Green Energy

Green Energy

An A-to-Z Guide

The SAGE Reference Series on
Green Society
Toward a Sustainable Future

DUSTIN MULVANEY, GENERAL EDITOR
University of California, Berkeley

PAUL ROBBINS, SERIES EDITOR
University of Arizona

Los Angeles | London | New Delhi
Singapore | Washington DC

Los Angeles | London | New Delhi
Singapore | Washington DC

FOR INFORMATION:

SAGE Publications, Inc.
2455 Teller Road
Thousand Oaks, California 91320
E-mail: order@sagepub.com

SAGE Publications Ltd.
1 Oliver's Yard
55 City Road
London EC1Y 1SP
United Kingdom

SAGE Publications India Pvt. Ltd.
B 1/I 1 Mohan Cooperative Industrial Area
Mathura Road, New Delhi 110 044
India

SAGE Publications Asia-Pacific Pte. Ltd.
33 Pekin Street #02-01
Far East Square
Singapore 048763

Publisher: Rolf A. Janke
Assistant to the Publisher: Michele Thompson
Senior Editor: Jim Brace-Thompson
Production Editors: Kate Schroeder, Tracy Buyan
Reference Systems Manager: Leticia Gutierrez
Reference Systems Coordinator: Laura Notton
Typesetter: C&M Digitals (P) Ltd.
Proofreader: Christina West
Indexer: Julie Sherman Grayson
Cover Designer: Gail Buschman
Marketing Manager: Kristi Ward

Golson Media
President and Editor: J. Geoffrey Golson
Author Manager: Ellen Ingber
Editors: Jill Coleman, Mary Jo Scibetta
Copy Editors: Anne Hicks, Barbara Paris

Copyright © 2011 by SAGE Publications, Inc.

Printed in the United States of America

Library of Congress Cataloging-in-Publication Data

Green energy : an A-to-Z guide / general editor, Dustin Mulvaney.

p. cm. — (The Sage reference series on green society: toward a sustainable future)
Includes bibliographical references and index.

ISBN 978-1-4129-9677-8 (cloth)—ISBN 978-1-4129-7185-0 (ebk)

1. Clean energy industries—Handbooks, manuals, etc. I. Mulvaney, Dustin.

HD9502.5.C542G74 2010 333.79'4—dc22 2011002804

11 12 13 14 15 10 9 8 7 6 5 4 3 2 1

Contents

About the Editors   *vii*

Introduction   *ix*

Reader's Guide   *xiii*

List of Articles   *xv*

List of Contributors   *xvii*

Green Energy Chronology   *xxi*

Articles A to Z   *1–464*

Green Energy Glossary   *465*

Green Energy Resource Guide   *475*

Green Energy Appendix   *481*

Index   *485*

About the Editors

Green Series Editor: Paul Robbins

Paul Robbins is a professor and the director of the University of Arizona School of Geography and Development. He earned his Ph.D. in Geography in 1996 from Clark University. He is General Editor of the *Encyclopedia of Environment and Society* (2007) and author of several books, including *Environment and Society: A Critical Introduction* (2010), *Lawn People: How Grasses, Weeds, and Chemicals Make Us Who We Are* (2007), and *Political Ecology: A Critical Introduction* (2004).

Robbins's research centers on the relationships between individuals (homeowners, hunters, professional foresters), environmental actors (lawns, elk, mesquite trees), and the institutions that connect them. He and his students seek to explain human environmental practices and knowledge, the influence nonhumans have on human behavior and organization, and the implications these interactions hold for ecosystem health, local community, and social justice. Past projects have examined chemical use in the suburban United States, elk management in Montana, forest product collection in New England, and wolf conservation in India.

Green Energy General Editor: Dustin Mulvaney

Dustin Mulvaney is a Science, Technology, and Society postdoctoral scholar at the University of California, Berkeley, in the Department of Environmental Science, Policy, and Management. His current research focuses on the construction metrics that characterize the life cycle impacts of emerging renewable energy technologies. He is interested in how life cycle assessments focus on material and energy flows and exclude people from the analysis, and how these metrics are used to influence investment, policy, and social resistance. Building off his work with the Silicon Valley Toxics Coalition's "just and sustainable solar industry" campaign, he is looking at how risks from the use of nanotechnology are addressed within the solar photovoltaic industry. Mulvaney also draws on his dissertation research on agricultural biotechnology governance to inform how policies to mitigate risks of genetically engineered biofuels are shaped by investors, policy makers, scientists, and social movements.

Mulvaney holds a Ph.D. in Environmental Studies from the University of California, Santa Cruz, and a Master of Science in Environmental Policy and a Bachelor's Degree in Chemical Engineering, both from the New Jersey Institute of Technology. Mulvaney's previous work experience includes time with a Fortune 500 chemical company working on sulfur dioxide emissions reduction, and for a bioremediation startup that developed technology to clean groundwater pollutants like benzene and MTBE.

Introduction

All human activities rely on transformations of energy. Energy, the ability to change a system, constitutes everything from human metabolism to how humans move about the planet. On Earth most of the energy we encounter originates from solar activity, with the exception of relatively smaller amounts of gravitational and internal heat. For most of human history, this solar energy was available in the form of biomass that humans directly ate, or used for heat, light, and cooking. But since the industrial revolution we have witnessed what the historian Lewis Mumford has called carboniferous capitalism; human civilization is drawing down the stocks of accumulated solar energy found in the form of fossil fuels such as coal, petroleum, and natural gas to supply our heat, electricity, and transportation fuels. This reliance on stocks of fossil fuel energy helped maintain the illusion that we were getting something for nothing.

But we now know that the extractions and transformations of fossil fuel energy come at a considerable cost and raise many questions about the sustainability of our energy supply as the conventional energy system confronts the realities of climate change, material limits, and environmental degradation. The use of fossil fuel energy is a major intervention in the carbon cycle as carbon (in hydrocarbons) moves from being sequestered in oil and gas pockets and coal seams below the Earth's surface to carbon dioxide (CO_2) in the atmosphere. This has led most scientists, including those represented on the Intergovernmental Panel on Climate Change (IPCC), to suggest that fossil fuel extraction is a significant component of anthropogenic climate change because of the role of CO_2 in causing the greenhouse affect. This, in turn, has forced many to suggest limits on the amount of CO_2 added to the atmosphere, where it is currently believed that up to an 80 percent reduction in global CO_2 emissions will be necessary to stabilize atmospheric CO_2 emissions—if that is even possible.

The depletion of fossil fuel stocks has led to calls to refocus energy generation on renewable energy sources in the face of impending scarcity and resource limitations. Claims about peak oil—Hubbert's Peak, for example—suggest that the increasing cost of extracting fossil fuel energy resources in more remote locations, from deeper and further afield, will raise the costs of extraction significantly, perhaps to the point at which it is not worth recovering. This implies that even if we do not "run out" of oil materially, we will run out from an economic perspective since the costs of extraction could cause fossil fuel prices to be greater than the prices for alternatives.

There are environmental and social impacts of fossil fuel extraction that go far beyond the impacts to the carbon cycle and fossil fuel scarcity. There are considerable impacts of land use change, air and water pollution, and toxic waste associated with our energy system. Take coal extraction as an example. Coalmining in West Virginia has dramatically altered the landscape as the "overburden" (hilltops that contain seams of coal) is blasted into streambeds to flatten the landscape, while massive surface mines in Wyoming, where larger seams of low sulfur coal are extracted, leave behind large contoured pits. Water is contaminated where piles of mine tailings leach toxic metals into groundwater. Air is polluted from coal combustion by among other things the cadmium and mercury found in coal, or the sulfur dioxide emissions that eventually are incorporated into the water cycle as acid rain. All while the fly ash left behind in giant ponds at coal power plants leaks lead, cadmium, chromium, mercury, arsenic, and other toxic metals from behind the impounded levy walls, threatening downstream community water resources and ecological habitats with steady seepage or catastrophic breaches.

Even some alternatives to fossil fuels could have vital impacts. Hydroelectric and nuclear energy sources, both of which do not emit CO_2 during operation, have other impacts that may fail to meet resource limitations and sustainability criteria. There are only so many places with large-scale hydroelectric capacity and where it has been adopted it is hindered by substantial social strife and ecological impacts. Likewise, uranium used in nuclear reactors is also in relatively short supply, and comes with enormous uncertainties about risks of accidents, costs, water pollution, waste disposal, and nuclear proliferation. While earlier times saw hydroelectric and nuclear energy sources as panaceas, they now face considerable scrutiny.

Our green energy future will rely on energy sources that do not utilize fossil fuels, that minimize environmental impact, and that replenish themselves by drawing on energy that presently flows from the sun or the Earth. Renewable energy makes up only a small percentage of the energy used today. While some statistical reporting includes large scale hydroelectric in the category of renewable, most limit the definition to small scale hydroelectric, solar, wind, wave, tidal, geothermal, and biomass sources. If we consider the stocks of fossil fuels buried beneath Earth's crust as the past products of photosynthesis, flows of solar, wind, and biomass energy are produced with current solar income. This notion that fossil fuel use is spending down the savings has led to widespread calls for more renewable sources of energy. Although many of these sources come with environmental and social tradeoffs, the switch to green energy is seen as a means to combat climate change, avoid peak oil, and lower toxic effluents and emissions, all while helping foster energy independence and revive the manufacturing economy.

But the key questions for green energy will be how to foster the innovations, policies, and behavioral transformations necessary to reduce the impacts of energy production and contract the carbon footprints of large energy consumers. Most recently this discussion has taken place through the United Nation's Framework Convention on Climate Change where the Kyoto Protocol and its potential successor—a multilateral agreement currently dubbed the Copenhagen Accord—aim to commit countries to binding emissions reductions. But it is not clear what exactly this will entail. Will it be a carbon tax or carbon trading and offsetting? What will be the role for nuclear, hydroelectric, and clean coal technology? One outcome of the December 2009 meeting in Copenhagen was an agreement to promote carbon sequestration through decreased deforestation and forest degradation. While this was an important step with multiple benefits, it does not address fossil fuel combustion. Reducing the carbon footprint of the global economy will require incentives for innovation in non-fossil fuel energy sources and efforts to reduce consumption by organizations, nations, and individuals.

This volume provides an overview of the social and environmental dimensions of our energy system, and the key organizations, policy tools, and technologies that can help shape a green energy economy. Each entry draws on scholarship from across numerous disciplines in the social sciences, natural and physical sciences, and engineering. The urgency of climate change helps underscore the importance of getting the right technologies, policies and incentives, and social checks and balances in place. The green energy challenge faced by human civilization will require many minds and a great effort on all fronts. We hope this collection of entries can provide those with an interest in green energy to participate in what will hopefully become an equitable and intergenerational conversation about the impacts of our energy consumption and how to make it cleaner and greener.

Dustin Mulvaney
General Editor

Reader's Guide

Energy Agreements and Organizations

Best Management Practices
CAFE Standards
Carbon Tax
Carbon Trading and Offsetting
Department of Energy, U.S.
Federal Energy Regulatory Commission
Feed-In Tariff
Intergovernmental Panel on Climate Change
Internal Energy Market
International Renewable Energy Agency
Kyoto Protocol
LEED Standards
Oil Majors
Organization of Petroleum Exporting Countries
Public Utilities
Renewable Energy Portfolio
World Commission on Environment and Development
Yucca Mountain

Energy Challenges

Arctic National Wildlife Refuge
Automobiles
Caspian Sea
Chernobyl
Chlorofluorocarbons
Climate Change
Combustion Engine
Dioxin
Exxon Valdez

Fossil Fuels
Global Warming
Greenhouse Gases
Insulation
Metric Tons of Carbon Equivalent (MTCE)
Mountaintop Removal
Natural Gas
Nitrogen Oxides
Nonpoint Source
Nonrenewable Energy Resources
Nuclear Power
Nuclear Proliferation
Offshore Drilling
Oil
Oil Sands
Oil Shale
Petroviolence
Sulfur Oxides (SOx)
Sustainability
Three Gorges Dam
Three Mile Island
Uranium
Volatile Organic Compound (VOC)
Waste Incineration

Energy Measurement

Exergy
Food Miles
Heating Degree Day
Hubbert's Peak
Radiative Forcing
Risk Assessment
Total Primary Energy Supply

Energy Solutions

Alternative Energy
Alternative Fuels
Appliance Standards
Berlin Mandate
Bicycles
Biodiesel
Biogas Digester
Biomass Energy
Carbon Emissions Factors
Carbon Footprint and Neutrality
Carbon Sequestration
Climate Neutrality
Coal, Clean Technology
Combined Heat and Power
Compact Fluorescent Bulb
Daylighting
Electric Vehicle
Energy Audit
Environmentally Preferable Purchasing
Environmental Measures
Environmental Stewardship
Energy Policy
Ethanol, Corn
Ethanol, Sugarcane
Flex Fuel Vehicles
Gasohol
Geothermal Energy
Green Banking
Green Energy Certification Schemes
Green Power
Green Pricing
Hydroelectric Power
Hydrogen
Innovation
Landfill Methane
Microhydro Power
On-Site Renewable Energy Generation
Plug-In Hybrid
Public Transportation
Recycling
Renewable Energies
Solar Energy
Solar Thermal Systems
Tidal Power
Wave Power
Wind Power
Wood Energy
Zero-Emission Vehicle

Energy Systems

Batteries
California
Columbia River
Electricity
Embodied Energy
Emission Inventory
Emissions Trading
Energy Payback
Energy Storage
Entropy
Flaring
Forecasting
Fuel Cells
Fusion
Gasification
Grid-Connected System
Heat Island Effect
Home Energy Rating
   Systems
Life Cycle Analysis
Metering
Photovoltaics (PV)
Pipelines
Power and Power Plants
Smart Grid
Solar Concentrator
Wind Turbine

List of Articles

Alternative Energy
Alternative Fuels
Appliance Standards
Arctic National Wildlife Refuge
Automobiles

Batteries
Berlin Mandate
Best Management Practices
Bicycles
Biodiesel
Biogas Digester
Biomass Energy

CAFE Standards
California
Carbon Emissions Factors
Carbon Footprint and Neutrality
Carbon Sequestration
Carbon Tax
Carbon Trading and Offsetting
Caspian Sea
Chernobyl
Chlorofluorocarbons
Climate Change
Climate Neutrality
Coal, Clean Technology
Columbia River
Combined Heat and Power
Combustion Engine
Compact Fluorescent Bulb

Daylighting
Department of Energy, U.S.
Dioxin

Electricity
Electric Vehicle

Embodied Energy
Emission Inventory
Emissions Trading
Energy Audit
Energy Payback
Energy Policy
Energy Storage
Entropy
Environmentally Preferable Purchasing
Environmental Measures
Environmental Stewardship
Ethanol, Corn
Ethanol, Sugarcane
Exergy
Exxon Valdez

Federal Energy Regulatory Commission
Feed-In Tariff
Flaring
Flex Fuel Vehicles
Food Miles
Forecasting
Fossil Fuels
Fuel Cells
Fusion

Gasification
Gasohol
Geothermal Energy
Global Warming
Green Banking
Green Energy Certification Schemes
Greenhouse Gases
Green Power
Green Pricing
Grid-Connected System

Heating Degree Day
Heat Island Effect
Home Energy Rating Systems
Hubbert's Peak
Hydroelectric Power
Hydrogen

Innovation
Insulation
Intergovernmental Panel on Climate
    Change
Internal Energy Market
International Renewable Energy Agency

Kyoto Protocol

Landfill Methane
LEED Standards
Life Cycle Analysis

Metering
Metric Tons of Carbon Equivalent (MTCE)
Microhydro Power
Mountaintop Removal

Natural Gas
Nitrogen Oxides
Nonpoint Source
Nonrenewable Energy Resources
Nuclear Power
Nuclear Proliferation

Offshore Drilling
Oil
Oil Majors
Oil Sands
Oil Shale
On-Site Renewable Energy Generation
Organization of Petroleum Exporting
    Countries

Petroviolence
Photovoltaics (PV)
Pipelines
Plug-In Hybrid
Power and Power Plants
Public Transportation
Public Utilities

Radiative Forcing
Recycling
Renewable Energies
Renewable Energy Portfolio
Risk Assessment

Smart Grid
Solar Concentrator
Solar Energy
Solar Thermal Systems
Sulfur Oxides (SO_x)
Sustainability

Three Gorges Dam
Three Mile Island
Tidal Power
Total Primary Energy Supply

Uranium

Volatile Organic Compound (VOC)

Waste Incineration
Wave Power
Wind Power
Wind Turbine
Wood Energy
World Commission on Environment and
    Development

Yucca Mountain

Zero-Emission Vehicle

List of Contributors

Auerbach, Karl
University of Rochester

Boslaugh, Sarah
Washington University in St. Louis

Boudes, Philippe
University Paris West Nanterre la Défense

Brahinsky, Rachel
University of California, Berkeley

Bremer, Leah
San Diego State University

Bridgeman, Bruce
University of California, Santa Cruz

Burns, William C. G.
Santa Clara University School of Law

Carr, David L.
University of California, Santa Barbara

Casalenuovo, Kristen
Virginia Commonwealth University

Chakraborty, Debojyoti
Independent Scholar

Chalvatzis, Konstantinos
University of East Anglia

Chatterjee, Amitava
University of California, Riverside

Chatterjee, Sudipto
Winrock International India

Coleman, Jill S. M.
Ball State University

De Lucia, Vito
Independent Scholar

de Souza, Lester
Independent Scholar

Deaner, Hugh
University of Kentucky

Downie, David
Fairfield University

Ehrhardt-Martinez, Karen
American Council for an Energy-Efficient Economy

Evans, Tina
Fort Lewis College

Evrard, Aurélien
Sciences Po Paris/Centre for Political Research (CEVIPOF)

Eysenbach, Derek
University of Arizona

Field, Chris
Carnegie Institution for Science

Finley-Brook, Mary
University of Richmond

Gabbard, R. Todd
Kansas State University

Gareau, Brian J.
Boston College

Golden, Elizabeth L.
University of Cincinnati

Goodier, Chris
Loughborough University

Gopakumar, Govind
Rensselaer Polytechnic Institute

Gray, Lindsay L.
Alexander Gorlin Architects

Gray, Steven A.
Rutgers University

Harper, Gavin
Cardiff University

Harrington, Jonathan
Troy University

Hurst, Kent
University of Texas at Arlington

Iles, Alastair
University of California, Berkeley

Isaacson, Marjorie
CNT Energy, Center for Neighborhood Technology

Isherwood, William
Chinese Academy of Sciences

Kakegawa, Michiyo
University of California, Santa Cruz

Keirstead, James
Imperial College London

Keith, Ron
George Mason University

Kinsella, William J.
North Carolina State University

Kline, Charles
University of Richmond

Kofoworola, Oyeshola Femi
University of Toronto

Kumar, P. Vijaya
Project Directorate on Cropping Systems Research, India

Mazze, Sarah
University of Oregon

McKechnie, Jon
Independent Scholar

McKinney, Vanessa
American Council for an Energy-Efficient Economy

Mudd, Gavin M.
Monash University

Norris, Timothy
University of California, Santa Cruz

Olanrewaju, Ajayi Oluseyi
Covenant University, Nigeria

Panda, Sudhanshu Sekhar
Gainesville State College

Papadakis, Maria
James Madison University

Pearce, Joshua M.
Queen's University

Pellegrino, Michael
Architect, AICP, NCARB

Phadke, Roopali
Macalester College

Price, Jessica
University of Wisconsin–Madison

Ren, Guoqiang
University of Washington

Santana, Mirna E.
University of Wisconsin–Madison

Schelly, Chelsea
University of Wisconsin–Madison

Sherman, Daniel
Independent Scholar

Smith, Susan L.
Willamette University College of Law

Star, Anthony
CNT Energy, Center for Neighborhood Technology

Valero, Alicia
Independent Scholar

Valero, Antonio
Independent Scholar

Vora, Rathin N.
University of Rochester

Vynne, Stacy
University of Oregon

Wang, Yiwei
University of California, Santa Cruz

Waskey, Andrew J.
Dalton State College

Whalen, Ken
*American University of
Afghanistan*

Williams, Akan Bassey
Covenant University

Winograd, Claudia
*University of Illinois at
Urbana-Champaign*

Zhang, Yimin
University of Toronto

Zimmermann, Petra
Ball State University

Green Energy Chronology

c. 500,000 B.C.E.: The first humans use fire.

12,000–6000 B.C.E.: During the Neolithic Revolution early humans learn to domesticate plants and animals, developing agriculture and beginnings of settlements in the Fertile Crescent. Previously gathered plants are sowed and harvested, while wild sheep, goats, pigs, and cattle are herded instead of hunted.

c. 6500 B.C.E.: The first known application of metalworking with copper begins in the Middle East.

4000–3000 B.C.E.: In a seemingly simultaneous innovation, fledgling civilizations in Europe and the Middle East use oxen to pull sledges and plow fields.

3200 B.C.E.: The wheel is used in Ancient Mesopotamia.

c. 3000 B.C.E.: Mules are used as cargo animals in the Middle East, fueling the earliest long-distance trade routes.

c. 3000 B.C.E.: Chinese, Egyptian, Phoenician, Greek, and Roman settlements use heat from the sun to dry crops and evaporate ocean water, producing salt.

1200 B.C.E.: Ancient Egyptians show knowledge of sailing, but Ancient Phoenicians become the first to efficiently harness the power of the wind, using early sailboats to develop an extensive maritime trading empire.

1000 B.C.E.: Egyptians use petroleum-based tar to help preserve the human body during the process of mummification.

1000 B.C.E.: The first known consumption of fossil fuels occurs in China. Coal is unearthed, and likely used to smelt copper in rudimentary blast furnaces.

600 B.C.E.: A rudimentary form of a magnifying glass is used to concentrate the sun's rays on a natural fuel, lighting a fire for light, warmth, and cooking.

200 B.C.E.: Greek scientist Archimedes is said to have used the reflective properties of bronze shields to focus sunlight and set fire to Roman ships, which were besieging Syracuse. In 1973 the modern Greek Navy recreated the legend, successfully setting fire to wooden boats 50 meters away.

100 C.E.: The Greeks invent the waterwheel.

100–300: Roman architects build glass or mica windows on the south-facing walls of bath houses, and other buildings, to keep them warm in the winter.

500: Roman cannon law, the Justinian Code, establishes "sun rights" to ensure that all buildings have access to the sun's warmth.

500–900: The first known windmills are developed in Persia; uses include pumping water and grinding grain.

700: In Sri Lanka, the wind is used to smelt metal from rock ore.

1088: A water powered mechanical clock is made by Han Kung-Lien in China.

1300s: The first horizontal axis windmills, shaped like pinwheels, appear in Western Europe.

1306: England's King Edward I unsuccessfully tries to ban open coal fires in England, marking an early attempt at national environmental protection.

1347–50: In just three short years the Bubonic Plague, a disease transferred from rats to humans, spreads across European Trade routes, engulfing the continent and decimating over one-third of the population.

1500s: Early in the century, Renaissance man Leonardo da Vinci proposes the first industrial applications of solar concentrators. He also is the first to describe a precursor to the water-driven turbine.

1582: Water wells pump the first water from the Thames River into London for irrigation.

1600s: The Dutch master drainage windmills, moving water out of low lands to make farming available. During the Protestant Reformation, they use windmill positions to communicate to Catholics, indicating safe places for asylum.

1601: William Gilbert publishes *De Magnete*, one of the earliest extensive works on electricity and magnetism.

1690: Progressive Governor William Penn requires that one acre of forest be saved for every five that is cut down in the newly formed city of Philadelphia.

1712: A piston-operated steam engine is built by Thomas Newcomen in England.

1752: Benjamin Franklin attaches a key to a damp kite string, and flies it into a thunder cloud. He observes sparks from the key running down the string to his hand, discovering that lightening is electrical in nature.

1767: Swiss scientist Horace de Saussure is credited with building the world's first solar collector, a simple glass box that traps sunlight. Sir John Herschel would elaborate on the invention, proving that solar collectors could heat water to its boiling point.

1769: An improved steam engine is patented by James Watt in England. Later he invents the rotary steam engine.

1786: Italian physician Luigi Galvani pioneers bioelectricity, the study of nerve cells passing signals to muscles via electricity, after his research with frog's legs and static.

1789: German chemist Martin H. Klaproth is credited with the discovery of the element uranium. The metal is later used in fission research and the development of nuclear weapons and reactors.

1803: American Robert Fulton builds the first steam-powered boat.

1808: Sir Humphrey David invents a battery-powered arc lamp. It is one of the first electric-powered lighting systems to be commercialized.

1816: Scottish clergyman Robert Stirling receives a patent for the first heat engine using a process that improves thermal efficiency, now called the Stirling Cycle. He calls his invention the Heat Economiser.

1818: The first steamship, the *Savannah*, crosses the Atlantic Ocean.

1820: English scientist Michael Faraday conducts experiments into electromagnetism, demonstrating that electricity can produce motion. He later invents the electric dynamo.

1821: The first U.S. natural gas well begins drilling in upstate New York.

1830: Steam-driven cars become commonplace on the streets of London.

1839: French physicist Edmund Becquerel discovers the photovoltaic effect, the process by which electromagnetic radiation (typically visible light) is absorbed by matter (typically a metal) which in turn emits electrons, creating electricity. The effect would later become the physics behind the development of a solar panel.

1841: Construction is completed on the Croton Aqueduct. For the next hundred years, the distribution system carries water 41 miles south to New York City using only the force of gravity.

1850s: Americans Daniel Halladay and John Burnham work to build and commercialize the Halladay Windmill, designed for use in the American west. Encouraged by their developments, they found the U.S. Wind Engine Company.

1855: English engineer Henry Bessemer revolutionizes the steel industry with the development of the Bessemer process, which uses oxidation to remove impurities from iron. The process makes steel significantly easier and quicker to produce, making its price comparable to iron and fueling expansive infrastructure construction in modern nations.

1860s: French mathematician August Mouchet first proposes the idea of a solar-powered steam engine. He later patents his invention.

1860: The first internal combustion engine is built by Etienne Lenoir in Belgium.

1862: U.S. President Abraham Lincoln creates the Department of Agriculture, charged with promoting agriculture production and the protection of natural resources.

1864: Scottish physicist James Clerk Maxwell demonstrates that electricity, magnetism, and light are all manifestations of the same phenomenon, the electromagnetic field.

1879: The U.S. Geological Survey is established, responsible for examining national geological structure and natural resources.

1879: American inventor Thomas Edison invents the incandescent electric light bulb. After repeated attempts with using metal, he finally succeeds using carbon-based filaments.

1880s: While working for the U.S. Wind Engine Company, Thomas O. Perry conducts extensive wind power experiments, attempting to build a more effective windmill. He invents a windmill that uses gears to reduce the rotational speed of the blades. Perry starts the Aermotor Company to sell his advanced windmill, which becomes indispensable for Midwestern farmers and ranchers who use windmills to pump water for livestock. Later, steel blades are developed for windmills. This innovation, combined with a significant amount of American families relocating west, makes windmills an essential source of power. Mills are used for pumping water, shelling corn, sawing wood, and milling grain.

1880: The world's first concrete-arch dam, the 75-Miles Dam, is built in the Australian city of Warwick.

1882: The first electric power stations of New York and London go on line.

1882: The world's first hydroelectric power station, the Vulcan Street Power Plant, is built in Appleton, Wisconsin.

1883: American inventor Charles Fritts builds the world's first working solar cell. It has an efficiency of about 1 percent.

1887: German physicist Heinrich Hertz discovers that ultraviolet light will lower the lowest possible voltage of a metal, causing an electric spark to jump between electrodes.

1888: Charles F. Brush adapts the first large windmill to generate electricity, in Cleveland, Ohio. Electricity generating mills are coined "wind turbines." General Electric later acquires Brush's company, Brush Electric.

1891: Baltimore inventor Clarence Kemp patents the first commercial solar water heater.

1896: The first offshore oil wells, built on wooden piers, begin drilling off the coast of California.

1896–1903: While studying phosphorescence, French physicist Antoine Henri Becquerel (son of Edmond Becquerel) discovers a chemical reaction that does not require an external source of energy, called radioactivity. He later wins the Nobel Prize for discovering radioactivity, along with Marie and Pierre Curie, who discover radium the same year. Their work, combined with numerous others, provides strong evidence that atoms have the potential to release immense amounts of energy.

1905: American Albert Einstein publishes his Theory of Relativity, revolutionizing human understanding of energy by unifying mass, energy, magnetism, electricity, and light. He also publishes his paper explaining the photoelectric effect. He later wins the Nobel Prize.

1908–17: Ernest Rutherford fires alpha particles at gold foil and discovers that atoms have a small charged nucleus that contains the majority of the atom's mass. He wins the Nobel Prize for his discovery. In 1917 he successfully splits an atom for the first time.

1920: The Federal Power Act establishes the Federal Power Commission, an independent agency responsible for coordinating federal hydropower development. The commission is later given authority over natural gas facilities and electricity transmission. It is eventually overtaken by the Department of Energy.

1929–70: Venezuela is the world's top oil exporter.

1930: General Motors and DuPont introduce Freon, synthetic chemicals widely used in air conditioners and refrigerators until the 1980s, when regulatory agencies ban their use because of harm caused to Earth's ozone layer.

1932: The nucleus of an atom is split for the first time in a controlled environment by physicists John Cockcroft and Ernest Walton under the direction of Ernest Rutherford.

1932: English scientist Francis Bacon develops the first practical fuel cell.

1936: Construction is completed on The Hoover Dam. The 725-foot arch is the world's largest electric-power generating station and the world's largest concrete structure. For a total cost of $49 million, it will eventually power the American west with more than four billion kilowatt-hours per year. Lake Mead, America's largest reservoir, is formed by water impounded from the dam.

1938: European physicist Otto Hahn and collaborators discover nuclear fission. Prompted by previous researchers, they bombard uranium with neutrons, bursting the atoms into trace amounts of the much lighter metal, barium, and releasing leftover energy in the process.

1939: In August a letter is drafted by Hungarian physicist Leo Szilard, signed by American Albert Einstein, addressed to U.S. President Franklin D. Roosevelt, advising him to fund nuclear fission research as means to a weapon, in the event that Nazi Germany may already be exploring the possibility. In September Germany invades Poland, beginning World War II. In October a secret meeting results the creation of the Advisory Committee on Uranium for the purpose of securing the element and using it in research to create an atomic weapon.

1940: The British MAUD (Military Application of Uranium Detonation) Committee is established for the purposes of investigating the possibility of using uranium in a bomb. The next year they publish a report detailing specific requirements for its creation.

1941: Plutonium is first identified by Glenn Seaborg, who immediately recognizes its potential in ongoing atomic weapon research. On December 7th, the largest attack on American soil occurs when Japanese war planes ambush the naval base Pearl Harbor. The next day the United States enters World War II by formally declaring war on Japan. In the following week, Germany and Italy declare war on America.

1941: The world's first megawatt-size wind turbine is made functional on a hill in Vermont, Grandpa's Knob. During World War II it is connected to the local electrical distribution system, providing electricity for the city of Castleton. The 1.25-megawatt turbine operates for approximately 1,100 hours before a blade failure occurs, caused by a lack of reinforcement due to war time shortages.

1942: The Manhattan Engineer District or the Manhattan Project is established by the Army Corps of Engineers, directed by physicist Robert Oppenheimer. The project makes the development of a nuclear weapon a top Army priority and begins to outline methods for construction, testing, and transportation of an atomic bomb.

1943: Italian geothermal fields produce about 132 megawatts of power.

1943: Under the growing threat that German scientists may be outpacing the Allies, U.S. President Roosevelt and British Prime Minister Winston Churchill sign the Quebec Agreement. Teams of British scientists join the Manhattan Project.

1945: In July the world's first nuclear explosion, the Trinity test, occurs in the desert of New Mexico. In August American forces detonate two atomic bombs on Japanese soil. The cities of Hiroshima and Nagasaki are effectively obliterated, losing 140,000 and 80,000 people, respectively. In the following week, Japan surrenders to Allied powers.

1946: After World War II, the U.S. Congress creates the Atomic Energy Commission for the purpose of regulating the development of nuclear technology, and states that the committee is to be placed under civilian, not military, control.

1947: Solar buildings in the United States are in high demand. Libbey-Owens-Ford Glass Company publishes the book, *Your Solar House*, profiling the nation's top solar architecture experts.

1947: Diesel/electric trains replace steam locomotives in the United States.

1954: The first Russian nuclear power plant opens.

1954: American photovoltaic technology makes a giant leap when scientists at Bell Labs develop the world's most efficient solar cell at 6 percent efficiency, enough power to run everyday electrical equipment.

1955: American Architect Frank Bridgers designs the world's first commercial office building with solar water heating and a passive design, the Bridgers-Paxton Office Building, in New Mexico.

1955–90: First proposed in 1955 as the Air Pollution Control Act and later as the Clean Air Act of 1963, the Air Quality Act of 1967, the Clean Air Act Extension of 1970, and the Clean Air Act Amendments of 1977 and 1990, the U.S. government enacts similar legislation regarding hazardous emissions into the atmosphere. Finally, met with criticism due to bureaucratic methods, the Clear Skies Act of 2003 amends much of the previous legislation. A significant amount of all the provisions are designed toward energy companies.

1959: NASA launches the *Explorer VI* satellite. It is built with an array of over 9,000 solar cells.

1960: The first solar powered coast-to-coast conversation takes places from New Jersey to California by the U.S. Army Signal Corps.

1962: Experimental developments into satellite communications between America and Britain, across the Atlantic Ocean, prove successful when *Telstar 1* is launched. The world's first working communication satellite delivers transatlantic phone calls, television pictures, and fax messages between the United States and England.

1964: NASA launches the first Orbiting Astronomical Observatory, powered by a one-kilowatt photovoltaic array.

1964: U.S. President Johnson signs the Wilderness Act into law. Over nine million acres of land are closed to excavation.

1967: The first commercial microwave, for use in homes, is introduced.

1970s: With help from the Exxon Corporation, American Dr. Elliot Berman designs a significantly cheaper solar cell, which equates to a cost of $20 per watt, instead of $100 per watt. Solar cells begin powering navigation and warning lights on offshore oil rigs, lighthouses, and railroad crossings. Domestic solar technology is considered a feasible alternative in remote rural areas, where utility grids are too costly.

1970: Earth Day is made a national holiday, to be celebrated April 22nd of each year. It is founded by U.S Senator Gaylord Nelson as an environmental teach-in. It is celebrated by many countries throughout the world.

1970: The Environmental Protection Agency (EPA) is created to enforce federal environmental regulations. The agency's mission is to regulate chemicals and protect human health by safeguarding air, land, and water.

1970: The National Oceanic and Atmospheric Administration is created for the purpose of developing efficient ways of using the country's marine resources.

1971: U.S. and British scientists begin development for the first wave energy system.

1972: The Institute of Energy Conversion is established at the University of Delaware.

1972: The U.S. Congress passes the Ocean Dumping Act, requiring companies to file for licenses from the Environmental Protection Agency to dump wastes into national waters. After heaps of medical waste wash up on the shores of New Jersey in 1988, Congress bans the dumping outright.

1973: The Arab Oil Embargo begins as a response to the U.S. decision to supply the Israeli military during the Yom Kippur War between Arab nations and Israel.

1974: The U.S. Department of Energy forms a branch dedicated to national research and development of solar energy, the Solar Energy Research Institute.

1974: After sharp increases in the price of oil from the Organization of the Petroleum Exporting Countries (OPEC) lead to a major American energy crisis, The Energy Reorganization Act is signed into law replacing the Atomic Energy Commission with the Energy Research and Development Administration, responsible for oversight of nuclear weapons, and the Nuclear Regulatory Commission responsible for commercial nuclear safety. The act also requires the future creation of the Strategic Petroleum Reserve, set to contain one million barrels of oil. Because of the same crisis, the U.S. government begins federally funding wind energy research through NASA and the Department of Energy, coordinated by the Lewis Research Center.

1975: In another response to the energy crisis, the Corporate Average Fuel Economy (CAFE) regulations are passed by the U.S. Congress, intending to improve the average fuel economy of consumer vehicles. In 2002 the National Academy of Sciences reviews the regulations, and finds they are responsible for a decrease in motor vehicle consumption by 14 percent.

1976: The NASA Lewis Research Center begins installing solar power systems across the world. They will provide power for lighting, water pumping, grain milling, electricity, refrigeration, and telecommunications. However, the project won't be complete for another 20 years.

1977: Still reeling from the oil crisis, the U.S. Department of Energy is created. The new department will coordinate several already established programs, assuming the responsibilities of Energy Research and Development Administration. The Energy Information Administration is responsible for independent energy statistics. The Office of

Secure Transportation provides secure transportation of nuclear weapons and materials. The Federal Energy Regulatory Commission is given jurisdiction over commercial energy including electricity and natural gas, as well as managing the Strategic Petroleum Reserve.

1978: In a final measure to avoid another energy crisis like the one started in 1973, the National Energy Act of 1978 is passed by the U.S. Congress. It includes a host of new statutes attempting to redefine how the country secures, consumes, and comprehends energy. The Public Utility Regulatory Policies Act (PURPA) promotes the greater use of renewable energy. The law regulates a market for renewable energy produces, forcing electric utility companies to purchase from these suppliers at a fixed price. Cogeneration plants become the industry standard. Another law enacted under the National Energy Act is the National Energy Conservation Policy Act, which requires utility companies to employ energy management strategies designed to curb the demand for electricity. Another law enacted gives an income tax credit to private residents who use solar, wind, or geothermal sources of energy. Also created is the "gas guzzler tax," which makes the sale of vehicles with a gas mileage below a specified EPA-estimated level liable to fiscal penalty. The Power Plant and Industrial Fuel Use Act and the Natural Gas Policy Act are also passed as part of the National Energy Act.

1978: Supertanker *Amoco Cadiz* runs aground of the coast of France, emptying the entire cargo of 1.6 million barrels into the water. It is the largest oil spill in history and causes an estimated U.S. $250 million in damages.

1978: The NASA Lewis Research Center installs a photovoltaic system on an Indian Reservation in Arizona. It provides power for pumping water and residential electricity in 15 homes until 1983, when grid power eventually reaches the village.

1979: A partial core meltdown occurs at Three Mile Island Nuclear Generating Station, releasing radioactive gases into the Pennsylvania air. An investigation later concludes that no adverse health effects will be perceptible to the community.

1980: The Crude Oil Windfall Profits Act creates what is technically an "excise tax" imposed on the difference between the market price of oil and a base price that is adjusted for inflation. It also increases tax credits for businesses using renewable energy.

1980: The American Council for an Energy-Efficient Economy is formed as a nonprofit organization. Their mission is to advance energy efficiency as a fast, cheap, and effective means of meeting energy challenges. The agency works on state and federal levels helping shape energy policy in favor of energy conservation, focusing on the end-use efficiency in industry, utilities, transportation, and human behavior.

1981: American engineer Paul MacCready, in addition to being credited with designing the first practical human-powered aircraft in 1977, invents the world's first solar-powered aircraft. He flies his *Solar Challenger* from France to England, across the English Channel.

1982: Adventurous Australian Hans Tholstrup drives the first solar-powered automobile. He travels a total of 2,800 miles. Tholstrup later founds a world-class solar car race.

1983: Worldwide, photovoltaic product exceeds 21.3 megawatts, a $250 million industry. Three out of four power plants in the United States still burn fossil fuels.

1985: To meet a drastic rise in demand for electricity, California installs enough windmills to exceed 1,000 megawatts of power, enough to supply 250,000 homes with electricity. By 1990 California is capable of more than 2,200 megawatts, about half of the world's capacity.

1986: The world's largest thermal facility is commissioned in Kramer Junction, California.

1986: The most significant nuclear meltdown in history occurs in Chernobyl, Ukraine. The entire area is subject to nuclear fallout. With some local residents unable to evacuate, generations of families suffer from intense radiation.

1987: The National Appliance Energy Conservation Act authorizes the Department of Energy to set minimum efficiency standards for space conditioning equipment and other appliances each year, based on what is "technologically feasible and economically justified."

1991: U.S. President George Bush announces that the Solar Energy Research Institute has been designated the National Renewable Energy Laboratory. Its mission is to develop renewable energy and energy-efficient technologies and practices, advance related science and engineering, and transfer knowledge and innovations to addressing the nation's energy and environmental goals by using scientific discoveries to create market-viable alternative energy solutions.

1992: Under President Clinton, the Energy Policy Act of 1992 is passed by Congress. It is organized under several titles enacting legislation on such subjects as: energy efficiency, conservation and management, electric motor vehicles, coal power and clean coal, renewable energy, alternative fuels, natural gas imports and exports, and various others. Among the new directives is a section that designates Yucca Mountain in Nevada as a permanent disposal site for radioactive materials from nuclear power plants. It also reforms the Public Utility Holding Company Act to help prevent an oligopoly and provides further tax credits for using renewable energy.

1992: Energy Star is established as a unified standard for energy-efficient consumer products. The Energy Star logo begins to appear on things like computers, kitchen appliances, laundry equipment, air conditioners, lighting, and various other energy saving products. Consumers who own Energy Star-endorsed products can expect a 20 to 30 percent reduction in energy usage.

1993: U.S. Windpower is one of the first companies to develop a commercially viable, variable-speed wind turbine over a period of five years. Funding for the project is supported by various utility companies, including the Electric Power Research Institute.

1993: Pacific Gas & Electric installs the first grid-supported photovoltaic system in California. In the same year the National Renewable Energy Laboratory completes construction of its Solar Energy Research Facility, immediately recognized as the most energy-efficient U.S. government building.

1993: The U.S. Green Building Council is founded as a nonprofit trade organization that promotes self-sustaining building design, construction, and operation. The council develops the Leaders of Energy and Environmental Design (LEED) rating system and organizes Greenbuild, a conference promoting environmentally responsible materials and sustainable architecture techniques.

1995: A U.S. Department of Energy program makes great advances in wind power technology. The new turbines can generate electricity with a cost of only $.05 per kilowatt hour.

1996: The U.S. upgrades a solar power tower into Solar Two, demonstrating that solar energy can be stored efficiently, resulting in power production even when the sun isn't shining. It sparks intense commercial interest in solar power towers.

1996: Facing major financial problems, Kenetech, the top U.S. producer of wind generators stops production and sells most of its assets.

1997: A national global climate change initiative, the Million Roofs Initiative, administered by the Department of Energy, is established, setting a goal of one million new solar energy systems to be installed in the United States by 2010. The initiative's objectives also include: evaluating greenhouse gas emissions, expanding energy options, creating energy technology jobs, removing market barriers, and generating grassroots demand for solar technologies.

1998: American scientist Subhendu Guha is the leading inventor of flexible solar shingles, a state-of-the-art technology for converting sunlight into electricity. The shingles operate on the same principles as conventional solar cells: sunlight falls on thin layers of silicon stimulating an electrical current for the building.

1999: In New York City, construction is completed on 4 Times Square. The building has more energy-efficient features than any other commercial skyscraper, including solar panels that produce part of the power.

2000s: The photovoltaic cell continues to be the primary source of power for U.S. space programs.

2000s: Alternative fuels for automobiles begin to enter the mainstream consumer market, including hydrogen and electric-powered vehicles.

2000: The Biomass Research and Development Board is created as part of a U.S. Congress act attempting to coordinate federal research and development of bio-based fuels obtained by living (as opposed to long dead, fossil fuels) biological material, such as wood or vegetable oils. Biofuel industries begin to expand in Europe, Asia, and the Americas.

2001: England's gas giant, British Petroleum (BP), announces the opening of a service station that features a solar-electric canopy. The station is the first in the United States (Indianapolis) and is a model that BP intends to use to revamp stations.

2002: Union Pacific Railroad installs blue-signal rail yard lanterns, which incorporate energy-saving light-emitting diode (LED) technology and solar cells, at a flagship rail yard in Nebraska.

2003: PowerLight Corporation installs the largest rooftop solar power system in the United States, a 1.18-megawatt system in at Santa Rita Jail in California. It uses 30 percent less electricity.

2005: The Energy Policy Act is passed by the U.S. Congress, signed into law by George W. Bush, making sweeping reforms in energy legislation, mostly in the way of tax deductions and subsidies. Loans are guaranteed for innovative technologies that avoid greenhouse gases, and alternative energy resources such as wind, solar, and clean coal production are given multimillion dollar subsides. For the first time wave and tidal power are included as separately identified renewable technologies. On the local level, individual tax breaks are given to Americans who make energy conservation improvements in their homes. However, total tax reductions greatly favor nuclear power and fossil fuel production, and the bill is later met with criticism. During the 2008 Democratic Primary, candidate Senator Hillary Clinton dubs it the "(Vice President) Dick Cheney lobbyist energy bill."

2007: Wind power accounts for about 5 percent of the renewable energy in the United States.

2007: The Energy Independence and Security Act of 2007 (originally named the Clean Energy Act of 2007) is passed by the U.S. Congress. Its stated purposes are "to move the United States toward greater energy independence and security, to increase the production of clean, renewable fuels, to protect consumers, to increase the efficiency of products, buildings and vehicles, to promote research on and deploy greenhouse gas capture and storage options, and to improve the energy performance of the Federal Government," as well as various other goals. Title I of the original bill is called the "Ending Subsidies for Big Oil Act of 2007." Included in the new provisions is a requirement of government and public institutions to lower fossil fuel use 80 percent by 2020. Also included is the repeal of much of the legislation included in the Energy Policy Act of 2005.

2008: In August the National Clean Energy Forum is attended by industry leaders, scientists, and policymakers at the University of Nevada, Las Vegas to discuss the future of green energy.

2009: Amid a global recession, the American Recovery and Reinvestment Act of 2009 is one of the inaugural acts signed by President Barack Obama. Otherwise known as the "stimulus package," it mostly makes provisions for job creation, tax relief, and infrastructure investment, but is also heavily focused on energy efficiency and science. Multibillion dollar funding is appropriated toward energy-efficient building practices, green jobs, electric and hybrid vehicles, and modernizing the nation's electric grid into a smart grid that uses digital technology to save energy. In the official seal of the act, an illustration of a fertile bright green plant is placed aside two grinding cogs.

ALTERNATIVE ENERGY

Alternative energy refers to renewable energy sources, which draw on current flows of energy that are constantly being replenished instead of accumulated stocks of fossil fuels that are drawn down over time. The availability of a particular form of renewable energy differs by region, and the potential to use it differs by the technologies available. Most of energy found on Earth ultimately comes from the sun (a less significant amount is derived from Earth's gravity and internal heat). Solar energy comes in many forms as direct heat evaporates water and drives the winds, while the sun's electromagnetic radiation can power photovoltaic and photo synthesis.

The sun provides the energy for plants to grow through photosynthesis where the plant coverts carbon dioxide gas into carbon biomass, which is stored in the structure of the wood or oil seed. When dry wood or biofuels are burned, carbon is consumed, releasing heat and carbon dioxide gas. Trees, grass, and other plants can be harvested and burned as firewood, wood pellets, wood chips, dried corn, biodiesel, and ethanol. Some plants are not burned in their entirety, but rather the oil extracted from them is burned, such as canola oil.

Two liquid forms of biomass are ethanol and biodiesel. Ethanol can be produced from the carbon-based feed stock such corn, wheat, barley, sugarcane, beets, waste fruits and vegetables, plant waste (straw) and agricultural

It can take as many as 200 large wind turbines like this one, shown under construction, to produce as much power as a fossil fuel plant.

Source: iStockphoto.com

1

residue. Dedicated cellulosic energy crops such as grasses, legumes, trees, sweet sorghum, Jerusalem artichoke, and sugar cane can be developed to thrive on land unsuitable for row crops, and yield more biomass per acre than crop residues. Biodiesel is derived from vegetable oils, greases, and animal fat residues from rendering and fish-producing facilities. It is produced by chemically reacting these oils or fats with alcohol and a catalyst to produce compounds known as methyl esters and the byproduct glycerin. It is simple and inexpensive to produce. Biodiesel performs as well as petrodiesel in terms of horsepower, torque, and haulage rates, and provides better lubricity (the capacity to reduce engine wear from friction) and reduced ignition knocking. It can also be used in furnaces for space heating. Like ethanol, it supports the farming community and contributes to life cycle energy reductions.

Passive solar space heating uses the sunlight to warm an interior space without the use of collector panels or piping. It requires certain considerations. First, orient the building in such a way as to capture as much sunlight as possible by placing the long axis of the building at a right angle to the true or solar north-south direction. Plant deciduous (leaf-falling) trees in a path between the summer sun and the house to provide shading. Plant evergreen trees on the north and east sides to provide a wind break during the colder months. If trees aren't available, place other structures (a garage or a barn) on the north and east sides, or locate your home on the side of a hill or rock outcropping. Secondly, weatherize your building by sealing and insulating, to block air and heat movement through the exterior of the building. Third, use windows to collect the sunlight into your building. Place the majority of the windows on the south side of the building, and place windows on either side of the building to align with the prevailing winds, for a cooling effect in the warm months. Limit the amount of windows on the north, northeast, and northwest sides of the house. Add sunscreen adjustable blinds, a controllable ventilation system, and sun shading, such as a roof overhang, over the east, west and south facing windows, to prevent the summer sun from heating your interior spaces. Apply an exterior cladding material that contains tiny perforations to allow outdoor air to travel through the exterior surface, applied on the east, south, or west walls (not the cold north wind wall). During the day, it absorbs the sun's energy, warming the air, which is drawn into the building's ventilation system. In the summer, it shades the interior wall and any hot air moving into it escapes outward through the perforations, cooling the wall(s).

Hydroelectricity is produced from the kinetic energy in moving water under the force of gravity. The sun evaporates water and forms clouds in the sky from which the water, in the form of raindrops, falls back to earth, forming a stream that runs downhill or a waterfall. The water enters the turbine heart of a generator and passes through a set of guide vanes, which force the water to hit the runner blades at the correct angle. Runner blades capture the kinetic energy and cause the generator to rotate, producing electricity. The water then exits out the draft tube, which is tapered to draw the water away from the blades, to increase efficiency. The pressure or head (determined by the vertical drop, and the volume of flow), the quantity of water flowing past a given point in a given period of time, determine the potential energy or power in water, expressed as horsepower or watts. At a high head site, a dam structure provides a means of inserting an intake pipe and strainer midway between the riverbed and the river surface.

The sun's energy causes the wind to blow, which moves the blades of a wind turbine (located on a tall tower and securely anchored to a concrete pad), causing a generator shaft to spin and produce electricity. A small change in wind speed causes a large change in electrical power output, so the blades must be located at a height that will offer the highest average wind speed and the best laminar flow, with the fewest obstructions.

Photovoltaic (PV) electric generators, made from transistors or integrated circuits, collect solar energy. When sunlight hits a photovoltaic panel, the sunlight is converted directly into electricity, which then powers the appliances, electronics, lighting fixtures, and other equipment that uses electricity to operate. Modules can be mounted to a fixed location pointing solar south, or to a tracking unit that automatically aims the panels directly at the sun. Unlike solar hot water panels, any interruption to the flow of sun energy onto the panel will disable the entire panel, so locate away from all obstructions and shadows.

Solar water heating (thermal) systems pre-heat the incoming water supply, reducing the fuel demand of the hot water heat. Solar collectors are best mounted on the roof at an angle approximately equal to your geographic latitude, facing solar south, but they can be detached from the house if necessary for optimum solar gain. They must have an unobstructed view of the sun between 9 a.m. and 3 p.m. Energy can be conserved by using the heat from your shower and sink (graywater) to your incoming cold water by using a simple heat recovery system and by using your graywater to flush your toilets as well. Active solar thermal systems typically use hydronic in-floor heating, where heated water (or antifreeze) is pumped through a series of flexible plastic pipes located under the flooring or embedded in the insulated concrete slab of the building. A zone thermostat turns a tiny pump on and off as heat is required in each zone.

Renewable heating fuels can be as diverse as firewood, wood chips or pellets, dried corn, peat, or cow dung. Catalytic (advance combustion) wood stoves and modern self-feeding pellet stoves, both with airtight burning chambers, are very efficient and clean-burning. There are many difference types and sizes of wood burners available, each with particular advantages and disadvantages.

Geoexchange technology, also known as geothermal, heat pump, and ground-source heating, offers a nearly endless supply of renewable energy stored underground. In winter, heat is drawn from below the surface of the earth through a series of pipes called a loop, an antifreeze solution circulates through the loop, carrying the warmth to a compressor and heat exchanger inside the building. The heat pump concentrates and transfers the energy via a hydronic heating system to heat the interior spaces. Electricity is needed to operate the compressor.

See Also: Alternative Fuels; Biodiesel; Biomass Energy; Electricity; Ethanol, Corn; Ethanol, Sugarcane; Geothermal Energy; Hydroelectric Power; Insulation; Nonrenewable Energy Resources; Photovoltaics (PV); Renewable Energies; Solar Energy; Wind Power; Wood Energy.

Further Readings

Flavin, Christopher and Nicholas Lenssen. *Power Surge: Guide to the Coming Energy Revolution*. New York: W. W. Norton & Company, 1994.

Goettemoeller, Jeffrey and Adrian Goettemoeller. *Sustainable Ethanol*. Maryville, MO: Prairie Oak Publishing, 2007.

Kemp, William H. *The Renewable Energy Handbook*. Tamworth, Ontario, Canada: Aztext Press, 2005.

Morris, Craig. *Energy Switch: Proven Solutions for a Renewable Future*. Gabriola Island, Canada: New Society Publishers, 2006.

Scheer, Hermann. *The Solar Economy: Renewable Energy for a Sustainable Global Future*. Sterling, VA: Earthscan, 2005.

Elizabeth L. Golden
University of Cincinnati

ALTERNATIVE FUELS

Standard fuel sources include nonrenewable energy sources such as petroleum-based products, natural gas, coal, propane, and nuclear power. Alternative fuels on the other hand refer to any non-conventional source materials that include biofuels (e.g., biodiesel, ethanol), hydrogen, wind power, hydroelectricity, geothermal energy, and solar power. The energy from alternative fuels may also be stored for later use using chemical storage systems (e.g., batteries, fuel cells). Alternative fuel sources are non-fossil fuel based and are an integral a part of a renewable and sustainable energy practice. Although fossil fuel substitutions have notable drawbacks, the development and implementation of alternative fuel technologies coupled with extreme conservation measures may provide industrialized nations a means to wean their historical dependence on fossil fuels without comprising economic prosperity.

The inexpensive fuels of the previous century are no longer and high fossil-fuel consumption among western nations shows trends of declining. In 2008, global oil consumption decreased by 0.6 percent, the largest decrease in nearly three decades that was driven primarily by 2 to 3 percent declines in traditionally major oil consuming nations such as the United States, Canada, and western Europe. However, the oil consumption decline of western nations is being countered by increasing consumption by developing world economies, particularly China and India. The world has already consumed about half of the recoverable fossil fuels since the modern industrial age, remarkable given fossil fuel development requires millions of years and specialized biological, geologic and climatic conditions. For instance, oil forms from the decomposition of marine life deposited in sedimentary basins that undergoes compaction from the increasing pressure of the overlying materials, thus trapping the hydrocarbons and the gas byproducts from the organic breakdown deep into the Earth's crust. The majority of oil and natural gas worldwide are found in sedimentary rock layers of the Cenozoic Era about 50 million years ago. The process of oil and natural gas creation took millions of years, while the process of extraction is on the order of a couple centuries.

Although uncertainty exists about when world oil production will peak and when viable oil reserves will cease, experts generally agree supplies will be severely limited within the next generation and a fossil fuel energy crisis may be unavoidable. The ratio of oil consumption to oil availability is nearly 3:1. Based on 2009 global consumption and production levels, British Petroleum (BP) estimates 42 years are left before current oil reserves are completely depleted. However, the BP estimate does not incorporate several key parameters that can influence oil consumption, such as: consumption pattern changes, including those stemming from population increases and conservation measures; alternative fuel usage; increased oil extraction costs from less accessible reserves; pricing and production strategies from oil producers (e.g., the Organization of Petroleum Exporting Countries, OPEC); or new oil reserve discoveries. Whether current oil reserves will be sufficient to meet demand for the next decade or several decades or whether world oil production has climaxed is largely immaterial. The global energy crisis will begin when oil production is growing more slowly than the amount demanded by the population coupled with volatile pricing. As such, energy consumption habits will alter as petroleum-based products gradually become more expensive and alternative energy solutions are sought.

Concerns over oil reserves and exorbitant prices in the early 21st century have led to increased consumption of other fossil-based fuels, particularly coal that has become the mainstay electrical energy source of developing world economies. As with oil, coal is derived from the decomposition of organic material on a geologic time scale. Coal forms from decaying plant material that accumulates near low-lying swamps during warm, humid climate conditions and then undergoes physical transformation as the material is subjected to increasing temperature and pressure below the surface of the Earth. The majority of the world's coal was formed during the Carboniferous period approximately 300 to 360 million years ago during a period of high sea-level and extensive tropical vegetation. Since coals are derived from carbon-based organisms, the combustion of coal produces large quantities of carbon dioxide (CO_2) as well as other air pollutants such as sulfur dioxide (SO_2), nitrogen oxides, and particulate matter. Anthracite, the highest grade coal, provides the highest energy content per unit mass while producing the smallest amount of pollutants compared with the lower and more abundant coal grades of bituminous, subbituminous, and lignite. Based on current consumption patterns, coal reserves will last at least five times longer than oil; however, coal, as with other fossil-based fuels is not a sustainable energy option.

In order to stave off the impending energy predicament, some have turned to increased fossil fuel extraction using unconventional methods and sources. Oil (or tar) sands are being exploited for bitumen, a viscous black oil that requires multiple extraction, separation, and dilution processes before the oil can be used as a viable fuel. Canadian oil sands, such as the Athabasca fields in Alberta, are particularly rich, producing nearly one million barrels per day with plans to triple that production. Oil sand extraction has significant environmental consequences (e.g., increased greenhouse gas emissions, diminished water quality, wildlife habitat loss) with minimal energy gains; two tons of tar sands are needed to produce one barrel of crude oil. Other unconventional sources include oil shales and liquefied coal, but the economic and environmental impacts are cost-prohibitive and widespread adoption is not likely. Consequently, fossil fuels are gradually being substituted with alternative energies.

Biofuels

Alternative energy fuels comprise renewable energy resources such as biofuels; solid, liquid or gaseous fuels derived from living or recently living organic material; or biomass. The energy from biomass is used as a fuel for domestic heating and cooking, running power generators for electricity production, and transportation fuels. Biofuels include long-exploited sources such as wood and grass that produce heat directly when burned.

Liquid biofuels are increasingly being developed as substitutes for gasoline for the transportation fuel market. The most touted substitutes are ethanol (also known as ethyl alcohol or grain alcohol) from fermented plant starches and sugar-based feedstocks and biodiesel from oil-based crops. In the United States, ethanol derived from corn is the leading fuel alternative, devoting 14 percent of the annual national corn crop for ethanol fuel production. Since pure ethanol provides only one-third of the energy produced from gasoline, ethanol is often blended with gasoline to increase the energy efficiency; a common ethanol blend for vehicles is E85, a high-octane mixture of 85 percent ethanol and 15 percent gasoline. Ethanol production and demand in the United States has increased significantly in the past decade, but gasoline consumption has paralleled those increases. Ethanol production technologies are inefficient, consuming as much energy during production as that is released during combustion. Corn-based ethanol currently meets only about 4 percent of the total U.S. fuel needs; distilling the entire corn crop would only satisfy less than one-fourth that demand. Ethanol from corn remains a viable alternative fuel in the United States due to the relative abundant corn supplies, an established infrastructure and, perhaps, more importantly, a 51 cents per gallon federal subsidy.

Similar energy efficiency problems—consuming almost as much energy as it releases—occur with ethanol originating from other plants. In non-grain crops, the material being converted to liquid fuel is the cellulose biomass, a plant material composed of complex sugar polymers found in agricultural and food processing byproducts (e.g., wood chips, corn husks, household garbage). Cellulose could be grown in greater quantities than grain with less environmental impact, but converting it into ethanol is more difficult than distilling the starches in corn because the chemical bonds holding the cellulose together must be broken first. This step requires additional energy, thus making cellulose-based ethanol no more energy efficient than corn ethanol. Energy is further expended during collection and transportation of the dispersed crop to a distillery for refinement, especially given that only 3 percent of the biomass eventually becomes ethanol. Although ethanol technologies may improve over time, the energy efficiency of ethanol is limited by the energy available from the input product.

In contrast to corn- and cellulose-based varieties, ethanol made from sugarcane can achieve a positive energy balance because sugar yields more calories per acre than any other crop. Sugarcane-based ethanol yields eight times more energy than the fossil fuel inputs to its production. Several low-latitude countries produce sugar ethanol on a large scale, using the non-sugar parts of the cane plant as fuel for the distillation process. In particular, Brazil devotes enormous tracts of land for sugar production that are used for generating 8 percent of the national total fuel supply, a remarkable percentage especially when considering the per capita energy usage by Brazilians is about one-fifth of that of Americans.

Another path to biofuels is growing oil crops and plant oil–based sources for use in diesel engines. Plant oil undergoes a process called esterification, joining two organic molecules together. Alcohol and a catalyst (a substance that enables a chemical reaction) convert the oil into an ester fuel called biodiesel. The process is more efficient than alcohol distillation, and many crops grow seeds that are rich in oil. Soybeans are a premium choice in the United States because they are grown on an industrial scale, but the amount of fuel obtainable per acre is even less than with corn ethanol, making the product viability difficult. In Europe, biodiesel stems chiefly from palm plantations in the tropics. The plantations are environmentally destructive and clearing land for them produces more carbon dioxide than the oil saves from not using fossil fuels. Other oil crops face similar problems, balancing energy acquisition with negative consequences such as environmental degradation and food production limitations. Biodiesel is also being made from oil-based plant substances that have been processed, such as vegetable and other cooking oils discarded by restaurants and food processors. As with other oil crop biofuels, the problem is scaling, as only a very small proportion of the world have the technological infrastructure to utilize biodiesel and the amount available would not be sufficient for residential and commercial energy needs.

Hydrogen

Hydrogen is a lightweight gas derived in large quantities from the chemical breakdown of hydrogen-based sources (e.g., water) or as a byproduct of other chemical processes. The two most common techniques for hydrogen production are steam production and electrolysis. The steam production method separates carbon atoms from methane (CH_4); however, the process results in the increase of greenhouse gas emissions linked to recent global climate change. Electrolysis involves the splitting of hydrogen from water, an expensive process without global climate implications. Although often publicized as the fuel of the future, hydrogen is technically not a fuel source but serves as an energy carrier for moving energy in a practical form between mediums. Hydrogen batteries or fuel cells are then used to produce electricity. Fuel cells have been primarily utilized for industrial applications (e.g., metal refinement, food processing) as well as for remote and portable electricity generation from powering portable electronic devices to space shuttles. Increasingly, hydrogen is being sought as a viable alternative fuel for vehicles.

While hydrogen has the advantage of producing considerably less pollution than fossil-based fuels, hydrogen has some significant downsides. Hydrogen is a particularly volatile substance, making transportation and storage difficult. Even though hydrogen has the highest energy content of any fuel by weight, the energy density is about a quarter that of gasoline. Hydrogen creation is expensive and the renewable energy sources that might produce hydrogen on a large scale, such as wind electricity for hydrolysis of water, can be used more efficiently to power motors directly.

Other Alternative Fuel Sources

The generation of electricity by multiple mediums (e.g., solar, water, wind) is another major alternative fuel group. In the transportation industry, gasoline-electric hybrid vehicles like the Toyota Prius are finding recognition as a means of fossil fuel energy conservation rather than complete fuel replacement. Petroleum-electric hybrid vehicles use highly efficient internal combustion engines, fueled by gasoline or diesel, to turn a generator that the powers the batteries and/or supercapacitors used for propulsion. Completely electric vehicle models powered by onboard battery packs have waxed and waned in popularity in the United States, but have seen a marked increase in adoption in Europe and Asia. A single battery charge can power electric vehicles for 40-50 miles, making them practical for the majority of routine daytrips, and can be recharged during the night when electricity usage is typically in lower demand. Other alternative transportation fuel development includes creating solar-powered vehicles powered by rooftop photovoltaic panels; however, current technology is only able to produce about one-tenth of one horsepower (i.e., less than the power generated from riding a bicycle).

Much of the most promising alternative fuel options come from electricity generation by renewable sources, primarily hydroelectric. The movement of water from higher to lower elevations, often using water held in a dam, creates kinetic energy as the water flows downstream. A hydroelectric power plant can convert this energy into electricity by pushing the water through hydraulic turbines that are connected to generators. In the western United States, huge dams produce large quantities of electric power, enough to drive, for instance, the San Francisco mass transit system. Hydroelectric power has also become an increasingly large fuel source for the developing world, most notably China that has undertaken some of the largest dam construction projects in the world (e.g., Three Gorges Dam). The environmental costs for such megadams are high, including ecological habitat loss and increased river sediment load that must be weighed against power generation by other means.

Electricity generated from solar energy has become more widespread in recent years with advances in photovoltaic (PV) technology. Photovoltaics are sunlight sensitive semiconductors composed of materials like silicon that are capable of converting sunlight into electricity. In addition to electricity conversion, PV technology is being used to develop ways to harness the solar energy into fuel cells. In terms of viability as an alternative fuel, solar energy–derived fuels remain relatively inefficient and expensive and, as a consequence, often require the support of government subsidy programs to sustain their development.

At present, solar panels require nearly three years to generate power equal to energy inputs made during the manufacturing process. Solar power is most economical on the smaller, residential scale (e.g., home water heating) in certain climate zones. Large-scale solar electricity plans are well into the future, and may include heat-based solar electric plants in the deserts that are still in their experimental phase.

Wind power has become highly developed in Europe and generates electricity less expensively than solar panels and has already become cost-competitive with fossil fuels in some locations. Modern wind power generation utilizes large three-four blade wind turbines (with heights 60 to 100 meters and blade spans of 20 to 40 meters or more) to harness the kinetic energy of the wind and convert it into mechanical energy for electricity. The maximum power generated by a single wind turbine is dependent on rotor diameter, tower height, and wind speed; for example, a rotor diameter of 40 meters and a maximum wind speed of 15 meters per second (or 33 miles per hour) can produce about 500 kilowatts. For large-scale power generation, multiple turbines are clustered together into wind farms. Approximately 200 of the largest wind turbines are needed to equal the output of a single fossil-fuel power plant, a figure reliant on moderate winds blowing.

See Also: Biodiesel; Ethanol, Corn; Ethanol, Sugarcane; Fossil Fuels; Hydroelectric Power; Hydrogen; Renewable Energies; Solar Energy; Wind Power.

Further Readings

Bourne, Joel. K., Jr. "Green Dreams." *National Geographic* (October 2007).
Hirsch, Robert L., Roger Bezdek, and Robert Bendling. "Peaking of World Oil Production: Impacts, Mitigation, & Risk Management." *U. S. Department of Energy Report.* www .netl.doe.gov (Accessed February 2009).
Kunstler, James H. *The Long Emergency: Surviving the End of Oil, Climate Change, and Other Converging Catastrophes of the Twenty-First Century.* New York, NY: Grove Press, 2005.
Lee, Sunggyu, James G. Speight, and Sudarshan K. Loyalka, eds. *Handbook of Alternative Fuel Technologies.* Boca Raton, FL: CRC Press, 2007.

Jill S. M. Coleman
Ball State University

Bruce Bridgeman
University of California, Santa Cruz

APPLIANCE STANDARDS

Appliance standards are federally mandated legislation establishing the energy efficiency regulations for assorted residential and commercial appliances. Appliance standards continue to be an important part of the larger policy approach to energy savings, requiring manufacturers, distributors, and retailers to produce energy efficient products or receive cumulative financial penalties for noncompliance. Although future energy savings and carbon emissions reductions can be achieved by way of numerous mechanisms, appliance standards have already played an important role in reducing both household and commercial energy consumption. Despite these past successes, however, the potential energy savings of new appliance standards continue to hold the opportunity for significant future energy savings as well. In other words, appliance standards represent a proven approach to reducing the energy needed to meet a variety of energy services such as refrigeration and cooking. Of particular note, appliances are the largest users of electricity in the average U.S. household and are responsible for roughly two-thirds of all electricity use in the residential sector. Household appliances include items like refrigerators, freezers, stoves, ovens, dishwashers, clothes washers, and clothes dryers, but they also include televisions, computers, and numerous other items. The broad range of appliances combined with the large number of U.S. households (approximately 114 million) make the management of appliance-related energy consumption a real challenge for U.S. policy makers and homeowners alike.

Refrigerators are generally the appliances that consume the most energy in the typical household, and currently there are more refrigerators than there are households in the United States. For example, in 2006 there were nearly 140 million refrigerators in use in the U.S. residential sector alone. Each of those refrigerators consumed an estimated 655 kilowatt-hours (kWh) of electricity per year. Total refrigerator-related energy consumption in that year was estimated at 91.7 billion kWh and was responsible for approximately 14 percent of total residential electricity use. Table 1 shows residential electricity end uses as a percentage of total residential electricity consumption.

Some of the best efforts to manage appliance-related energy consumption have been achieved through appliance standards. Between 1987 and 2000, standards-based electricity savings were estimated at approximately 2.5 percent and U.S. carbon emissions savings at approximately 2 percent (based on federal appliance efficiency standards implemented since 1987). In other words, total electricity savings between 1987 and 2000 totaled an estimated 88 trillion kWh per year, and estimates of continued energy savings through 2010 and 2020 are projected to save 6.9 percent and 9.1 percent of total U.S. electricity use in those years, respectively. Despite these achievements, there continue to be significant levels of unrealized energy savings that could be attained through the implementation of standards on those appliances for which standards aren't currently imposed, and through the implementation of stricter standards on appliances for which existing federal standards are now ready for updating. For example, new federal standards on furnace fans, central air conditioners, refrigerators, and other appliances could provide an additional 180 trillion kWh of electricity savings and 2.3 quads of primary energy savings annually by 2030.

Ultimately, the goal of appliance standards is to encourage the continuous transition away from relatively energy-inefficient appliances toward more energy-efficient appliances. By setting increasingly stringent standards, federal

Table I   Residential Electricity Consumption by End Use

End Use/Appliance	Percentage Electricity Consumption
Air Conditioning	16.0
Space Heating	10.1
Water Heating	9.1
Appliances (total)	64.8
Refrigerator	13.7
Lighting	8.8
Clothes Dryer	5.8
Freezer	3.5
Dishwasher	2.5
Other Appliances	30.5
TOTAL	100

Source: Energy Information Administration 2001, Residential Energy Consumption Survey.

and state agencies have the ability to encourage appliance manufacturers to continue to innovate and to provide consumers with products that use fewer energy resources. Moreover, standards ensure that efficiency improvements are built into all new products and aren't limited strictly to premium products. They also reduce the need for consumers to become "experts" in the energy attributes of competing products. As a result, all buyers are afforded a higher minimum efficiency. In short, each new advancement in appliance standards and technologies allows consumers to enjoy the same energy services (or better services) while consuming less energy to achieve them.

Minimum efficiency standards are generally set at levels designed to eliminate a certain proportion of the current, less-efficient products from the market. Efficiency standards provide manufacturers with the freedom to design their products any way they want, as long as they meet or exceed specified efficiency levels. Historically, standards have been a preferred approach, as they have been shown to be both relatively effective and easy to implement. Moreover, according to the Collaborative Labeling and Appliance Standards Program, programs that are well designed and implemented are hugely cost-effective and also very good at "limiting energy growth without limiting economic growth." Standards programs also tend to be favored because they focus on changing the behavior of a relatively limited number of manufacturers rather than the behavior of the larger number of appliance consumers.

History

The first appliance standards were implemented in Europe in the 1960s. According to Steven Nadel, "the first confirmed account of appliance standards was in 1966, when the French government adopted refrigerator standards." However, much of the initial effort to implement standards resulted in fairly weak legislation that was poorly implemented and had only a limited effect in terms of appliance energy consumption. In the United States, standards were first discussed as a policy tool after an electrical blackout darkened the Northeast in 1965. Following growing concern about how to provide adequate power generation, California provided authorization to the California Energy Commission in 1974 to set appliance efficiency standards. The first standards took effect two years later in both California and New York.

Voluntary standards were considered at the federal level around the same time; however, it wasn't until a growing number of states had adopted standards on one or more products (in the mid-1980s) that mandatory federal standards were revisited as a policy option. After more than 10 years, the United States passed the National Appliance Energy Conservation Act in 1987, specifying standards on many major appliances and requiring the Department of Energy to periodically review and revise minimum efficiency standards. Additional national standards were enacted in 1988 and 1992. More recently (since 2001), a variety of states have ramped up their efforts to impose even more stringent efficiency standards. These states include California, Maryland, Connecticut, New Jersey, Arizona, Washington, Oregon, Rhode Island, New York, and Massachusetts. (See the American Council for an Energy-Efficient Economy State Energy Efficiency Policy Database for current state information: http://www.aceee.org/energy/state/index.htm.) In addition, recent action at the federal level has included the adoption of the Energy Policy Act of 2005, which (among other requirements) set new federal efficiency standards for 16 products and called for the establishment of new standards for five additional products.

Table 2 Savings From Federal Appliance and Efficiency Standards

Year Enacted	Standards	Electricity Savings (TWh/yr)			Carbon Reductions (million metric tons)			Net Benefit ($bil) Through 2030
		2000	2010	2020	2000	2010	2020	
1987	NAECA	8.0	40.9	45.2	3.7	10.0	10.1	46.3
1988	Ballasts	18.0	22.8	25.2	4.4	5.0	5.0	8.9
1989/91	NAECA*	20.0	37.1	41.0	4.8	8.1	8.1	15.2
1992	EPAct	42.0	110.3	121.9	11.8	27.5	27.9	84.2
1997	Refrige/freezer*	0.0	13.3	28.0	0.0	2.9	5.5	5.9
2000	Room AC*	0.0	1.3	2.1	0.0	0.3	0.4	0.6
2001	Ballasts*	0.0	6.2	13.7	0.0	1.3	2.7	2.6
2001	Clothes washer*	0.0	8.0	22.6	0.0	2.2	5.4	15.3
2001	Water heater*	0.0	2.5	4.9	0.0	1.4	2.2	2.0
2001	Central AC&HP*	0.0	10.7	36.4	0.0	2.3	7.2	5.0
2005	EPAct 2005	0.0	14.7	53.0	0.0	3.7	11.5	47.5
	TOTAL	88	268	394	25	65	86	234
	% of projected U.S. use	2.5%	6.9%	9.1%	1.7%	3.6%	4.4%	

Source: Nadel et al., 2006.

*Indicates standards updates.

Note: NAECA = National Appliance Energy Conservation Act; AC&HP = air conditioners and heat pumps.

Key Players

State governments, nonprofit organizations, and appliance manufacturers have been among the key players in instigating, defining, and advancing appliance standards in the United States.

Among the government actors, states have continued to play a critical role in pushing the frontier of appliance standards and motivating action at the federal level. Notably, it was the growing number of state standards and concerns about the possibility of future federal standards that led appliance manufacturers to work with nonprofit organizations and other energy efficiency advocates (including the American Council for an Energy-Efficient Economy) to develop a consensus proposal on federal standards.

The Effect

Standards have attained some success and have had a tremendous effect on U.S. energy consumption. Even appliance manufacturers concede that "standards have driven the development of high efficiency [refrigerator] components such as high [energy efficiency ratio] compressors, adaptive defrost controls, and low-wattage fan motors that are currently used exclusively for the U.S. marketplace." In addition to energy savings, efficiency standards also provide the opportunity to save money, create jobs, limit global warming, and protect public health. More efficient appliances mean lower energy bills (all else equal), and reduced demand helps lower power rates across the board. In addition, businesses are likely to add jobs to meet the increased demand for energy-saving products. Of course, increased efficiency also reduces carbon emissions and air pollution from power plants, which contribute to public health problems like asthma. The savings from federal appliance and equipment efficiency standards are summarized in Table 2 (p. 11).

The Way Forward

Upgraded national efficiency standards offer the potential for significant growth in standards-related efficiency savings. Future policies need to focus on broadening the scope of appliance standards and on making existing standards increasingly strict, so as to maximize the technological capacity of existing efficiency technologies. In many ways, California is setting the pace by implementing standards for about 20 products not covered by federal standards. The National Association of Regulatory Utility Commissioners has also adopted resolutions in support of both upgraded national efficiency standards and expanded state efficiency standards.

See Also: Electricity; Energy Audit; Energy Payback; Energy Policy; Power and Power Plants.

Further Readings

Energy Information Administration. "End-Use Consumption of Electricity 2001." From the Residential Energy Consumption Survey. http://www.eia.doe.gov/emeu/recs/recs2001/enduse2001/enduse2001.html (Accessed February 2009).

McInerney, E. and V. Anderson. "Appliance Manufacturers' Perspective on Energy Standards." *Energy Build,* 26:17–22 (1997).

Nadel, Steven. "Appliance and Equipment Efficiency Standards." *Annual Review of Energy and Environment,* 27:159–192 (2002).

Nadel, Steven, et al. "Leading the Way: Continued Opportunities for New State Appliance and Equipment Efficiency Standards." Washington, DC: American Council for an Energy-Efficient Economy, 2006.

Nadel, S. and D. Goldstein. "Appliance and Equipment Efficiency Standards: History, Impacts, Current Status, and Future Directions." Proceedings of the ACEEE Summer Study on Energy Efficiency in Buildings, 2.163–2.172. Washington, DC: American Council for an Energy Efficient Economy, 1996.

Wiel, S. and J. McMahon. "Energy Efficiency Labels and Standards: A Guidebook for Appliances, Equipment and Lighting." LBNL-45387. Washington, DC: Collaborative Labeling and Appliance Standards Program, 2001.

Karen Ehrhardt-Martinez
American Council for an Energy-Efficient Economy

ARCTIC NATIONAL WILDLIFE REFUGE

The Arctic National Wildlife Refuge (ANWR) is a federally protected conservation reserve of 19 million acres in the northeast corner of Alaska. The refuge is home to communities of Inupiat and Athabascan peoples as well as a rich array of wildlife in a pristine wilderness. At issue is a 1.5-million-acre section of coastal plain known as the 1002 Area, or 1002. Political disputes over 1002 have been ongoing for over 30 years and center on whether or not Congress should permit oil drilling in this ecologically significant tract. Proponents of drilling assert that tapping oil reserves in 1002 would reduce U.S. dependence on foreign oil, enhance U.S. energy security, and buffer future gasoline price increases. They also believe that drilling can be done in an environmentally sensitive manner. Opponents of drilling assert that the prospective oil reserves in 1002 represent a negligible proportion of U.S. oil consumption and would have no meaningful effect on world oil prices, U.S. gasoline prices, or U.S. energy security. They also argue that the environmental and ecological impacts on 1002 specifically, and ANWR generally, would be devastating. In the ANWR debates, the positions are largely polarized between those who promote oil drilling and those who want the coastal plain to be permanently protected, intact wild lands. The issue of whether oil development can proceed in an environmentally responsible manner is generally not part of the national argument.

History and Natural Resource Management

Present-day ANWR has its roots as the Arctic National Wildlife Range, an 8.9-million-acre region created by Department of Interior in 1960 under the administration of Dwight Eisenhower. During the 1950s, prominent conservationists and the National Park Service urged preservation of this tract because of its value as undisturbed wilderness, the scope of its unique wildlife, its cultural heritage, and its recreational potential. Secretary of Interior Fred Andrew Seaton issued Public Land Order Number 2214, creating the Arctic National Wildlife Range in 1952.

The Arctic National Wildlife Range became ANWR in 1980, with the passage of the Alaska National Interest Lands Conservation Act of 1980 (ANILCA; P.L. 96-487). ANILCA increased the size of the range to nearly 18 million acres and designated three of ANWR's rivers as "wild" and all but a portion of the original range as "wilderness." The remaining acreage of ANWR is designated as "minimal management," including 1002. ANWR was expanded again in 1983 and 1988, bringing its total size to about 19.3 million acres, roughly equivalent to the size of South Carolina. ANWR contains the largest designated wilderness region in the U.S. National Wildlife Refuge System and also includes coastal lagoons and barrier islands on its northeastern shores. ANWR is managed by the U.S. Fish & Wildlife Service of the Department of Interior.

The Wilderness Act (P.L. 88-577) and the Wild and Scenic Rivers Act (P.L. 90-542) afford extraordinary protection to land formally designated as wilderness and to rivers designated as wild. Under these designations, land management goals focus on preserving the natural conditions and wilderness character of the region, including ecological processes. Human recreational use and visitation is not prohibited but is highly regulated; commercial development, permanent human structures, and settlements are not allowed. The "minimal management" classification also involves the maintenance of natural conditions and resource values. ANWR itself contains no roads or trails and is not accessible by motor vehicle; visitors typically fly to the refuge, although it is accessible by boat and on foot. In section 1003 of ANILCA, Congress specifically prohibited oil and gas production, leasing, or exploration leading to production of oil and gas from ANWR without an act of Congress.

Ecological Significance

The ecological and biological significance of ANWR is considerable. It is an ecologically intact, undisturbed wilderness that lies entirely above the Arctic Circle and contains a continuum of six ecological zones. As the only conservation area in the circumpolar north to include the complete range of North American arctic ecosystems, it is unique. Primarily permafrost, ANWR contains arctic tundra, glaciers, and boreal forest. There are no known exotic (introduced) species in the refuge.

Biodiversity within ANWR is high. It is home to all three species of bears in North America and provides denning sites for polar bears. Nearly 200 bird species have been observed within the refuge, and it is a major migratory pathway for waterfowl. Many charismatic species of birds and mammals live in ANWR, including the golden eagle, snow geese, swans, cranes, seals, whales, the arctic fox, lynx, Dall sheep, wolves, black bears, polar bears, and caribou; altogether, 45 species of land and marine mammals inhabit the refuge, several of which use it as critical seasonal breeding habitat. The tundra that supports much of this wildlife is a fragile ecosystem of braided rivers, mosses, small shrubs, sedges, and other vegetation. At higher elevations, poplar groves, stands of spruce, and the boreal forest support seasonal migrations of regional caribou herds and birds from Central and South America. Iynx, wolverines, and other large predators are permanent residents of the forest.

Although 1002 represents less than 10 percent of ANWR, it contains most of two ecological zones within the refuge—the coastal plain and arctic foothills. Because 1002 is so close to both the mountains and the coast, the geographic compactness of these varied habitats has given rise to a greater amount of biodiversity than other tracts of comparable size on the North Slope in Alaska. Several species of birds, muskoxen, polar bears, and the

Porcupine caribou herd also rely on 1002 for breeding grounds or as migratory way stations. Preservation of habitat for these species, including the ability to move freely throughout their range, is required for their survival as well as for the overall integrity of ANWR's ecosystems.

Settlements and Land Rights

Settlements and land rights in ANWR are somewhat complicated. ANWR is not strictly uninhabited. Over 11,000 acres in ANWR are privately owned, and some landowners have erected cabins. A little over 175,000 acres was conveyed through ANILCA and the Alaska Native Claims Settlement Act (P.L. 92-203) to corporations owned by native peoples.

There are two human settlements located in or adjacent to ANWR. Kaktovik, an Inupiat Eskimo village of approximately 150 people, is located on the northern coast on Barter Island. Arctic Village, an Athabascan Gwich'in settlement, also comprising about 150 individuals, is on the Chandalar River on lands that border the south-central portion of the refuge. The Inupiat and Athabascan peoples have lived in the region for generations, and more than 300 archaeological sites have been found within ANWR.

The Kaktovik Indian Corporation owns approximately 92,000 acres of surface lands, all within the boundaries of ANWR. Under the Alaska Native Claims Settlement Act of 1971 and ANILCA, the Kaktovik Indian Corporation has surface rights to conveyed lands located within the refuge, but subsurface ownership is retained by the federal government. Under these conditions, it is possible for the U.S. government to lease or exchange subsurface rights to parties other than the Inupiat peoples. If Congress authorized it, oil drilling could thus take place on Inupiat lands without their permission because of the legal limitations of surface rights.

1002

The region of greatest political focus within ANWR is an area of coastal plain approximately 1.5 million acres in size, located along the Beaufort Sea. This portion of ANWR was in the original Arctic National Wildlife Range and is essentially the only part of the original range that was not designated as wilderness by ANILCA in 1980. Instead, Section 1002 of ANILCA identified this tract for further study before Congress could designate it as wilderness or authorize oil development. This tract of land is now conventionally known as 1002 and is classified as a minimal management area for the purposes of natural resource conservation. The studies mandated by Congress included a complete inventory of fish and wildlife, an assessment of potential oil resources contained by the area, and an analysis of the possible effects of oil and gas exploration and development on natural resources.

This area is on the North Slope of Alaska—an oil-rich region extending north of the mountainous Brooks Range to the Arctic Ocean. In the early 1920s, 23 million acres of the North Slope were set aside as Naval Petroleum Reserve No. 4 (now the National Petroleum Reserve—Alaska) to preserve a supply of oil for national security needs. The National Petroleum Reserve—Alaska is larger than ANWR and is technically the largest area of undisturbed public land owned by the United States. In 1968, oil was discovered on Alaskan state land in Prudhoe Bay, the largest oil field in North America. Other oil discoveries were made on Alaska's North Slope, giving rise to the expectation that oil would also be found within 1002.

The political tensions between wilderness preservation and commercial oil interests emerged with the discovery of oil at Prudhoe Bay in 1968, intensified with the completion of the Trans-Alaska Pipeline system at the end of the 1970s, and became fully formed in 1986. At that time, the U.S. Fish & Wildlife Service drafted a report recommending that ANWR's coastal plain be opened for oil and gas development, a move fundamentally opposed by conservation organizations and many members of the House of Representatives. A bill authorizing drilling in ANWR was expected to pass readily in the Senate in 1989 when the *Exxon Valdez* oil spill occurred, effectively ending the momentum for such an authorization. Since that time, conflicting legislative initiatives authorizing drilling in ANWR, banning drilling in ANWR, and designating 1002 as wilderness have regularly been introduced in Congress but have failed to be enacted in any version. The division is largely (but not entirely) partisan, with Democrats against drilling and Republicans for.

Because of changes in oil equipment and technology that make development more compact, with an associated smaller "footprint," in 2001 and 2003, the House of Representatives passed bills (never enacted) that would have opened the refuge to oil development but limited development to 2,000 surface acres of the coastal plain within ANWR. The 2,000-acre development concept was reiterated by President George W. Bush in 2005, considered by Congress again in 2006, and has been a source of additional controversy within the ANWR drilling debates for two reasons. First, the legislative language has been ambiguous about the spatial distribution of the 2,000 acres within the coastal plain. Critics of the approach argue that a consolidated tract of 2,000 acres would not be able to fully exploit oil within 1002, and that under the language of the bill, the 2,000 acres could be spread out across the 1.9-million-acre ecozone in a vast network of exploration, drilling rigs, roads, and pipelines, a scenario worse, in many ways, than the presence of a single contiguous oil field. Second, the legislation did not address the status of Native lands within the refuge—thousands of acres of which could potentially be open for development—in an unrestricted manner—if drilling were permitted in ANWR.

Broadly speaking, those in favor of oil development in ANWR are the state of Alaska, its citizens, and its elected officials; commercial oil interests; individuals and organizations advocating oil security and independence; and (cautiously) the Kaktovik Indian Corporation. In general, most of these stakeholders stand to gain economically—through tax, lease, or sales revenues; profits; and jobs—from oil development. Other stakeholders may not have a direct financial interest but advocate for greater oil independence and security in the United States through a greater domestic supply of petroleum. More generalized perceived benefits from oil development include lower oil prices, spillover effects of jobs throughout the United States (not just in Alaska), and an extension of the economic viability of the Trans-Alaska Pipeline System.

Formal opposition to drilling in the refuge relies heavily on nonprofit environmental preservation organizations. These institutions argue that conservation of a unique American wilderness outweighs the uncertain, if not outright limited, projections of oil reserves in ANWR. Two key studies, those of the National Academy of Sciences and the U.S. Geological Survey, are often cited in discussions of potential harms from oil development in the refuge. Based in part on observations of damage created by existing North Slope oil fields, ecological harm could result from (1) noise from drilling and seismic exploration, (2) damage to permafrost from construction equipment for roads and oil fields, (3) oil spills, (4) the presence of industrial settlements, (5) obstacles to freedom of movement for herding animals, (6) accidental kills of wildlife during construction, (7) the presence of roads and road construction, (8) the presence of low-flying aircraft, and (9) saline water spills from excavation and drilling.

Documented biological changes from oil development on the North Slope include damage to tundra, decline in species populations, and displacement of species. With respect to ANWR, concern over the loss of calving grounds for the Porcupine caribou in 1002 is the most prominent issue discussed, but loss of denning habitat for polar bears, displacement of bowhead whales, a reversal in the successful reintroduction of muskoxen, and loss of habitat for roughly 135 species of migratory birds are also frequently mentioned as concerns about opening the refuge for oil development.

Effects on human communities are also of note. The Inupiat on the north coast have cultural and subsistence ties to bowhead whales, and the Gwich'in in the southern region depend on the caribou for their subsistence and cultural way of life. Indeed, the Arctic Village is on one of the main Porcupine caribou seasonal migration routes. Changes in whale or Porcupine caribou dynamics are likely to be of direct consequence not just to the two villages located in or next to ANWR but also to the extended community of several thousand Inupiat and Gwich'in in the region. In general, however, the Gwich'in are strongly against drilling in the refuge, whereas the Inupiat are more generally in favor.

Oil Reserves in ANWR

To understand the estimates about potential oil recovery within ANWR, it is important to distinguish between prospective resources, proven reserves, technically recoverable reserves, and economically recoverable reserves. A prospective oil resource is one that we believe to exist in a certain location using a variety of scientific indicators but has not been physically proven to exist through testing and other exploratory measures. Such measures could include core drilling, seismic exploration, and other geological techniques. Proven reserves are those that have been physically demonstrated to exist and have some credible, quantitative estimate as to the total amount of oil in place. Oil resources, both prospective and proven, are usually characterized as technically recoverable (we have the technology to get it out of the ground) or as economically recoverable (we can get it out of the ground in a cost-effective manner). To be technically recoverable, reserves rely on the state of technology for extracting oil deposits in a variety of geological conditions. Economically recoverable reserves reflect a subtle interaction between the technology required to extract the crude oil and the prevailing market price for petroleum. If market prices are too low, it is not cost-effective to recover reserves that are otherwise technically accessible.

In ANWR, oil reserves are prospective and are based on 1998 U.S. Geological Survey estimates. The survey estimates that the technically recoverable, prospective reserves in ANWR's coastal plain range from 5.7 billion to 16 billion barrels of oil, with an average estimate of 10.4 billion barrels. This is equivalent to 4–13 percent of the estimated amount of all technically recoverable, undiscovered reserves in the remainder of the United States. It also represents about two years or less of total U.S. domestic consumption of petroleum products: Americans consumed just over 7.5 billion barrels of petroleum products in 2007.

Recent analysis by the Energy Information Administration sheds light on the economic role of this oil over the next 20 years. Oil reserves are not instantaneously available: oil fields and transportation networks take time to construct, and there are physical daily limits to extraction and shipping. As a consequence, we should expect that any oil development from the refuge's estimated reserves will enter the market about a decade after development begins and in a flow over several years, not immediately and all at one time.

Because of these time constraints, the relatively small volume of oil reserves predicted for ANWR, and the large volume of U.S. oil consumption, the Energy Information Administration analysis suggests that U.S. oil import dependency would decline modestly only for about four years, about 14 years after oil development starts in the refuge. Effect on peak global price would be realized about eight years after production starts and would have a marginal decrease in price of 41 cents per barrel if ANWR's reserves are on the low end, and a marginal decrease in price of $1.44 per barrel if ANWR's reserves are on the high end.

See Also: *Exxon Valdez*; Offshore Drilling; Oil; Oil Sands; Oil Shale.

Further Readings

Arctic Circle World Wide Web Project. "Arctic Circle." http://arcticcircle.uconn.edu/ (accessed April 2009).

Congressional Research Service. *Oil and Gas Leasing in the Arctic National Wildlife Refuge (ANWR): The 2,000 Acre Limit.* Washington, DC: CRS Report RS22143, 2006.

Energy Information Administration. *Analysis of Crude Oil Production in the Arctic National Wildlife Refuge.* Washington, DC: U.S. Department of Energy, May 2008.

National Research Council. *Cumulative Environmental Effects of Oil and Gas Activities on Alaska's North Slope.* Washington, DC: National Academies Press, 2003.

U.S. Fish & Wildlife Service. "Arctic National Wildlife Refuge." http://arctic.fws.gov/ (accessed April 2009).

U.S. Geological Survey. *Arctic Refuge Coastal Plain Terrestrial Wildlife Research Summaries,* Biological Science Report. USGS/BRD/BSR-2002-2001. Washington, DC: U.S. Geological Survey, 2002.

Maria Papadakis
James Madison University

AUTOMOBILES

There are more than 250 million automobiles in the United States, and many models have been large and fuel-inefficient in recent decades.

Source: iStockphoto.com

Automobiles, collectively, remain one of the largest consumers of petroleum products in the world. According to 2007 estimates, there are about 806 million cars and light trucks in operation around the world, with about 250 million of these vehicles located in the United States. Forecasters are predicting that by 2020, 1 billion cars will be on the road in the world. At present, these vehicles consume about 260 billion gallons of petroleum-derived fuels every year. Petrol-fueled vehicles may be very efficient, but the environmental impacts and the reality of limited nonrenewable resources have forced researchers to consider alternative fuels.

Environmental Effects

The ubiquity of gasoline (petrol) engines has had profound effects on the natural environment. In the first instance, drilling for crude oil, from which gasoline is derived, has left a trail of tarnished ecological landscapes and political and economic situations that call into question justice and peace.

Emissions from gasoline-powered automobiles contain solid particulates, sulfur oxides, nitrogen oxides, hydrocarbons, carbon monoxide, and carbon dioxide (CO_2). Hydrocarbons, by-products of unburned fuel, have been reduced through regulations governing tighter-sealed fuel caps and more efficiently burning fuel. Solid particulates, nitrogen oxides, sulfur oxides, and carbon monoxide are all contributors to local smog pollution. The first attempt to control these emissions came in the form of catalytic converters, an engine exhaust component that rendered vehicle emissions less toxic. In the United States, catalytic converters were first installed on vehicles in 1975. Today, many state jurisdictions require vehicles to pass a "smog check," or assessment of their vehicles emissions.

However, CO_2 remains difficult to control. Gasoline produces about 156 pounds of CO_2 for every 1 million British thermal units of energy. CO_2 has been identified as a greenhouse gas and is seen as a driver of climate change. At present, the only means of decreasing CO_2 output is through the design of more fuel efficient engines or the use of an alternative fuel that emits less CO_2. Fuel-efficiency standards, commonly measured in miles/gallon, have not increased significantly, despite strong political efforts. In the United States, the popularity of large, fuel-inefficient cars and the increased vehicle weight from mandated safety features have held fuel efficiency standards in relative check for decades, despite improving technology. Furthermore, the U.S. automobile industry has been successful in opposing national standards that would require greater fuel efficiency by reasoning that the cost of research would ultimately hurt their viability.

Electric Vehicles

Electric cars are the oldest alternative-fuel vehicle, with variations dating back to the early 1800s. In fact, in the earliest days of automobiles, electric cars were seen as having numerous advantages over the complicated gears of a gasoline engine and the cumbersome warm-up time of steam engines. Electric cars, however, depend on battery technology to store energy. Thus, their range is limited by the size and efficiency of battery technology. As road networks grew and internal combustion engines grew more sophisticated, electric vehicles became reserved for golf carts and novelty vehicles.

By 1990, however, electric cars experienced a revival. In response to regulations mandating "zero emission" vehicles, automakers produced a variety of electric cars, from the novel General Motors EV-1 to an electric version of the Toyota Rav-4. The revival was short-lived. Electric vehicles were deemed experimental and were reclaimed by manufacturers when their leases expired, regardless of owner satisfaction. Several explanations have been put forth as to "who killed the electric car," ranging from consumer confidence to a petroleum lobby conspiracy.

Most recently, Tesla Motors, a California-based start-up, has unveiled prototypes of a high-performance roadster that it hopes to take into production. Electric car advocates hope that a high-performance luxury car that can be charged at home will remove the impression that electric vehicles are ill-suited for daily use.

Gas-Electric Hybrids

At this time, the most visible representation of an alternative fuel vehicle is one that is not actually an alternative fuel. The gas-electric hybrid vehicle is a traditional automobile with an electric battery. The battery augments the gasoline engine, essentially allowing the vehicle to turn off the gas engine. Some hybrid drivetrains are configured to run on either the internal combustion engine or the electric motor, depending on the vehicles' needs. Other drivetrains use the electric motor to augment the combustion engine, essentially providing increased fuel economy. Electric motors typically run off energy stored in recyclable batteries. These batteries can be charged by the combustion engine, through the vehicle's physical operation (energy captured from braking, for example), or through plug-in charging. Hybrids can provide up to a 40 percent increase in fuel efficiency.

Hybrid technology does not break the dependence on nonrenewable fossil fuels; its environmental effects come in the form of increased fuel efficiency, yet it still produces emissions. Thus, these automobiles are seen as an important bridge to moving from gas-only vehicles to emerging technologies that may come to market in the coming decades. As of 2009, the Toyota Prius leads the hybrid wave of automobiles, with several manufacturers offering hybrid versions of traditional vehicles.

Biofuels

Another alternative to traditional gasoline engines is the substitution of biofuels. Biofuel vehicles can operate on a fuel source derived from renewable, living organisms. Typically, biofuels still release harmful emissions into the atmosphere through combustion, although not at the rate of fossil fuels.

It has become something of a legend to attribute biodiesel to Rudolph Diesel, the inventor of the first diesel engine. Indeed, Diesel's first engine was powered by peanut oil—a biofuel. Before his death, Diesel had been known to be harshly critical of the emerging geopolitics of fossil fuels and saw the potential of his engine to be a part of a peaceful, agrarian revolution. Yet it was not until 1937, long after his death, that Diesel's vision was realized when G. Chavanne, a Belgian scientist, patented the production of diesel fuel from vegetable oil.

Biodiesel can be produced from a variety of oils, or "feedstocks." Most commonly, it is derived from virgin vegetable oils derived from soybeans or rapeseed. It may also be processed from waste vegetable oil (used in cooking) and animal fat (captured either from cooking or processing). Not all biodiesel is the same. Although different feedstocks can yield the same performance, "cloud points" may differ. A cloud point is the temperature at which pure biodiesel (B100) begins to gel. These cloud points limit the use of pure biodiesel anywhere but in the warmest of climes. More often, biodiesel is mixed with petro-diesel (B20 would be 20 percent biodiesel), and/or a warming system might be employed to keep an automobile's tank and fuel lines above the cloud point.

Biologically derived ethyl alcohol (ethanol) is another biofuel. Ethanol is produced by processing sugar derived from common crops, such as sugar cane or corn. Ethanol may be blended with gasoline and run in traditional petrol engines. Ethanol-blended fuel has a higher octane than regular gasoline and thus allows an engine to run at greater thermal efficiency. Ethanol has been widely integrated in Brazil, where sugarcane crops provide most of the fuel source. In the United States, sorghum (corn) is the crop of choice. Flex fuel vehicles are traditional automobiles that have engines designed to sense the petrol/ethanol blend ratio and make adjustments to ensure efficient operation.

Although many see the advantages of using cleaner-burning biofuels in vehicles, there is considerable debate over the ultimate environmental impact. Although tailpipe emissions are reduced, a tremendous amount of energy and resources may go into the growing and harvesting of feedstocks for these fuels. In addition, ethical questions have been raised over the use of "food for fuel."

Other Technologies

Through the years, alternative fuel trends have come and gone. Hydrogen fuel cells, which promise to produce electricity through a chemical processing of hydrogen and oxygen reaction, are all but finished. Although it offered considerable promise as an abundant resource with no emissions, the technology proved too difficult to harness, and the U.S. federal government killed research funding in 2009. Compressed Natural Gas, consisting primarily of methane, is an easy substitute in gasoline combustion engines. However, although it is the cleanest-burning fossil fuel, it is still a fossil fuel and a nonrenewable resource. In the United States, propane is the third most popular fuel source for vehicles; however, it is not widely used in passenger automobiles. Other proposals, including using algae-derived biodiesel or compressed air vehicles, remain in infant stages.

Conclusion

Automotive history is littered with tales of magic fuels that will outperform gasoline and break society from its petroleum addiction. However, even the most promising of current proposals give pause. Although tailpipe emissions may be cut, the effect of converting and storing alternative fuels leaves an ecological footprint somewhere. Political debates over these issues, as well as concerns over social justice, economic effects, and the effectiveness of technology, leave profound questions on the future of alternative fuel vehicles.

See Also: Alternative Fuels; Biodiesel; Electric Vehicle; Ethanol, Corn; Ethanol, Sugarcane; Fossil Fuels; Zero-Emission Vehicle.

Further Readings

Bethscheider-Keiser, Ulrich. *Future Cars: Biofuel, Hybrid, Electrical, Hydrogen, and Fuel Economy in All Shapes and Sizes*. Ludwigsburg, Germany: National Book Network, 2008.
Vaitheeswaran, Vijay and Carson, Iain. *Zoom: The Global Race to Fuel the Car of the Future*. New York: Twelve Books, 2008.
Westbrook, Michael H. *The Electric Car: Development and Future of Battery, Hybrid and Fuel-Cell Cars*. London: Institution of Electrical Engineers, 2001.
Worldwatch Institute. *Biofuels for Transport: Global Potential and Implications for Energy and Agriculture*. London: Earthscan, 2007.

Derek Eysenbach
University of Arizona

B

BATTERIES

Many electric devices need to be plugged into electricity in order to run. Batteries allow electricity to be stored in a convenient, affordable way that can travel with devices. In this manner, people can use electric devices in an expanding variety of venues. As computing technology advances, more and more personal hand-held devices are available, each needing a lightweight, rechargeable, long-lasting, and affordable battery.

Using batteries, machines, appliances, and other forms of technology need not be plugged into electricity in order to run. Theoretically this capacity would limit the amount of electricity used per person per day. With rechargeable batteries powering many wireless devices, electricity is still used to recharge the batteries, often on a daily basis. Additionally, spent batteries are often not disposed of properly, resulting in contained chemicals leaching into land and water, adding to environmental harm.

Batteries have been around for over two centuries; Italian physicist Alessandro Volta (1745–1827) invented the first one in 1800. The modern battery uses an updated design; batteries in Volta's original style are called voltaic batteries. In fact, Volta's model was impractical for the amount of current needed to run most machines. Several designs were used until the 1950s, when the contemporary alkaline battery emerged. In 1955, the Canadian engineer Lewis Urry (1927–2004) was working for the Eveready Battery Company (now known as the Energizer Battery Company), and discovered a way to optimize the

This researcher at the U.S. Department of Energy's Sandia National Laboratories is shown at work on a lithium-ion battery being developed for use in hybrid vehicles.

Source: U.S. Department of Energy, Sandia National Laboratories

previously infeasible alkaline battery. His invention was first marketed in 1959 and continues to be a staple battery design.

In the 1970s, a new design on an old battery made it more environmentally friendly. Rechargeable batteries were at the time made with nickel and cadmium. The cadmium was later replaced with a metal hydride, which was less toxic to the environment. More recently, scientists have developed improved lithium and lithium-ion batteries, which are much lighter in weight and therefore ideal for small, portable hand-held devices. Another specialized battery is the lithium iodide battery, which has a relatively long lifespan; these batteries are useful in implanted cardiac pacemakers.

Lithium batteries also were the initial choice for electric cars. The promise of lithium batteries is their ability to store large amounts of power; however, their drawback is the time it takes to recharge them. Modern designs are emerging that would allow the batteries to recharge more quickly; an allied drawback would be a need for specialized charging stations, akin to gas stations, that could supply enough power quickly enough for the newer batteries. Another obstacle that electric car battery engineers must overcome is the limiting factor of battery size, and therefore the weight of the vehicle.

The design behind a basic battery is elegantly simple. There are two poles—positive and negative in charge. Electrons are negatively charged and are stored at the negative pole. Opposite charges naturally attract; therefore, if there is a route to the positive pole they will flow in that direction, causing a current of electricity in the opposite direction. The current can be utilized to power a device, such as a laptop computer, handheld wireless device, digital camera, or automobile. The device to be powered is called a load, and it is part of the circuit between the negative and positive poles of the battery. If the negative and positive ends were connected with a conductive wire, without a load, the electrons would flow quickly from negative to the positive end, and the battery would run out of charge rapidly. In chemical terms, the battery would reach *equilibrium*. In common terms, it would be *dead*. Due to the large current flow, connecting the positive and negative ends of a powerful battery without a load in between is dangerous.

Different batteries contain different chemicals, and therefore pose various risks to handlers or to the environment. The common, non-rechargeable alkaline battery typically contains either potassium hydroxide or sodium hydroxide. Either chemical can cause burns if the battery is old and the chemical starts to leak out. Lead-acid (car) batteries also contain strong skin and eye irritants; if enough contact is made between the sulfuric-acid-filled electrolyte solution and the eyes, blindness can result. Other, less common battery types also contain irritants. Furthermore, several styles of lithium and lithium-ion batteries can release toxic gases if improperly disposed of.

It is important to dispose of batteries properly, to protect the environment as well as anyone who may come into contact with a spent battery. In an effort to curb the negative impact of disposed batteries, local, state, and federal regulations that guide safe disposal are made available to consumers. Battery recycling is a process by which most, if not all, materials from a battery are harvested and reused. For example, the plastic casing and lead of lead-acetate batteries can be reused for new lead-acetate batteries. The sulfuric acid in these batteries can be neutralized and purified to convert it to clean water, or can be converted to sodium sulfate for use in industry. Metals from nickel-cadmium, nickel metal hydride, or lithium-ion batteries are separated at high temperatures and re-used in other industries as well.

Solar energy can also be stored in a form of battery, called a photovoltaic cell.

See Also: Alternative Energy; Automobiles; Electric Vehicle; Electricity; Photovoltaics (PV).

Further Readings

Buchmann, I. *Batteries in a Portable World: A Handbook on Rechargeable Batteries for Non-Engineers*. Richmond, Canada: Cadex Electronics, Inc., 2001.

Engardio, P. "The Electric Car Battery War." *BusinessWeek*. (February 12, 2009).

Hurd, D. J. *Recycling of Consumer Dry Cell Batteries* (Pollution Technology Review, No. 213). Norwich, NY: William Andrew, 1994.

Linden, D. and T. B. Reddy, eds. *Handbook of Batteries*. New York: McGraw-Hill Professional, 2001.

van Schalkwijk, W. *Advances in Lithium-Ion Batteries*. New York: Springer, 2002.

Claudia Winograd
University of Illinois at Urbana-Champaign

BERLIN MANDATE

The Berlin Mandate is a decision reached in 1995 at the first Conference of the Parties to the United Nations Framework Convention on Climate Change (UNFCCC) that established a negotiating mandate for adopting specific biding commitments to reduce greenhouse gas emissions. These negotiations led to the 1997 Kyoto Protocol.

Faced with growing concern over the effects of anthropogenic greenhouse gas emissions, the United Nations General Assembly in 1990 authorized formal international negotiations aimed at creating a framework convention on climate change. Important to this decision was the First Assessment Report of the Intergovernmental Panel on Climate Change (IPCC). In 1988, governments of the world had requested that the United Nations Environment Programme and the World Meteorological Organization create the IPCC to provide independent scientific advice on climate change. The IPCC was asked to prepare, based on the best available scientific information, comprehensive and authoritative reports on climate change and its effects. The first assessment report of the IPCC served as the basis for negotiating the UNFCCC.

Following two years of negotiations, governments signed the UNFCCC during the 1992 Earth Summit in Rio de Janeiro. The convention sets the overall framework for global, intergovernmental efforts to address the immense challenges posed by climate change and enjoys near-universal membership, with 192 countries having ratified it.

The convention recognizes that the climate system is a shared resource, the stability of which can be affected by industrial and other emissions of carbon dioxide and other greenhouse gases (GHGs). Its stated objective (Article 2) is the stabilization of greenhouse gas concentrations in the atmosphere at a level that would prevent dangerous anthropogenic interference with the climate system. Under the convention, all governments agree to gather and share information on greenhouse gas emissions, national policies, and best practices; launch national strategies for addressing greenhouse gas emissions and adapting to its expected effects, including the provision of financial and technological support to developing countries; and cooperate in preparing for adaptation to the effects of climate change.

Article 4 of the UNFCCC required the parties to "adopt national policies and take corresponding measures on the mitigation of climate change, by limiting its anthropogenic emissions of greenhouse gases and protecting and enhancing its greenhouse gas sinks and reservoirs." In addition, developed-country parties agreed to take measures aimed at reducing greenhouse gas emissions to 1990 levels. However, as no binding emission levels or timetables were set, many nations declined to make independent national commitments that they believed would harm their economies in the short term.

As a result, many nations that signed the UNFCCC believed that its provisions were only a first step and were, on their own, inadequate to generate major reductions in GHG emissions. Many countries within the European Union and the Alliance of Small Island States, which face severe threats from climate change, began pushing for stronger commitments in climate meetings immediately following the Earth Summit.

These meetings acted as preparatory meetings for the first Conference of Parties (COP-1), which convened formally in Berlin in March 1995. Delegates had several major goals to accomplish, including an initial review of the adequacy of the UNFCCC commitments in Article 4. These discussions were informed by the 1995 release of the IPCC's *Second Assessment Report*, which provided greater clarity and understanding of human effects on the climate system.

Following difficult negotiations, COP-1 agreed that the existing UNFCCC commitments were inadequate to prevent dangerous climate change, but it could not agree on specific reduction targets or timetables. The European Union supported establishing substantial reduction targets on industrialized countries immediately, but the United States, Japan, Canada, Australia, and New Zealand opposed binding emissions reductions that would begin before the year 2000. Reducing emissions, they argued, would require adopting new economic measures to facilitate fundamental changes in energy use, and they said that they were not yet in a position to take those steps. Thus, COP-1 created the Berlin Mandate, which acted as a formal agreement that such commitments were necessary (even if they could not be agreed upon yet) and established a framework to negotiate, by the end of 1997, quantitative limits on GHG emissions to be taken after the year 2000.

Joint implementation (JI) was another major issue in Berlin. JI would allow a country with an emission reduction or limitation commitment under the future agreement to earn emission reduction units from an emission-reduction or emission-removal project in another participating nation. JI proponents argued that the mechanism would offer parties a flexible and cost-efficient means of fulfilling a part of their commitments, and the host party would benefit from foreign investment and technology transfer. Opponents, who included many developing nations, thought JI provided a way for industrialized nations to avoid reducing GHG emissions within their own borders, as financing reduction in poorer countries would cost less. The debate in Berlin ended with a decision to launch a pilot program that allowed industrialized countries to develop projects in both industrialized and developing nations, hoping to generate more support for JI before the anticipated protocol was completed in 1997.

Other accomplishments of COP-1 included determining the location of the permanent secretariat in Bonn, Germany, as well as establishing the convention's official subsidiary bodies. However, the meeting is best known for the Berlin Mandate, in which the world agreed on the necessity for binding reductions of GHG emissions, rather than the vague phrasing of the UNFCCC, and to seek agreement on the first set of such reductions by the end of 1997. This mandate was fulfilled with conclusion of the Kyoto Protocol in December 1997.

See Also: Climate Change; Global Warming; Greenhouse Gases; Intergovernmental Panel on Climate Change; Kyoto Protocol.

Further Readings

Arts, Bas and Wolfgang Rüdig. "Negotiating the Berlin Mandate: Reflections on the First Conference of the Parties to the UN Framework Convention on Climate Change." *Environmental Politics*, 4 (Autumn 1995).

Chasek, Pamela, et al. *Global Environmental Politics*, 5th ed. Boulder, CO: Westview, 2010.

Downie, David, et al. *Climate Change*. Santa Barbara, CA: ABC-CLIO Contemporary World Issues Series, 2009.

Framework Convention on Climate Change Conference of Parties. "Report of the Conference of the Parties on Its First Session, Held at Berlin From 28 March to 7 April 1995." United Nations Document FCCC/CP/1995/7/Add.1, 6 June 1995. http://unfccc.int/resource/docs/cop1/07a01.pdf (Accessed March 2009).

David Downie
Fairfield University

BEST MANAGEMENT PRACTICES

Work is the ethical, efficient, and effective application of energy to productive ends. Management is the organization and direction of work. Best management practices is the search for the most ethical, efficient, and effective means for directing energy to productive ends.

Individuals and groups engage in work that seeks to produce something. However, the "ends" of the work of an individual may be other than an end product. For example, the quality of a painting or the making of a model airplane may not be the goal for an individual as much as the occupation of his or her time. Or, for individuals in a group, it may be that the sharing between people is more important than the quality of the tasks that occupy them. Although these ends may satisfy human needs, they are not the practices that focus on achieving the highest-quality end product with the least amount of resources and effort.

Best management practices is a search for the "best" method for accomplishing goals. The best practice depends on the task or set of tasks that one or more persons have to perform. Some of these may be repetitive and some may be unique. For example, operations management deals with the orders for goods that the marketing department generates. The goods may simply have to be gathered to be delivered, or in more complicated situations, it may be necessary to first manufacture the goods or organize the service to be delivered. In both of these situations, the practices used are usually routine activities.

In contrast to operations management, project management deals with unique situations. Projects are often never-before-attempted undertakings that have beginning, middle, and end stages that are carried out to achieve a goal. When a project is defined, it is a problem scheduled for solution. Project managers often face unique problems because they are doing work never previously attempted. As a result, project managers seek to solve problems by bringing together money, manpower, and materials to do or build something. Projects themselves may be small—writing a book or building a house; medium—making a new model tank or enormous "elephants"; or large—constructing Hoover Dam. In both the ordinary activities of operations management and the unique problems project managers often face, there is currently a widespread search for the best practices for accomplishing any task or set of tasks.

In many ways, the search for the best practice resembles the scientific management work of Frederick Winslow Taylor (*The Principles of Scientific Management*, 1911); Henry Laurence Gantt (Gantt charts); Frank and Lillian Gilbreth, who were made famous in *Cheaper by the Dozen*; and others. These pioneers sought to reduce what had previously been work performed on a craft or traditional jobs basis to a series of steps that was the least number identifiable for accomplishing a job. With stopwatches, movie analysis of work, and other techniques such as charts of productivity, they were able to greatly increase the efficiency of U.S. industry. Their ideas spread rapidly into many parts of the world by the 1920s, where the term "Taylorism" was often used to describe and to denounce them. Opposition arose in several ways, but a key complaint was that by reducing work to a series of repetitive steps, the individual was dehumanized because the creative process had been engineered out of the job by the time-motion experts.

The assumption used in best management practices is that there is one best way to accomplish a task or job. Practices to accomplish tasks or jobs include processes, methods, techniques, activities, incentives, rewards, or any other skill that may be used. The best practice is the one or the set that delivers the most "bang for the buck" to achieve the desired outcome. The best practice is often innovative and dynamic. At the least, it needs to be adaptable, capable of being replicated by others in many different circumstances, and capable of continuous improvement. "Continuous improvement" means that the practice being used is assumed to be imperfect but capable of being improved until the steps to achieve the end goal have been reduced to the least number of steps possible at the lowest cost.

In work situations, it is common for the way the job was first accomplished to become the practice of those who are faced with repeating the job. However, even if the best practice available at the time a task or job was initially attempted, it is possible for workers and managers to fall into routines. This is a natural response to daily activities or even hectic workloads. The routines are efficient but may become less effective, leading to production falling below the best practices available. As a result, continuous improvement is a program for being attentive to new best practices or to improving the methods, techniques, skills, processes, or other ways for doing a task or job.

The way to identify the best practice or practices is to hunt for those producers who are most effective and efficient at accomplishing a task. Their practice is then borrowed. Another way is to treat the task or job in the way in which project management organizes its work into an organized work breakdown structure. The work breakdown structure is like a hierarchical organization chart of personnel in an organization, but with the elements of the task or job put into their logically subordinate rankings. Other techniques used include benchmarking, detailed audits of operations and strategic management, computerized auditing, and management tools.

Best management practices is being used in the energy industry, where the search is often for best practices adaptable to unique projects. To engage in best practices, it is necessary for careful documentation to be kept of both practices and experimental changes for improvement. Some have developed energy efficiency best practices programs. In the electricity industry, companies using best practices include Efficiency Vermont, Florida Power and Light, Pacific Gas and Electric, Energy Trust of Oregon, Xcel Energy, and others. The oil industry, in cooperation with the Bureau of Land Management, is engaged in joining the best industry practices with the best environmental protection practices.

Other areas of the energy industry are partnered with the U.S. government in training that seeks to promote the best environmental practices with the best of industry practices. This is an outgrowth of implementing ISO 14000.

The term *best practices* has become a buzzword in many quarters. Others have sought to use the term *good practices* because for many, "best practices" simply means standardization or rules that are followed mechanically. For others, it connotes finality, universality, and obedience to the practice that is viewed as "best." The reality is, however, that what is best practice for one may not be best practice for others. Therefore, care is needed in adopting practices.

See Also: Environmental Stewardship; Innovation; Sustainability.

Further Readings

Friedman, Thomas L. *Hot, Flat, and Crowded: Why We Need a Green Revolution—and How It Can Renew America.* New York: Farrar, Straus & Giroux, 2008.

Global Association of Risk Professionals. *Foundations of Energy Risk Management: An Overview of the Energy Sector and Its Physical and Financial Markets.* New York: John Wiley & Sons, 2008.

Institute for Prospective Technological Studies. *Reference Document on Best Available Technologies in the Large Volume Organic Chemical Industry.* Seville, Spain: European Commission, Joint Research Centre, IPTS, 2003.

Kanter, Moss. *Best Practices: Ideas and Insights From the World's Foremost Business Thinkers.* New York: Basic Books, 2003.

Shilling, Melissa. *Strategic Management of Technological Innovation.* New York: McGraw-Hill Companies, 2006.

Andrew J. Waskey
Dalton State College

BICYCLES

Impending energy shortages require alternative transportation modes, and the bicycle is ideally equipped to fill some of the gaps. Healthy, convenient, and relatively fast, urban bicycle use can be significantly enhanced with inexpensive infrastructure improvements such as bicycle lanes isolated from motor traffic, separate bicycle signals, and covered and lighted bikeways. Attitudes must also change to make cycling a widely accepted, green alternative.

One component of the future transportation mix that will probably grow in importance is the humble bicycle—human-powered, but amazingly efficient. Two-legged locomotion places humans among the least efficient animals, but humans on bicycles go farther per calorie of energy than any other animal except soaring birds (essentially solar-powered flight). On level ground, a cyclist can move about three times as fast as a pedestrian with the same effort. This is the promise of cycling as transportation, but it comes with some reservations.

This wide lane of bicycle commuters in Beijing, China, runs beside lanes for the city's growing numbers of private cars and trucks.

Source: iStockphoto.com

History

The bicycle is a surprisingly recent invention. After numerous experiments in the 18th and 19th centuries, bicycles similar to modern models appeared in the 1880s. Predecessors of the modern (or safety) bicycle were not seen as practical means of transportation and were used primarily for short, recreational rides. The popular all-metal high wheel (or ordinary) bicycles of the 1870s were luxury items afforded only by the wealthy and the perilous design often limited usage to younger men. Practical bicycles depended on two inventions of the modern industrial era: the ball bearing and the pneumatic tire. Together with steel fabrication methods, these inventions made bicycles light, comfortable, and affordable to almost everyone. However, they were also key steps to the automobile, which was invented only a few years later.

The 1890s were the golden era of cycling, taking place between the invention of the modern bicycle and the introduction of affordable automobiles, when the bicycle dominated the popular imagination. Bicycles enabled workers to live farther from their places of employment, resulting in the development of near-in suburbs around many cities. They made possible weekend jaunts into the countryside for urban dwellers, while creating demand for paved roads. Together with the new urban and interurban electric trolleys (trams), an efficient, pollution-free transportation system took form.

With the rise of the automobile, and later the superhighway, however, bicycles gradually slipped into subsidiary roles as sports equipment and children's toys and were not taken seriously for transportation in the United States, though in parts of Europe they remained an important transportation option.

Function

The bicycle is an intimate interaction of machinery with human physiology. Muscles are strongest near the middle of their length range; on a properly adjusted bicycle, the knee and ankle are not quite fully extended at the bottom of a pedal stroke, and the knees not too bent at the top of the stroke, so that the leg muscles—the largest and most powerful muscles in the body—are at their strongest. Pedaling with the toe allows the muscles surrounding the ankle to participate in propulsion. Bicycle shoes that clip onto the pedals allow the rider to perform a power stroke on both the up and the down phases of pedal motion; for commuters who prefer ordinary shoes, toe clips (small cages for the toe of the shoe) perform that function. This innovation helps make cycling more efficient than walking, where the lifting phase of the stride performs no motive work. Coasting requires no work at all, unlike walking, where even downhill segments require effort. A century of refinement has made the modern bicycle light, flexible, and easy to ride at any speed, with gearing schemes to keep the muscles working at the same speed and torque (twisting power) on all but the steepest hills, while allowing higher speeds on level or down-sloping terrain.

Bicycles as Transportation

To affect society's energy mix, bicycles must be used widely for transportation, not just recreation. They are ideal for short-distance commuting and running errands: 22 percent of commutes in the United States and 35 percent in Canada are less than 5 kilometers (km), well within comfortable bicycle range. At a reasonable 20 km/hour (12 miles/hour), this means about a 15-minute ride. Bicycles offer transportation on demand—no walking to and waiting for a bus or tram. They go where the commuter wants to go, not where an established route directs. The consistent, mild exercise is better for health than powered transport. The bicycle is also inexpensive, at about 2 percent of the cost of an automobile. Maintenance is similarly inexpensive, though bicycle parts are by nature fragile because they must be light. Parking is usually right outside the door of the destination, making cycling faster than driving for many urban applications.

Despite these advantages, bicycles do not dominate urban transport. Here the limitations of bicycle travel become clear. First, cycling is not for everyone. The handicapped, aged, or others who cannot physically handle the demands of the bicycle need alternate modes of transportation, and once those alternatives are in place, others will be tempted to use them. Riding is unpleasant and often dangerous in cold weather, in rain, and at night—the bicycle is a fair-weather friend. In cities such as San Francisco or Pittsburgh, hills are a significant barrier. The speed of bicycles also limits them to short-distance trips, which is a barrier in sprawling American cities such as Los Angeles or Dallas. A rider produces about one-tenth of one horsepower, or one-eighth in short bursts, which is inadequate to move loads much heavier than a briefcase.

If bicycles are to become significant contributors to a green energy mix, green public policy should minimize the bicycle's inherent limitations. Some limitations are easily overcome; others require more investment or a longer time frame. Perhaps most important is better rights-of-way. Most city planning considers bicycles more an annoyance than a transit alternative. Bike lanes, where they exist, are usually stripes painted on the sides of traffic lanes, between moving and parked vehicles, offering no protection from either. The few dedicated bicycle rights-of-way are usually designed for recreation (often in parks), rather than transportation.

For guidance on better alternatives, planners can look to Europe. A main artery in Munich has two-way bicycle lanes on each side, above the curbs and separated from sidewalks. Cyclists are protected from moving vehicles by the lines of parked cars. Separate traffic signals regulate bicycles, and parking racks surround subway entrances. Because the system is heavily used, drivers are aware of the stream of cyclists at intersections. Along Munich's river, bikeways detour under bridges to eliminate most intersections, creating a bicycle autobahn. Such facilities lead to extensive bicycle use even in large cities. The Netherlands, famous for its bicycle density, has not only urban bicycle lanes but also intercity paths made exclusively for bicycles. Such facilities are inexpensive compared to motor vehicle highways: the lanes are narrower, and they are built to carry 200 pounds of vehicle and rider rather than 3,000 pounds or more of automobile and driver.

In addition to well-designed bike paths that separate bicycles from moving motor vehicles, facilities such as secure parking and showers encourage riding. Some employers, including many universities, already offer such facilities. Bicycle racks in front of buses, along with spaces in urban rail transit cars, help to eliminate the distance restrictions of bicycle commuting while maintaining door-to-door convenience.

Beyond the current infrastructure, new facilities can compensate for other disadvantages of cycling with existing technology. To counter dark and rain, lighted and covered bikeways can make cycling pleasant at all times and in all weather except extreme cold, at very low cost compared with that of building new highways. Eventually, urban areas must be reconfigured to be more compact.

Some of the factors limiting bicycle use are psychological rather than physical. Cycling must become an accepted transit alternative, not seen as mere sport or exercise or as something only for children too young to drive. Convenient, safe rights-of-way are only part of this transformation. The enforcement of traffic rules regarding cyclists might seem unfriendly to cyclists, but in the long run it contributes to cycling being taken seriously and improves safety. Cyclists are often seen without helmets, riding on the wrong side of the street, riding at night without lights, weaving through traffic, or drifting through stop signs. The latter problem, caused partly by irresponsibility and partly by the reluctance of riders to give up their hard-won momentum for regulations designed for motor vehicles, can be ameliorated by substituting yield signs for stop signs in many locations. For the rest, education and experience are the only alternatives.

With enlightened but inexpensive public policy, the bicycle can be made into a sustainable, green transit mode for a substantial fraction of transit needs. Bicycles are not completely free of fossil fuel requirements—their manufacture requires fossil fuels, and paving rights of way also requires fuels—but those needs are minimal compared with those of automobiles.

See Also: Embodied Energy; Environmental Measures; Sustainability.

Further Readings

Bijker, Wiebe E. *Of Bicycles, Bakelites, and Bulbs: Toward a Theory of Sociotechnical Change.* Cambridge, MA: MIT Press, 1997.
Herlihy, David V. *Bicycle: The History.* New Haven, CT: Yale University Press, 2004.
Whitt, Frank Roland and David Gordon Wilson. *Bicycling Science*, 2nd ed. Cambridge, MA: MIT Press, 1982.
Wilson, S. S. "Bicycle Technology." *Scientific American.* v.228/3 (1973).

Bruce Bridgeman
University of California, Santa Cruz

BIODIESEL

Developed as an alternative to traditional fossil fuels, biodiesel is a renewable biofuel variety derived from vegetable oils, animal fats, and used cooking oils. Biodiesel is currently produced by transesterification, a chemical process developed more than 150 years ago that extracts biodiesel from these oils and fats. Biodiesel fuel comes in several blends (e.g., B5, B10, B20, B99) that indicate the percent of biodiesel within the blend of conventional diesel fuel. The B20 (i.g., 20 percent biodiesel and 80 percent diesel) blend is the most common, providing the best benefit-cost ratio. Although traditional petroleum diesel still fuels more than 99 percent of diesel engines, biodiesel production worldwide has increased over 30 percent annually between 2002 and 2008 and over 50 percent between 2005 and 2007. An estimated 11.1 million tons of biodiesel were produced in 2008, primarily by the United States and European countries. Developing nations are also ramping up production, and it is expected that Brazil

The world's biodiesel production increased over 50 percent between 2005 and 2007, which may lead to increased access to the fuel, such as this biodiesel pump installed beside a gasoline pump.

Source: iStockphoto.com

will be one of the top producers by 2015. However, increasing biodiesel production is causing concern among some scientists who fear that extracting fuel from food crops may not be as environmentally beneficial as previously extolled and may lead to issues with food security.

In Europe, rapeseed is the main feedstock used in biodiesel production, whereas in the United States, soybean is the primary feedstock. Other common feedstocks are mustard seed, palm oil, and waste cooking oil. Two main criteria used to evaluate the costs and benefits of biodiesel production are (1) net energy value (NEV), which measures energy efficiency as a ratio of energy output to energy input, and (2) greenhouse gas (GHG) emission benefits, which measure biodiesel GHG emissions as a percentage of petroleum diesel GHG emissions. Both of these criteria can be measured through life cycle analysis studies. Although there is a substantial body of life cycle analysis studies for biodiesel, there is minimal consensus over the results. The conflicting results are largely an effect of the assumptions made in the models (e.g., land use change, agricultural practices, and input calculations). In addition, there is an inherent problem in comparing current technologies such as transesterification of vegetable oil with proposed technologies such as algal biodiesel production. Estimates for the NEV of biodiesel range from 1.2 to 3.5 (soybean), showing a net energy benefit. A NEV of 1.2 for soybeans can be interpreted as requiring 4.3 hectares of soybeans to fuel one car for one year. For GHG emissions, the estimates range from a 40 percent (soybean) to 50 percent (rapeseed) improvement over petroleum diesel fuel emissions. However, there is an open debate about whether to include land cover change in the calculations, which may increase the final GHG emission estimates of biodiesel. A recent analysis of biofuels, including ethanol, has shown negative GHG benefits when land cover change is included, but no calculations have been made specifically for biodiesel.

The recent push for more biodiesel production is a direct result of policies implemented by Europe and the United States to promote increased usage of biofuels. Production of biodiesel in Europe started in the early 1990s with revisions to the Common Agricultural Policy, which promoted the development of biofuels. In the United States, Title IX (the energy section) of the Farm Security and Rural Investment Act of 2002 (the Farm Bill) established new programs and grants to support development of biorefineries, promoted programs to raise public awareness of biodiesel fuel use, and assisted farmers and small business in purchasing renewable energy systems. The Farm Bill was followed by the Energy Policy Act of 2005, which included various incentives, such as tax relief, loans, and grants, directed toward developing renewable energy alternatives. In January 2007, the state of California (USA) introduced the Low-Carbon Fuel Standard (LCFS), the first clean energy policy designed to establish a GHG standard for transportation fuels and promote alterative fuel sources. By 2020, the LCFS is expected to reduce carbon content of all passenger vehicles by 10 percent and triple California's renewable fuels market, including biodiesel. The Californian LCFS is expected to be the model for future U.S. GHG emission standards.

In early 2007, the European Commission published its "Renewable Energy Roadmap," recommending that biofuels constitute at least 10 percent of diesel and petroleum fuel consumption by 2020. In the same year, the U.S. Congress passed the Energy Independence and Security Act, which included legislation to increase the capacity of renewable fuel production from 4.7 to 36 billion gallons by 2020. However, in early 2008, the European Commission started to back away from the targets it had given previously and began to consider adding high standards of "sustainability" for biofuel sourcing and manufacturing. Similarly, meeting the U.S. target of higher renewable fuel production will require drastically increasing soybean acreage and supplementing domestic supplies with imported biodiesel from countries such as Brazil.

The increased push for biofuel production and usage in the United States and abroad has led to some controversial subsidy agreements colloquially known as "splash and dash." In the United States, fuel suppliers receive a subsidy for every gallon of biodiesel blended with conventional diesel. Biodiesel imported from abroad is given a "splash" of regular diesel and makes a "dash" for external fuel markets, such as Europe. Fuel suppliers thus profit from the subsidy, acquiring millions of dollars in tax credits for a purpose not originally intended.

The future of biodiesel is focused on developing second-generation feedstocks (from nonfood crops), inventing more efficient extraction methods, producing higher NEV and lower GHG emissions, and improving the cost competitiveness of the fuel. One promising second-generation biodiesel feedstock crop being explored is jatropha (a small tree indigenous to the tropics), which is considered an improvement over traditional feedstocks because it grows very quickly and in marginal lands. However, there is concern that planting crops such as jatropha and oil palm throughout large areas in African and southeast Asian countries will lead to deforestation, biodiversity loss, and large carbon emissions resulting from land conversions. Other alternatives being investigated include methods to extract biodiesel from wet biomass (which is less energy intensive than current methods) and cultivating feedstocks that can grow in salt water (e.g., sea asparagus).

With second-generation biodiesel still in development, many private investors are already looking to algae as a feedstock. Algae are considered a third-generation feedstock because they produce far higher yields and require lower inputs than first- and second-generation feedstocks. The U.S. Department of Energy estimates that one acre of algae can produce over 30 times as much energy as one acre of soybeans. Biodiesel from algae was studied intensely by the National Renewable Energy Laboratory from 1978 to 1996, but high costs associated with bringing production to an industrial scale and difficulty transferring results from laboratory into field conditions eventually brought the research to a standstill. Despite recent developments, much of the progress in second- and third-generation biodiesel production has only produced results in a laboratory setting and suffers from technological hurdles when practiced on a larger scale.

At present, feedstocks supplying biodiesel overlap significantly with food crops, which may pose challenges for food security in developing nations. When prices of feedstocks, such as soybeans, increase in response to the growing biofuel market, people who depend on these crops for protein may no longer be able to afford them. The introduction of second- and third-generation feedstocks may ease the effects of biodiesel production on food security, but it still relies heavily on monoculture agriculture. Thus, the potential benefits of producing renewable biodiesel must be weighed against its possible economic, ecological, and social effects on the place where it is produced. Nevertheless, supportive legislation for renewable energy has spurred large amounts of private and public funding for expanded biofuel development, and biodiesel will almost certainly be a part of our energy future.

See Also: Alternative Fuels; Biomass Energy; Energy Policy; Life Cycle Analysis.

Further Readings

Crane, David and Brian Prusnek. "The Role of a Low Carbon Fuel Standard in Reducing Greenhouse Gas Emissions and Protecting Our Economy" (January 2007). State of California Office of the Governor. http://gov.ca.gov/index.php?/fact-sheet/5155 (Accessed August 2009).

Fargione, Joseph, Jason Hill, David Tilman, Stephen Polasky, and Peter Hawthorne. "Land Clearing and the Biofuel Carbon Debt." *Science.* v.319/5867 (2008).

Hill, Jason, Erik Nelson, David Tilman, Stephen Polansky, and Douglas Tiffany. "From the Cover: Environmental, Economic, and Energetic Costs and Benefits of Biodiesel and Ethanol Biofuels." *Proceedings of the National Academy of Sciences.* v.103/30 (2006).

National Renewable Energy Laboratory. "A Look Back at the U.S. Department of Energy's Aquatic Species Program: Biodiesel from Algae." www.nrel.gov/docs/legosti/fy98/24190 .pdf (Accessed March 2009).

Rajagopal, Depak and David Zilberman. "Review of Environmental, Economic and Policy Aspects of Biofuels." The World Bank Policy Research Working Paper No. 4341 (September 1, 2007). http://econ.worldbank.org (Accessed March 2009).

Runge, C. Ford and Benjamin Senauer. "How Biofuels Could Starve the Poor." *Foreign Affairs.* v.86/3 (2007).

Sagar, Ambuj and Sivan Kartha. "Bioenergy and Sustainable Development?" *Annual Review of Environmental Resources.* v.32 (2007).

Yiwei Wang
Michiyo Kakegawa

Timothy Norris
University of California, Santa Cruz

BIOGAS DIGESTER

Biogas is the gas produced from the rotting of organic matter in conditions that favor the growth of anaerobic bacteria. A number of bacteria flourish in the absence of light and oxygen. These bacteria digest natural organic material in swamps, bogs, and other marshy areas, and they also digest organic material in mammal digestive systems and in sewage systems. This digestion, or fermentation, produces a breakdown of organic compounds from complex to simple. A by-product is gas, much of which is methane, the simplest of organic compounds.

Biogas digesters, or more precisely, anaerobic digestion plants, are engineered plants for processing organic material in bulk to produce biofuels. The term *biodigester* can also be applied to the reactor vessel in which the process of anaerobic decomposition occurs. The reaction vessel itself is usually called the fermenter. These vessels produce more gas and less heat than does aerobic "rotting."

The process of anaerobic digestion was given a fictional depiction in the third of the cult Road Warrior movies *Mad Max, Beyond Thunderdome* (1985), starring Mel Gibson and Tina Turner. Gibson's character goes to Barter Town, where there are electrical lights run by biogas. The source is pig manure that produces methane.

Real biogas digesters deal with the fact that there two kinds of biogas producers. The fermentation type uses anaerobic digestion of biodegradable materials such as vegetation, plant material from fields, manure, sewage, green waste (flowers, green trimmings, commercial food waste), municipal waste, and crops grown specifically as energy-producing materials.

When biogas is mixed with oxygen, the gas can be burned for fuel. The gas produced by a biogas digester system can be part of a waste management system. The heat produced by burning biogas can be used with a heat engine, which converts thermal energy to mechanical energy or electrical energy. It can also be compressed and used to power motor vehicles.

Biogas is a renewable energy source. In many places, such as the southern United States, wind power will not generate enough energy to be a viable source. Nor are there any sources of geothermal power. Even solar power is not as practical as some have thought, considering the hot humid climate of the region—similar to that of southeastern China. The clouds and humidity block sunlight, making solar power a less effective system than is the case in desert or arid regions. What is produced, however, in hot and humid regions are large quantities of biomass. This natural resource is being considered for use as a biogas material. Biofuel crops can also easily be grown and can also be genetically engineered to produce vegetation that is richer in organic compounds that will produce methane.

Because biogas digesters can be used at very low cost on farms and in rural communities, new types of biogas digesters have and are being developed. Some digesters specialize in using biomass feedstocks of varied types. For some, the volume of pig, cow, or horse manure is so large that the supply more than meets the demand. For example, cow manure, when biodegraded in a biogas digester, produces a gas that is 55–65 percent methane, with traces of hydrogen and nitrogen, and 30–35 percent carbon dioxide. In other cases, forest materials or tree trimmings are available in large quantities. The key is to have a steady supply of a biodegradable feedstock that can be used by the biogas digester for which it was designed.

Biogas digester plant designs are focused mostly on the type of organic waste to be converted into biogas. Other engineering design features include the temperature of operation and the cost of materials for construction. Waste can be digested in one stage or in two stages. The bulk of the methane produced would be produced in the first stage. The second stage produces more biogas, but at a slower rate and with a lower volume of gas produced.

A batch biogas digester can usually be operational after a few weeks. If the biogas digester is located in a warm climate, little gas will be needed for the fermentation process.

However, if the facility is in a colder climate, it will probably be necessary to divert portions of the biogas production for use in warming the reaction system.

One advantage of biogas digester systems is that they aid in getting rid of organic waste in a clean and efficient manner. The ancient practice of burning animal dung, whether buffalo chips in the Old West or cow patties in India today, has never been as clean as a biogas digester system. Visitors to India can see a haze at certain times of the year resulting from burning cow dung. The smoke causing the haze is an eye irritant, and it is unhealthy for anyone to be exposed to it for a long time. Another advantage to using biogas digesters is the "waste product." The end product of the fermentation slurry is a valuable organic fertilizer; large quantities of cow and sheep manure are sold commercially to gardeners in the United States.

A disadvantage of such systems is a "rotten egg" smell resulting from the production of hydrogen sulfide, which can occur if the system is not properly operated. Another disadvantage is that operators have to be on guard against infection by the pathogens that will naturally be present.

Several biogas digester programs are being operated by the Dairyland Power Cooperative in La Crosse, Wisconsin, using cow and pig wastes. Small biogas digesters also have been built in the Philippines, and more are being prepared.

See Also: Alternative Energy; Biodiesel; Biomass Energy; Renewable Energies.

Further Readings

Christensen, T. H. and R. Cossu. *Landfilling of Waste Biogas*. London: Taylor & Francis, 1996.

Demuynck, M. and E. J. Nyns, eds. *Biogas Plants in Europe: A Practical Handbook*, Vol. 6. New York: Springer-Verlag, 2007.

Deublien, Dieter. *Biogas From Waste and Renewable Resources: An Introduction*. New York: John Wiley & Sons, 2008.

Taylor, Glenn. *Biogas!* Bloomington, IN: iUniverse, 2007.

van Buren, Ariane ed. *A Chinese Biogas Manual: Popularizing Technology in the Countryside*. Rugby, UK: Practical Action, 1998.

Andrew J. Waskey
Dalton State College

BIOMASS ENERGY

Biomass energy refers to energy derived from organic materials of living or recently expired plants and animals. Sources of organic materials (or biomass) include wood, agricultural crops and residues, domestic food waste, industrial and municipal organic wastes, and livestock manure. Biomasses are carbon-based compounds composed primarily of carbon, oxygen, and hydrogen (the three chemical constituents of carbohydrates); nitrogen and other substances (e.g., alkali earth metals) in small quantities may also be found. In effect, the carbohydrates are the organic contents that comprise biomass and contain the stored energy from the sun. The stored energy in the carbohydrates is released

These large pieces of wood waste will be broken down into wood chips for biomass fuel production.

Source: iStockphoto.com

upon death of the plant or animal. Biomass decomposition processes, including burning, produce energy that can then be used for electricity generation and transportation fuels.

Despite the differing categories and characteristics of biomass materials, the developments of conversion technologies are on the rise. These technologies can be categorized as thermal (e.g., combustion), chemical, or both (e.g., anaerobic digestion). As a result of the growing awareness of biotechnology, countries such as the United States have been able to employ biomass energy in the generation of electricity for large communities. Biomass generation electricity can come from engines run by special-purpose turbines that use the methane gas (biogas), a byproduct from the decomposed organic matter of plants and animal wastes.

Biomass energy is considered a renewable and sustainable energy source. Organic materials are readily available and easily replenished. Biomass resources can be classified as primary, secondary, or tertiary biomass sources depending on their processing level. Primary biomass resources are organic materials that are produced directly from photosynthesis (e.g., forests, agricultural crops). Biomass that has been processed either through physical (e.g., rice husks), chemical (e.g., liquids from pulp extraction), or biological (e.g., animal and human waste) means are considered a secondary source. Tertiary biomass sources are post-consumer residues and waste products, such a cooking oils, animal fats, and industrial wastes. Whether from a primary, secondary, or tertiary sources, biomass provides energy and waste maintenance.

Although biomasses release carbon dioxide (CO_2) when burned or undergoing decomposition, the CO_2 released is the same amount the plant materials had used up through their cycles of photosynthesis. Biomass processes do not alter the overall CO_2 content of the atmosphere but only work to maintain it. The CO_2 expelled during biomass combustion or decay is an essential part of the carbon cycle.

The dependence on biomass energy for daily energy needs is not actually new. For thousands of years, people have burned wood for heating and cooking, and even now, wood is still relied on in many developing countries. In the United States, technologies exist to convert wood and wood waste to electricity. This bioelectricity is then consumed by various businesses and industries. Just like other forms of energy, biomass energy can be converted from one form to another. During the process of combustion, biomass energy can be converted to heat; in addition, wood waste can be burned to produce steam for making electricity. Biological tissues and some domestic waste can be made to undergo controlled (or monitored) decomposition to produce methane, which in turn can be used to drive turbines for electricity production; crops can be fermented to produce ethanol for transportation, and vegetable oils and animal fats can be blended synthetically to produce biodiesel.

Statistically, biomass sources provide about 3 percent of all energy consumed in the United States. In 2002, about 40 percent of all renewable energy consumed in the United States was from biomass. U.S. electricity generation from biomass, excluding municipal solid waste, represents about 11 percent of all generation from renewable sources. Globally, biomass is capable of meeting about 14 percent of the world's energy needs. For instance, biomass energy can be a solution to Africa's current energy development problems. Although traditional biomass energy sources (e.g., wood burning for cooking) have been used for centuries, the traditional techniques employed are not necessarily sustainable, safe, or environmentally friendly. The introduction of modern biomass energy technologies (e.g., ethanol production) can provide stand-alone energy generators for industries, localities, and rural communities not connected to national grids.

The Environment and Biomass Energy

Biomass energy has some significant environmental benefits over fossil fuels as an energy source. Like fossil fuels, biomass can pollute the air with CO_2 and carbon monoxide (CO) gases when burned as they are both derived from the biological materials of plants and animals, however the degree of pollution is much less than that of fossil fuels. Although fossils fuels offer high energy density (at many times more than per the same quantity of biomass), the combustion process of organic materials millions of years old produces much larger quantities of CO_2 as well as nitrous and sulfur oxides. The significant addition of CO_2 from fossil fuel burning during the age of industrialization has coincided with recent global surface warming trends and regional climate changes. Biomass combustion on the other hand does not produce more carbon dioxide than was taken out of the atmosphere for plant growth nor the high levels of the sulfur-based compounds that are capable of causing acid rain when mixed with rain water.

Unlike fossil fuels, biomass energy sources are renewable and are sustainable using proper management practices. Biomass can be harvested as part of a constantly replenished crop, resulting from woodland cultivation, including coppicing (a tree pruning method) and commercial agriculture. Resource management can sustain current atmospheric CO_2 levels by encouraging new plant growth able to intake CO_2 during photosynthesis while negating the CO_2 emitted during biomass energy conversion processes (e.g., combustion). In a way, biomass energy then has the potential of reducing greenhouse gas emissions.

Increased reliance on biomass energy sources can also reduce foreign exchange expenditures that are channeled toward purchase of petroleum-based products from other countries. Industrially, the use of biomass as energy sources will eliminate wastage, as the majority of biological waste from food and beverage production, for instance, can be recycled into new energy sources for powering the production machines. This invariably will aid waste management, lead to enhancement of waste-to-wealth technology and indirectly reduce production cost. Consequently, instead of organic materials (e.g., paper mill residues, municipal waste, sawdust) put into landfills, these materials can be recycled with appropriate technology to produce gas or liquid fuels for domestic and industrial electric power generation or transportation.

The transportation industry is increasingly utilizing the environmental advantages of biomass energy in the form of biofuels, fuels that are composed of at least 80 percent biomass. Examples of biofuels include ethanol, biodiesel, and syngas. The ethanol from biomass fermentation derived from sugars found in many plants (e.g., sugar cane, corn), can be used to drive motor vehicles, a fuel that produces less CO_2 per volume compared with fossil fuels. Biodiesel from biomass materials is a renewable, biodegradable, and non-toxic fuel that can be used instead of conventional diesel from petroleum-based products. It results in less pollution than petroleum diesel and has a solvent-cleansing ability that loosens and dissolves sediments in storage tanks. Biodiesel is significantly cleaner when burned, producing fewer air pollutants such as particulates, carbon monoxide, air toxics, and unburned hydrocarbons. Although biodiesel emits a slightly higher amount of nitrogen oxides (about 10 percent higher than petroleum diesel), biodiesel produces less soot, smells better, and harmful sulfur compound byproducts are relatively minimal. A partial combustion of biomass can also produce a biofuel known as syngas, a carbon monoxide and hydrogen mixture, that can be used directly as fuel source or as a synthetic gas mixed with other chemicals to produce diesel substitutes.

Using biomass as an alternative source of energy is a unique way to manage a great deal of domestic and industrial waste, some of which could constitute serious environmental risks. During the biological decomposition of organic wastes, methane is released into the atmosphere. This gas is colorless and odorless and is the principal constituent of natural gas (75 percent). It burns in a sufficient amount of oxygen to form carbon dioxide and water vapor. Methane, CO_2, and $H_2O_{(g)}$ are all greenhouse gases (GHG) (atmospheric gases that absorb radiation in the thermal infrared spectrum from the sun and the earth and reemit this long-wave radiation to the atmosphere. These gases stabilize the mean surface temperature of the earth; however, minute changes in GHG concentrations can significantly impact global climate, creating an enhanced greenhouse effect). As a greenhouse gas, it is regarded as very strong, having a global warming potential of 23, meaning that methane has 23 times more global warming ability than carbon dioxide over a hundred-year period and has 20 percent of the total radiative forcing of the net irradiance of all greenhouse gases in their combined effects.

However, when biomass decomposition is controlled in such a way that the NH_4 can be harvested to produce useful energies, the gas does not contribute to an increase in atmospheric methane concentration. Instead, the methane gas byproduct can be consumed or converted to useful energy products (e.g., cooking gas). Electricity can also be generated using methane-powered turbines that then can be used to power several industrial and production processes. Thus, biomass-derived methane using modern biomass energy technologies can be commercially useful and not necessarily lead to the adverse environmental effects that traditional biomass energy extraction methods may produce.

In addition to methane and carbon dioxide, other gases released during biomass combustion include sulfur dioxide (SO_2) and nitrogen oxides. Although produced in significantly smaller quantities than their fossil fuel counterparts, the levels of these gases released vary widely depending on the contents of sulfur and nitrogen in the biomass materials. When allowed to interact directly with the atmosphere, these gases can react with water to form acidic precipitation and pollutants, such as smog. Acid rains have the ability to corrode concrete and other building materials (including paint), damage forests and crops, alter soil composition, and negatively affect aquatic life. High levels of SO_2 in the atmosphere can cause health problems for people too. For example, repeated exposure to higher SO_2 concentrations can cause heart disease and temporary breathing difficulty, especially for asthma patients. Nitrogen oxides, on the other hand, have various effects depending on its compound. Exposure to nitrogen dioxide (NO_2), for example, can cause irritation of the eyes, nose, and throat, and can lead to respiratory infections, particularly in children. Nitrous oxide (N_2O) is a GHG with a large global warming potential (> 200) and a stability period of 150 years; in other words, nitrous oxide has a long life cycle once introduced into the atmosphere and can make a large contribution to the enhanced greenhouse effect. Hence, controlled and efficient combustion of these sulfur- and nitrogen-containing biomasses, most especially in the presence of coal, is needed to lower the emission potential of these gases.

Generally, unlike resources produced from petroleum products, biomass energy resources require much less energy for their extraction, processing, and disposal. As a consequence, their production and processing also generally mean lower carbon dioxide and other GHG emissions into the atmosphere. Biomass energy can provide a renewable, sustainable, and green alternative to fossil-based fuels provided modern technologies and sustainable land management practices are employed.

See Also: Climate Change; Fossil Fuels; Global Warming; Renewable Energies.

Further Readings

German Solar Energy Society (DGS). *Planning and Installing Bioenergy Systems: A Guide for Installers, Architects and Engineers.* London, UK: James and James, 2005.

Hoogwijk, Monique, Ándre Faaij, Richard Van den Broek, Göran Berndes, Dolf Gielen, and Wim Turkenburg. "Exploration of the Ranges of the Global Potential of Biomass for Energy." *Biomass and Bioenergy.* v.25/2 (2003).

Kyritsis, S. P. Helm, A. Grassi, and D. Chiraramonti, eds. *First World Conference on Biomass for Energy and Industry.* London, UK: James and James, 2001.

Smil, Vaclav. *Energy at the Crossroads: Global Perspectives and Uncertainties.* Cambridge, MA: The MIT Press, 2005.

United Nations Foundation. "Biomass Conversion Technologies." *Sustainable Energy Development in UEMOA Member Countries.* http://www.globalproblems-globalsolutions-files.org/gpgs_files/pdf/UNF_Bioenergy/UNF_Bioenergy_5.pdf (Accessed August 2009).

Ajayi Oluseyi Olanrewaju
Covenant University, Nigeria

CAFE Standards

After the first Arab oil embargo in 1973, Congress enacted the Energy Policy and Conservation Act (EPCA) in 1975, codified at 49 USC 32901-32919, charging the National Highway Traffic Safety Administration (NHTSA) in the Department of Transportation with establishing Corporate Average Fuel Economy (CAFE) standards for vehicle manufacture fleets and the U.S. Environmental Protection Agency (EPA) with measuring vehicle fuel efficiency to reduce American dependence on foreign oil. CAFE standards are the sales-weighted harmonic mean, expressed in miles per gallon (mpg), of a vehicle manufacturer's fleet of automobiles and lightweight trucks under 8,500 pounds gross vehicle weight. Starting with the vehicle model year (MY) 2011, sports utility vehicles, minivans, and all other light trucks (i.e., those vehicles that weigh 8,500–10,000 pounds gross vehicle weight) must comply with CAFE standards.

NHTSA is required to set CAFE standards at the maximum possible level given technological feasibility, economic practicality, effects of other legal standards on fuel economy, and national need to conserve energy. The federal CAFE standards preempt any state and local laws directly regulating fuel economy, whether those laws are aimed at vehicle manufacturers or vehicle purchasers. For example, the federal district court held in *Metropolitan Taxicab Board* that New York City regulations specifying minimum fuel economy for New York City taxicabs were preempted.

Over the course of the last 30 years, federal fuel economy standards have not increased as rapidly as the technological ability to increase fuel economy. From MY 1978 to MY 1985, CAFE standards for passenger cars increased from 18 to 27.5 mpg, whereas the standards for light trucks rose from 17.2 to 19.5 mpg. For four years, MY 1985–89, NHTSA reduced the passenger car's standards 1–1.5 mpg under pressure from the vehicle manufacturers, while light truck standards remained at 20.5 mpg. During MY 1986–2007, light truck standards gradually increased from 20 to 22.2 mpg. Passenger car standards in MY 1990 were raised to 27.5 mpg, where they stagnated through 2007.

NHTSA in 2006 issued standards for MY 2008–11 that fundamentally changed the approach to creating CAFE standards. The agency decided to set light truck standards according to the size of the truck's "footprint" in square feet. This footprint approach would have allowed manufacturers selling more large trucks to meet less stringent standards, rather than setting an absolute limit on average fuel economy for the manufacturer's fleet.

In response to a petition for review by the Center for Biological Diversity, the Ninth Circuit Court of Appeals in April 2007 overturned the MY 2008–11 standards and remanded the standards to NHTSA. The Ninth Circuit determined that the light truck standards failed to address light trucks between 8,500 and 10,000 pounds, were not consistent with the factors set in the EPCA, and violated the National Environmental Policy Act because of inadequate environmental impact assessment, including setting a zero dollar value on global warming effects from light trucks.

In December 2007, Congress enacted the Energy Independence and Security Act. The act requires NHTSA to gradually increase the combined passenger car and light truck standards, reaching 35 mpg by 2020. In MY 2021–30, CAFE standards must be set at the maximum feasible fuel economy based on available technology, without regard to other factors. If NHTSA issues standards on the basis of "footprints" or other mathematical formulas, each manufacturer must meet the higher standard of either 92 percent of the projected average of all manufacturers or 27 mpg. Flex fuel vehicles will not be allowed to take extra credit for possible use of E85 (a fuel blend comprising 85 percent ethanol and 15 percent gasoline), but manufacturers will be given extra credit for biodiesel-ready vehicles. NHTSA must also set standards for medium- and heavyweight trucks.

Shortly after taking office in January 2009, President Barack Obama ordered NHTSA to propose MY 2011 rules meeting the 35-mpg level—nine years earlier than the Energy Independence and Security Act requires and 4 mpg higher than the MY 2011 rules previously proposed by the Bush administration in response to the Ninth Circuit remand.

NHTSA has always been reluctant to set higher CAFE standards because of fears that the standards would force vehicle manufacturers to make smaller cars, which were believed to be less safe. However, studies in recent years indicate that better safety engineering of small cars has improved their safety, and conversely, light trucks have a poor safety record. Thus, there is no longer a significant relationship between vehicle weight and safety, and CAFE standards need not be kept artificially low to protect safety.

EPCA gives NHTSA exclusive authority to regulate vehicle fuel economy through the CAFE standards. However, during the last five years, states have become increasingly impatient with the federal government's failure to address global warming and to regulate greenhouse gas emissions such as vehicle carbon dioxide (CO_2) emissions. In an attempt to force a federal response to the global warming problem, California issued air pollution rules regulating vehicle CO_2 emissions. California is the only state allowed by the Federal Clean Air Act to issue vehicle air pollution regulations more stringent than the uniform national vehicle emission regulations, but California's regulations become effective only if the EPA grants California's request for a waiver of the uniform national standards. The Bush administration in 2008 declined to grant California a waiver; however, the Obama administration in January 2009 ordered the EPA to grant California's waiver application. When the EPA grants California a waiver, other states are allowed to opt in and impose California standards on vehicles sold in their state, rather than apply the uniform national standards. Thus, when California's CO_2 vehicle emission standards become effective, they will apply in all states that have adopted California's CO_2 vehicle emission rules.

Vehicle manufacturers have sought to block California's and other states' CO_2 vehicle emission rules by arguing that EPCA preempts state fuel economy standards, and thus CAFE standards issued under EPCA are the only standards that can limit fuel economy. They reason that because CO_2 vehicle emission standards can only be met at this point by adjusting fuel economy, EPCA preempts the state CO_2 vehicle emission standards. Federal district courts in both California and Vermont, in the *Central Valley* and *Green Mountain* cases, have concluded that EPCA does not preempt California's CO_2 vehicle emission standards. In addition, as CAFE standards increase, the vehicle manufacturers' legal argument that state vehicle CO_2 emission regulations are inconsistent with federal CAFE standards may simply become irrelevant.

See Also: Automobiles; California; Combustion Engine; Flex Fuel Vehicles; Gasohol; Plug-In Hybrid; Zero-Emission Vehicle.

Further Readings

2006 Light Truck CAFE Standards for 2008-2011 Model Years, http://www.nhtsa.dot.gov/ staticfiles/DOT/NHTSA/Rulemaking/Rules/Associated%20Files/2006FinalRule.pdf (Accessed February 2009).

Center for Biological Diversity v. National Highway Traffic Safety Administration, No. 06-71891 (9th Cir.), overturning 2006 light truck CAFE standards. http://www.ca9 .uscourts.gov/datastore/opinions/2007/11/14/0671891.pdf (Accessed February 2009).

Central Valley Chrysler-Jeep Inc. v. Goldstone, No. CV F 04-6663 (E.D.Cal. Dec. 11, 2007), upholding California CO_2 vehicle emission standards. http://www.climatelaw.org/cases/ country/us/case-documents/us/us/Ishii%20Order.pdf (Accessed February 2009).

Energy Policy and Conservation Act, automobile fuel economy provisions. http://www.nhtsa .dot.gov/nhtsa/Cfc_title49/ACTchap321-331.html (Accessed February 2009).

Green Mountain Chrysler-Plymouth-Dodge-Jeep v. Crombie, No. 05-cv-302 (D. Vt. Sept. 12, 2007), *upholding Vermont's adoption of California's CO_2 vehicle emission standards.* http://www.climatecasechart.com/ (Accessed February 2009).

Metropolitan Taxi Board of Trade v. City of New York, No. 08 Civ. 7837 (PAC)(S.D. NY Oct. 31, 2008), overturning New York City requirements for taxicab fuel economy. http://www.nytimes.com/packages/pdf/nyregion/city_room/20081031_Crottydecision.pdf (Accessed February 2009).

NHTSA CAFE home page. http://www.nhtsa.dot.gov/portal/site/nhtsa/menuitem.43ac99aefa8 0569eea57529cdba046a0 (Accessed February 2009).

Susan L. Smith
Willamette University College of Law

CALIFORNIA

As the U.S. state with the largest population and most economically productive economy, California is important to green energy policy and programs. The state competes with entire nations to rank among the 10 largest economies globally, and though the state's per capita energy consumption is low in comparison with other U.S. states, California is so large that its total greenhouse gas emissions are ranked 18th globally.

California has a history of pace setting in many realms other than energy (two prominent examples include Hollywood and tax policy). Its leadership on energy-related policy is particularly notable, however, and its successes and mistakes are influential and highly scrutinized.

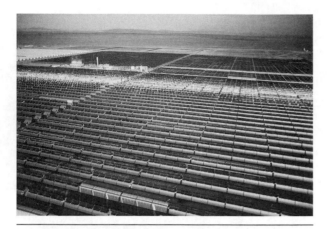

The panels in this vast field of solar panels in a California desert move to follow the sun's path during the day. California's solar installations made up 58 percent of the U.S. solar market in 2007.

Source: iStockphoto.com

The California clean air and water acts, for example, led the nation, influencing decades of research and development in the pollution-creating energy sphere. The state, nevertheless, continues to be an uneven and contradictory home for green energy; powerful detractors of green power have emerged in California, alongside its proponents. In fact, California's successful green energy efforts likely would not exist if it were not for the environmental problems caused by the energy sector in the past.

The California mythmakers and boosters have long insisted that the state is a place of nearly supernatural beauty and resources. Generations of migrants and immigrants have been lured to the state with the promise of luxury and riches in the form of gold, jobs, housing, and fame. California historian James C. Williams traces California's long economic boom story to its prodigious reliance on energy technologies to fuel the instant cities of the 19th and early 20th centuries and to power successive waves of industry. This is evident in investments in mining technologies, the path-breaking development zones of Los Angeles' suburbia, the shipbuilding and other military complexes of the World War years, the wildly productive agricultural industry, and the computer technology and Internet booms centered in Silicon Valley.

Many others, including Donald Worster, have also written about the importance of energy-related technologies in transforming California's desert and mountain landscapes into some of the most productive agricultural and urban spaces in the world. The story of the creation of the state's water infrastructure—which reaches through deserts, across the Sierra Nevada mountain range, and into neighboring states—is particularly notable on this point.

One key to California's need for energy is in its consistent growth through much of the 20th century. Beginning with World War I, and accelerating during and after World War II, the state experienced a steady and dramatic population boom, accelerating the cumulative effects of industrial and residential energy use.

With growth came environmental pollution and degradation, and pro–green power rules and regulations emerged in response to California's high energy consumption and played a role in inspiring the general rise of U.S. environmentalism in the 1960s and 1970s. The burgeoning environmental movement, with key epicenters in California, began targeting the energy sector as a primary site of human-made pollution and degradation.

Out of this milieu, many nationally significant proconservation and pro–alternative energy activist movements emerged, particularly from the San Francisco Bay Area. They, along with public and private sector researchers and investors, have since been further motivated by the mounting climate change crisis and by the energy crises of the 1970s and early 2000s to transform energy development. Efforts have included legislation limiting emissions and encouraging the use of solar and wind energy, which played a role in inspiring hundreds of residentially based experiments in off-the-grid living. More centralized efforts, such as government support for energy conservation and renewable energy standards, have been roadmaps for other states and for national policy (under certain administrations).

Environmental Roots

Similar to civil rights and other movements typically associated with the 1960s in the popular imagination, California's environmentalism has long historical roots, going back to John Muir's early-20th-century fights to save the Hetch Hetchy Valley from flooding to build a major electricity-producing dam. In the post–World War II years, as car traffic increased and industrialization expanded, Californians agitated for the first anti-smog law in the country in 1959, several years before the first federal Clean Air Act's passage.

The movement's impact, however, snowballed in the 1970s, around the time of the 1973 oil embargo by the Organization of Petroleum Exporting Countries (OPEC), which sent gas prices skyrocketing. The embargo highlighted the United States' dependency on OPEC countries, inspiring calls for energy independence—and sometimes for green energy—that have been echoed through debates over American involvement in the Middle East ever since.

In the years leading up to the embargo, California researchers had already been sounding the alarm about the state's unsustainable growth and the need for cleaner energy. Fights had been breaking out across the state over power plant sitings, as citizens organized themselves into increasingly powerful blocks, pushing for the passage of the California Environmental Air Quality Act in 1970, which gave the public an important tool to use in challenging the energy companies. Other fights, over the effects of offshore drilling, inspired a series of protective laws and led to passage of the 1972 Coastal Zone Conservation Initiative, mandating attention to environmental effects before certain energy projects could move forward. California pioneered large-scale energy regulation by creating the California Energy Commission in 1974 with a mandate to regulate environmental aspects of the state's energy system.

The state also became a hot spot for advocacy for specific green technologies, such as solar water heaters, inspiring pro-solar policies under President Jimmy Carter that were later repealed by President Ronald Reagan (Reagan, similar to pro–solar water heater policy, was a product of California). Championed by then-governor Jerry Brown, who filled top positions with renewable energy advocates, some of the world's largest solar installations were erected in the Mojave Desert near Barstow (containing, in 1982, more than 90 percent of the world's solar thermal capacity). The world's largest (in 1985) wind facility also was built at Altamont Pass, east of San Francisco.

In 1989, the state had another global first when Sacramento residents—customers of the publicly owned Sacramento Municipal Utility District—voted to close the nuclear Rancho Seco Power Plant.

As the OPEC crisis faded into memory, however, and the push for deregulation of energy gathered its own power, many of California's impressive green power achievements and policies, including some of its massive solar installations, were dismantled. However, the new awareness of the effects of power plants on urban populations in the 1990s and the energy crisis of 2000 that made Enron's power manipulations famous inspired renewed calls for conservation and green energy. Even after the setbacks for green power, California far outpaces many other states in areas such as solar electricity, for example. In 2007, the state's solar energy installations represented 58 percent of the nation's grid-connected solar market (this does not include off-the-power-grid solar panels on homes, which are harder to track). The outcome of the conservation efforts of the 1970s and 1980s has also lingered: California Energy Commission figures show that the state's trend of keeping per person energy consumption far below national rates continued through 2008.

Energy Future

More recently, increased attention to the climate crisis has ratcheted pressure on the state to force the private power companies to go green. Under California's Renewable Portfolio Standard—a set of state goals first enacted in 2002 and updated in 2006—the major energy providers in the state are required to raise the percentage of renewable energy sources in their portfolios by about 1 percent a year, with the goal of reaching 20 percent by 2010 and 33 percent by 2020. According to state regulators, just 12.7 percent of the energy produced by the three primary private utility companies came from renewable sources in 2007, according to the following breakdown: geothermal, 47.93 percent; wind, 19.04 percent; biomass, 14.32 percent; small hydroelectric, 11.12 percent; biogas, 4.73 percent; and solar, 2.86 percent.

Many other related policy moves are merging various strands of the green power movement. One example of this is in the efforts of the Oakland-based group Green For All, which advocates for building the green power economy as a means for lifting underemployed groups out of poverty through "green-collar" jobs. Both the California Renewable Portfolio Standard concept and the green-collar employment concepts have been prominent in debates over the federal economic stimulus package under Obama's administration, and Obama's environmental advisers—many of whom are from California—may be using California policies as models.

See Also: Energy Policy; Renewable Energy Portfolio; Solar Energy; Wind Power.

Further Readings

Berman, Daniel M. and John T. O'Connor. *Who Owns the Sun? People, Politics, and the Struggle for a Solar Economy.* White River Junction, VT: Chelsea Green, 1997.

Goodell, Jeff. "Look West, Obama." *Rolling Stone* (February 4, 2009).

McWilliams, Carey. *California, the Great Exception.* Berkeley: University of California Press, 1998.

Rosenfeld, Arthur H. "Energy Efficiency in California." California Energy Commissioner presentation (18 November 2008). http://www.energy.ca.gov/2008publications/CEC-999-2008-032/CEC-999-2008-032.pdf (Accessed January 2009).

Sherwood, Larry. *IREC Solar Market Trends August 2008.* Interstate Renewable Energy Council, 2008. http://irecusa.org/fileadmin/user_upload/NationalOutreachPubs/IREC%20Solar%20Market%20Trends%20August %202008_2.pdf (Accessed February 2009).

Smil, Vaclav. *Energy at the Crossroads: Global Perspectives and Uncertainties.* Cambridge, MA: MIT Press, 2003.

Walker, Richard. *The Country in the City: The Greening of the San Francisco Bay Area.* Seattle: University of Washington Press, 2007.

Williams, James C. *Energy and the Making of Modern California,* 1st ed. Akron, OH: University of Akron Press, 1997.

Worster, Donald. *Under Western Skies: Nature and History in the American West.* New York: Oxford University Press, 1992.

Rachel Brahinsky
University of California, Berkeley

CARBON EMISSIONS FACTORS

Human activities such as conversion of land and combustion of fossil fuels contribute to greenhouse emissions. Among these, carbon emissions from fuels represent 75–80 percent of the global warming emissions. If this trend continues, the consequences for the global climate could be enormous. Carbon emissions factors (CEF) or carbon dioxide (CO_2) emission factors have been established to estimate how much CO_2 is emitted to the atmosphere when greenhouse-emitting materials, such as fossil fuels, are used to power human activities (e.g., industries, commercial enterprises, transportation, and generation of electricity). The CEF is used to relate the relative intensity of an activity (e.g., the amount of coal burned) to emission values in carbon dioxide equivalence (CO_2e), a quantity that indicates the global warming potential of each greenhouse gas (GHG) using carbon dioxide as the GHG standard. The use of CEF has helped to quantify and report emissions of different sources in a unifying way. These CO_2 estimates have led to the implementation of regulations, the forecasting of future emission scenarios, and the rising of public awareness about global climate change issues.

The United Nations Intergovernmental Panel on Climate Change estimates that current emissions need to be cut by 80 percent to avoid further climate change consequences. Actions required to deal with this problem include implementing better measures to quantify direct and indirect emissions. CEFs are thus revised periodically to improve measures of CO_2 emissions.

CEFs estimate direct and indirect sources of CO_2 emissions. When measured at the consumer end, they are considered direct sources, because the emissions could be attributed to a particular activity. Direct sources of CO_2 emission include combustion of fuels, industrial processes, solvents, agriculture, land use changes, and waste. Direct sources potentially contribute to the following greenhouse gases: CO_2, methane, and nitrous oxide. In contrast, indirect sources of carbon emissions are derived from carbon monoxide, nonmethane volatile compounds, and sulfur oxides. Indirect sources or contributors to carbon emissions are difficult to track. To date, only a fraction of the total—direct and indirect—carbon emissions have been quantified.

The following are examples of the U.S. average CO_2 emissions factors by units of energy for 2007:

- Gasoline (U.S. gallon): 19.56 lbs. (8.89 kg)
- Diesel (U.S. gallon): 22.38 lbs. (10.17 kg)
- Liquefied petroleum gas (LPG; U.S. gallons): 12.81 lbs. (5.82 kg)
- Natural gas (million Btu): 117.08 lbs. (53.21 kg)
- Electricity (kWh): 1.27 lbs. (0.58 kg)

The consumption of fossil fuels is the largest contributor to global carbon emissions, and hence are the primary focus when discussing carbon emission factors. A fuel CEF is the CO_2 emitted by that fuel during its conversion to units of energy, mass, or volume. A formula to calculate CO_2 emission factor consists of the "input + use" multiplied by the "estimated fuel emission factor" (CO_2 emitted by fuel type). For simplicity, CEF assumes an oxidation factor of 99 for all fuels except natural gas and liquid petroleum gas (for which the oxidation factor is given at 99.5 percent). Sources for the calculation of coefficients for each CO_2 emissions factor vary with fuel type and thus could be influenced by the geographical origin of the raw material. In the absence of a fuel CEF, a rough estimate of carbon emissions can be made considering that for every pound of carbon burned, 3.7 lbs. of CO_2 are released to the atmosphere. In the United States, CEFs for fuels are expressed in terms of the energy content of coal as pounds of carbon dioxide per million British thermal units. However, international systems report CEF as million tonnes (Mt) and kilotonnes (kt) of carbon or as ratios of these units.

To calculate CEFs from fuels derived from coal, coal carbon content coefficient and heat content are required. These coefficients depend on the physical characteristics of particular coals. Coal types such as anthracite, bituminous, subbituminous, or lignite differ in their carbon emissions. Within a coal type, geographic location may be a source of difference in carbon content. Thus, corrections for carbon properties and region help to decrease CEF uncertainties. These corrections are known as fixed CEFs and can be used to calculate emissions by consumption, which are variable.

Currently, the largest sources of global carbon emissions (in terms of annual Gigaton [Gt] carbon output) arise from the combustion of solid, liquid, and gaseous fuels, comprising approximately 35 percent, 36 percent, and 20 percent of the global emissions, respectively. Examples are coal-carbon, liquefied petroleum gas, natural gas, motor gasoline, and diesel fuel. Per fuel-consumer category contribution of CO_2 emissions are, in decreasing order: petcoke and petcoke-cement; domestic consumption (anthracite carbon); agriculture; and other bituminous-based coal activities.

By consumer sector, generation of electricity from coal plants is by far the major contributor to CO_2 emissions and accounts for approximately 42 percent of global CO_2 emissions (in terms of annual Gt carbon output). Other consumer sectors contributing high CO_2 emissions are transportation, industrial processes, residential-commercial consumption, agriculture, landfills, wells, bunkers, and nonfuel hydrocarbons. In the United States, the U.S. Department of Energy estimates the average household energy consumption produces 36,000 lbs. (16,364 kg) of CO_2 annually, excluding the use of cars. By source, the average U.S. household contributes approximately 512 lbs (233 kg) of CO_2 from liquid petroleum gas (LPG), 13,532 lbs. (6,151 kg) from power, and 8,442 lbs. (3,837 kg) from natural gas.

Various scientific reports suggest that carbon emissions are not only underestimated but also increasing. They attribute the increase to the consumption of fuel for energy by countries such as China and India. Most of China's emissions arise from the manufacturing of electronics and other technology products for international markets. Some developed countries have decreased their carbon emissions, but critics argue that these countries have simply transferred their emissions to other countries. At this time, China and the United States remain the largest contributors to greenhouse emissions.

Although CEF calculations contain several layers of uncertainty (e.g., measures at the consumer end rather than the source), these measures have proven valuable for people, governments, and regulators. Aided by emission calculators and programs, people can obtain information to link their emissions and the emissions of their countries or regions to climate change policy decisions. This information could help consumers to decrease their carbon emissions or shift toward cleaner fuels (e.g., solar or wind energy).

Other potential uses of the CEF include international economic trades and environmental policies. For example, to offset their emissions, individuals, companies, or nations can pay carbon credits to other parties. These actions may help distribute global resources and protect natural resources. Because CEFs estimate potential contribution of different sectors to CO_2 emissions, these factors also contribute to the establishment of regulations that aim to decrease global emissions. Changes in consumption patterns could significantly decrease carbon emissions. A recent report suggests that converting 50 percent of all lights to compact fluorescent bulbs could reduce CO_2 emissions by 36 percent. Solutions like this are practical and can be implemented at personal or institutional levels.

In summary, CEFs have increased people's and countries' accountability for their role in producing and contributing to greenhouse gases. The use of CEFs to monitor greenhouse emissions also has led to local and global climate regulations. Currently, carbon emissions factors provide values that underestimate the total emissions and more precise measurement is needed. Better tracking of greenhouse gases and improved regulations applying CEF could decrease atmospheric pollution and humanity's contribution to global climate change.

See Also: Carbon Trading and Offsetting; Emissions Trading; Global Warming; Greenhouse Gases.

Further Readings

AEA. "Greenhouse Gas Inventories for England, Scotland, Wales and Northern Ireland 1990–2006." AEAT/ENV/R/2669. Issue 1, 2008. http://www.airquality.co.uk/reports/cat07/0809291432_DA_GHGI_report_2006_main_text_Issue_1r.pdf (Accessed March 2009).

Greenhouse Gas Protocol Initiative. "Calculation Tools." www.ghgprotocol.org/calculation-tools/all-tools (Accessed March 2009).

International Energy Agency (IEA). *CO$_2$ Emissions From Fuel Combustion, 2008 Edition.* Paris, France: OECD Publishing, 2008.

Miller, Peter. "Energy Conservation." *National Geographic*, 215/3 (2009).

Raupach, Michael R., Gregg Marland, Philippe Ciais, Corinne Le Quéré, Josep G. Canadell, Gernot Klepper, and Christopher B Field. "Global and Regional Drivers Accelerating CO$_2$ Emissions." *Proceedings of the National Academy of Sciences*, 104 (2007).

Stoyke, Godo. *Home Energy Handbook. Carbon Buster's.* Gabriola Island, Canada: New Society, 2007.

U.S. Environmental Protection Agency (EPA). "Greenhouse Gas Emissions." www.epa.gov/climatechange/emissions/index.html (Accessed February 2009).

U.S. Environmental Protection Agency. "Unit Conversion, Emission Factors, and Other Reference Data November 2004." http://www.epa.gov/climatechange/emissions/state_energyco2inv.html (Accessed February 2009).

Chris Field
Carnegie Institution for Science

CARBON FOOTPRINT AND NEUTRALITY

"Carbon footprint" and "carbon neutrality" are concepts proposed in response to the climate change challenges humans are facing today. Climate changes, which refer to long-term changes in average weather in a specific region, have been affected by human activities in an unpredictable way. Global warming, with observed temperature increases all over the world and more frequent droughts, floods, and other forms of extreme weather, has been attributed to the increase of greenhouse gases (GHGs) in the atmosphere, which primarily absorb and emit radiation with a certain energy that has thermal heating effects. Typical GHGs, as mentioned in the Kyoto Protocol, include carbon dioxide (CO_2), methane, nitrous oxide, hydrofluorocarbons, perfluorocarbons, sulfur hexafluoride, and so on. It is widely known that these GHGs have a negative effect on our living environment. Among these GHGs, CO_2 makes up the majority of the gases that are affected by human activities; to simplify understanding and analysis of the situation, other GHGs are usually converted into carbon dioxide equivalent (CO_2e) according to their global warming potentials on a time scale of 100 years. Research has shown that before the Industrial Revolution in Europe (around 1750), the concentration of CO_2 in the atmosphere was 280 parts per million (ppm); the overall amount now has increased to 390 ppm CO_2e. The trend is ongoing, and the concentration increases at 1.5–2 ppm annually. The remarkable changes seen in the level of GHG contents and their effects on the environment have drawn worldwide attention. The scientific term *carbon footprint* is used to represent the total amount of GHGs that are caused directly or indirectly by an individual, organization, event, or product. Carbon footprints can be labeled as primary or secondary. The primary footprint measures direct emissions of GHGs that are discharged in a way fully controlled by humans or organizations. Examples of primary footprint are using coals to generate electricity and heat and burning gasoline to power vehicles for transportation. The secondary footprint measures indirect emissions of GHGs throughout the entire life cycle of an individual, organization, product, or event. For instance, an individual who purchases a laptop is indirectly responsible for the emissions and the computer's carbon footprint, which is the combined emissions throughout the life cycle of the product, including preparation of raw materials (silicon, plastics, etc.), fabrication of electronic components, product fabrication, transportation and distribution, and emissions from the use of electrical power. The laptop manufacturer is indirectly responsible for the carbon emissions formed when making and transporting raw materials (purification of silicon, production of plastics, etc.) or components (chips, screen, light-emitting diodes, software, etc.) and when transporting and distributing the laptop to consumers. The chip manufacturer who uses all the electronic components such as transistors, resistors, and capacitors is considered indirectly responsible for the carbon emissions from their suppliers. By the same token, downstream companies in manufacturing are indirectly responsible for the emissions from upstream companies by using their products and services directly.

It is obvious that the accurate calculation of emissions is a complicated task that requires a systematic knowledge of the product, from the raw materials to any intermediate products, to the final products; from production, to transportation, to distribution to end users. This process is usually so demanding that it is almost impossible to produce the full carbon footprint that covers the whole life cycle of a product. Therefore, various methodologies have been established to quantify carbon emissions; however, the results obtained are based on different principles and are rarely comparable with each other. A more basic approach, which simplifies calculation by merely counting the direct emissions and electricity consumption, is widely practiced and accepted. Emissions from major sources, including direct emissions from on-site fuel consumption to run the facility and owned transportation and indirect emissions from electricity usage, will be quantified; emissions from uncontrollable sources will be excluded from calculations. These data are easily obtained from utility meters and from the volume or mass of input materials within the organization. Using the published standard emission conversion factors provided by the Department for Environment, Food and Rural Affairs (Defra) of the United Kingdom, the emissions from fuel, electricity, and transportation can be converted into the emission of CO_2 or its equivalent (CO_2e). On the basis of the carbon emission data, a full carbon footprint can be produced by providing the following information in the report: (1) define the methodology employed in carbon emission calculation; (2) specify the scope of coverage as to emissions from which entities will be included; (3) collect emission data and calculate carbon footprint; (4) verify results by an independent third party, if possible; and (5) disclose the carbon footprint report in relevant media. Once the basic carbon footprint has been revealed, actions can be taken to optimize emission management and finally reduce the emissions rationally. These steps include setting emission goals, identifying the opportunity for reduction, evaluating reduction options to maximize environmental and financial benefits, and finally, implementing reduction strategies and making improvements according to the feedback. In the most "green" cases, if an organization sequesters and releases equal amounts of CO_2, or its exhaust can be offset by purchasing shares from other organizations, it is said that carbon neutrality is achieved within that organization.

The term *carbon neutrality* is commonly used to represent the condition at which a net zero carbon emission is reached. Actual net zero carbon emission is hard to achieve, as an organization will, to some extent, release GHGs; it is common to compensate for these carbon emissions with carbon reductions achieved elsewhere to achieve carbon neutrality. The compensation strategy involved in zeroing net carbon emission is commonly referred to as carbon offset, which represents a viable approach for an organization to define itself as carbon neutral. To facilitate the exchange of rights to carbon emissions, official markets have been established to help those who have extra emission quotas to make a profit and for those who need to, to increase their emissions allowances. This process is called emissions trading and was motivated by the Kyoto Protocol, which sets legal goals and timetables for the reduction of GHGs of industrialized countries.

Emissions certificates are issued to participants, and they are allowed to trade in both mandatory and voluntary markets. For example, on the large scale, a nation that fails to meet its reduction obligation can enter the mandatory markets to purchase emission certificates to compensate for their treaty shortfalls. On the voluntary markets, companies and organizations are offered different options to trade their emissions certificates. In North America, the Chicago Climate Exchange and over-the-counter market have been established to help members trade their emissions shares to accomplish their carbon reduction obligations. These mechanisms have played important roles by providing incentives to companies that adopt advanced technology to reduce their carbon emissions and even have extra emissions quotas to sell them and make extra profits; in contrast, those companies that failed to assume their reduction obligations may pay for extra emissions allowances. The loss of economic interest will serve as an impetus for the companies to advance their emissions reduction technologies and facilities.

See Also: Alternative Fuels; Carbon Trading and Offsetting; Climate Change; Fossil Fuels; Life Cycle Analysis.

Further Readings

International Energy Agency (IEA). *CO_2 Emissions From Fuel Combustion, 2008 Edition.* Paris, France: OECD Publishing, 2008.

Intergovernmental Panel on Climate Change. *Fourth Assessment Report.* Geneva: Intergovernmental Panel on Climate Change, 2007.

UK Carbon Trust. "Carbon Footprinting." http://www.carbontrust.co.uk (Accessed May 2009).

Guoqiang Ren
University of Washington

CARBON SEQUESTRATION

Carbon sequestration is the process through which carbon dioxide (CO_2) from the atmosphere is absorbed by trees, plants, and crops through photosynthesis and stored as carbon in biomass (tree trunks, branches, foliage, and roots) and soils or aquatic vegetation. The term *sinks* is also used to refer to forests, croplands, and grazing lands and their ability to sequester carbon. Agriculture and forestry activities can also release CO_2 to the atmosphere. Therefore, a carbon sink occurs when carbon sequestration is greater than carbon releases over some time period.

These men are cutting blocks of peat from a bog in Ireland. Peat plays a significant role in carbon sequestration, accounting for 25–30 percent of the world's stored carbon.

Source: iStockphoto.com

Forests and soils have a large influence on atmospheric levels of carbon dioxide—the most important global warming gas emitted by human activities. Tropical deforestation is responsible for about 20 percent of the world's annual CO_2 emissions. Therefore, agricultural and forestry activities can both contribute to the accumulation of greenhouse gases in our atmosphere and be used to help prevent climate change by avoiding further emissions and sequestering additional carbon. Sequestration activities can be carried out immediately, appear to present relatively cost-effective emission reduction opportunities, and may generate environmental cobenefits

At the global level, the estimates are that about 100 billion metric tons of carbon over the next 50 years could be sequestered through forest preservation, tree planting, and improved agricultural management. This would offset 10–20 percent of the world's projected fossil fuel emissions.

The Carbon Cycle

Carbon is held in the vegetation, soils, and oceans. The global carbon cycle involves carbon flows among the various systems—terrestrial, atmospheric, and oceanic. Biological growth captures carbon from the atmosphere and distributes it within the terrestrial system, and decomposing vegetation and respiration release carbon back into the atmosphere. Annual plants have a cycle that includes growth during some parts of the year and death and decomposition during others. Thus, the level of atmospheric carbon increases in the northern hemisphere in the winter and decreases in the summer. (Because of its much greater landmass, the northern hemisphere has more vegetative activity than the southern hemisphere and therefore dominates this cycle.)

Carbon and Forests: An Overview

The process of photosynthesis combines atmospheric carbon dioxide with water, subsequently releasing oxygen into the atmosphere and incorporating the carbon atoms into the cells of plants. In addition, forest soils capture carbon. Trees, unlike annual plants that die and decompose yearly, are long-lived plants that develop a large biomass, thereby capturing large amounts of carbon over a growth cycle of many decades. Thus, a forest ecosystem can capture and retain large volumes of carbon over long periods.

Forests operate both as vehicles for capturing additional carbon and as carbon reservoirs. A young forest, when growing rapidly, can sequester relatively large volumes of additional carbon that are roughly proportional to the forest's growth in biomass. An old-growth forest acts as a reservoir, holding large volumes of carbon even if it is not experiencing net growth. Thus, a young forest holds less carbon, but it is sequestering additional carbon over time. An old forest may not be capturing any new carbon but can continue to hold large volumes of carbon as biomass over long periods of time. Managed forests offer the opportunity for influencing forest growth rates and providing for full stocking, both of which allow for more carbon sequestration.

Agricultural and Forestry Practices Sequester Carbon

There are three general means by which agricultural and forestry practices can reduce greenhouse gases: (1) avoiding emissions by maintaining existing carbon storage in trees and soils; (2) increasing carbon storage by, for example, tree planting and conversion from conventional to conservation tillage practices on agricultural lands; and (3) substituting biobased fuels and products for fossil fuels, such as coal and oil, and energy-intensive products that generate greater quantities of CO_2 when used.

There are a number of activities that could result in an increase in forest and forest-related carbon compared with the base situation. These include reducing deforestation, expanding forest cover, expanding forest biomass per unit area, and expanding the inventory of long-lived wood products inventory. Each of these is discussed below.

Reduction of Tropical Deforestation

From a pragmatic perspective, the most straightforward approach seems to be that of reducing deforestation. Tropical deforestation is driven primarily by the conversion of forests to agricultural uses. If this trend could be slowed, stopped, or reversed, less carbon would be released into the atmosphere, and the forests would become a net sink for carbon. Tropical deforestation has been recognized as a problem for several decades; however, programs to mitigate this problem have been notable for their lack of success. Most data indicate that the level of tropical deforestation has been relatively high and essentially constant for at least the past two decades. Furthermore, the value of the forest for carbon sequestration and other environmental values is typically ignored. In addition, tropical forests are often on a country's frontier: an area where government control is limited, property rights are weak, and law enforcement is sporadic.

Forest Expansion

Any expansion of global forests implies the capture of atmospheric carbon. Forests have been expanding in the northern hemisphere as marginal agricultural lands have reverted to forest. In Europe, for example, the reversion began as early as the beginning of the 19th century, and in New England, it began perhaps in the middle of the 19th century. However, it was probably not until the latter part of the 20th century that the reversion in the northern temperate forest generated an overall net expansion of forest area as agriculture declined in many regions and the lands reverted to forests, often through natural regeneration. In many cases, however, afforestation has been a result of conscious human effort. Whatever the cause, an expansion of the land area in forest means additional carbon sequestration. Net carbon sequestration occurs if new areas are converted to forests and sequestration occurs more rapidly than losses, some of which may be occurring elsewhere.

Carbon in Wood Products

Finally, harvested wood that is converted into long-lived wood products provides an additional stock of captive carbon. Wood products do not last forever. However, the global inventory of wood products increases when more products are added to the inventory than are removed from the destruction and dissipation of some products. As the wood products' inventory stock increases, more carbon is held captive, or sequestered, in that stock. Globally, the total stock of wood products appears to be increasing gradually.

Sequestration as an Opportunity and Option for Mitigation

Well-designed, land-based climate change mitigation activities are therefore an essential component of climate change mitigation. International efforts in climate change mitigation under the Kyoto Protocol offer three market-based mechanisms, among which the Clean Development Mechanism is of particular importance to the developing world. Under the Clean Development Mechanism, a developed country buys carbon credits generated from a project (including forestry) in a developing country. These credits are used in offsetting the emissions target allotted to the developed country.

Such projects can bring sustainable livelihoods to local people through the diversification of agriculture, soil and water protection, direct employment, the use and sale of forest products, and ecotourism. In the process, communities can also build their capacity to adapt to the effects of climate change. Through effective planning and implementation, all of these positive outcomes can be achieved cost-effectively. Forestry Clean Development

Mechanism projects are challenging in the sense of development and execution, but they offer biodiversity and community benefits that no other carbon credits scheme can offer.

Managing Terrestrial Carbon in Peat Lands

Globally, the extent of peat lands is approximately 400 million hectares. They occur in arctic, boreal, temperate, subtropical, and tropical zones and cover over 120 countries. Peat comprises layers of partly decomposed plant materials deposited over 5,000–10,000 years. Peat lands have a high water table and slow decomposition rate. They are extensively found in regions with high rainfall and/or low temperatures. Therefore, peat lands play a critical role in water management. Given their depth and extent, peat lands also play a significant role in carbon storage and sequestration. It is estimated that 25–30 percent of all terrestrial carbon is in peat lands. This is equivalent to 550,000–650,000 million tons of carbon dioxide, 75 percent of carbon in the atmospheric carbon, or 100 years of fossil fuel emissions. Therefore, peat lands have helped prevent global warming over the past 10,000 years by absorbing over 1,200 billion tons of CO_2.

As far as climate change is concerned, conservation and rehabilitation of degraded peat lands are both urgent and strategic. However, no market-based mechanisms have yet been envisaged to achieve this important objective. As a consequence, limited public funding has to be stretched to enhance the capacity of stakeholders and raise public awareness.

Therefore, carbon sequestration is the best option to mitigate global climate change, as it offers a cost-effective option for emission reduction and also comes with a host of environmental and community benefits.

See Also: Carbon Trading and Offsetting; Climate Change; Emissions Trading; Global Warming; Greenhouse Gases; Intergovernmental Panel on Climate Change; Kyoto Protocol.

Further Readings

Guidebook for the Formulation of Afforestation and Reforestation Projects Under the Clean Development Mechanism. ITTO Technical Series 25. Yokohama: International Tropical Timber Organization, 2006.

Intergovernmental Panel on Climate Change. *Fourth Assessment Report.* Geneva: Intergovernmental Panel on Climate Change, 2007.

Intergovernmental Panel on Climate Change. *Special Report on Land Use, Land Use Change, and Forestry.* Cambridge: Cambridge University Press, 2000.

Pearson, Timothy, et al. *Sourcebook for Land Use, Land-Use Change and Forestry Projects.* Washington, DC: BioCarbon Fund, World Bank, 2006.

Debojyoti Chakraborty
Independent Scholar

CARBON TAX

To date, the forefront of international and domestic climate change policy making has been on a mechanism known as "cap-and-trade." Under the cap-and-trade approach, economy-wide or sector "caps" are established for greenhouse gas emissions and tradable allowances (the right to emit a ton of greenhouse gases) to greenhouse gas emission sources or fuel distributors. This approach is the centerpiece of the Kyoto Protocol, the European Union's Greenhouse Gas Emissions Trading System, and in the United States, several regional programs, including the Regional Greenhouse Gas Initiative in the Northeast, the Western Climate Initiative, and the Midwestern Regional Greenhouse Gas Reduction Accord.

However, the consensus of most ecological economists, many policy makers, and even some in the energy sector, including Exxon's chief executive, Rex Tillerson, is that a radically different approach is needed. They suggest that an imposition of a carbon tax would be a superior method for effectuating substantial reductions in carbon dioxide, the greenhouse gas that has accounted for approximately 90 percent of increased climate forcing in recent years.

How a Carbon Tax Works

A carbon tax is a tax on the carbon content of fossil fuels and is effectively a tax on the carbon dioxide emitted when fuels are burned. Carbon taxes are easy to calculate because the precise carbon content of every form of fossil fuel is well known, and most fuels emit carbon dioxide in direct proportion to their carbon content. Because the carbon content of natural gas, oil, and coal vary, a carbon tax would be based on British thermal units rather than volume or weight. Thus, coal would be taxed somewhat more heavily than petroleum products, and much more than natural gas. Some proposed carbon taxes also account for the impact of other non-carbon greenhouse gases (e.g., methane) produced from fossil fuels, often using the relative global warming potential of a gas to determine the carbon dioxide content equivalent (CO_2e) in metric tons.

A carbon tax could be imposed anywhere in the chain from the point of ultimate fuel combustion ("downstream"), such as on gasoline for vehicles or natural gas for heating, or at the point of extraction ("upstream"), or where fuel is imported. Many advocate the latter approach on the grounds of administrative ease. For example, in the United States, imposition of an upstream carbon tax would mean that virtually all fossil fuel emissions could be covered by imposing the regime on fewer than 3,000 entities. Most carbon tax proposals are designed to be revenue neutral, meaning that the vast majority of the tax revenues would be returned to the public to minimize or totally eliminate the potential regressive effects of the scheme.

The Rationale for a Carbon Tax

A carbon tax is often characterized as a "Pigouvian tax," named after the 20th-century economist Arthur Pigou, who argued that taxes could be levied to correct the negative externalities associated with market activities rather than to merely raise revenue. A negative externality occurs when an economic actor's activities impose an economic cost on third parties, but the actor does not fully pay this cost. A quintessential example is a steel plant pumping pollutants into the air. Although the firm has to pay for electricity, labor, and materials, it is not assessed for the negative costs imposed on those living near the plant, such as higher medical expenses, loss of income, and diminution of the quality of life. Ecological economists have catalogued a host of negative externalities associated with fossil fuel production, including damages associated with climate change, such as adverse effects on agricultural production, health costs, damages associated with violent weather events, and the effects of rising sea levels on coastal regions. Indeed, many ecological economists believe that the failure to account for the externalities of fossil fuel use constitutes the great market failure in the world today.

A carbon tax that accurately reflected the negative external costs associated with gasoline, electricity, oil, coal, and other fuels could contribute to efforts to address climate change in several ways. First, a carbon tax would raise the price of fossil fuels, encouraging utilities, businesses, and individuals to reduce consumption and increase energy efficiency. For example, a recent study by Tufts University economist Gilbert Metcalf concluded that a carbon tax of just under $17 per ton of carbon dioxide would nearly double the average price of coal, the most carbon intensive of fossil fuels, leading to fuel substitution and increases in efficiency that would ultimately reduce demand by 32 percent. Second, a tax would send accurate price signals for carbon that would level the playing field between carbon-based fuels and clean renewable energy sources, energy efficiency programs, demand-side management, and other technologies that could radically reduce greenhouse gas emissions. Finally, a carbon tax would create a profit niche for environmental entrepreneurs to find ways to deliver lower-carbon energy at competitive prices.

The Potential Advantages of a Carbon Tax Over a Cap-and-Trade Approach and Some Responses

Proponents of a carbon tax contend that such a scheme would be much easier to implement administratively than a cap-and-trade scheme. Because the types of fuel and amounts purchased are precisely tracked by most industrial sources, it is argued that instituting a carbon tax would require little, if any, additional reporting and could be easily calculated. Moreover, existing revenue collection and enforcement mechanisms could be used in administering the tax.

In contrast, it's contended by carbon tax supporters that establishment of a cap-and-trade program is an extremely complex process, requiring rigorous new systems for compliance reporting, audits, and verification. The administrative simplicity of a carbon tax means that it could be implemented much more quickly in a country such as the United States—a critical consideration, as time is now of the essence in combating climate change. However, in 2006 the state of California was able to pass the Assembly Bill 32 (AB 32): Global Warming Solutions Act that establishes a statewide carbon emission level cap, calling for 1990 greenhouse gas emission levels to be reached by the year 2020.

Opponents of a carbon tax question whether a carbon tax would necessarily prove to be more administratively simple. They point to the complexity of tax codes throughout the world and express concern that a carbon tax scheme could be laden with loopholes and exceptions that would render it far from simple to administer. The fact that tax rates might have to be adjusted from time to time would further increase administrative burdens. Proponents of a carbon tax argue that these taxes pose less potential for political manipulation and fraud. The protracted negotiations process for cap-and-trade systems provides ample opportunities for stakeholders to engage in rent-seeking activities to maximize their profits or market share, or to undercut competitors. Carbon tax supporters argue that this can undermine the integrity and efficiency of climate change programs. For example, largely as a result of lobbying by industrial interests, the European Union, in formulating its Emission Trading System, set the initial emissions cap so high that most regulated entities were able to comply with their obligations without making substantive reductions in greenhouse gas emissions. There is also substantial evidence of fraudulent transactions in cap-and-trade systems. For example, a recent review in the United Kingdom concluded that at least half of the claimed emissions reductions in a recent government-sponsored auction may not have been real. In contrast, proponents of a carbon tax argue that the transparency and relative simplicity of carbon taxes afford far fewer opportunities for manipulation and fraud.

However, opponents of a carbon tax assert that carbon taxes may be equally susceptible to political manipulation. It is by no means clear that the political will exists to establish a carbon tax rate that is sufficiently high to substantially reduce greenhouse gas emissions, or that would not be pervaded with loopholes. For example, when the United Kingdom imposed a climate tax on industrial carbon emissions nine years ago, industrial interests were able to obtain widespread exemptions that have severely undercut the program's effectiveness in reducing emissions.

A third argument in favor of a carbon tax is that cap-and-trade schemes are prone to high price volatility, because although supplies of credits are fixed, demand can vary considerably as a result of several factors, including changes in energy demand and fuel-price fluctuations. For example, trading prices for carbon dioxide in the European Union Emission Trading System have fluctuated by more than 50 percent in a single year. In the U.S. sulfur-emissions trading program, volatility has been even more dramatic, with prices varying from a low of $70 per ton in 1996 to $1,500 per ton in late 2005. Volatility of this magnitude can impede investments in projects and programs that could substantially reduce emissions, such as clean technology development and energy efficiency initiatives, because of unpredictable returns on investment. In contrast, carbon tax proponents contend that such taxes establish a well-defined price for carbon, and although the tax may rise over time, this increase would be known in advance, facilitating a stable investment environment for businesses and utilities.

Supporters of the cap-and-trade approach question the effects of carbon market volatility. They point out that volatility is pervasive in many commodity markets, and yet businesses manage to cope. Moreover, although the sulfur dioxide cap-and-trade system in the United States has, as previously noted, been subject to large swings in credit prices, it has been by almost all accounts a rousing success.

A final argument put forward by carbon tax supporters is that it makes sense to allow national greenhouse emissions to vary from year to year because prevailing economic conditions significantly affect costs of emissions abatement. Unfortunately, most cap-and-trade systems do not provide business entities with such flexibility. In contrast, a carbon tax would permit business entities to pay more tax and abate emissions less in periods when abatement costs are unusually high, and vice versa when abatement costs decline. As a consequence, one study has estimated that a carbon tax can reduce the societal costs of combating climate change by approximately 80 percent below that of a cap-and-trade approach.

Opponents of a carbon tax also contend that because taxes only indirectly reduce emissions through price, we cannot easily ascertain the appropriate level of tax to induce changes in energy demand and consumption to ensure that emissions decline far and fast enough to avoid dangerous climate change. In contrast, some argue that cap-and-trade systems provide better prospects for meeting scientifically prescribed targets because they directly regulate greenhouse gas emissions by placing constraints on quantity.

See Also: Climate Change; Energy Policy; Greenhouse Gases.

Further Readings

Environmental and Energy Study Institute. "A National Carbon Tax: Another Option for Carbon Pricing." Available online at www.eesi.org/120908_tax (Accessed January 2009).

Parry, Ian and William A. Pizer. "Emissions Trading Versus CO$_2$ Taxes Versus Standards." *Backgrounder* (September 2007).

Redburn, Tom. "The Real Climate Debate: To Cap or to Tax?" *New York Times* (November 2, 2007).

Weisbach, David, et al., "Carbon Tax vs. Cap-and-Trade." *Bulletin of the Atomic Scientists* (September 2008). www.thebulletin.org/web-edition/roundtables/carbon-tax-vs-cap-and-trade (Accessed January 2009).

William C. G. Burns
Santa Clara University School of Law

CARBON TRADING AND OFFSETTING

Energy from carbon-based fuels has been essential in the industrialization of society. The emissions from the use of these fuels include several greenhouse gases, and carbon dioxide is the principal component of the various greenhouse gases emitted. A market-based method to control carbon dioxide emissions into the environment has resulted in the development of commodified emissions rights and systems for trading in these rights. Emission trading markets that focus on carbon dioxide and carbon dioxide equivalents are known as carbon markets. In emissions markets, permitted quantities of emissions below a defined ceiling or cap are tradable and can be sold to others whose emissions are higher than the cap. In carbon markets, tradable emissions are denominated in tonnes or metric tons of carbon dioxide or carbon dioxide equivalents. Markets may be regulated by governments or nongovernmental organizations or be self-regulated by the participants. National authorities can operate regulated markets under their own initiative as well as implement international agreements such as the Kyoto Protocol. The Kyoto provisions at Article 17 include regulated carbon trading between signatories to the agreement as well as offsets between signatories and nonsignatory countries. Under Kyoto, offsetting can be through the Clean Development Mechanism and Joint Implementation mechanisms. In addition, nongovernmental markets operate within several jurisdictions, and several transnational brokerages also facilitate offsets globally.

With industrialization and newer technologies, the pace of energy demand and usage has been increasing dramatically and continues unabated as different populations participate in the globalizing economy. With a history of reliance primarily on fossil fuels, increasing demand and irregular supply of fossil fuels has led to volatile prices and consideration of alternative fuels. In addition to industrialization, the stability or security of energy supplies is important for everyone as an essential feature of life. With limited global fossil fuel sources distributed unevenly and differently from processing facilities and end users, access is a security concern.

Nonfossil fuels are now commercially reasonable in some contexts and are being used. For instance, in Europe, fuel for the manufacture of cement consists largely of wastes from sources such as chemical production, sewage sludge, and tires. Biomass, methanol, ethanol, and propane are also significant fuel sources in some parts of the world. Fossil fuels and some nonfossil fuels are carbon based. In large part, emissions from energy generation and use are associated with the increased presence of atmospheric carbon dioxide, other gases involving carbon, and carbon-based particulate matter such as soot.

Eighty percent of the greenhouse gases from industrialized jurisdictions such as Europe can be traced to energy usage. Residential and transportation sectors contribute most of the balance of the carbon in greenhouse gases. Other emergent industrializing jurisdictions are expected to contribute to the growing greenhouse gas emissions and environmental toxicity from anthropogenic carbon. For a net reduction in atmospheric carbon, an increase in carbon sinks or reduction in emissions—or both—is required. As the economy evolves, efforts to control carbon emissions may lead to a redistribution of the number, location, and sources of carbon dioxide emissions without any net reduction in emissions. Emissions trading in general, and carbon trading in particular, operates to redistribute or rebalance emissions and reductions. These trading systems may also be effective in moderating net carbon emissions into the atmosphere.

Trading

Control of greenhouse gas emissions, and hence atmospheric carbon, is established through various regimes. These can range from regulatory provisions made domestically in countries as well as by agreement between countries and between human and legal persons. The United Nations Framework Convention on Climate Change and the Kyoto Protocol are examples of agreements between countries that have provisions that are implemented in domestic legislative regimes. Market-based controls have also been designed and implemented in several jurisdictions, including the United States and Canada, independent of the Kyoto Protocol. Nongovernmental organizations have also established emissions trading markets in several jurisdictions, as seen in the exchanges owned by Climate Exchange PLC.

For the purposes of carbon markets, the Kyoto Protocol recognizes atmospheric carbon resulting from land and land use and emissions. Currently tradable commodities in carbon markets therefore include emissions units themselves, removal units based on land use including forestry, emission reduction units in Joint Implementation projects, and certified emission reductions in the Clean Development Mechanism.

All trading markets for emissions require that a limit, ceiling, quota, or caps for emissions into the environment be established either by regulation or by agreement of market participants. Internationally, authorized emission amounts are allocated to countries or regions for further distribution domestically or locally. End users of emission units may be obligated to retain a minimal amount of units or "cushion" above the actual usage. This cushion is to protect against being without units if required. Emission units above the cushion amount and up to the cap would then be unused and transferable at the discretion of the holder of the units.

In a carbon trade, the vendor or the seller provides tradable units to the purchaser, who requires them either because the purchaser is a source of emissions that may exceed its emissions limit or to bank them for future use or trade. Decisions about the quantities of carbon units to be produced and traded then become management decisions subject to cost-benefit analyses and a determination to either reduce emissions or purchase tradable balances from other parties, whichever is more commercially reasonable under the circumstances.

In addition to direct emission trades, it is also possible to trade in futures and emission-based derivative products. As the volume and interest in carbon markets grows, products and strategies similar to those used in market trading and financing can be deployed.

Beyond regulated and nongovernmental trading systems, other sources of emissions are also known to offer tradable credits. These participants, who are not part of established emission reduction trading systems, offer discrete emission reductions. Other equivalent terms used are voluntary compliance, open-market trading, or verified emission reductions. Of interest here is that, as with nongovernmental markets, the participants voluntarily enter into a trading mechanism without legal requirement. In some instances, regulatory and nonregulatory offsetting could be designed as discrete emission reduction mechanisms.

Offsetting

Offsetting was intended to support the introduction of new clean energy projects and technologies in industrializing economies. Under the Kyoto Protocol, offsets may be either compliance-based Clean Development Mechanisms or Joint Initiative mechanisms involving at least one self-regulated or voluntary market party. Nongovernmental organizations and independent self-regulated persons may also design and trade units at their own risk and discretion.

Under the Clean Development Mechanism, it is possible for a party to the Kyoto Protocol to obtain certified emission reductions when a qualified emission reduction project is implemented in a jurisdiction that is not subject to the protocol. Certified emission reduction units are available to be applied against the Kyoto Protocol targets. Under the joint initiative mechanism, the parties must be identified in Annex II of the Kyoto Protocol, and the project must provide reductions or sinks that are additional to what would have occurred without the initiative. An example of an independent brokerage service in trades of carbon credits is the World Bank, which may charge a 13 percent commission on each trade.

For a purchaser, there are two steps to obtaining an offset. The purchaser first determines the required amount of offset for the purchaser's emissions by determining their intended carbon footprint and computing the carbon equivalent. Next, the purchaser locates a vendor who will agree to implement measures to achieve the same required amount of reduction in carbon emissions to the environment. The purchaser pays the vendor to implement the reduction. If the transaction proceeds and the required measures are implemented, the intended result is for the participants to generate no more than zero net emissions into the atmosphere. In an offset transaction, funds from purchasers of carbon equivalents serve to finance operations that minimize environmental degradation or effect atmospheric carbon neutrality. For the vendor, the same transaction can be used for capital improvements that otherwise may not have been implemented.

At the international level, emission quotas are allocated to individual countries or to regions such as the European Union. These allocations include all sources of emissions over which each sovereign country has jurisdiction. The process of allocating quotas to individual operators tends to address point sources more easily than nonpoint and mobile sources. As a result, point sources of emissions may bear a greater share or burden of the costs. For individual enterprises, there is a similar issue as to the assignment of emissions to various segments of the facility.

To allocate emissions within the scope of a facility—or even an enterprise—requires management design and execution as a course of doing business. Management can achieve intended emission targets by configuring the various sources of emissions to offset reductions of emissions in some spaces to balance increased emissions in others. Alternatively, management can determine that it is not commercially reasonable to reduce emissions and can elect to purchase credits instead, which may also qualify as offsetting in some trading markets. Further options include international offsets, as well as offsets by carbon recapture and sequestration.

As with the World Bank, private brokers operate to facilitate offset transactions and can function at various levels of complexity, from retail transactions involving personal activities to large, industrial-scale offsets.

Information systems can be used to map and manage the carbon footprints of facilities, which can be linked through software to carbon trading markets. Technological solutions can be used to flatten the supply/demand curve. Some jurisdictions have deployed "smart meters" involving electricity supply meters that can record consumption by variable, including time of day and season of the year. The information from such meters permits energy suppliers to vary rates and supplies to retail and commercial customers for supply reasons. It is even possible to assign to suppliers remote control of noncritical individual appliances at user facilities. Such supply-based remote controls can be used to redistribute supply when demand from certain critical applications requires it.

Market trading mechanisms can produce net overall reductions in emissions by deploying declining ceilings or allocations or quotas on emissions. In a regulated carbon market, taxes on trades can be varied to adjust incentives to reduction as circumstances require and can be used on all trades as they settle. Brokerage fees and other transaction costs can have a similar effect as taxes.

Offsets can be effectively used to limit total emissions within regions where sources of emissions are being added. On a global scale, reductions in emissions in some areas can be used to introduce other sources without exceeding the applicable limit to emissions and balancing the overall global anthropogenic atmospheric carbon burden.

New technologies can achieve reduced carbon emissions, and alternative sources of energy can be valuable in maintaining energy supplies while advancing socially valuable products and services.

Continuing Issues

Carbon occurs in all organic materials including living organisms, is ingested as food, and is widely used in human products and production processes. The distinction between the use of carbon in products and in production processes is not always a simple matter. For instance, carbon-based products such as plastics derived from fossil fuels may be used directly or indirectly to generate energy. Carbon for human use other than for direct ingestion as food is obtained from fossil or nonfossil fuels including oil, coal, gas, and—more recently—from methanol, ethanol, propane, biomass, and oil sands or bitumen. All these sources, when appropriately fractionated and processed, may be used directly, as fuels, or indirectly, as in the generation of electricity. Because carbon is involved in all of human life, carbon trading could potentially be similarly integrated into human activities. In addition to controlling its distribution in the environment, carbon trading may have complex and pervasive effects on human life.

The immediate concern about carbon trading arises out of changes in the anthropogenic or human use and control of carbon. Atmospheric carbon dioxide accounts for 80 percent of the contribution to global warming from current greenhouse gases. Carbon is also present in the atmosphere as carbon monoxide, methane, and chlorofluorocarbons at concentrations that are two to six orders of magnitude lower than that of carbon dioxide.

The proportion of gases to each other does not directly correspond to the effect each has on climate change. With a potential for global warming 3.7 times greater than carbon dioxide per mole, even trace amounts of methane can have a greater effect than the same amount of carbon dioxide. As a consequence, the total contribution of carbon in all its forms in the atmosphere is significant. Carbon trading can be used to recognize the presence of all forms of carbon in environments at levels that are directly or indirectly toxic to life on the planet, including carbon-based life-forms such as human beings. Denominations other than carbon dioxide, such as carbon equivalents, could be designed to account for the trace greenhouse gases.

Introducing technological solutions may not always have the intended benefits. The valuation of a solution considered from the context of the globalized economy appears to favor capital-intensive systems over labor-intensive solutions. Valuation of a piece of equipment may be more easily visible than the work of poor, unemployed persons who may be providing the required service at lower cost. Introduction of a new facility in such a context may produce a more favorable balance of carbon credits, which represents financial gain. These credits can be sold pursuant to the Clean Development Mechanism.

As previously noted, offsets or trading may involve redistribution, and not a reduction in toxicity resulting from emissions. An example is the introduction of an incinerator to process and retrieve emissions in an offset agreement. The operation of this technologically green energy source may eliminate employment for hundreds of persons whose only source of income is the collection and sale of recyclable and reusable materials from municipal waste dumps. In addition, should the incinerator waste be unregulated, the net effect on environmental toxicity is questionable.

Unauthenticated claims, applying past achievements for future credit, unverified implementation, enforcement of agreements, limited audit capabilities, and redistributing funding and equipment can result in the destruction of existing efficiencies and exacerbate atmospheric and social problems rather than alleviating them. Even where an offset operates as intended, there are no provisions by which the gains introduced in a project will be retained. For example, if a forest is planted as a sequestration project in an offset, there are no assurances that the forest will not be subsequently clear-cut without further compensatory actions or enforceable consequences.

As with nuclear fission reactors, sources of energy that may be solutions to carbon emission reductions can incur costs that may not always be acceptable. Concerns with nuclear energy sources include security, safety, and nonproliferation during their operation; the dismantling of installations when decommissioned; and the management of waste for a thousand years or more. Further research on existing technology and alternative energy sources may produce more effective solutions to reducing greenhouse gas emissions.

See Also: Alternative Energy; Alternative Fuels; Automobiles; Biodiesel; Biogas Digester; Biomass Energy; Carbon Emissions Factors; Carbon Footprint and Neutrality; Carbon Sequestration; Carbon Tax; Chlorofluorocarbons; Climate Change; Climate Neutrality; Coal, Clean Technology; Combined Heat and Power; Combustion Engine; Department of Energy, U.S.; Electricity; Embodied Energy; Emission Inventory; Emissions Trading; Energy Audit; Energy Storage; Fossil Fuels; Fuel Cells; Fusion; Gasohol; Greenhouse Gases; Intergovernmental Panel on Climate Change; Internal Energy Market; Kyoto Protocol; Landfill Methane; Metric Tons of Carbon Equivalent (MTCE); Natural Gas; Nonpoint Source; Nonrenewable Energy Resources; Oil Sands; Oil Shale; Petroviolence; Plug-In Hybrid; Public Transportation; Volatile Organic Compound (VOC); Waste Incineration; Wood Energy.

Further Readings

Climate Exchange plc. http://climateexchangeplc.com (Accessed February 2009).
"Energy Insights Predicts Climate Change Issues to Drive Increased Tech Investment in 2008." *Consulting-Specifying Engineer*, Supplement, 43 (March 2008).

Holtcamp, Wendee E. "An Off-Setting Adventure." *The Environmental Magazine*, 18/6 (November/December 2007).

Hopkin, Michael. "Emissions Trading: The Carbon Game." *Nature*, 432/7015 (November 18, 2004).

International Carbon Action Partnership. http://www.icapcarbonaction.com (Accessed February 2009).

Lashof, Daniel A. and Dilip R. Ahuja. "Relative Contributions of Greenhouse Gas Emissions to Global Warming." *Nature*, 344 (April 1990).

U.N. Intergovernmental Panel on Climate Change. "IPCC Special Report on Carbon Dioxide Capture and Storage." http://www.ipcc.ch/ipccreports/srccs.htm (Accessed February 2009).

U.S. EPA Climate Change. http://epa.gov/climatechange/index.html (Accessed February 2009).

Willson, Richard W. and Kyle D. Brown. "Carbon Neutrality at the Local Level: Achievable Goal or Fantasy?" *Journal of the American Planning Association*, 74/4 (September 2008).

Wysham, Daphne. "Carbon Market Fundamentalism." *Multinational Monitor*, 29/3 (November/December 2008).

Lester de Souza
Independent Scholar

CASPIAN SEA

The Caspian Sea, or as some insist on calling it, "Caspian Lake" (as it is landlocked), is the world's largest inland body of water, but its 144,000 square miles of space and water may no longer be its claim to geographical fame. What has become more substantial, at least to the countries of Russia, Azerbaijan, Iran, Turkmenistan, and Kazakhstan that rim this once-pristine waterscape, is the large body of liquid and gaseous fossil fuels waiting to emerge from beneath the seafloor, perhaps enough to eventually rename this portion of the Earth's surface the "Caspian Deposit."

The 13th-century traveler Marco Polo called it a "fountain from which oil springs in great abundance," indicating that it may not have always been so pristine, as crude oil and water were mixing many years before oil and gas replaced wood and coal as the world's main natural energy resources—a process beginning in the late 19th century. Polo's observation sprang from a visit to Baku, the capital city of Azerbaijan, which is today—with 2 million people—the largest populated area on the Caspian. What the traveler did not know is just how much oil lay beneath the water and drainage basin, the latter spreading out 400 square miles from its 4,400-mile fickle shoreline—the Caspian's waters rise and fall intermittently, and no one knows why, though one theory points to the unusually high salinity of its water as the cause. By the time the Industrial Revolution was in full swing in the West, the world's first offshore oil wells poked pipes deep into the seafloor at Bibi-Heybat Bay, just a few miles east of Baku. By 1900, Baku became famous as the "Black Gold Capital," with its 3,000 wells producing half the world's supply of oil.

It is said that at each end of every rainbow is a pot of gold. The history and geography of the Caspian's relationship with nonrenewable energy resources can be likened to a rainbow. Radiating from the pots of black gold are the extraordinary hues of oil exploitation that color the shoreline, from where the great Volga River, flowing south from the legendary Russian city of Kazan, reaches into the sea, to the massive oil field at Tengiz in the northeast, down to Krasnovodsk, Turkmenistan, then west across the sea to Baku, then up along the western coast back to the Volga. Below the east–west transverse is the southern Caspian, most of which is controlled by Iran but,

Large numbers of oil derricks near Baku, Azerbaijan, are tapping some of the massive oil reserves in the Caspian Sea.

Source: iStockphoto.com

ironically, contains few deposits of oil, therefore remaining undeveloped. Here and in the Garabogaz Bay, the water glows turquoise blue.

Soviet Caspian

Along the shoreline, seeping oil and Caspian waters have mixed freely for centuries, creating pools of dark green sludge—always the bane of local farmers. The number of pools greatly expanded with the Sovietization of the region, which is also when the Caspian rainbow really began to shine.

When the Bolsheviks gained control of Russia in 1917, they inherited an empire consolidated over the centuries by czarist monarchies. The territory of that empire included all the independent countries bordering the Caspian, each a Republic of the Soviet Union except for Iran. This gave the Soviets a complete monopoly on the natural resources in the region. Soviet economic development policies included an acceleration of industrialization that quickened the need for an increased supply of energy resources. Though Soviet industrialization depended mostly on coal as an energy resource, oil soon began to figure in more prominently, as it is dense with hydrocarbons and easily transported, making it the most efficient of fossil fuel resources. The Caspian Sea, of course, became the central focus of exploitation, particularly the Volga estuary, extending about 50 miles from the Caspian and Tengiz in Kazakhstan.

In addition to the flourish of Soviet oil derricks, canals, dams, and floodgates that now tinted the Caspian came the industrial and agricultural elements and acids flowing down the Volga and finally into the sea. Sometimes the chemical soup would render colorful algae blooms. Much of the time, the estuary's thick sieve of flora filters the water, freeing it of contaminents before it reaches the sea. Accessorizing many of the derricks, even today, are pink ponds mixing water and salts, the latter a by-product of oil extraction. In 1985, the oil field at Tengiz was in flames: The ponds of oil fueled the fires that lit the

Caspian sky for a year. The heat was so intense it melted and melded the white sand, turning it to blue glass.

Post-Soviet Caspian

With the fall of the Soviet Union and the consequential crash of the Russian economy, the Caspian region suffered dereliction for some years—but not for too long. The West's appetite for oil is unquenchable, and with the geopolitical insecurities of the Persian Gulf states, the main suppliers of oil to the West, its attention turned toward the Caspian. Using new survey technologies, the U.S. Energy Information Administration estimated that below and around the sea are proven reserves of 17–33 billion barrels of light, sweet crude oil, as opposed to heavy, sour oil, which is more expensive to refine into automotive gasoline and jet engine fuel and more polluting when burned. Possible oil reserves were large—145 billion barrels, which is more than what lies beneath Iran and Iraq combined. Moreover, atop the oil sit 236 to 337 trillion cubic feet of proven reserves of natural gas, with 317 trillion cubic feet in possible reserves. Until recently, gas was burned at the well because it was too expensive to store and transport, given its high flammability. Recent environmental concerns have made natural gas an important source of nonrenewable energy because when burned, it emits less pollutants into the atmosphere. The former Soviet republics, where most of the deposits are now located, turned to Western oil companies, forming consortiums with the major players in the global oil industry, thus beginning the Caspian oil and gas boom of the mid-1990s.

Predictions about the amount of oil and gas beneath the Caspian region bubbled, and with that, the prodding, piping, and tapping took place in earnest. Much of it resulted in the formation of mud "volcanoes" as brown slush oozed from potential wells, enhancing the hue of sludge already flowing from sewer pipes and factory drains and adding yet another color to the Caspian rainbow. Some of the productive wells almost bathe in red-liquid sulfur, again a by-product of oil extraction, its color coordinating with the eerie orange of the fantastic salt lakes cultivated when salt was currency in the region.

Investments by Western oil companies have not made life much better for the majority of citizens living in the countries possessing the Caspian pots of black gold, especially for those who dwell along its edge, where one might expect living conditions to improve substantially. But this is an old story. In peripheral regions rich in natural resources valued by the West, poverty, disease, corruption, and environmental degradation are the usual motifs in accounts of resource extraction. Much of the money earned by Caspian countries, which is estimated to be in the hundreds of billions of dollars, is funneled to a few and spent abroad. Nepotism prevents local native businesses from satisfying the technological, material, and service needs of the new oil industry. The lack of an entrepreneurial spirit and business culture in these ex-Soviet republics also hinders regional development.

"Great Game"

Today, a "great game" is playing out that will change the political and physical landscape of the region in years to come. In fact, the current president of Russia has not ruled out the use of violence to resolve the serious issues that are the source of competition between relevant states.

The first issue involves disputes over who has rightful claim to portions of the Caspian basin, and therefore the reserves of fuel beneath the surface. At one time, the issues involved two states: the Soviet Union and Iran. Treaties were signed and agreements reached over fishing rights, but not necessarily over rights of drilling for gas and oil. Today the issue is more complicated, as now five countries border the "Sea"—an important geographical classification because it makes applicable to this body of water the United Nations Conventions on the Laws of the Sea. Still, such measures for delineating state boundaries along the world's marine coastlines has not made the issue of sovereignty over Caspian space any less contentious.

The other issue is the European Union's fear of becoming dependent on Russia for its energy needs, not because of the enormous amount of oil and gas on Russian territory but, rather, Russia's incipient monopoly on the transport of fuel into Europe via pipelines leading out of the Caspian region, into Eastern Europe, and finally to the largest economies in Central and Western Europe. These pipelines will carry oil and gas not just from proven reserves in Russia but from central Asia, the Middle East, and the Caucasus. Because of geopolitical concerns, European consortiums are now competing with Russia's nationalized oil company—Gazprom—to sign on countries of transit so that the lines linked will be the shortest, safest, and ultimately most cost-effective routes into Europe.

Over the Caspian Rainbow

At each end of the Caspian rainbow sits a pot of black gold worth hundreds of billions of dollars. But the spectacular hues and hydrocarbon wealth found there have spoiled its waters, making them less and less habitable for the marine life that is a vital renewable food resource for many who live by the sea. One kind of fish in particular has been affected most by the excesses of oil and gas production—or rather, no other Caspian fish has drawn so much attention to the negative results of this excess than the grand sturgeon, which produces world-famous, highly desired, and expensive black, gold, brown, white, and yellow caviars, yet another aspect that gives the Caspian its geographical fame.

See Also: Fossil Fuels; Natural Gas; Offshore Drilling; Oil; Pipelines.

Further Readings

Cullen, Robert. "The Rise and Fall of the Caspian Sea." *National Geographic,* 195/5 (1999).
Galpin, Richard. "Energy Fuels New 'Great Game' in Europe." *BBC News* (9 June 2009). http://news.bbc.co.uk/2/hi/europe/8090104.stm (Accessed June 2009).
Joyner, Christopher C. and Edmund A. Walsh. "The Caspian Conundrum: Reflections on the Interplay Between Law, the Environment and Geopolitics." *International Journal of Marine and Coastal Law,* 21/2 (2006).

Ken Whalen
American University of Afghanistan

CHERNOBYL

One source of energy that does not produce greenhouse gases is nuclear power. Nations with burgeoning populations are turning to nuclear power as an efficient, clean way to produce vast amounts of power that can support their electricity needs. Although nuclear power plants do provide clean energy, they come with a risk. If an accident should occur at the power plant that causes a meltdown, or breach in radiation safety, tremendous amounts of nuclear radiation could theoretically be released into the surrounding environment, injuring all life forms including people and plants, as well as contaminating structures and cloud formations, leading to nuclear rain.

Just such a disaster occurred at the Chernobyl nuclear power plant on April 26, 1986. The power plant was located immediately outside the city of Pripyat in the Ukraine, which was then part of the Soviet Union. Although, in general, nuclear power plants are considered safe producers of energy, this disaster cast doubt and fear over the safety of nuclear energy. Most doubts will likely be found in post-Soviet production of nuclear energy, rather than nuclear energy production in other nations, but strict regulations are an outcome of the disaster. At the time, whereas Western power plants typically protected their radioactive cores in strong containment buildings made from steel or concrete, Soviet power plants were not routinely built with such protection.

An initial nuclear reaction escalated out of control, causing a steam explosion in reactor 4 of the Chernobyl power plant. The extreme conditions that resulted caused a second explosion, this time chemical. The force of this second explosion tore off the reactor's roof, and the radioactive core was exposed to the environment. Not only was a large amount of radioactive material released, but the extremely hot core was exposed to the oxygen of the surrounding air, causing it to burn, and releasing even more radioactivity in the smoke. The plumes of smoke carried radioactive debris to the atmosphere, from which it traveled to nearby nations and even, to some extent, to the U.S. eastern seaboard. It is estimated that the lethal fallout totaled 300 times the amount released by the atomic bombing of Hiroshima, Japan, on August 6, 1945. The main radioactive isotopes released by the Chernobyl disaster were cesium-137, strontium-90, iodine-131, and plutonium. It is difficult to determine the precise amount of damage caused by the Chernobyl disaster because, at the time, record keeping was censored both in what was written and in what was released by Soviet authorities. At this time, scientists are examining what data there are in an attempt to better understand the effects of radiation contamination in the atmosphere, as well as on human and other health.

Historical accounts maintain that the plant had not followed the proper safety precautions and had not passed regulated safety tests but had been recorded as being safe by its director, so as to save employees' jobs and bonus pay. In addition, an experiment had been planned to test the safety precautions, and it was this experiment that ironically led to the steam explosion and subsequent disaster. Two people died immediately from the explosion, but many more deaths were eventually caused by the radioactive fallout.

The city of Chernobyl is about 9 miles (nearly 15 kilometers) southeast of the power plant and is generally unaffiliated with the power plant, as most power plant employees and their families resided in Pripyat. Nonetheless, it does lie within the Zone of Alienation (aka the Chernobyl Zone), which is a circle of 19 miles' (30 kilometers) radius around the Chernobyl power plant that was evacuated after the disaster. The zone is generally uninhabited, save for security personnel and some scientists; some people live there for short durations, some people resisted the evacuation mandate, and sometimes tourists come for a day trip from Kiev, the Ukraine capital.

See Also: Alternative Energy; Green Power; Nuclear Power; Nuclear Proliferation; Power and Power Plants; Three Mile Island; Uranium.

Further Readings

Hiroshima International Council for Health Care of the Radiation-Exposed. "Global Radiation Exposure." http://www.hiroshima-cdas.or.jp/HICARE/en/10/hi04.html (Accessed May 2009).

International Atomic Energy Agency. "Environmental Consequences of the Chernobyl Accident and Their Remediation: Twenty Years of Experience." Vienna: IAEA, 2006.

Mould, R. F. *Chernobyl Record: The Definitive History of the Chernobyl Catastrophe.* London: Taylor & Francis, 2000.

Stone, R. "Inside Chernobyl." *National Geographic Magazine* (April 2006).

Claudia Winograd
University of Illinois at Urbana-Champaign

CHLOROFLUOROCARBONS

Chlorofluorocarbons, or CFCs, are part of a group of chemical compounds known as haloalkanes. They contain chlorine, fluorine, and carbon and are extremely stable compounds that can be separated by ultraviolet radiation. They are known for a variety of uses, including as flame retardants, fire extinguishers, refrigerants, propellants, and solvents. Because of their ozone-depleting potential, CFCs were subject to a phase-out in production and consumption under the Montreal Protocol.

The CFC story begins with chemist Thomas Midgley, Jr., who in 1930 developed the CFC commonly known as Freon for General Motors. This CFC was considered a safe alternative for uses in heat pumps, refrigerators, and other appliances. CFCs also gained widespread use in aerosols and asthma inhalers. Some common CFCs and their uses include:

- CFC-12 for early home refrigeration and wartime insecticide sprays
- CFC-13 for commercial cooling and refrigeration
- CFC-11 for domestic toiletries and cleaning products in aerosols

By the early 1970s, 200,000 metric tons of CFCs were used in aerosols each year in the United States alone.

By the 1970s, it became increasingly evident that CFCs have a major side-effect: they deplete the stratospheric ozone layer. An article published by Frank Sherwood Molina and Mario Rowland in *Nature* in 1974 strongly suggested that CFCs persisted long enough in the atmosphere to reach the stratospheric ozone layer and deplete it through a chemical reaction that releases chlorine atoms that then bond with oxygen atoms in ozone molecules, thus destroying the ozone molecule. As a consequence, CFCs were set for phase-out under the provisions of the Montreal Protocol, the international treaty designed to protect the ozone layer, which was signed in 1987.

Before the ratification of the Montreal Protocol, public concern in the industrialized world regarding the negative effect of CFCs gave rise to action by corporations and governments. States within the United States enacted bans (either upheld by legislation or "voluntarily" enforced) on CFC aerosols, passed a labeling law for CFC-containing products, and passed other bills restricting CFC aerosol use. By 1975, CFC aerosol sales had plummeted in the United States, and by 1978, all CFC aerosols were banned in the United States, with medical essential uses (such as CFCs for metered dose inhalers) remaining exempt.

Chemical companies have worked on alternatives to CFCs. Hydrochlorofluorocarbons were a popular replacement to CFCs that broke down in the atmosphere quicker than CFCs. Hydrofluorocarbons also could be used in other applications and broke down even faster. However, because of their global warming potential, in 2007, parties to the Montreal Protocol accelerated the agreed-upon phase-out of hydrochlorofluorocarbons from 2030 to 2015. Chemicals that do not deplete the ozone layer have subsequently been put into production, making CFCs—chemicals that have extremely effective applications but extremely harmful effects on the natural environment—closer to obsolete.

See Also: Climate Change; Global Warming; Greenhouse Gases.

Further Readings

U.S. Environmental Protection Agency. "Ozone Depletion Glossary." http://www.epa.gov/
    ozone/defns.html (Accessed February 2009).
van der Leun, Jan C. "The Ozone Layer." *Photodermatology, Photoimmunology &*
    Photomedicine, 20 (2004).

Brian J. Gareau
Boston College

CLIMATE CHANGE

Climate is the average of the weather conditions described through variability in temperature, precipitation, and wind over a period of time. This time period may range from months to millions of years. The climate system is a complex interactive system consisting of the atmosphere, land surface, snow and ice, oceans and other bodies of water, and living things.

Changes in climate have occurred in Earth's history in the past, but the rate of change in the past 30–40 years has snowballed into a major global debate and concern. The past 10,000 years in Earth's history has been a period of unusual stability that favored human civilization, but anthropogenic activities, particularly after the industrial revolution in the 1970s, have lead to a warming of Earth's atmosphere by 0.75 degrees Celsius compared with the preindustrial level. The rise in Earth's temperature by 2050–70 is projected by different models to range between 1.5 and 4.5 degrees C. This will have a grave effect on the ecology and socioeconomics of nations. The effect, however, will not be equal but will be greater on the poorer and marginal sections of human society—they will find it rather more difficult to adapt to the changed climate scenario. Through a global convention, the Kyoto Protocol, many nations of the world have come together and acted to limit the levels of the greenhouse gases (GHGs) in the atmosphere—major causes of global warming—and to build resilience through appropriate strategies.

The Earth's warming is mostly attributed to increasing levels of GHGs in the atmosphere. To better understand this process, one has to understand the "greenhouse effect." Greenhouses are glass house structures used for raising plants in which the temperature inside the glass houses gets raised as a result of an inherent property of the glass to trap heat. Thirty percent of the light from the sun that reaches the top of the atmosphere of the Earth is reflected back, of which roughly two-thirds is reflected as a result of clouds and small particles in the atmosphere known as "aerosols." Light-colored areas of the Earth's surface—mainly snow, ice, and deserts—reflect the remaining one-third of the sunlight. The energy that is not reflected back to space—approximately 240 watts per square meter—is absorbed by the Earth's surface and atmosphere. The Earth must radiate the same amount of energy back to the atmosphere, but it does so by emitting by long-wave radiation (infrared radiation). The Earth's surface is warm because the blanket of water vapor and carbon dioxide (CO_2) in the atmosphere reflects the radiation back to the Earth. This is a natural greenhouse effect. Gases such as CO_2, methane, oxides of sulfur and nitrogen (SO_X and NO_X), and chlorofluorocarbons are greenhouse gases, and their increase in the atmosphere leads to an increase in global temperatures. Clouds, too, have a blanketing effect similar to greenhouse gases; however, the effect is offset by their reflectivity.

The amount of CO_2 in the atmosphere has increased by about 35 percent in the industrial era. This increase has been primarily a result of an increase in fossil fuel use and the destruction of forests. Human beings have altered the composition of the atmospheric chemistry and changed many of the intricate cycles of the biosphere, geosphere, and atmosphere. It is, therefore, important to develop an understanding of the history of the science of climate change.

In 1896, the Swedish scientist S. Arrhenius proposed a relationship between atmospheric CO_2 and temperature, indicating that water vapors and CO_2 have the capacity to absorb infrared radiation and thus help maintain the average temperature of the planet at 15 degrees Celsius. He also calculated that the doubling of CO_2 levels would cause the temperatures to surge by 5 degrees Celsius. Later, he, along with Thomas Chamberlain, was among the first to claim that human activities could warm the earth by adding CO_2 to the atmosphere.

The development of infrared spectroscopy in the 1940s furthered the understanding of absorption of infrared radiation by CO_2. Gilbert Plass in 1955 concluded that adding more CO_2 to the atmosphere would intercept infrared radiation that is otherwise lost to space. Before the 1950s, people seemed unconcerned about these findings because they believed that oceans were great carbon sinks and would automatically absorb CO_2 and nullify the pollution caused by excessive release of CO_2. Advancement of scientific research on the subject led to an understanding that CO_2 has an atmospheric lifetime of approximately 10 years, and only one-third of the anthropogenic CO_2 is absorbed by the oceans. Increased levels of CO_2 also led to gradual acidification of the oceans, thereby reducing their ability to absorb atmospheric CO_2. In the early 1960s, Charles Keeling, using the most modern technologies, produced concentration curves for atmospheric CO_2 in Antarctica and Mauna Loa. These curves were considered "icons" in global warming science.

The global warming theory did not remain unopposed. Research on ocean sediments indicated a global cooling trend. Michael Crichton's best-seller *State of Fear* cautioned the world about our limited understanding of the Earth's processes. Through the novel, the author questioned the credibility of temperature data sets across the globe. The mismatch of rise in temperature and lowering of levels of CO_2 during the period 1940–70 remain unaddressed. Critics of global warming pointed out that the theory of global warming does not take into account periodic solar activity or appearances of solar spots in the sun, which are also correlated to Earth's surface temperature. The concentration of aerosols, which shoot up during biomass burning, also leads to cooling, although their residence time in the atmosphere is very short.

Climate Change as a Global Issue

The sharp upward trend of global annual mean temperature since the 1970s–80s led to the emergence of climate change as a global issue. The United Nations Environment Programme and the World Meteorological Organization led to the formation of the Intergovernmental Panel on Climate Change (IPCC) in 1988. This organization tries to predict the effect of the greenhouse effect according to existing climate models and literature. Comprising more than 2,500 scientific and technical experts from divergent research fields, IPCC is referred to as the largest scientific cooperation project in world history. It has a mandate to assess the scientific information relevant to human-induced climate change, its effects, and the options for adaptation and mitigation.

The IPCC is organized into three working groups plus a task force on national GHG inventories. Each of these four bodies has two cochairmen (one from a developed country and one from a developing country) and a technical support unit. Working Group I assesses the scientific aspects of the climate system and climate change; Working Group II addresses the vulnerability of human and natural systems to climate change, the negative and positive consequences of climate change, and options for adapting to them; and Working Group III assesses options for limiting greenhouse gas emissions and otherwise mitigating climate change, as well as economic issues.

The mandate of the IPCC is to provide in regular intervals an assessment of the state of knowledge on climate change. The IPCC also prepares special reports and technical papers on topics where independent scientific information and advice is deemed necessary for the preparation of assessment reports. The IPCC *First Assessment Report* was published in 1990. This report confirmed that human activities were affecting the climate. The report and the Second World Climate Conference, held the same year, helped focus attention on climate change and the need for international action. It was in this regard that the United Nations General Assembly in December 1990 established the Intergovernmental Negotiating Committee for a Framework Convention on Climate Change. This committee had the mandate to prepare an effective intergovernmental treaty on the climate change issue. After five brainstorming sessions held by the committee between February 1991 and May 1992 on the issues of emission reduction targets, binding commitments, timetables, financial mechanisms, technology transfer, and the "common but differential responsibility" of developed and developing countries, it finally came out with a text document—United Nations Framework Convention on Climate Change (UNFCCC)—in May 1992 at the Earth Summit in Rio di Janeiro. The ratification process started immediately the next month. In March 1994, the UNFCCC entered into force after the 50th ratification to it. Since 1994, the parties to UNFCCC, called the Conference of Parties, meet annually to review the progress and to consider further action.

The *Second Assessment Report* was out by 1995; it provided key input to the negotiations, which led to the adoption of the Kyoto Protocol to the UNFCCC in 1997. The Kyoto Protocol is an amendment to the UNFCCC. The countries that ratify it commit to reducing CO_2 and five other GHGs or to engage in emissions trading if they maintain or increase the emissions of these gases. The treaty was negotiated in Kyoto, Japan, in December 1997 and came into force on February 16, 2005, with 161 countries parties to it. All parties to UNFCCC can sign or ratify the Kyoto Protocol. Most of the provisions of the Kyoto Protocol apply to developed countries listed in Annex 1 of UNFCCC. Under the protocol, industrialized nations are required to reduce their emissions of GHGs by 5.2 percent compared with the year 1990 by the period 2008–12. It also reaffirms the principle that developed countries have to pay for and supply technology to other countries for climate-related studies and projects. In spite of a very high per capita emission of GHGs, the United States has refused the sign the protocol, and the issue has been debated globally.

The *Third Assessment Report* was completed in 2001. It was submitted to the 7th Conference of the Parties to the UNFCCC and proved to be an important and well-acknowledged reference for providing information for deliberations on agenda items of the Conference of the Parties. The *Fourth Assessment Report* is the latest available report in the series.

As a commitment to the implementation of the UNFCCC, signatories to the Kyoto Protocol are mandated to submission to National Communications to the IPCC. These communications are windows to apprise the world of the nation's attempt to undertake assessments of the effects of climate change on its resources and ascertain both how vulnerable the nation is to vagaries of climate change and what should be the adaptation strategies. Different counties are at different stages in placing their National Communications documents. Countries like India and China have prepared a National Climate Change Action Plan.

The Evidence Mounts

Scientific evidence supporting global warming is now reported to be accumulating. Melting of ice caps in the poles, receding of the glacial towers, rising of sea levels, bleaching of corals, and changes in phenological patterns in many plants, such as early fruiting and flowering, are just a few. Acidification of ocean water is a cause of concern because many of the marine species build their shells and skeletons from calcium carbonate. With more acidic water, it is difficult for organisms like coral and giant clams to mobilize calcium carbonate from the ocean water. A compilation of the effect of climate change is available in the two edited volumes on climate change and biodiversity by Dr. Thomas Lovejoy of the H. John Heinz III Center for Science Economics and Environment in Washington, D.C. In 2006, Al Gore, former vice president of the United States, and the IPCC jointly received the Nobel Peace Prize. Paramount Pictures cast Gore in the film *An Inconvenient Truth*, which in no time became a powerful medium to teach and educate the general public about the science of climate change.

Projections through modeling techniques have been made at global and regional scales to predict the effects of climate change for different scenarios. The projections at the global scale show that a large proportion of the world's population is vulnerable to climate change. This vulnerability to climate change is not distributed evenly. The countries that have benefited immensely through development produce a larger share of GHGs, and they are also the ones that would be the least affected. The effect will be most severe on the poor countries of the world including some island states and African nations. Together, these nations form a group of 100 countries with a population of about 1 billion people, but with CO_2 emissions of about 3.2 percent of the global total. Mitigation measures to keep emissions at lower levels and adaptation to climate change are the strategies to address the issue of climate change. Most of the predictions are based on very coarse models with very high levels of uncertainty. The predictions are also for long time periods.

Decisions and policy makers, however, require more precise predictions to know the possible effects of climate change to design measures to counter these possible effects.

See Also: Energy Policy; Global Warming; Greenhouse Gases; Intergovernmental Panel on Climate Change; Kyoto Protocol.

Further Readings

Crichton, Michael. *The State of Fear*. London: Harper Collins, 2005.

Dodman, Tim, et al. "Building Resilience." In *State of the World 2009. Into a Warming World*. Worldwatch Institute. http://www.worldwatch.org (Accessed January 2009).

Le Treut, Herve, et al. "Historical Overview of Climate Change." In *Climate Change 2007: The Physical Science Basis*. Contribution to Working Group I to the Fourth Assessment Report of the Intergovernmental Panel on Climate Change. Cambridge: Cambridge University Press, 2007.

Lovejoy, Thomas. "Climate Changes Pressures on Biodiversity." In *State of the World 2009. Into a Warming World*. Worldwatch Institute. http://www.worldwatch.org (Accessed January 2009).

Lovejoy, Thomas and Lee Hannah. *Climate Change and Biodiversity*. New Haven, CT: Yale University Press, 2005.

State of the World 2009. Into a Warming World. Worldwatch Institute. http://www .worldwatch.org (Accessed January 2009).

Sudipto Chatterjee
Winrock International India

CLIMATE NEUTRALITY

Climate neutrality, also sometimes referred to as "carbon neutrality," is the aim of developing solutions that have no net carbon output to the atmosphere. Solutions are deemed be "climate neutral" if they are determined to have negligible or zero effect on the Earth's climate.

Reasons for Climate Neutrality

Before the advent of industrialization, the level of carbon dioxide in the Earth's atmosphere was 280 parts per million. Industrialization, the use of fossil fuels, and our burgeoning energy use have changed the balance of gases in our atmosphere. The level of carbon dioxide in the Earth's atmosphere is now around 380 parts per million. The amount of carbon dioxide in the atmosphere is one of a number of what are known as "climate forcings." Carbon dioxide levels are known as an "anthropogenic factor," as human activities have affected the level to some degree.

Other forcings include the level of solar radiation and deviations in the Earth's orbit. As carbon dioxide emissions are a climate forcing we can control, many would advocate the development of "zero-carbon" solutions or "climate neutral" solutions to control this factor's contribution to climate change. Some would argue that we need to "stabilize" the level of carbon dioxide in the atmosphere, and others would go further by arguing that we need to reduce it from current levels.

There are various "feedback mechanisms" in the natural world that have the potential to amplify the effect of this initial forcing once a "tipping point" is crossed. Climate scientists are concerned that if we exceed a certain level of carbon concentration in the atmosphere, the effects of its effect on the Earth's climate might be irreversible.

The Multiscalar Nature of Climate Neutrality

We can look at climate neutrality on a number of different scales, looking at the whole picture—with the "global system" working toward a goal of climate neutrality, or individual projects or businesses aiming for a zero-carbon target to ensure that they are benign and do not contribute to climate change.

Achieving Climate Neutrality in Practice

Climate neutrality is primarily achieved through the substitutions of fossil fuel–intensive energy generation technologies with zero-carbon alternatives—namely, renewables. This is complemented by a package of efficiency measures to reduce demand for energy. A number of groups have proposed national alternative energy strategies that are "climate neutral." One of the most ambitious is Zero Carbon Britain, which is a UK energy strategy that looks at how the United Kingdom could meet its energy needs in a wholly climate neutral manner.

Included Technologies

There is fierce debate about the definition of "carbon neutrality" and what technologies can fall under the banner of being considered carbon neutral. However they are stored, they should not fall under the heading of climate neutral solutions.

Although some would argue that nuclear power produces no carbon emissions at the point of use, we must look at the problem holistically and understand that carbon emissions are produced at other stages of the nuclear fuel cycle (facility manufacture or waste transportation), so some would argue that this is not a climate neutral solution. Also, nuclear power produces a range of radioactive wastes, and we still do not have a solution to safely storing them for the long term.

A range of technologies known as "carbon capture and storage" have been proposed, where carbon emissions from the burning of fossil fuels are captured and sequestered deep underground in stable geological formations. In many instances where this technology has been proposed, it is as part of an "enhanced oil recovery" scheme, whereby the piped carbon dioxide is used to force additional oil from deep underground. In this respect, looking at the solution holistically, it cannot be considered climate neutral, as the effect of the carbon emissions from burning the recovered oil far outweighs any benefit from sequestering the carbon. Critics of carbon capture and storage have argued that such solutions are vulnerable to shifts in geology that could release large amounts of carbon dioxide at some point in the future. It has also been proposed that a reduction of atmospheric carbon levels can be achieved by absorbing carbon from the atmosphere (e.g., using natural processes such as photosynthesis or a range of next-generation technologies that are at an early stage of development).

Geo-Engineering Technologies

Geo-engineering is the proposed science of deliberately manipulating the Earth's climate to counteract any climatic change from greenhouse gas emissions. There are a number of technologies that could work in conjunction with climate neutral solutions to reduce the levels of atmospheric carbon and work to mitigate the effects of the carbon that is present.

- Carbon dioxide air capture
  - ○ Biomass energy with carbon capture and sequestration
  - ○ "Artificial trees"
  - ○ Scrubbing towers
  - ○ Biochar
- Ocean iron fertilization
- Solar radiation management
  - ○ Stratospheric sulfur aerosols
  - ○ Cloud reflectivity enhancement

Carbon Dioxide Equivalence

It is important to view the problem of greenhouse gas emissions holistically. There are other substances in our atmosphere that contribute to the phenomenon of climate change, including, but not exclusively, methane, nitrous oxide, hydrofluorocarbons, perfluorocarbons, and sulfur hexafluoride. Often, when looking at the problems of these gases' activity in the atmosphere, we assign a "carbon equivalence" value that reflects how active the gas is as an agent of climate change in the atmosphere compared with carbon dioxide. For example, we can say that methane is 21 times more active as an agent of climate change than carbon dioxide.

Carbon Trading

Carbon trading is seen as a market-based way to collectively reduce carbon emissions. In this scheme, a cap is chosen for annual carbon emissions, and then permits are issued to countries, industries, and individuals to allow them to release carbon, up to the permitted amount. There is significant criticism of trading systems as a method for achieving climate neutrality, as markets are open to manipulation and distortion. The alternative to this is a "gate cap" on carbon-emitting fuels, whereby extraction licenses would be issued for fossil fuels.

A carbon offset allows an organization or individual that is conducting a carbon-emitting activity to offset the effect of their carbon emissions by purchasing an "offset," or funding an activity that is deemed to remove an equivalent amount of carbon from the atmosphere. Carbon offsetting has come under criticism as a solution to climate neutrality, as it is said that it provides a "license" for poor environmental behavior. The efficacy of offsets has also come under criticism.

Direct/Indirect Emissions

One of the challenges in developing climate neutral solutions is that there is often a lot of "embodied carbon" in materials and goods, where the carbon emission has occurred at some other place from where the goods or service originated. To this end, carbon accounting needs to take into consideration the embodied carbon in materials, goods, and services so that the direct and indirect emissions resulting from providing a product or service can be taken into account.

See Also: Carbon Footprint and Neutrality; Carbon Trading and Offsetting; Emissions Trading; Greenhouse Gases.

Further Readings

Bookhart, Davis. "Strategies for Carbon Neutrality." *Sustainability: The Journal of Record*, 1/1 (February 2008).

Helweg-Larsen, Tim and Jamie Bull. *Zero Carbon Britain*. Powys: Centre for Alternative Technology, 2007.

Kirby, Alex and United Nations Environment Programme. *Kick the Habit: A UN Guide to Climate Neutrality*. Geneva: United Nations, 2008.

Worth, Dan. "Accelerating Towards Climate Neutrality With the U.S. Government Stuck in Neutral." *Sustainable Development*, (2005).

Gavin Harper
Cardiff University

COAL, CLEAN TECHNOLOGY

Coal is a combustible rock used as a fossil fuel. It is either black or brown. It is composed of carbon, hydrogen, and oxygen. The quantities of hydrogen and oxygen in coal determine which one of the four types it is: lignite, subbituminous, bituminous, or anthracite.

Most of the world's coal supplies were formed during the Carboniferous Period, which was about 300 million years ago. It was an age in which the Earth was warmer, with huge quantities of tropical plants growing in shallow swamps and seas. As the plants died, they settled to the bottom of the swamp or shallow sea, where the matter was compacted into peat bogs. Numerous tree and fern leaves—the remains of leaves or branches from the ancient swamps—have been found petrified in coal mines. Eventually, after a few hundred years, the peat layers were covered with sand and mud.

As the overburden increased, the compacting organic matter was pressed into a harder and dryer form known as lignite, which is the next stage in the formation of coal. Lignite is a sedimentary, dark brown type of coal. There are large formations of lignite in the United States, especially west of the Appalachian Mountains. It is usually strip mined and burned in electric power–generation steam plants. It is the youngest type of coal and is usually only a million years old. Studies have shown that it took seven feet of vegetable matter to make one foot of coal.

With increasing pressure from the overburden, lignite turned into subbituminous coal. Harder than lignite, subbituminous coal is sedimentary coal. As the overburden pressure increased over the long passage of centuries, some beds of subbituminous coal were turned into metamorphic bituminous coal. Most of the big deposits of subbituminous coal are in the Powder River region of northeastern Wyoming and southeastern Montana.

Cannel coal, a type of bituminous coal, is also called candle coal because it can be lighted like a candle; it is usually classified as a form of oil shale. It contains a fair amount of hydrogen and burns cleanly. It was used widely before the Industrial Revolution and was also used to make coal gas for burning but was replaced by other resources.

Earthquakes buried some coal deposits as they were tilted by the Earth's movements. Miners in shaft coal mines in places such as Pennsylvania have dug at a slant 1,000 or 2,000 feet deep into the mountains to dig out the coal that was pressed between layers of sandstone or other rock formations. Usually anthracite and bituminous deposits are the deepest deposits; however, in some places, earthquakes and erosion have moved these deposits to the surface.

The most expensive coal is anthracite, which is a metamorphic form of coal. It is also the oldest, having been buried the longest. This means that anthracite deposits have been subjected to the most pressure. Some anthracite deposits are over 400 million years old. Anthracite is almost pure carbon because most of the hydrogen and oxygen have been squeezed out of it.

Seams or beds of coal range in thickness from one inch to hundreds of feet. Because bituminous and anthracite coal are the most compact types of coal, these seams are usually less thick than the beds of lignite or subbituminous coal. Some coal deposits began as peat bogs that were covered with overburden. Later, a second or even a third swamp arose on top of the older one(s) to form a second or third coal deposit.

Coal reserves, that is, coal that can be economically mined, are estimated on a global basis to be about 660 billion tons. The United States in 2008 was estimated to have about 480 billion tons of coal reserves. Along with Canada, the usage is about a billion tons a year, mostly as fuel in steam-generating plants. Europe (including Russia), China, and South Asia have most of the remaining reserves. Coal fields are areas in which there are large and abundant deposits of coal available for mining. Bituminous coal forms the largest portion of the coal reserves and is the most widely used.

What is not known and is likely to be the subject of exploration and exploitation in the coming decades is how much coal is in the sea beds; some undersea mining has already taken place. It is also likely that India, China, and other countries will increase their use of coal as the cost of petroleum increases.

How Coal Is Mined

Historically, coal mining took place in shafts that used the room-and-pillar system. Eventually, the coal in the rooms cut into the coal bed is removed, allowing pillars to remain, supporting the roof of the mine. When the end of the coal seam is reached, the miners back out of the mine, removing the coal pillars and allowing the roof to collapse.

Surface mining of coal is almost always strip mining. This process strips away the overburden to expose the coal beds in areas that are relatively flat. In areas on hillsides or mountainsides, augers are used for digging out the coal. If the overburden is shallow, it is bulldozed; however, if it is much thicker, it is blasted into spoil that can be removed by giant earthmoving machines that may be as large as a 20-story building. Large coal-digging machines scoop up the coal and load it into large trucks. These trucks then usually haul the coal to railroad sidings, where the coal is dumped into coal cars that transport it to power plants.

The overburden or spoil is piled away from the cut that has exposed the coal seam. It is dumped into a spoil bank. In the past, the spoil was allowed to simply lie untreated where water could leach unwanted minerals into groundwater. The institution of environmental regulations since the adoption of the Surface Mining Control and Reclamation Act of 1977 (SMCRA) mandated that miners restore the spoil to the mine sites with plantings of vegetation. This restores the land to a natural state. To do this, the earthmoving machine digs a second cut and piles it into the cut from which the coal was dug. These operations create a series of parallel spoil banks that can be leveled when the mine is closed.

Strip mining methods vary according to the lay of the land. On level land, strip mining is area mining. In hilly land, it is contour mining. In contour mining, the stripping follows the contour of the mountain. A very controversial form of strip mining in mountainous areas is mountaintop mining. It involves removing a mountain's top, which is the overburden, and then removing the coal. This type of mining has been very controversial in areas of the Appalachian Mountains where it has been practiced. At times, as much as 1,000 feet (300 meters) of overburden are removed to get to the coal. The overburden is then dumped into valleys or "hollers" beside the mountain. When the mining is finished, the remains of the mountain are leveled to match the spoil in the adjacent valley. This method disturbs local watershed and leaves the mountain looking like a plateau.

Environmental concerns about strip mining coal arose from the great increase in strip mining beginning in the 1930s. The first coal was mined in Colonial America in 1740. Until the 1930s, coal was almost always shaft mined. The economics and politics of coal mining were shaped by the political culture of the 19th century. Among the prominent corporate cultural attitudes was attention to business interests, with little regard for miners' safety. The struggle of miners to form unions was at times met with strong-arm tactics and resulted in violence on a number of occasions.

Demand and Regulation

Demand for coal during World War II allowed environmental and miner concerns to be ignored despite numerous state laws. Variances in state laws allowed mining companies at times to avoid regulation. When strip mining reached 60 percent of total coal production in 1973, ignoring the environmental effect of strip mining was no longer prudent. Despite laws proposed during the energy crisis of the 1970s, legislation was proposed that would require environmental regulation enforcement despite the additional costs it would mean for mining companies, and ultimately for consumers.

SMCRA uses a cooperative federalism approach that engages both federal and state laws. The law sets standards of performance that miners must meet during mining operations and afterward when the mined land is reclaimed. To mine, a SMCRA permit is required. To obtain a permit, the environmental conditions must be described and a plan presented for land reclamation. To mine, companies must post a bond, which is a deposit or guarantee of money that will be sufficient to restore the land in case the company refuses to do so or is unable to do so because of a financial reversal or for other reasons. Inspections are carried out to see that standards are being met, and if not, "notices of violation" can be issued to require mine operators to fix the problem in a timely manner of face fines or an order to cease mining.

The regulations arising from SMCRA prohibit surface mining in wilderness areas or in national parks. However, national forest lands and lands under the control of the Bureau of Mines or other agencies can be opened for mining. Citizens can challenge proposed strip mining in areas where they believe the mining will cause too much environmental damage.

An Abandoned Mine Land fund was created under SMCRA to pay for the cost of cleaning up mines abandoned before the adoption of SMCRA. The fund gains its monies from a tax of 35 cents per ton on surface-mined coal and 15 cents per ton on shaft-mined coal. Half of the funds collected in each state can be awarded to the state government for its own reclamation programs. The other half is used to cover the cost of emergencies, landslides, coal mine fires, or urgent cleanup projects. Some of the money is used to pay homeowners for damage caused by shaft mine collapses beneath their homes. These mine shaft collapses are often in mines long abandoned that cause land subsidence that knocks houses off of their foundations or causes them to be out of level.

When coal is mined and sent straight to a customer, it is run-of-mine coal. However, the two largest users of coal—coke operations making iron and steel and most coal-fired steam-generating plants—have higher standards for the coal they burn. This means that mining companies have to clean their coal before shipping it. Cleaning coal is an operation performed in a preparation plant. If the mine is large enough, it may have its own preparation plant on the mine's site.

Processing

Preparation plants use a variety of machinery for removing impurities from the coal. Usually these are dirt and rock, but other impurities must also be removed. The processes wash the coal and separate it from the impurities. Sorting also takes place so that the coal is separated into different-size pieces. New clean coal technologies also include washing the coal with chemicals to remove minerals and other impurities

After washing the coal in preparation plants, it goes through a drying process and then is shipped, usually by rail in "unit trains." These are long coal trains that carry only coal from the mine to the power plant and then return for another load. If a unit train has 100 coal cars, then it is carrying approximately 10,000 tons of coal to a power plant or to some other user such as a coke works.

Slurry pipelines that transport coal crushed into a powder and then mixed with water have been built. These usually deliver the coal slurry to a power plant, but the water is removed before the coal is burned to send the water back to the mine for another load of coal. The Black Mesa pipeline in Nevada and Arizona is one such system.

Coal is a dirty fuel. In the 19th century, smoking chimneys were taken as a source of prosperity. However, when it burns, coal creates soot, which is mostly carbon and coal ash that often contains mineral compounds. These compounds are usually aluminum, iron, silicon, calcium, and other heavy metals. Burning coal also dirties the air with ash and with gases including sulfur dioxide and nitrogen oxides.

Sulfur in coal often comes in mineral forms. One common kind is iron pyrite, commonly known as "fool's gold" from its close appearance to gold nuggets. Other forms of sulfur are sulfur organic compounds in the coal. When coal is burned, sulfur is released, often forming sulfur dioxide that combines with water in the atmosphere to form sulfuric acid. A powerful acid, sulfur dioxide is the source of damage to trees because the sulfuric acid is absorbed into rain that falls as "acid raid." There are low-sulfur coal deposits, but these types of coal are more expensive and are used for special purposes.

The nitrogen-oxygen compounds formed when coal is burned include nitrous oxide, which is a greenhouse gas. Nitrous oxide is able to absorb much more radiant energy than carbon dioxide, so increases of nitrous oxide in the atmosphere from burning large quantities of dirty coal pose a climate change threat. In addition to these ecological dangers, the pollution caused by open burning of coal is a threat to human health. The pollutants can cause both short-term illnesses from respiratory irritations and long-term problems that can threaten or reduce life.

Coal fueled the Industrial Revolution and the great expansion of industrial societies during the 19th century. It was the fuel used in factories, by railroad engines, by steamships, and for heating homes and buildings. It is the source for making coke, which is used to make steel. Burning coke provides the heat necessary for melting iron and limestone. It removes the oxygen from the iron and provides a way to remove gases from the molten steel. Coke has become a substitute for coal in many countries where it is illegal to have a visible column of smoke rising from the chimney. A by-product of coke manufacture is coal gas, which was a widely used gas until being replaced by natural gas. However, the new clean coal technologies are working with new ways to gasify coal.

At the end of the 19th century, coal began to be supplanted as the fuel for transportation and for heating by the development of gasoline engines and other inventions that used petroleum and natural gas. Today, coal's energy competitors include energy from temperature inversion systems, wave power, tidal power, oil shale, tar sands, solar power, nuclear power, wind power, thermal power, hydroelectric power, and renewable biomass fuels, as well as traditional gas and oil resources.

In the latter half of the 20th century, the use of coal as a fuel for home heating declined. Other users also moved to other resources that did not seem to be as dirty and old-fashioned as coal. It seemed that coal was on the way out, but by the beginning of the new millennium, coal production had doubled over the 1970 level of consumption. Geologists are currently searching widely for new deposits of coal. Advances in seismic methods have enabled us to focus the search for minerals, including coal. Likely prospects are investigated using test drilling.

Coal is the source of numerous products that could allow it to be a substitute for petroleum. Hundreds of plastics are made from coal, as are roofing tar and many other products including cosmetics and artificial sweeteners.

New Clean Technologies

Instead of open burning of coal, new clean coal technologies are focused on much more efficient use or on reducing the environmental effect of coal production, on electrical generation, and on reducing smoke-stack emissions to zero.

Energy-efficiency technologies including the use of fluidized bed combustion boilers in coal-fired steam plants are achieving energy captures at a rate of 50 percent rather than the older one-third rate—most of the heat produced by coal has simply been lost as waste heat. A similar method is being used in some systems that power locomotive engines.

New clean coal technologies also are reducing smoke stack emissions, which reduces negative environmental effects such as acid rain. Technologies handle the flue gases by using steam, which removes the sulfur dioxide and also permits the capture of carbon dioxide, which can be stored underground or pumped into the ocean, where it is neutralized. Removing all carbon dioxide from the coal plant smoke stacks addresses the concern that many have about carbon dioxide as a source of climate change.

Whether the coal industry and its consumers can advance these new technologies at a rate that will be economical and still prevent the threat of governmental policies being instituted is a public issue. The cost of electricity is affected by cost increases resulting from factors affecting the coal industry. The imposing of laws that satisfy ideological visions of great improvements in the natural environment may be accompanied by serious reductions in living standards for many people.

Among the new clean coal technologies available are methods for dewatering lignite coal. This operation improves the burn and increases the calories of heat produced. The heat increase improves steam efficiency for generating electricity.

A similar technique removes minerals from coal before it is burned. When burned, the minerals are a pollutant; this is especially the case when sulfur is part of its composition. A clean technology would be to add to remove sulfur or other minerals from the flue gas with chemicals.

Burning coal with oxygen instead of air to reduces nitrogen dilution and enhances CO_2 capture. The last stage of clean coal technologies will be to capture and store CO_2 generated for sequestration or subsequent use in oil fields to gasify sluggish underground oil flows.

The world's first "clean coal" power plant was opened in September 2008 in Germany. The plant is owned by the German government, which has the will to bear the high cost of seeking to reduce coal-smoke emissions. Plans in the United States are for the development of near-zero atmospheric emissions coal plants.

The administration of President Barack Obama strongly supports facing issues involving global warming. Obama's record is one of support for coal and for new coal technologies. In 2005, he supported President George W. Bush's Energy Policy Act, which included $1.8 billion in tax credits for clean coal technology development. He helped to introduce the Coal-to-Liquids Fuel Promotion Act of 2007, which provided tax incentives for investing in new coal technologies that are more efficient and emit significantly lower amounts of carbon dioxide and other gases. He has also promoted government aid for the use of technologies that capture coal gas from syngas made from coal. However, he also strongly supports a system of carbon "cap-and-trade" credits, which is believed by some to spell the end of the coal industry. Obama has said that the cost of using a coal-fired plant with significant carbon dioxide emissions will bankrupt them. However, he is in favor of clean coal technologies, if they can be developed.

Coal-fired electricity plants are the biggest customers for coal. These plants are the second-largest source of carbon dioxide emissions in the United States, and the imposition of greenhouse gas emission taxes has the potential for raising significantly the price of electricity, whose cost will be borne by utility customers. Whether such taxes will increase the rate of discovery of clean coal technologies or whether it will be a boon to alternative energy sources is not yet known.

See Also: Alternative Energy; Climate Change; Fossil Fuels; Global Warming; Greenhouse Gases; Mountaintop Removal; Power and Power Plants; Sulfur Oxides (SO_x).

Further Readings

Ackerman, Bruce A. A. and William T. Hassler. *Clean Coal/Dirty Air: Or How the Clean Air Act Became a Multibillion-Dollar Bail-Out for High-Sulfur Coal Producers.* New Haven, CT: Yale University Press, 1981.

Eichengreen, Barry. *The European Economy Since 1945: Coordinated Capitalism and Beyond.* Princeton, NJ: Princeton University Press, 2006.

Flavin, Christopher and Nicholas Lessen. *Power Surge: Guide to the Coming Energy Revolution.* New York: W. W. Norton, 1994.

Freese, Barbara. *Coal: Human History.* New York: Perseus Distribution, 2002.

Goode, James B. *Cutting Edge: Mining in the 21st Century.* Ashland, KY: Jesse Stuart Foundation, 2003.

Goodell, Jeff. *Big Coal: The Dirty Secret Behind America's Energy Future.* New York: Houghton Mifflin Harcourt, 2007.

Hinde, John R. *When Coal Was King: Ladysmith and the Coal-Mining Industry on Vancouver Island.* Vancouver: University of British Columbia Press, 2004.

Lee, Sunggyu, et al., eds. *Handbook of Alternative Fuel Technologies.* Oxford: Taylor & Francis, 2007.

Lorde, Audre. *Coal.* New York: W. W. Norton, 1996.

Moroney, John R. *Power Struggle: World Energy in the Twenty-First Century.* Westport, CT: Greenwood, 2008.

Morris, Craig. *Energy Switch: Proven Solutions for a Renewable Future.* Minneapolis, MN: Consortium Book Sales & Distribution, 2006.

Probstein, Ronald F. R. and Edwin Hicks. *Synthetic Fuels.* Mineola, NY: Dover Publications, 2006.

Romey, I, G. Imarisio and P. F. M. Paul, eds. *Synthetic Fuels From Coal: Status of the Technology.* London: Graham & Trotman, 1988.

Shayerson, Michael. *Coal River.* New York: Farrar, Straus & Giroux, 2008.

Speight, James G. *Synthetic Fuels Handbook.* New York: McGraw-Hill, 2008.

Wiser, Wendell H. *Energy Resources.* New York: Springer-Verlag, 1999.

Woodruff, Everett B., et al. *Steam Plant Operation.* New York: McGraw-Hill Professional, 2004.

Andrew J. Waskey
Dalton State College

COLUMBIA RIVER

Eternalized through the songs of American folk singer Woody Guthrie, the Columbia River still looms large in the American imagination as a symbol of political might, engineering prowess, and more recently, ecological restoration.

In his compelling 1995 book *The Organic Machine*, historian Richard White navigates the complicated history of human settlement along the Columbia River. The Columbia is the largest river in the Pacific Northwest, with a drainage basin of 258,000 square miles. The river traverses 1,243 miles from British Columbia through the states of Washington and Oregon. Archeologists trace Native American settlement in this region back 11,000 years, although American and British explorers did not arrive in the region until the late 18th century. Lewis and Clark's famous expedition reached the Pacific Northwest and the Columbia basin in 1805. Earnest plans to develop this frontier did not begin until the arrival of the Northern Pacific Railroad in the late 1800s.

At nearly a mile in length, the Grand Coulee Dam on the Columbia River is the largest dam in the United States and can produce 6.5 million kilowatts of electricity.

Source: Wikipedia/Gregg M. Erickson

The Columbia River has been long considered the powerhouse of the Pacific Northwest because of its immense hydroelectric capacity. Today, there are over 400 small, medium, and large dams within the Columbia basin. The main stem of the Columbia River has a total of 14 dams (3 in Canada and 11 in the United States). The main-stem dams, and several on the neighboring Snake River, operate navigation locks to transport barge traffic from the ocean inland to Idaho.

The Grand Coulee Dam is the best-known landmark on the Columbia River. Built in 1942 and measuring one mile long and 330 feet tall, the Grand Coulee Dam is a marvel of early-20th-century engineering. In addition to producing up to 6.5 million kilowatts of power, the dam irrigates over half a million acres of farmland and provides abundant wildlife and recreation areas. Lake Roosevelt, the reservoir at the Grand Coulee Dam, stretches 151 miles to the Canadian border.

Western water historians have documented the immense political wrangling that occurred behind the scenes to get the Columbia dams authorized and built. In his 1986 epic book *Cadillac Desert*, Marc Reisner described these as "go-go" years, a period when America's technological supremacy made a project as immense and implausible as Grand Coulee achievable in just five years. This was largely because of the political will of the U.S. Congress, which voted 77 separate authorizations between the years 1928 and 1956 to support water development in the American West.

The Grand Coulee was originally built to irrigate the desert regions of the Pacific Northwest, particularly eastern Washington State. This goal changed after the onset of World War II. By the mid-1940s, the dam was producing abundant and surplus electricity. This cheap energy was appropriated to serve wartime industrial operations. Energy-intensive industries, like the Aluminum Company of America, became established in the Columbia basin to provide the raw materials necessary for manufacturing World War II planes and ships.

The Columbia also provided electricity and water for the Hanford site. First established as part of the Manhattan Project, the Hanford site was the first full-scale plutonium production plant in the world. In the Columbia River Risk Assessment Project, University of Washington scholars document that eight out of the nine reactors at Hanford were "single-pass reactors." This meant that water from the Columbia was pumped directly through the reactors and returned to the river after a short time in cooling ponds. This environmental contamination was kept secret from the public. Over Hanford's 40 years of operation, populations were unknowingly exposed to radiation when they ate local fish or recreated in the river. In February 1986, the U.S. Department of Energy was forced to release 19,000 pages of documents that revealed this radioactive contamination. In 1988, Hanford was designated as four separate Superfund sites. When complete in 2011, the cost of toxic cleanup at this site will exceed $12 billion.

The Columbia basin is home to many species of anadromous fish—those that ascend rivers to spawn. The Columbia is particularly known as a vital habitat for salmon species, including chinook, sockeye, steelhead, coho, and chum. The status of Pacific Northwest salmon has precipitously declined over the last hundred years as a result of habitat modification. The U.S. National Park Service has reported that of the 400 or so stocks that existed along the West Coast, 100 have already gone extinct, and another 200 are considered to be at risk of extinction.

Dam development has exacerbated the salmon problem by creating impenetrable barriers. Several ecological engineering solutions have been devised to help restore salmon. These include fish transport using barges and trucks, construction of fish ladders along dams, and artificial propagation. Salmon recovery efforts are incredibly complex because they involve several layers of local, state, and federal bureaucracies. Restoration efforts must also balance the need for electricity production with endangered species protection.

The Columbia River is currently the site of far-reaching experiments in river renewal through dam removal. The largest dam removal projects in the Columbia basin involve the decommissioning of the Elwha and Glines Canyon Dams as part of the Elwha Ecosystem Restoration Project. Removal of these dams will allow migratory salmon to spawn uninterrupted upstream. Many ecologists are studying the implications of salmon recovery for nutrient cycling on these rivers.

See Also: Hydroelectric Power; Microhydro Power; Nuclear Power.

Further Readings

Center for Columbia River History. http://www.ccrh.org/ (Accessed February 2009).
Columbia River Risk Assessment Project. http://www.tag.washington.edu/projects/hanford .html (Accessed February 2009).
Harden, B. *A River Lost: The Life and Death of the Columbia*. New York: W. W. Norton, 1996.
Lichatowich, J. *Salmon Without Rivers: A History of the Pacific Salmon Crisis*. Boston: Island, 1999.
Reisner, M. *Cadillac Desert: The American West and Its Disappearing Water*. New York: Viking, 1986.
Taylor, J. *Making Salmon: An Environmental History of the Northwest Fisheries Crisis*. Seattle: University of Washington Press, 1999.
U.S. Bureau of Reclamation. "Grand Coulee Dam." http://www.usbr.gov/pn/grandcoulee (Accessed February 1999).
White, R. *The Organic Machine: The Remaking of the Columbia River*. New York: Hill and Wang, 1995.

Roopali Phadke
Macalester College

COMBINED HEAT AND POWER

Combined heat and power (CHP) technologies refer to one type of distributed electricity generation that involves the strategic placement of power generation equipment in close proximity to the site of demand, so as to provide an on-site (or a relatively proximate) supply of energy resources. CHP is considered to be particularly advantageous because it increases the energy efficiency of electricity production in two ways. First, because electricity production is on-site, distributed generation reduces the large proportion of transmission line losses that are inherent to the traditional system of electricity generation. In addition, CHP systems also offer greater efficiency because they are able to simultaneously capture both the electrical and the thermal outputs associated with power generation.

Importantly, CHP systems (also known as cogeneration) generate electricity (and/or mechanical energy, sometimes referred to as shaft power) and thermal energy in a single, integrated system. The integrated system differs markedly from standard, centralized approaches, in which electricity is generated at a central power plant while on-site heating and cooling equipment are separately employed to meet heating and cooling demands. As opposed to the traditional approach, the CHP system creates efficiencies by recovering the thermal energy that is generated in the electricity production process. Because CHP captures the heat that would be rejected in the traditional approach, the total efficiency of these integrated systems can exceed 80 percent. (Conventional electricity generation converts only about a third of a fuel's potential energy into usable energy.)

CHP systems are not "one size fits all." Instead, they include diverse configurations of energy technologies that can vary significantly depending on the specific end-user needs for heating and/or cooling energy and mechanical and/or electrical power. In other words, CHP is not a specific technology but, rather, an application of technologies to meet multiple end-user needs. And the technologies involved have continued to evolve over time. Among the newer technologies involved in CHP systems are new generations of turbines, fuel cells, and reciprocating engines that have been developed through intensive, collaborative research, and technology demonstrations by government and industry. Advanced materials and computer-aided design techniques have also dramatically increased CHP system efficiency and reliability while also reducing the implementation and adoption costs associated with the conversion to a CHP system.

In terms of atmospheric emissions, the significant increases in efficiency associated with CHP have consistently resulted in both lower fuel consumption and reduced emissions. This preventive approach to emissions control differs dramatically from traditional pollution control mechanisms that achieve pollution control through flue gas treatments and that often further reduce efficiency and useful energy output. Comparatively, CHP systems are both more economically productive and more environmentally friendly.

Although CHP systems are gaining in popularity, they do not represent a new approach to meeting energy service demands. In fact, CHP technologies have a long history. Perhaps most notably, the first commercial power plant in the United States—the famous Edison Pearl Street station, built in 1882 in New York City—is often credited as the first CHP plant. However, later government regulations aimed at promoting rural electrification also promoted the construction of centralized plants managed by regional utilities and discouraged systems of decentralized power generation. It was not until 1978 that the Public Utilities Regulatory Policies Act was passed, encouraging utilities to pursue a more decentralized approach. Subsequently, cogeneration rose to approximately 8 percent.

At present, CHP electric generation provides approximately 56,000 megawatts of operation in the United States, up from less than 10,000 megawatts in 1980. The chemical, petroleum refining, and pulp and paper industries widely use CHP in their processes, and recently, the food, pharmaceutical, and light manufacturing industries have also begun to use CHP. Similarly, university campuses have adopted CHP systems for their ability to condition the indoor air of many of their building on campus and provide useful, low-cost electricity. Estimates from 2007 indicate that CHP:

- provided 329 billion kilowatt-hours of electricity—8 percent of all electricity produced in the United States;
- had an average system efficiency of 68 percent, with some new systems exceeding 90 percent; and
- emitted on average one-tenth of the nitrogen oxides (NO_x) per kilowatt-hour of average utility grid electricity.

In recent years, CHP has become an important element of the national energy debate, and the United States has taken some preliminary steps toward developing policies to promote CHP by establishing a national target. The next step will involve translating these efforts into concrete policies and programs at both the federal and state levels to overcome existing hurdles and barriers.

Although CHP is a promising and highly efficient use of limited fossil fuel resources, it is also important to recognize that major barriers do continue to hinder its widespread adoption. First, interconnection standards are often onerous and costly for CHP units, and utilities often require unnecessary and expensive equipment for the CHP unit to connect to the grid. Second, the rates that utilities are willing to pay for excess CHP generation are low, and existing utility customers who install a CHP unit may also be charged an exit fee designed to recoup any investments the utility may have put into serving that customer. Third, there is no current system that recognizes the high energy efficiency rates, and therefore lower emissions, that CHP units achieve. A set of policies, similar to emissions credits, could account for these emission benefits. In addition, CHP units, like all capital investments, have set depreciation schedules for tax purposes. At present, however, the set of depreciation schedules for CHP units can range anywhere from 5 to 39 years, and they often fail to adequately reflect the useful lives of these devices. Finally, many energy facilities managers are simply unaware of the benefits of CHP units.

CHP systems provide an important alternative for generating the energy services demanded in the residential and commercial sectors, and in the face of existing energy constraints and the challenges of addressing global climate change, investments in these technologies are likely to expand. However, to be successful, a variety of persistent barriers and constraints will need to be addressed.

See Also: Electricity; Power and Power Plants; Public Utilities.

Further Readings

American Council for an Energy-Efficient Economy. "Combined Heat and Power: The Efficient Path for New Power Generation." http://www.aceee.org/energy/chp.htm (Accessed January 2009).

Elliot, Neal and Mark Spurr. *Combined Heat and Power: Capturing Wasted Energy.* Washington, DC: American Council for an Energy-Efficient Economy, 1999.

Vanessa McKinney
Karen Ehrhardt-Martinez
American Council for an Energy-Efficient Economy

COMBUSTION ENGINE

There are two types of combustion engine—internal and external. The internal combustion engine generates energy from the reaction of a fuel with an oxidizer, which causes an explosion. The reaction products are gases, and the reaction itself generates intense heat that causes the gases to expand at an extraordinarily rapid rate. The expanding gases push against a piston, which is then forced to move, and the result is mechanical energy that has been harvested from the chemical energy within the fuel and the oxidizer. Typically, gasoline fuel reacts with oxygen-containing air in an internal combustion engine of an automotive vehicle. Fuels are commonly fossil fuels; examples include gasoline, diesel gas, natural gas, and petroleum. These engines can, however, be converted to run on "greener fuels" such as biodiesel, ethanol, hydrogen gas, or methanol. For a greater amount of power, the oxidizer can also be nitrous oxide.

The four-stroke reciprocating internal combustion engine is so called for using a four-stroke combustion cycle, which was first designed by a German engineer, Nikolaus Otto (1832–91), in the year 1867, and is sometimes called the Otto cycle. The four strokes are intake, compression, combustion, and exhaust. Everything occurs in the same cylinder, in the same space, but in chronological sequence. To power a vehicle as large as a car, engines need multiple cylinders. Typical cars have four, six, or eight cylinders. Arrangement of the cylinders gives the engine its name: inline, V, or flat. Four-cylinder engines are typically in the inline formation, whereas six-cylinders are arranged in a V shape (e.g., a V6 engine). American autos have historically been made with V6 engines, with the belief that the increased power was necessary, but Asian cars have been successful in the American market with more compact, less emissions-generating, and more fuel-efficient four-cylinder engines. This success has inspired American auto makers to consider the four-cylinder engine as well. In addition, the traditional V6 engine has been updated by engineers to rival four-cylinders in fuel efficiency.

A parallel evolution of car engines began in the late 19th century, when German-French engineering student Rudolf Diesel (1858–1913) studied combustion engines and how fuel-inefficient they were. He determined to create an engine that used more fuel for energy instead of wasting it as excess heat. He called his new engine the "combustion power engine" (today called a Diesel engine) and submitted his first patent in 1892. This new engine works quite similarly to the way a typical internal combustion engine works, with the same four strokes. The difference is that in a traditional internal combustion engine, fuel is mixed with air, then compressed and ignited. In a diesel engine, the air is compressed first, which heats it up, and then when the fuel is added, it ignites. Thus, a diesel engine does not have spark plugs, and the compressed air forces the oxygen particles closer together, ready to ignite the fuel more efficiently. Today, diesel fuel is thicker and oilier than gasoline. It is also easier to refine, and therefore it is generally cheaper to purchase. The diesel engine is louder than a gasoline engine and is thus not as attractive for personal vehicle use; because of its increased efficiency—and therefore power—it has been used primarily in larger commercial vehicles, including trucks. An environmental advantage to diesel engines is that most can run smoothly on biodiesel without any modifications.

Cars are typically run by four-stroke reciprocating internal combustion engines. Two-stroke engines are useful for lower-power devices such as lawnmowers and mopeds. Other types of internal combustion engine include the HEMI engine, first used in a car in a 1948

Jaguar. A rotary engine is not typically used in cars, although it has been seen in some models since the 1960s. At this time, Mazda is engineering rotary engines for its cars that are fuel-efficient and can run on alternate fuels such as hydrogen. Hydrogen as a fuel is clean because the exhaust is water vapor. It is a difficult fuel to burn comfortably in typical internal combustion engines, however, because of its extreme ignitability, often leading to uncontrolled—and thus inefficient—combustion, but the separated intake and combustion chambers in a rotary engine circumvent this problem.

Internal combustion engines can power machines of all sizes and are sometimes used in handheld tools instead of batteries. The most common internal combustion engine encountered in everyday equipment uses reactions in sequence; however, more powerful internal combustion engines are constantly reacting. The latter type of internal combustion engine is found in jet and rocket engines. Variations of the internal combustion engine that can provide still more power are superchargers and turbochargers.

External Combustion Engine

External combustion engines use combustion of a heat source externally to heat up an internal fluid energy source. This internal source can be gas (Stirling's engine) or steam (steam engine). Although larger and clunkier than internal combustion engines, these external combustion engines are more environmentally friendly and are currently being developed by engineers to help reduce the carbon footprint of the world's machinery.

See Also: Alternative Fuels; Automobiles; Biodiesel; Carbon Footprint and Neutrality; Electric Vehicle; Ethanol, Corn; Ethanol, Sugarcane; Flex Fuel Vehicles; Hydrogen; Oil.

Further Readings

Grosser, M. *Diesel: The Man & the Engine.* New York: Atheneum, 1978.
Hege, J. B. *Wankel Rotary Engine: A History.* Jefferson, NC: McFarland, 2006.
Lumley, J. L. *Engines: An Introduction.* Cambridge: Cambridge University Press, 1999.
Stone, R. *Introduction to Internal Combustion Engines.* Warrendale, PA: SAE International, 1999.
Pulkrabek, W. W. *Engineering Fundamentals of the Internal Combustion Engine,* 2nd ed. Upper Saddle River, NJ: Prentice Hall, 2003.

Claudia Winograd
University of Illinois at Urbana-Champaign

COMPACT FLUORESCENT BULB

The compact fluorescent bulb (CFB), also known as the compact fluorescent lamp, compact fluorescent tube, or compact fluorescent light, is a fluorescent lightbulb designed to replace traditional incandescent bulbs in use in everything from everyday lamps to commercial lighting and even public and city lights. These lights use much less energy per light produced than traditional lightbulbs, lowering usage costs as well as greenhouse gas

Compact fluorescent bulbs like this one, with its characteristic spiral design, need as little as 25 percent of the energy used by incandescent light bulbs.

Source: iStockphoto.com

production. CFBs were developed in response to a need for cheaper energy that was more environmentally friendly.

The 1973 oil crisis had widespread effects on Western energy technology. It inspired motor companies to revisit flex fuel technology and also gave impetus to companies such as General Electric to develop new lightbulbs that would use less power. An engineer with General Electric, Edward E. Hammer, developed the prototype for the modern CFB shortly thereafter, in 1976. Hammer's design was economically unfeasible to make at the time, so it was put on hold. Nearly 20 years later, other companies began to produce CFBs, and the first one was sold in 1995. Despite the fact that General Electric did not make Hammer's first CFB, his prototype is currently housed in the Smithsonian Institution in Washington, D.C. The CFB has a characteristic spiral tube design: It is a glass tube that contains mercury and gases, emitting light. It still takes more energy initially to make a CFB than an incandescent bulb, but the energy saved by the CFB compared with that used by an incandescent bulb is greater.

There are two principle types of CFBs: integrated and nonintegrated. Nonintegrated CFBs require special housing units and therefore are more expensive and require some work to install, but yield the best energy-saving results. Most people are more familiar with integrated CFBs. These bulbs have been manufactured to fit into the same electrical fixtures as traditional incandescent bulbs to maintain convenience for the consumer. Excitingly for energy-saving enthusiasts, CFBs can be used in solar-powered fixtures to provide clean energy streetlamps and other illumination devices.

The purchase price of CFBs is higher than that of traditional incandescent bulbs, but they last far longer and use much less energy over their lifetime than do incandescents. Because they use less power to output the same amount of light, CFBs generate significantly less greenhouse gases over their lifetime. Interestingly, the life span of a CFB can be decreased by quick usage times. The ideal operating time for a CFB to promote its long life is at least 15 minutes. In addition, although CFBs can be fitted into a light fixture with a dimming capability, these bulbs are not recommended for dimming purposes because this also reduces their life span. As well, CFBs take longer to reach full output than incandescent counterparts—from several seconds to three minutes. Although this delay is not strongly perceptible to the human eye, further development is required before CFBs are ideal for motion-sensitive lighting arenas. Because standard CFBs may work best upright—that is, with their fluorescent tube pointing upward—there are special models for use in recessed lighting fixtures.

According to Energy STAR, part of the U.S. Department of Energy, "[a]n ENERGY STAR qualified compact fluorescent light bulb . . . will save about $30 over its lifetime and pay for itself in about 6 months. It uses 75 percent less energy and lasts about 10 times longer than an incandescent bulb."

One minor drawback is the small amount of ultraviolet light emitted by the CFBs, which could irritate the skin of people with certain medical conditions. Most people, however, will not be harmed by a CFB. There may be a slight health risk involved with lengthy exposure at a distance closer than 20 centimeters (approximately eight inches), but even then, only certain CFBs with minimal coating on the fluorescent tube pose a risk.

A second minor drawback is that toward the end of a CFB's life span, it may begin to emit slightly less light. However, the decreased output of light is, in most cases, negligible to the human eye.

CFBs are fluorescent bulbs, however, and therefore contain mercury in their fluorescent tubes. This mercury content concerns environmentalists, as it can be released in landfills into the surrounding land, and hence into the water supply. Many state and local waste-dumping sites provide options for these mercury-containing CFBs, but many still do not. Convenient, proper disposal methods are still needed in many other countries, as well.

There is also some global health concern over factory workers making these bulbs. CFBs may be made in nations with lower standards for health and safety of factory workers; therefore, laborers may be exposed to unsafe levels of mercury during the production of these bulbs.

Although initially more expensive to purchase than a traditional incandescent bulb, over five years, the money saved in electricity bills by using these bulbs far outweighs the initial investment. Large buildings with vast square footage of ceiling lights can benefit significantly from installing CFBs in place of traditional lighting, and although initial models of CFBs gave off a cold-feeling, cool-colored light that was unpopular for bedrooms and other home uses, modern models are ranking equal to or higher than their incandescent counterparts.

See Also: Alternative Energy; Appliance Standards; Department of Energy, U.S.; Environmentally Preferable Purchasing; Flex Fuel Vehicles; Greenhouse Gases; Oil; Solar Energy.

Further Readings

Berardelli, P. "Lasers Make Lightbulbs Shine Brighter." *Science.* (June 2009).

Held, G. *Introduction to Light Emitting Diode Technology and Applications.* Boca Raton, FL: Auerbach, 2008.

Icon Group. *The 2009–2014 World Outlook for Incandescent Light Bulbs.* San Diego, CA: ICON Group International, Inc., 2008.

Icon Group International. *The 2009 Report on Compact Fluorescent Lightbulbs: World Market Segmentation by City.* San Diego, CA: ICON Group International, Inc., 2009.

Masamitsu, E. "The Best Compact Fluorescent Light Bulbs: PM Lab Test." *Popular Mechanics.* (May 2007).

Stemp-Morlock, G. "Mercury: Cleanup for Broken CFLs." *Environmental Health Perspectives* (September, 2008). http://www.energystar.gov (Accessed June 2009).

Claudia Winograd
University of Illinois at Urbana-Champaign

DAYLIGHTING

Daylighting is the practice of using natural daylight to illuminate the interiors of buildings, reducing the need for artificial light and, as a result, increasing the energy efficiency of those buildings. Though primarily thought of as a source of illumination, daylighting can also serve as an adjunct to heating and cooling systems, further reducing a building's artificial energy requirements. Often a necessary element of premodern building design, and an aesthetic element in the late 19th and 20th centuries, daylighting as implemented in contemporary architecture is used as an innovative solution to energy efficiency and sustainability problems.

The use of daylight as a primary or complementary means of illumination for the interior of buildings is, of course, not a modern phenomenon. Premodern architecture often used natural light for illumination. This usage was not always driven strictly by necessity, as might be thought, but also could be both functionally and aesthetically innovative. The Pantheon in Rome, dating from the 2nd century, is an innovative design, incorporating what might be the world's most famous skylight. It has a simple, yet elegant design—the only source of interior illumination is the daylight that shines in through the oculus, essentially a simple hole in Pantheon's dome. Paris' Sainte Chapelle, built around 1240, makes extensive use of daylight as an aesthetic element of its Gothic design. Natural light is used throughout the chapel—most strikingly in the upper chapel, where stained glass windows surrounding the entire upper chapel diffuse and color sunlight passing through to the interior.

The use of daylight as an architectural design element became more prominent beginning in the 19th century, with the widespread availability of steel and glass. Building designs, once more restricted because of the structural limitations of traditional building materials and cost factors, started making greater use of daylight as the use of glass became more commonplace. In 1851, Sir Joseph Paxton's Crystal Palace ushered in this new era of architectural design. The Crystal Palace, essentially an enormous greenhouse, was originally designed to house Great Britain's Great Exhibition. According to Hermione Hobhouse, in her book on the Crystal Palace and the Great Exhibition, it was constructed from iron, wood, and 896,000 square feet of glass.

Since Paxton's Crystal Palace, the modern era of architectural design has been replete with examples of structures incorporating natural illumination in their design. Some notable examples include the following:

- Bibliotheque Nationale de France, Paris, France, Henri Labrouste, 1868
- Pennsylvania Station, New York City, McKim, Meade, and White, 1910
- Bauhaus, Dessau, Germany, Walter Gropius, 1925
- The Glass House, New Caanan, Connecticut, Philip Johnson, 1949
- Crown Hall, Chicago, Illinois, Ludwig Mies van der Rohe, 1956
- The East Building, National Gallery of Art, Washington, D.C., I. M. Pei, 1978

Although the use of daylight as an illumination source was a design element in all of these buildings, it is likely the architects of these buildings incorporated the use of daylight principally as an aesthetic element. Energy efficiency and environmental sustainability were perhaps secondary benefits but were probably not principal concerns for the architects, as at the time these buildings were built, these issues were not principal concerns for society as a whole.

Today, concerns about global warming, pollution, and environmental sustainability have fostered "greener" building designs, which strive to be more environmentally friendly. The American Institute of Architects now promotes architectural designs that are energy efficient, environmentally sustainable, and as carbon neutral as possible. As society's environmental awareness has increased, the design and construction of greener buildings has also increased. Contemporary architectural design now frequently incorporates design elements focusing on greater energy efficiency and environmental sustainability, and the use of daylight is more than an aesthetic element.

Daylighting is one of the principal tools used in green building design. The use of daylighting in a building's design has many benefits.

- Daylighting increases energy efficiency by reducing need for artificial light sources.
- Daylighting can also reduce the need for artificial heat, though this might be offset by the need for greater use of artificial cooling in warmer months.
- Because daylight is available during common working hours, the period of highest electricity demand, its use reduces strain on the electricity grid. Reduced demand during peak hours means a reduced need for building more power plants—principal sources of greenhouse gases and other pollutants.
- Daylighting reduces the life cycle cost of a building by reducing the usage of lights, fixtures, and other related items, thus extending the usable life of those products and reducing the maintenance and replacement costs.
- Daylighting reduces overall building operating costs through reduced use of artificial energy consumption.
- Daylight helps improve working and learning conditions. Studies have indicated that greater use of daylighting techniques in workplaces has improved worker productivity and satisfaction, and its use in schools has led to improvement in student scores.

Daylighting solutions can be as simple as the installation of skylights or windows, allowing more light into a building, but contemporary green building design often employs a wider array of daylighting strategies. These include:

- Site orientation: A building must be properly oriented to efficiently employ daylight.
- Climate studies: It is important to understand the availability and direction of daylight for the region where a building is constructed.

- Toplighting: Efficient use of toplighting allows daylight into interior spaces and provides indirect natural illumination throughout the building.
- Exterior and interior shading devices: These help reduce heat gain as well as increase diffusion of light.
- Glazing: Window-glazing enhancements, such as films and coatings, optimize the light allowed into a space, absorbing and reflecting excess light and harmful radiation.
- Interior reflectance: The higher the reflective quality of interior surfaces, the greater the penetration of daylight into the building; however, good daylighting techniques employ indirect light, so interior reflectance should be employed strategically.
- Lighting integration: Lighting can be integrated with daylighting to raise and lower artificial light as necessary without human intervention, preserving function illumination levels.
- Heating, ventilation, and cooling (HVAC) integration: Daylighting can be used to offset heating and cooling costs. Smart heating, ventilation, and cooling systems integrated into construction complement the use of daylighting in building design.

Daylighting by definition is as simple as employing windows and skylights in a building, but daylighting used efficiently in contemporary green building design requires innovative construction techniques complemented with integrated artificial lighting and heating, ventilation, and cooling systems to realize its greatest energy efficiency gains.

See Also: Alternative Energy; Green Energy Certification Schemes; LEED Standards; Sustainability.

Further Readings

American Institute of Architects. "50to50: Sustainability 2030." http://www.aia.org/SiteObjects/files/50to50_20071212.pdf (Accessed January, 2009).

Ander, Gregg D. "Daylighting." *The Whole Building Design Guide, 2008* http://www.wbdg.org/resources/daylighting.php (Accessed January, 2009).

ArchitectureWeek. "Great Buildings Collection." http://www.greatbuildings.com (Accessed January, 2009).

Heschong, Lisa. "Daylighting in Schools: Reanalysis Report." California Energy Commission, October 2003. http://www.newbuildings.org/downloads/FinalAttachments/A-3_Dayltg_Schools_2.2.5.pdf (Accessed January, 2009).

Hobhouse, Hermione. *The Crystal Palace and the Great Exhibition of 1851: Art, Science and Productive History. A History of the Royal Commission for the Exhibition of 1851.* New York: Continuum Press, 2004.

New Buildings Institute. "Lighting." http://www.newbuildings.org/lighting.htm (Accessed January, 2009).

Romm, Joseph J. and William D. Browning. "Greening the Building and the Bottom Line." The Rocky Mountain Institute, 1998. http://www.rmi.org/images/other/GDS/D94-27_GBBL.pdf (Accessed January, 2009).

Ron Keith
George Mason University

Michael Pellegrino
Architect, AICP, NCARB

DEPARTMENT OF ENERGY, U.S.

The U.S. Department of Energy (DOE) is a cabinet-level department in the executive branch of the U.S. government. Its mission is to advance the national, economic, and energy security of the United States; to promote scientific and technological innovation in support of that mission; and to ensure the environmental cleanup of the national nuclear weapons programs.

The department's strategic goals to achieve its mission are designed to deliver results along five strategic themes: energy security, nuclear security, scientific discovery and innovation, environmental responsibility, and management excellence. Within these themes are 16 strategic goals that the DOE uses to achieve its mission and its vision. Its vision is to achieve near-term results that promote its strategic goals.

The DOE was created in 1977 by the Department of Energy Organization Act. Before its creation, energy regulation was handled by many different agencies that were spread across the federal bureaucracy. Departments at that time with energy responsibilities included the departments of Agriculture and the Interior. Agencies included the Federal Power Commission, the Atomic Energy Commission, the Federal Energy Regulatory Commission, and the Nuclear Regulatory Commission. In addition, energy policy historically had assigned control of most natural resources used to produce energy to the states that regulated energy production under their specific laws, regulations, and judicial decisions.

The energy crisis of the mid-1970s necessitated the creation of a more unified approach to energy policy. The energy crisis was caused by the oil embargo in October 1973 put in place by the Organization of Petroleum Exporting Countries (OPEC). Most of the members of OPEC are Arab countries. They used oil as a political weapon to punish the United States and Western European countries for supporting Israel in the 1973 Arab–Israeli War. The embargo created high energy costs, gasoline shortages, long lines at gasoline stations, and other problems including inflation. To meet the challenge, Congress began to focus on developing an energy policy that would create energy reliability, economic stability, national security, and a cleaner environment.

The first responses to the crisis by presidents Richard Nixon and Gerald Ford were aimed at price and supply control. The responses were limited because it was not clear whether the embargo was a permanent threat to vital national interests or just a passing event. By the time of President Jimmy Carter's administration, political opinion had reached the point at which many agreed with President Carter that responding to the energy crisis was the "moral equivalent of war." OPEC's control of oil supplies was creating enormous problems with inflation resulting from the high cost of oil. OPEC's price for oil prompted a massive transfer of wealth to the Middle East, with little the West could do to stop it.

President Carter requested that Congress adopt the National Energy Act. It comprised five major pieces of legislation. Altogether, the act created a system of regulation of traditional energy fuels focused on conservation. In 1977, coordination of these regulations was assigned to the DOE by the Department of Energy Organization Act (P.L. 95-91, 91 Stat. 565). The act received strong approval from the members of Congress because its legislative investigation had found that increasing shortages of nonrenewable energy resources were likely to occur in the near future. This was at a time when the United States and the West were engaged in a Cold War against communism, led by the Soviet Union, making the United States vulnerable to foreign interests or opponents because of its dependence on foreign oil and other energy supplies. The vulnerability presented such a significant threat to national security that only a strong national energy program could meet the challenge. To meet national security energy needs, the fragmented federal energy policy that had historically evolved needed to be replaced with a proactive national energy program that was integrated and coordinated.

The creation of the DOE as a cabinet-level department was the result of the need for an industrial society like the United States to deal systematically with its energy needs. As a consequence, the DOE was instituted as the agency with management responsibility for information collection, policy planning, coordination, and program administration. To streamline the administrative organization of the DOE, the Economic Research and Development Administration and the Federal Energy Administration were abolished, and their responsibilities and powers were transferred to the DOE.

The DOE organization act renamed the Federal Power Commission the Federal Energy Regulatory Commission; it was placed in the DOE organization chart for purposes of information but remained an independent regulatory agency. The Department of the Interior, the Department of Housing and Urban Development, the Department of the Navy, Department of Commerce, the Interstate Commerce Commission, and other agencies all transferred their legal responsibilities for various energy regulations to the DOE. The Energy Information Administration, the Economic Regulatory Administration, and the Office of Energy Research were established as DOE agencies.

DOE Responsibilities

To fulfill its mission of advancing a coordinated national energy policy, the DOE was given responsibility for biannually reporting a National Policy Plan to Congress. The plan covers major energy information on production, consumption, and conservation. It also presents strategies for energy security and makes recommendations for action.

Since October 1977, when the DOE first began to operate, it has sought to fulfill its mandate by putting together numerous energy, scientific, technology, and defense programs in a coordinated manner. These programs have also included nuclear weapons testing. In the years since it began, the DOE has had to shift its focus from time to time as the energy needs of the country have changed. Among these changes is a turn to a growing investment in scientific and technological research and development to reduce the nation's growing dependence on foreign energy resources.

Under the leadership of President George W. Bush two new laws were enacted: the American Competitiveness Initiative (ACI) and the Advanced Energy Initiative (AEI). These two programs were begun in the hope that through science, technology, and engineering, answers can be found to the energy needs of the country and to the energy needs faced by the world. The initiatives seek to promote scientific innovation and technology development. They also give the DOE authority to encourage competitive energy markets, whether domestic or international, in the hope that market forces will open the way for innovations that will answer energy needs through private energy investments.

The AEI and ACI include government support for the development of energy sources—such as wave energy technology—that are today experimental. The research grants for experimental technologies to companies seeking to exploit these technologies are in the American tradition of using government to support the general welfare with public investment that can then be exploited by private investors. Research in agriculture at state and private colleges has yielded advances in science and technology that have led to the creation of many new companies and the improvement of the lives of individuals and families, as well as, through their labors, increased tax revenues for both federal and state governments.

The ACI program increased government spending for the scientific and technological education of students who are the thinkers, inventors, and business entrepreneurs of the future. Investing in the human capital of the United States by increasing educational opportunities for American youth was expected to equip them with the knowledge and skills needed to compete effectively in the global economy that emerged at the beginning of the 21st century.

The AEI initiative seeks to increase U.S. energy security. This involves a two-pronged approach: increasing domestic energy resources and restraining demand to reduce dependency on imported oil and natural gas. Restraining demand means increasing energy efficiency in the transportation and electric grids. Both goals of increasing energy efficiencies in the transportation and electric grid and increasing energy resources are dependent on research and development. Because the consumption of energy affects the environment, research and development are also concerned with finding and building new systems of clean technologies.

The development of alternative fuels and sources was the subject of the Energy Security Act adopted by Congress in 1980. It comprised six major acts: the U.S. Synthetic Fuels Corporation Act, Biomass Energy and Alcohol Fuels Act, Renewable Energy Resources Act, Solar Energy and Energy Conservation Act, Solar Energy and Energy Conservation Bank Act, Geothermal Energy Act, and Ocean Thermal Energy Conversion Act. This battery of laws engaged the nation in seeking energy resources from solar, geothermal, oil shale, tar sands, and other sources.

Developing Renewable Energy Technologies

Today the AEI is spending monies on developing renewable energy technologies such as biomass, wind, and solar energy. Another area of investigation is hydrogen research and development. Hydrogen has the potential of being used as a fuel in automobiles. It may also be the fuel for the electrical grid of the future if the technological difficulties in capturing energy from nuclear fusion can be overcome. AEI also instituted the Global Nuclear Energy Partnership to increase the use of nuclear power as a clean, safe, and affordable energy resource.

The United States has a large portion of the world's coal reserves. Burning coal dirties the air with ash and with gases, including sulfur dioxide and nitrogen oxides. Sulfur dioxide combines with water in the atmosphere to form sulfuric acid, which falls as "acid raid." Low-sulfur coal is more expensive and is usually used for special metallurgical purposes.

Burning coal at electric steam–generating plants supplies the United States with a large share of its electricity. In addition to smokestack emission of soot, ash, carbon dioxide, and sulfur compounds, nitrogen-oxygen compounds are also emitted. Nitrous oxide is a greenhouse gas that absorbs a much larger quantity of radiant energy than carbon dioxide. Increases in nitrous oxide in the atmosphere from burning large quantities of dirty coal pose a climate change threat.

Burning the large resources of coal in the United States also has environmental consequences that arise from mining. However, it is in the interests of energy security that the abundant coal reserves be tapped. The AEI under President Bush sought to accelerate the development of clean coal technology. This type of technology seeks to eliminate virtually all of the emissions that are created at coal-fired steam plants. The AEI has been able to build a near-zero atmospheric emissions coal plant.

The DOE has the ultimate responsibility for nuclear resources for both civilian and military use by the United States. As steward of the U.S. nuclear weapons stockpile, it oversees the cleanup of the environmental consequences of the development and testing of nuclear weapons. Its responsibility for nuclear safety includes not only the nation's weapons program but also other government-sponsored nuclear energy research. Research in experimental or pilot reactors has been conducted at many sites, including campuses of many of the nation's universities. After decades of experimentation, these facilities age and are decommissioned. However, dangerous radioactive materials cannot simply be bulldozed—they must be removed and placed into long-term storage.

The DOE develops rules for nuclear safety to protect the health and safety of workers and the general public. It promulgates rules and enforces them as well. Its safety concerns also include maintaining U.S. nuclear stockpiles as a nuclear deterrent to rogue states, protecting them from capture by terrorists but also seeking international agreements and actions that will prevent nuclear proliferation.

The administration of President Barack Obama, supported by Democratic Party majority control of Congress, has promoted billions of dollars in spending for energy conservation and efficiency. On March 26, 2009, DOE Secretary Steven Chu announced plans for spending $3.2 billion on energy efficiency and conservation projects in cities, counties, states, territories, and Native American tribes across the United States. The money is to be issued in the form of block grants that give the receiving government more latitude in spending than do project or formula grants.

The Energy Information Administration is the agency in charge of supplying information about energy and is a part of the federal government's ongoing quest to have the facts needed for energy sufficiency and security. It maintains up-to-date information on energy for every state. It also maintains information on the pipeline infrastructure for both gas and petroleum products. The information on pipeline flows was useful in the aftermath of Hurricane Katrina in 2005. It was able to show that price gouging did not occur and to have the facts in hand to keep the public informed against wild rumors.

A wide variety of scientific groups, environmental interest groups, and engineering organizations such as the American Society of Heating, Refrigerating and Air-Conditioning Engineers and the High-Performance Commercial Green Building Partnership, as well as others, are actively seeking to inform the DOE of matters that will aid in the development of a green society. Some of these groups' proposals are aimed at requiring future buildings to be much greener than they have been in the past. This means that instead of building as cheaply as possible and installing inefficient energy systems for lighting, heating, and cooling, as well as regulating equipment, future buildings will be constructed with energy-efficient systems. The initial costs will be higher, but the cost of operating office buildings, skyscrapers, warehouses, dormitories, public buildings, homes, or other facilities will be much cheaper because less energy will be needed.

The DOE is promoting zero-net energy commercial buildings. These buildings will use efficient technologies and be capable of generating on-site energy that would replenish energy taken from the electricity grid. The goal for these buildings is 2025.

See Also: Alternative Energy; Coal, Clean Technology; Energy Policy; Environmental Stewardship; Federal Energy Regulatory Commission; Greenhouse Gases; Nuclear Proliferation.

Further Readings

Byrnes, Mark E., et al. *Nuclear, Chemical, and Biological Terrorism: Emergency Response and Public Protection*. New York: Taylor & Francis, 2003.

Margulies, Phillip. *Department of Energy*. New York: Rosen Publishing Group, 2006.

National Renewable Energy Laboratory, U.S. Department of Energy, et al. *A Manual for the Economic Evaluation of Energy Efficiency and Renewable Energy Technologies International Law & Taxation*. Washington, D.C.: Government Printing Office, 2005.

National Research Council, National Research Council Staff, Board of Radioactive Waste Management Staff. *Long-Term Institutional Management of U.S. Department of Energy Legacy Waste Sites*. Washington, D.C.: National Academies, 2000.

Tuggle, Catherine and Gary Weir. *The Department of Energy*. New York: Facts on File, 1989.

U.S. Department of Energy. *Dictionary of Environment, Health and Safety*. Boca Raton, FL: CRC Press, 1992.

U.S. Department of Energy. *2009 Energy Data Book Series: Biomass Energy Data Book: Ethanol, Biodiesel, Biorefineries, Feedstocks, Oilseeds, Mill Wastes, Pellet Fuels, Tertiary Biomass Feedstocks, Urban Residues*. Washington, D.C.: Government Printing Office, 2009.

U.S. Department of Energy Handbook. Washington, D.C.: International Business Publications, 2006.

Westerfield, Jude S., et al. *Tribal Cultural Resource Management: The Full Circle to Stewardship*. Lanham, MD: AltaMira, 2002.

Williams, Walter Lee. *Determining Our Environments: The Role of Department of Energy Citizen Advisory Boards*. Westport, CT: Greenwood, 2002.

Andrew J. Waskey
Dalton State College

DIOXIN

Dioxin refers to a large group of halogenated organic compounds that are polychlorinated dibenzodioxins. They are commonly called "dioxins" because they all contain the p-dioxin skeleton consisting of two benzene rings joined by two oxygen bridges. A variable number of chlorine atoms are attached to the benzene ring at any of eight possible sites on the benzene rings. The locations of the chlorine attachments impact the toxicity of the dioxin. The most infamous of the dioxins is 2,3,7,8 tetrachlorodibenzo-p-dioxin (TCDD), which was a contaminant in Agent Orange, which was used as an herbicide in the Vietnam War.

Dioxins are produced in a variety of chemical reactions. In the 1980s, nearly 80 percent of dioxins entering the atmosphere were the result of burning organic materials in the presence of chlorine. Because chlorine is present in many settings, combustion of various kinds produced a small amount of dioxins that, in the aggregate, created significant pollution and environmental burden. Coal-fired utilities and municipal waste incinerators were two large sources of environmental dioxins. Diesel truck exhaust, metal smelting, application of sewerage sludge, burning of trash, and burning of treated wood were other significant sources. To some extent, new emission requirements implemented in the United States over the past two decades have significantly reduced, but not eliminated, dioxin emissions. At this time, in the United States, only 3 percent of all dioxin emissions come from these traditional sources. However, burning of residential trash, especially if not equipped with emission control methods, continues to be a large source of dioxin production in the United States. In other countries, burning of a variety of organic materials for heating, cooking, or power continues to be a source of dioxins. Other sources of dioxins include paper mills that use chlorine bleaching in producing paper, production of polyvinyl chloride plastics, and manufacture of pesticides. Cigarette smoke produces a small amount of dioxins.

Although great strides have been made in reducing the production of dioxins, there still remains a great deal of dioxins produced decades ago because dioxins decompose very slowly in the environment. Dioxins are also very widespread because when released into the air, as in burning, they can travel long distances. In water, they settle into sediments, where they can be ingested by fish or other aquatic organisms and travel long distances both geographically and up the food chain.

Dioxins are absorbed through dietary intake of fat and are stored in fatty tissues. Ninety-five percent of the human intake of dioxins occurs through dietary intake of animal fats. Only a small amount comes from breathing air containing dioxins or from skin contact. One way to reduce intake of dioxins is to reduce intake of animal fat.

Dioxins are not readily metabolized or excreted in the human body. Half-life in humans ranges from eight years to over a hundred years, depending on the specific dioxin. Dioxins cross the placenta, and thus pose risk to developing fetuses. Dioxins are excreted in human milk, and thus also can be toxic to developing infants.

Not all dioxins have the same toxicity. The 2,3,7,8 Tetrachlorodibenzo-p-dioxin (or TCDD) form is the most toxic. Toxicity of a given dioxin is measured in terms of its relationship to this form. The toxicity is measured in Toxic Equivalent by relating to the 2,3,7,8 TCDD form and summing the toxicities. Thus, a dioxin with half the toxicity of 2,3,7,8 TCDD would add a half Toxic Equivalent to the overall toxicity of the mixture.

There are several fairly well-agreed-upon toxic effects of dioxins. One is chloracne, which is a particularly severe skin disease with acne-like lesions. Chloracne occurs with high levels of dioxin exposure. A particularly notable case was the poisoning of a Ukrainian politician, Victor Yushchenko, in 2004. Chloracne can also be seen in workers in some chemical plants in which dioxin levels are high. Other well-agreed-upon health effects of high levels of dioxin exposure include other skin rashes, skin discoloration, excessive body hair, and possibly mild liver damage.

Cancer is a much-debated toxicity of dioxins. There are several studies that do suggest that workers exposed to high levels of dioxins at their workplace have a higher incidence of cancer. Animal studies have shown a cancer risk. The International Agency for Research on Cancer considers 2,3,7,8 TCDD to be a known carcinogen. However, despite no clear consensus on the cancer risk of most other dioxins, all dioxins have been considered to be cancer-causing by many scientists and advocates. These advocates are often very vocal in their beliefs.

A major concern regarding dioxins is their ability to act as "endocrine disruptors." This has been the concern of a number of jurisdictions including the European Union. Concern that dioxins can cause endocrine problems such as thyroid disease, endometriosis, and immune compromise has been the source of much discussion and strong feelings among certain advocate groups. Studies of Vietnam-era military personnel involved with handling Agent Orange have shown a possible increase in the risk of diabetes. Further, these endocrine-disruptor effects are felt to affect reproductive capability and the development of infants. California lists dioxins in its Proposition 65, which is a list of such agents. However, the science is far from settled on these issues, with emotions running high on both sides.

Several well-known dioxin exposure cases have received much publicity. Times Beach, Missouri, was the site of roads coated with dioxin (and other chemicals)-containing materials causing dioxin contamination of the surrounding soil. Eventually, the U.S. Environmental Protection Agency called for the evacuation of the area. Love Canal in Niagara Falls, New York, was another site of dioxin (and other chemical) contamination resulting in evacuation of the residents of the area.

The importance of dioxin in the framework of "green energy" is twofold. First, as traditional methods of energy production look at more efficient ways of producing energy, care must be taken to continue to control the emissions of dioxins in the combustion process. Second, alternative sources of energy generation that do not involve combustion (e.g., solar, wind, and tidal) should help to further reduce emissions. However, in considering such alternatives, it is important to be aware of the potential generation of other pollutants or by-products of the materials needed to produce such forms of energy. Even if all industrial sources of dioxins can be eliminated (which is not a realistic expectation), there will remain the production of dioxins through natural processes such as forest fires. The potential for human exposure also will still remain from the "reservoir" sources in the environment.

Unfortunately, the environmental burden of dioxins will be with us for generations to come because of their slow degradation in the environment. A concern that needs careful attention is remediation. Because of the high emotions and concerns regarding the risk of dioxins, there is a tendency to call for removal of the dioxin-containing material from particular areas. However, this may actually release dioxins that are trapped in sediment back into the food chain.

See Also: Alternative Energy; Coal, Clean Technology; Power and Power Plants; Waste Incineration.

Further Readings

American Chemistry Council. "Dioxin in Depth." www.dioxinfacts.org (Accessed January 2009).

Committee on the Implications of Dioxin in the Food Supply, National Research Council. *Dioxins and Dioxin-like Compounds in the Food Supply: Strategies to Decrease Exposure.* Washington, DC: National Academies Press, 2003.

National Center for Environmental Assessment, U.S. Environmental Protection Agency (EPA). "Dioxin and Related Compounds." http://cfpub.epa.gov/ncea/cfm/recordisplay .cfm?deid=55264 (Accessed January 2009).

Pesticide Action Network (PAN). "Dioxin (2,3,7,8-TCDD)." www.pesticideinfo.org/Detail_ Chemical.jsp?Rec_Id=PC35857 (Accessed January 2009).

U.S. Food and Drug Administration. "Questions and Answers about Dioxins." www.fda.gov/ Food/FoodSafety/FoodContaminantsAdulteration/ChemicalContaminants/DioxinsPCBs/ ucm077524.htm (Accessed January 2009).

Karl Auerbach
University of Rochester

ELECTRICITY

Electrical energy is so fundamental to daily living in the postindustrial world that it is impossible to imagine having a similar quality of life without it. Lighting that allows us to see through darkness, refrigeration that preserves the food we eat, heating and air-conditioning that increase our comfort level indoors; today these processes are made exceptionally convenient through the consumer availability of electrical energy. Consider further the broad applications in digital media: televisions, computers, cellular phones, MP3 players, kitchen appliances, the digital components in automobiles, and even the vehicles themselves. According to the Energy Information Administration, electrical energy is projected to remain the fastest-growing end-form of energy used worldwide through 2030. With society in the present and foreseeable future relying so heavily on electrical energy, the methods and resources chosen to generate it have vast consequences for environmental preservation.

The electrical nature of matter is well understood. This knowledge enables diverse applications of electrical energy. Integration of green electrical generation into the existing infrastructure is happening slowly, but governments must offer financial incentives to make renewable resources economically

Coal-fired plants like this one in the Midwest generated as much as 49 percent of the total electricity supply in the United States in 2006.

Source: iStockphoto.com

105

competitive with fossil fuels. New technologies for electrical generation aim to reduce greenhouse gas emissions and curb reliance on fossil fuels. Conserving electrical energy is a practical and simple method for consumers to contribute to these goals at home.

Major Concepts

The industry uses "electricity" interchangeably to refer to two very different concepts: electrical energy and electrical power. To prevent confusion, the terminology will be defined here from a scientific standpoint, and only these unequivocal definitions will be applied. Electricity is the field of knowledge that encompasses all phenomena arising from the presence of electric charge. Energy, measured in the SI unit joule (J), is the capacity for a system to do work to change the physical state of itself or of another system. Energy can be neither created nor destroyed but can be transferred from system to system or converted between different forms within a system when work is performed. Power is the rate at which work is done or energy is transmitted, measured in J per second, called watts (W).

There are several different measures of energy used by the industry and its analysts. Electric utilities often bill customers in kilowatt hours (kWh), which is the amount of power used for a given time period. As one can see from the definition of power, this is equivalent to the electrical energy consumed. North American analysts measure the energy content of resources used for electrical, heating, and transportation purposes in quadrillions of the nonmetric British thermal unit (Btu). Public utilities sell natural gas to consumers by volume in therms or ccfs, which are 100 cubic feet—equivalent to 100,000 Btu. The choice of units used depends on the resources being compared, regional norms (metric versus nonmetric), and the total amount in question; whichever unit gives the amounts from 1 to 100 is generally preferred. For instance, 1 J and 1 Btu are very small amounts in terms of average household usage, which is why consumers are billed in kWh and therms (1 kWh = 3,600,000 J and 1 Btu = 1,055 J). In the winter, the average household may consume 1,000 kWh of electrical energy per month and 50 therms of natural gas toward thermal energy for gas-powered heating, cooking, and hot-water systems.

All forms of energy can be classified in one of two ways: kinetic energy resulting from rotational, vibrational, or translational motion (i.e., thermal, acoustic) or potential energy stored within a system by virtue of its position as a whole or the relative positions of its particle constituents (i.e., gravitational, elastic, electrical, chemical). The total mechanical energy of a system is the sum of its kinetic and potential energies at the macroscopic level, considering the system as a whole. At the microscopic level, the total internal energy of a system is the sum of its thermal energy resulting from the motions of its atom or molecule constituents, its chemical energy resulting from the elasticity of its molecular bonds, and its electrical energy resulting from the distribution of its charge.

Electrical energy is the capacity of a system to move charged particles. Just like mass, charge is a fundamental property of matter because it is an inherent characteristic of protons and electrons. Charge is measured in SI units of coulombs (C), but it is often quantized as integer multiples of the elementary charge $e = 1.602 \times 10^{-19}$ C possessed by the electron and proton, by convention, $-e$ and $+e$, respectively. Although the convention is arbitrary, the positive and negative signs denote the general rule for the interaction of charge: Like charges repel each other, and opposite charges attract. Defined in terms of this rule, electrical energy is the amount of work required to separate two opposite charges by an infinite distance or to bring together two like charges that are initially separated by an infinite distance (infinity being the reference distance for zero energy in both cases). The electrical energy of a system then depends on its net charge, which is the sum difference of the number of its electrons and protons multiplied by e as well as the distribution of the charged particles within it. Sometimes it is more useful to talk about the electrical potential, or voltage, which is the electrical energy per unit of charge moved, measured in J/C, or the volt (V).

Electrical Usage

Electrical energy is the most usable form of energy for consumers because it is extremely flexible. It can easily be converted into other forms, as with the electric motor. Using the principle of electromagnetic induction, electric motors act as reverse generators to convert electrical energy into mechanical energy or motive power. Electric motors are used in many household appliances such as fans, refrigerators, hair dryers, electric heaters, blenders, stove range hoods, washing machines, and clothes dryers. National power grids supply these stationary electric motor applications with a steady external energy supply. Electric vehicles use an electric motor adapted for portability, which requires an equally portable power source such as a battery.

The use of batteries enables the effective storage of potentially large amounts of electrical energy. The voltaic cell or battery is an electrochemical device that stores chemical energy and converts it directly into electrical energy through a series of reactions involving the transfer of ions between two electrolytes. Rechargeable batteries such as those found in laptop computers and electric vehicles have reversible chemical reactions; thus, by supplying electrical energy to the cell, the battery's full storage capacity can be restored. Industrial-sized battery systems provide a way for intermittent renewable energies such as wind and solar to better integrate with the existing on-demand infrastructure. The series of 20 sodium-sulfur batteries at Xcel Energy's wind power facility in Minnesota is capable of storing 7 megawatt/hours (MWh) of electrical energy and releasing it immediately into the national grid. Industrial battery systems are not yet economically practical for large-scale power storage, however. The most cost-effective method widely used today to store energy from the grid is the pumped storage hydroelectric power plant.

The efficient transmission of electrical energy over thousands of miles makes the decentralized locations of power stations economically feasible. The power loss of electrical transmission through wires is proportional to the square of the current. Thus, to minimize loss, the amount of current must be as small as possible. Because current is inversely proportional to voltage, this means the voltage must be extremely high. However, residences and commercial buildings require a low voltage for their energy needs (110/230 V, depending on the country). The development of alternating current (AC) voltage and the invention of the transformer solved this problem. Step-up and step-down transformers respectively increase and decrease AC voltage. Transformers must operate at a standard frequency to avoid energy losses and device damage. Power station generators are synchronized to rotate with the country's standard AC frequency. In the Americas, the standard frequency is 60 Hz (3,600 rpm); for most other countries, the standard is 50 Hz (3,000 rpm). On the United States' power grid, transformers step up the voltage to as high 765 kV for long-distance transmission with minimal power loss.

Electrical energy drives many applications, both man-made and natural. The electrical nature of matter is inherent in the charge of electrons and protons that make up the atomic structure. The applications of electrical energy are possible because of the ability to transfer this charge between and within systems. Transmission of electrical energy via current occurs when electrons or positive ions drift through conductive wires in one collective direction. In the absence of an external power supply, the average drift of free electrons or free positive ions within a material is zero. Power delivered from nonelectric energy resources generates electrical energy for consumer and commercial applications. Lighting became the first consumer application of electrical energy in the 1870s, with the invention of the incandescent light bulb. Telecommunications systems use electrical energy to transmit sound or acoustic energy through wires over vast distances. Natural systems exhibiting electrical energy include species of marine animals such as electric eels and catfish, as well as the atmosphere, with the spectacular phenomenon of lightning.

Infrastructure and Policy

The electrical energy infrastructure is well developed for the integration of renewable energy technologies. The electrical energy industry is separated into two markets: power stations sell generated energy wholesale to retail power providers, known as utilities, who control distribution to consumers. Utilities are responsible for demand-side management and the maintenance of power grids—thousands of miles of transmission lines connecting power stations to consumers. National grids enable power stations to use a diversity of resources and generation schemes to increase the capacity and reliability of the electrical energy supply. Power stations can be built where renewable resources are located, cutting capital costs and carbon emissions related to transport while boosting local economies. Viable business models for renewable integration have already been achieved for wind farms, accounting for 30 percent of new capacity installed in the United States in 2007 and 40 percent in Europe, according to the European Wind Energy Association.

Wholesale price is inherently the main factor for utility companies in deciding which resource-derived electrical energy to purchase. Utilities purchase electrical energy generated from various power stations and then sell it en masse to customers at flat rates. Today, coal-fired plants offer the most cost-effective energy in the United States, accounting for 49 percent of net electrical generation in 2006, with the more fuel-efficient, lower-capital-cost natural gas in second place at 20 percent. Nuclear power is considered an alternative to fossil fuels because of its lower greenhouse gas emissions, so incentives have made it the third most popular resource at 19 percent. In contrast, renewable-derived electrical energy from wind, hydro, solar, geothermal, and biomass accounted for only 9 percent of the total electrical energy generated in the United States that year.

Government policy is necessary to make renewable and emissions-reducing technologies economically feasible. Renewable power station implementation depends on the financial incentives offered through government policies that offset the capital costs of their more expensive technologies. Their zero fuel costs alone are not enough to make a renewable generation scheme competitive with a plant using fossil fuel technologies. Governments tax power stations for greenhouse gas emissions and air pollution while offering incentives for the use of alternative energies and emissions-reducing technologies. Energy policy in the United States authorizes loan guarantees for the use of emissions-reducing technologies such as nuclear reactors, clean coal, chemical scrubbers, and renewable energies. Subsidies and tax credits are offered to power stations that generate electrical energy from hydropower, wind power, geothermal energy, and solar power to make the electrical energy from renewable resources more competitive with fossil fuels. In keeping with the Kyoto Protocol, countries such as Sweden, the Netherlands, and Italy have also introduced a carbon tax to further encourage the adoption of emissions-reducing technologies at fossil fuel plants. Green certification schemes enable conscientious consumers to purchase only electrical energy generated from renewable resources, further encouraging development of those technologies.

Generation Alternatives and Efficiency

Power stations generate electrical energy through multiple conversions between other forms of energy such as chemical, nuclear, thermal, mechanical, and gravitational potential. The general conversion is thermal to mechanical to electrical, but the operation can vary greatly depending on the resource used. Almost all power stations use turbines connected by a common shaft to a generator to convert mechanical energy directly into electrical energy via the principle of electromagnetic induction. Renewable resources like wind power, wave power, and hydropower flow through the turbines to generate mechanical energy directly. The chemical energy of fossil fuels and biomass and the nuclear energy of uranium are converted into thermal energy, which drives the turbines through the heating of steam or gas. Geothermal and solar thermal collectors provide thermal energy directly.

Fossil fuels are the largest contributors to electrical energy generation because of their high energy content, but existing schemes can integrate biomass for economic and environmental benefit. The process of combustion—burning with oxygen—converts the chemical energy of fossil fuels and biomass. The resulting thermal energy, referred to as the fuel's energy content, is the difference between the chemical energy of the initial fuel and oxygen and the chemical energy of the resulting steam and emissions: coal, 8.4 MWh/ton; petroleum, 12 MWh/ton; and natural gas, 15.3 MWh/ton. Compare these numbers with the energy content of a biomass such as air-dried wood (20 percent moisture content), 4.2 MWh/ton, and it is easy to see why fossil fuels provide two-thirds of the world's electrical energy demands. However, waste biomass from agriculture and lumber industries and municipal solid waste (garbage) are attractive alternative resources for combustible fuel power plants because of their economic benefit. Because it needs to be disposed of anyway, waste biomass's fuel cost is often negative, and governments offer financial incentives for using it. Power plants typically use biomass in cofiring schemes that require only trivial alterations of existing equipment. Cofiring a fossil fuel with biomass reduces certain greenhouse gas emissions such as sulfur dioxide. Emissions from the major greenhouse gas carbon dioxide are still an issue with all combustible fuels.

Greenhouse gases are an inevitable product of electrical generation involving combustible fuels. The chemical products resulting from combustion are oxide compounds of each element in the fuel: steam from hydrogen, carbon dioxide from carbon, and oxides of any impurities. Knowing the chemical composition of a particular fuel, one can calculate the amount of carbon dioxide produced burning it per MWh of energy content: coal, 288 kg/MWh; petroleum, 252 kg/MWh; natural gas, 180 kg/MWh; and air-dried wood, 299 kg/MWh. The release of carbon dioxides into the atmosphere from fossil fuel combustion for electrical generation contributed 2,516 million metric tons or 35 percent of the total anthropogenic greenhouse gas emissions in the United States in 2007. Other greenhouse gases released by U.S. electrical power plants in 2007 include 9 million tons of nitrous oxide, which creates smog and leads to respiratory illness, and 3.65 million tons of sulfur dioxide, the leading cause of acid rain. To reduce greenhouse gas emissions while still using steam engine generators for electrical generation, the external heat source must come from an alternative or renewable resource.

Alternative external heat sources for steam engine generators that reduce emissions include nuclear power, geothermal power, and solar collectors. Nuclear power stations generate thermal energy from fission inside nuclear reactors. Nuclear power accounted for 19 percent of the electrical energy consumed in the United States and 14 percent world-wide in 2007. There is great controversy over the economic practicality, environmental effect, and social welfare of using nuclear power, however, so its popularity as a resource for electrical generation is waning. Certain countries such as New Zealand have banned nuclear technology altogether. The benefits of renewable energies such as geothermal and solar thermal are much less controversial. They provide the external heat source safely, directly, and emissions-free. There are no fuel costs, and minimal employees are needed for operation—only the start-up capital and maintenance costs are required for the collecting equipment, which are offset by tax credits and subsidies. Concentrated solar thermal collection using parabolic troughs is a growing industry in Southern California, where it generates more than 350 megawatts of electrical power for the grid during daylight hours of peak demand.

Combustible fuels can be conserved by increasing the thermal efficiency of the generation design with combined cycle or cogeneration systems. Thermal power stations that use the combustion of coal or petroleum and all nuclear power stations generate the mechanical energy needed to operate the generator with the steam-driven turbines of Rankine engines. This thermodynamic cycle usually has 20 percent to 40 percent efficiency, meaning 80 percent to 60 percent of the heat generated is lost as waste. For thermal power stations using natural gas, the gas-driven turbine of the Brayton cycle is substituted for the steam turbine. With both thermodynamic cycles, the final process expels excess thermal energy into the atmosphere. This so-called waste heat can be put to use to increase the overall efficiency of the gas engine by using it to preheat the air about to enter the combustion chamber. More often, power stations use the waste heat of the Brayton engine as the external heat source for the Rankine engine. Power stations that employ this method are called combined cycle systems (up to 60 percent efficiency). Combined heat and power or cogeneration plants (up to 85 percent efficiency) use this waste heat for hot water production for district heating or directly for residential or industrial heating at locations near the plant.

Consumer Conservation

Consumers can deter greenhouse gas emissions by reducing demand with efficient electrical appliances or by not using high-energy appliances at all. Drying clothes by air instead of using an electric clothes dryer can save up 1,500 kWh of annual household energy consumption. On the basis of guidelines set by the U.S. Department of Energy and Environmental Protection, Energy Star ratings help consumers choose which products, from household appliances to heating and cooling systems, are the most energy efficient. Consumers reduced greenhouse gas emissions equivalent to those of 27 million cars with the use of Energy Star products in 2007 alone. Computers are another potential energy-saving device. Models such as the Dell Studio Hybrid and the Lenovo Think Centre operate on minimal power consumption—44 and 58 watts, respectively.

Green certificates give consumers the opportunity to encourage the integration of renewable energies into the existing electrical infrastructure. They also allow renewable energy generators to sell their electrical power at a competitive market value. For instance, Dominion Virginia Power offers two different options for residential and business customers to buy "green power." The first method specifies that Dominion will purchase green certificates equal to 100 percent of the monthly electrical energy used by the customer for an additional 1.5 cents per kWh. For the average household consumption of 1,000 kWh per month, 100 percent renewable energy would only cost an additional $15 per month under this scheme. Alternatively, Dominion customers may specify the amount of green certificates purchased by Dominion on their behalf in two-dollar increments, with each increment equivalent to 133 kWh of 100 percent renewable-derived electrical energy. Several organizations give consumers the ability to purchase green certificates separately if their power provider does not offer such a scheme. The extra revenue generated from green certificates covers the above-market costs and also encourages further development of electrical generation from renewable resources.

Many electrical utility companies enable their residential and commercial customers to contribute to renewable-technology development and to reduce energy use. These initiatives include offering discounts toward the purchase of efficient appliances and compact fluorescent light bulbs, increasing public awareness of energy-saving methods, and giving consumers the option to purchase electrical energy generated by renewable resources through green certificates. Utility websites often contain links to home energy use calculators. With the input of household size, heating/cooling system, installed appliances, and previous utility bill, customers can see how much electrical energy they use and learn ways to reduce consumption with programmable thermostats and proper insulation.

Heating the ambient air accounts for over 50 percent of an average household's energy requirement. If the heating system is electrical, just lowering the thermostat from 72 degrees to 68 degrees saves 4,800 kWh annually—the equivalent of reducing carbon dioxide emissions by 5,560 pounds and saving the consumer anywhere from $360 to $600. Natural gas heating systems are much more efficient, but the same reduction can save 208 therms per year, equivalent to 2,600 pounds of carbon dioxide emissions and from $260 to $430 off the annual gas bill. Keeping the thermostat at 68 degrees and turning an electric heating system off overnight when sleeping can reduce usage by 7,290 kWh, cutting carbon dioxide emissions by 8,456 pounds annually. For a natural gas heating system, the same method would decrease annual carbon dioxide emissions by 3,943 pounds.

See Also: Coal, Clean Technology; Electric Vehicle; Hydroelectric Power; Photovoltaics (PV); Power and Power Plants.

Further Readings

Boyle, Godfrey, ed. *Renewable Energy: Power for a Sustainable Future,* 2nd ed. New York: Oxford University Press, 2004.

Energy Information Administration. "Electricity Data, Electric Power Capacity and Fuel Use, Electric Surveys and Analysis." http://www.eia.doe.gov/fuelelectric.html (Accessed January 2009).

Energy Star. "ENERGY STAR Qualified Products." http://www.energystar.gov/index .cfm?fuseaction=find_a_product (Accessed January 2009).

European Wind Energy Association. "Press Release." http://www.ewea.org/fileadmin/ewea_ documents/documents/press_releases/2008/EWEC_press_release_day_2.pdf (Accessed January 2009).

Gibilisco, Stan. *Teach Yourself Electricity and Electronics,* 4th ed. New York: McGraw-Hill, 2006.

Smil, Vaclav. *Energy in Nature and Society: General Energetics of Complex Systems.* Cambridge, MA: MIT Press, 2008.

Kristen Casalenuovo
Virginia Commonwealth University

Electric Vehicle

First introduced during the mid-19th century, electric vehicles refer to any mode of transportation that utilizes an electric motor, rather than the internal combustion engine, for propulsion. Current electric vehicles include commercial transport systems such as buses, trains, and trolleys. Most progress taking place today is in the development of technologies for consumer electric vehicles such as cars and sports utility vehicles. All major automotive companies including General Motors, Toyota, Mitsubishi, Renault, and Volkswagen are pouring millions of dollars into research for electric models, most of it toward battery and fuel cell technology. Electric vehicles are nearly silent, reducing urban noise pollution; do not waste energy while idling; and are emissions-free if they are charged with only

renewable-derived electrical energy. With 27 pounds of carbon dioxide released for every gallon of fuel burned, consumers who replace their gas-powered vehicles with a full electric model could lower their carbon footprint by about five tons per year. The major obstacles to the embrace of electric vehicles are their low mileage range and the battery systems responsible for their expensive price tags. Battery and fuel cell technologies are constantly being updated to overcome these problems. Any situation in which daily travel is limited, such as urban driving, can be accommodated with current electric vehicles. Infrastructure projects to accommodate long-range driving and government incentives reduce the electric vehicle's cost for consumers.

Infrastructure for electric vehicles is under development in certain areas. This electric vehicle recharging station is in Portland, Oregon.

Source: iStockphoto.com

Under the Hood

From the exterior, the only noticeable difference from their gasoline cousins is that electric vehicles are nearly silent in operation. Underneath the hood is quite another story. A direct current or alternating current (AC) electric motor powered by a controller replaces the gasoline engine. Electric trains often use direct current motors, whereas most consumer vehicles use AC motors.

Although an AC installation is a more complex, expensive design, it is capable of greater power and often has a regenerative system that recharges the batteries during braking. Regenerative brakes are especially useful for start-and-stop urban driving, where they can significantly reduce the energy requirement of a trip. The muffler, catalytic converter, tailpipe, and gas tank are all replaced by an array of batteries or a fuel cell and a charge hook-up. An electric water heater is added for heating and a vacuum pump for power brakes. Smaller electric motors may also be added to power any other systems that used to be powered by the engine, such as the water pump, power steering pump, and air conditioner. Any gasoline-powered vehicle can be converted into an electric vehicle by applying these changes.

Much like the traditional setup, drivers administer the amount of power delivered to the motor with an accelerator pedal. Then, in an electric vehicle, the potentiometers and controller handle the rest. The controller sends voltage to the motor in the amount determined by the signal sent by the potentiometers. The signal strength sent to the controller by the potentiometers depends on how hard the accelerator pedal is pressed by the driver. There are two potentiometers for safety, in the event where one fails in the full-on position. The signal is sent to the controller only when both signals from each potentiometer match. On the basis of the signal strength, the controller delivers a certain amount of voltage to the motor from the batteries. The reason the motor is so quiet is because of the pulsed current sent by the controller. Whether direct current or AC, the current is pulsed by the controller's transistor to be sent to the motor in discrete packets. The motor vibrates at the same frequency of the pulses, which can easily be set outside the range of human hearing at 15,000 Hz or above. For an AC electric motor, the controller has three transistors to pulse the current in three phases and three more to reverse the polarity of the voltage for each phase.

Vehicles that use both an electric motor and a gasoline-powered internal combustion engine are called hybrids. Regular hybrid vehicles improve gas mileage and recharge their batteries through regenerative braking systems. Hybrid vehicles reduce idle emissions by shutting off the internal combustion engine when stopped in traffic. Plug-in hybrid vehicles (PHEVs) can be externally charged and operated in charge-depleting mode to improve the fuel economy by as much as 20 percent. The advantage of PHEVs over full electric vehicles is that they solve the problem of the mileage range and battery expense while still reducing noise pollution and greenhouse gas emissions. However, neither hybrids nor PHEVs fulfill the zero-emissions goal toward a sustainable transportation system; only a full electric vehicle has the potential to accomplish that.

The Battery Problem

Battery technology ultimately determines the limitations of the electric vehicle. Batteries are the most expensive component in the vehicle, are often physically heavy and bulky, and their storage capacity affects the mileage range of a full charge. For every kilowatt-hour (kWh) of energy stored, a car can be propelled about four miles. Current lead-acid battery systems can cost up to $2,000, need full replacement every three to four years, weigh about 1,000 pounds, require four to 10 hours for full charge, and have a capacity of 12–15 kWh, which is about 55 miles on a full charge. Some electric vehicles such as the Saturn EV-1 use nickel metal hydride (NiMH) battery systems, which last for 10 years and have twice the capacity. These battery systems have a price tag to match, however, costing anywhere from $20,000 to $30,000. General Motors owns patents on the proprietary NiMH traction battery technology, so its widespread use in electric vehicles may never be realized.

Automotive companies are racing to develop advanced lithium-ion (Li-ion) battery systems because of their extremely high energy density, which is twice that of NiMH. Li-ion batteries are already well developed for applications in laptop computers and cellular phones. For the same weight as the lead-acid systems described above and a much heftier price tag (over $50,000), the 375-volt Li-ion battery system in the Tesla Roadster stores up to 53 kWh of energy in a 3.5-hour charge—enough for the car to travel up to 244 miles. Tesla's system delivers up to 200 kW of electric power, enabling acceleration from 0 to 60 miles per hour in four seconds and a top speed of 125 miles per hour. The PHEV Chevrolet Volt's Li-ion battery, in comparison, is rumored to cost $10,000 for 16 kWh of storage. Lithium-ion batteries only last two to three years and degrade even faster under high temperatures. Their lifetimes can be increased with advanced cooling systems and by introducing variants on the battery's chemistry. Tesla's batteries last for five years or 100,000 miles.

Battery systems on electric vehicles could become obsolete with adequate development of fuel cell technology. Batteries are ultimately material intensive, so while weaning countries off a dependence on oil, they may ignite the consumption of another finite substance such as lithium. Fuel cells are lighter, smaller, and can be instantly recharged. Fuel cells powered by hydrogen are being combined with battery storage to eliminate the need for an external charge and increase the mileage range. A kilogram of hydrogen has the energy equivalent of a gallon of gasoline, and the only emissions are water vapor. Chevrolet rolled out a test fleet of hydrogen-powered Equinox Fuel Cell vehicles in early 2008 with 35-kW NiMH battery packs. With 4.2 kg of compressed hydrogen, the Equinox vehicles can cover 160–200 miles before needing more fuel. An infrastructure of hydrogen fueling stations and affordable fuel cell technology need to be implemented before fuel cell electric vehicles can be mass-produced.

Infrastructure

The infrastructure is not here yet for electric vehicles to flourish, but it may be well on its way. For long-range driving, battery exchange stations could solve the problem of long charge times. For such a plan to work, batteries and electric vehicles must be standardized, ownership of the batteries must be settled, and an expensive infrastructure would be required. Silicon Valley startup Better Place is building an electric car infrastructure of battery swapping stations and charging spots in the Hawaiian Islands, to be completed by 2012. Their other infrastructure projects are taking place in Israel, Denmark, Australia, and California. Better Place will own the batteries and sell standardized vehicles to consumers supplied by Renault-Nissan, Subaru, and possibly more partners in the future. Better Place sells vehicles to consumers minus the price of the battery systems (for $20,000 versus $60,000) in an effort to make the vehicle's prices competitive with those of gas vehicles. Consumers can recharge their vehicles through their home's power supply or use their subscription plans to purchase electrical mileage from the charging or battery swapping stations.

Tax incentives and electric city buses offer consumers an affordable way to support the electric vehicle economy. The U.S. Congress passed legislation for a tax credit in October 2008 toward any plug-in electric vehicle. The total amount ranges from $2,500 to $15,000 depending on the kWh capacity of the battery and the weight of the vehicle. Denmark charges no tax on electric vehicles versus a 180 percent tax on gas vehicles. However, with the Tesla Roadster starting at $109,000 and the PHEV Chevrolet Volt projected to cost $40,000, consumers may not be ready to buy a personal electric vehicle just yet. Trolleys and electric buses are rolling in cities across the globe in the United States, China, Holland, Austria, Switzerland, and Australia. The Tindo is an all-electric 27-passenger city bus that provides free service to the public in Adelaide, Australia. It receives its charge from an AU$550,000 solar photovoltaic system on the roof of the Adelaide Central Bus Station. A full charge gives Tindo (which is aboriginal for sun) a range of 200 kilometers with enough energy left over to power its air-conditioning system.

See Also: Automobiles; Batteries; Flex Fuel Vehicles; Fuel Cells; Hydrogen.

Further Readings

Better Place. "Our Bold Plan." http://betterplace.com (accessed January 2009).
Larminie, James and John Lowry. *Electric Vehicle Technology Explained*. Chichester, UK: Wiley, 2003.
Leitman, Seth and Bob Brant. *Build Your Own Electric Vehicle*. 2nd ed. New York, NY: McGraw-Hill, 2008.
Students Making Advancements in Renewable Transportation Technology. "SMARTT Challenge." http://smarttchallenge.com (accessed January 2009).

Kristen Casalenuovo
Virginia Commonwealth University

EMBODIED ENERGY

The production of goods and services requires energy inputs at each stage of the process, including resource extraction, transportation, manufacturing, retail, operation, and disposal. The energy involved can be divided into two components: direct energy and indirect energy. Direct energy consists of inputs consumed onsite in the production of a good or service, such as natural gas or electricity. Indirect energy includes that associated with materials and services at other stages of the product life cycle. For example, indirect energy inputs may encompass energy involved with resource extraction, manufacture of raw materials, operation and maintenance of the end product, disposal, and transportation between each stage. Cumulatively, this indirect and direct use of energy is referred to as embodied or embedded energy. Multiple models exist for calculating embodied energy, and the results are often used as a means of evaluating the environmental impacts of building materials, food products, energy sources, and other goods. Applications of this form of accounting lack uniformity, however, as different boundaries and methodologies lead to varied results.

Methodologies

The methodologies for calculating embodied emissions differ in the scale of the data they incorporate, particularly in terms of the boundaries selected and data specificity. Without international consensus in this area, a wide range exists for the embodied energy assigned to any specific material or process. There are three primary methodologies for calculating embodied energy: input-output models, process analysis, and hybrid models.

Input-output models are considered comprehensive, yet imprecise. The method derives from economist Wassily Leontief's input-output tables used for economic planning worldwide and similarly relies on assumptions that entire sectors of the economy are homogenous, in this case, in their energy use. Critiques of this method hone in on the highly aggregated inputs, which are representative of national averages and do not allow for variances in energy inputs that arise on a case-by-case basis.

A less complete but more precise methodology is process analysis. Process analysis does not rely on the input-output tables used for the former method but, rather, on tailored data specific to the process under analysis. However, energy inputs more than two stages back are difficult to include and are often left out. Indirect energy embedded in materials is typically excluded as well, as is energy lost in energy conversion processes in the form of waste heat.

As such, the two models are sometimes melded in a methodology termed *hybrid analysis*, which includes case-specific data while retaining some of the comprehensiveness of input-output models. Hybrid models use input-output tables to identify and analyze all of the major energy pathways but then substitute data for specific materials when possible.

Boundaries

Additional controversy exists around the boundaries selected for the indirect energy inputs. Many calculations do not include energy required for labor and government services, for example, as well as other aspects of the economic system such as marketing and catering. Some call this omission a significant methodological error that complicates attempts to rank the sustainability of goods and services.

The most comprehensive boundary condition goes from the extraction of raw materials to the end of a product's life, at disposal. This boundary condition is often termed *cradle to grave*. Less-broad boundary conditions assess solely the energy used from the extraction of raw materials to the point when it leaves the factory gate—cradle to gate—or to arrival at the point of use—cradle to site.

Applications

The analysis of embodied energy, also referred to as life cycle energy, has been applied in depth to buildings, households, energy itself, and food among other fields. Embodied energy in building materials is a key consideration for green builders and generally makes up a greater portion of total energy use in energy-efficient buildings. Of primary importance are the energy intensity of material production as well as the mass of the material needed. For example, aluminum and cement have high embodied energy, because of the high temperatures required for manufacturing the two materials. Heavier materials and those with greater mass require greater quantities of fuel for transportation. Recycling and reuse of materials can decrease embodied energy in building materials.

Scholars investigating the energy requirements of households look at their direct energy use, such as electricity and motor fuels, as well as their indirect energy use, made up of the energy embodied in goods and services consumed by households. Research shows indirect energy making up the majority of household energy consumption and embodied energy playing a major role in studied nations' total energy consumption. Food has been identified as a key leverage point at which embodied energy can be reduced at the household level, in part because food is consumed in large amounts over a short duration of time as compared with other goods like vehicles and electronics.

In the cases of both food and fuel, the total energy inputs, or embodied energy, can be contrasted with the energy output of the final product, yielding a ratio referred to as the energy return on investment. For example, the energy inputs needed for corn-based ethanol are much higher than those of cellulosic ethanol, yet the former is the manufacturing process that is more developed. Early research brought the energy payback period for renewable energy technologies like solar photovoltaic panels into question, as photovoltaic plants require a significant amount of energy for the concrete, metal, and human labor involved in construction and maintenance, leading to a low energy return on investment. However, later studies put that controversy to rest, with conservative estimates that the energy payback period for home systems is approximately two to eight years.

International Implications

Although the environmental impacts of goods and services have been studied extensively since the 1970s, the topic has remained of great interest in part because of the environmental effects of international trade. A large body of research asserts that decisions around firm location are driven by differences in labor, capital, and materials. Along with these differences come varied manufacturing methods, economic efficiency, and emissions associated with local energy production. Nations will tend to produce those goods that they have a competitive advantage in producing, often moving the production of pollution-intensive goods to developing nations. International attempts to control greenhouse gas emissions at the national level, such as the Kyoto Protocol, impart additional importance to the location where energy is used and the associated greenhouse gas emissions are released. When one country produces goods with high embodied energy for another nation, the potential exists for "carbon leakage" from developed to developing nations.

Embodied energy has the potential to serve as a practical indicator of environmental impact and a tool for decision making. However, inconsistencies in boundaries and imperfections in methodologies detract from the overall usefulness of the concept.

See Also: Emissions Trading; Energy Payback; Life Cycle Analysis; Sustainability.

Further Readings

Bankier, Colin and Steve Gale. "Energy Payback of Roof Mounted Photovoltaic Systems." http://www.energybulletin.net/node/17219 (Accessed June 2006).

Thoramark, Catarina. "A Low Energy Building in a Life Cycle—Its Embodied Energy, Energy Need for Operation and Recycling Potential." *Building and Environment.* v.37 (2002).

Treloar, Graham. "Extracting Embodied Energy Paths from Input–Output Tables: Towards an Input–Output-based Hybrid Energy Analysis Method." *Economic Systems Research.* v.9/4 (1997).

Sarah Mazze
University of Oregon

EMISSION INVENTORY

An emission inventory is a database recording the quantities of pollutants emitted into the atmosphere. In the United States, under the provisions of Title 40 Protection of Environment Part 51 legislation, pollutants are greenhouse gases (GHGs), of which six are recognized as "criteria" air pollutants. These six are particle pollution or particulate matter, ground-level ozone, carbon monoxide, sulfur oxides, nitrogen oxides, and lead. The inventory specifies the source and destination for the pollutant at or above threshold quantities, whether it is released intentionally or not. In expanded form, such inventories can include information on the sources and the transfer of the identified substances either into the local environment or outside the boundaries of the reporting operator for recycling, waste treatment, or other business purpose. Where the inventories include release and transfer data they may be termed *pollutant release and transfer registers*. Mandatory databases are maintained by authorities in several jurisdictions. Inventories may also be appropriate to satisfy independent industry certification requirements such as the International Organization for Standardization 14064 regime and voluntary registries like the California Climate Action Registry.

The existence of GHGs, including criteria air pollutants, is acknowledged as significant not only to global warming but also to its local effects such as health and property losses. In its April 2, 2007, decision, in the case of *Massachusetts et al. v. Environmental Protection Agency et al.* (No. 05-1120) 415 F. 3d 50, the U.S. Supreme Court noted that it was undisputed by the parties in the case that GHG emissions originating in human actions are a cause of global warming and found that the Environmental Protection Agency (EPA) has the authority and obligation to regulate GHG emissions under authority of the Clean Air Act 42 U.S.C. §7401 (1970 as amended). In Chapter 85 of the same Clean Air Act, the Public Health and Welfare provisions also address air pollution prevention and control. The formation and maintenance of an emissions inventory database or register then is a means of recognizing environmental effects of criteria air pollutants.

The U.S. EPA's National Emission Inventory (NEI) database is updated every three years and contains annual information about point, nonpoint, and mobile sources that emit criteria air pollutants and their precursors, as well as about hazardous air pollutants. The primary sources of information for the NEI database are current or previous emissions inventories reported from state and local environmental agencies, as well as its own databases. These EPA databases include the Toxic Release Inventory database, databases developed from Maximum Achievable Control Technology programs, databases from the electricity generators' reports to the Emission Tracking System/Continuous Emissions Monitoring data and Department of Energy fuel use data, estimates from the Federal Highway Administration, and the EPA's own computer modeling of emissions from road use and non–road use sources. Information from the NEI database is also published by the EPA's Emission Factor and Inventory Group.

In Mexico, there were three phases to the NEI program. Phase 1 began in 1995 with initial development and implementation of the methodology manuals and training. Phase 2 established the database for the six northern states. Following the expansion of the database to include all 32 states, the phase 3 report for 1999 was released in October 2006. The Canadian equivalent registry is known as the National Pollutant Release Inventory, first published by authority of the Canadian Environmental Protection Act in 1995, with data reported for 1993. In Australia, the National Pollutant Inventory was officially available in 2000.

Regional approaches to the formation of GHG emission inventories have also been used as a result of local cooperation transcending political borders. The Mexican NEI phase 2 priority to the region straddling its northern international border is a modest example of transborder cooperation. Similarly, the New England Governors and the Eastern Canadian Premiers meetings since 1973 represent sustained regional cooperation across another international border. The European Union also uses a regional approach to emissions inventories.

With the initial premise of the inventory being the community's right to know, maintaining currency, access, and transparency in the inventory is crucial. Because such inventories do not provide for any consequences other than for failure to report, their utility is in promoting awareness that may then affect other actions. If the principle that "what gets measured gets managed" holds, the databases themselves can also be deployed for business purposes and may be cross-referenced or used in whole or in part with other databases to be subject to regulatory regimes that could include provisions for enforcement.

An inventory of all sources, quantities, sinks, and transfers of identified substances could be used by management to improve business efficiencies. A routine record of such movements, whether intentional or not, could also be used in litigation as a business record and as a defense to potential claims. Recommended features of an inventory of criteria air pollutants should include relevance of the information in a form designed to support decision making, completeness, consistency over time, transparency of assumptions and references, and accuracy.

An accurate inventory can be demonstrative of GHG levels at a facility, which may be an advantage in claiming credits under regimes where there are fees or caps to emissions. Another context in which an inventory record is valued is in an emissions trading region. An emission inventory can be as versatile an instrument as the users of the information are capable of.

Because perception is a significant factor in society, the accuracy of the reported data when compared with data obtained by independent monitoring can affirm or degrade the credibility of the reporting operator in the perception of an observer. Independent monitoring routinely conducted by the U.S. EPA through its Particulate Matter Supersites and other programs can validate data reported from various sources. The data can also be audited as provided in Environment Canada's National Pollutant Release Inventory process. Over time, with the multiplication of monitoring agencies and cross-referencing of increasingly sophisticated databases, in addition to business and regulatory disclosures, the imperative for an operator to comply with reporting obligations becomes a function of doing business and an indicator of good citizenship in society and not just a legal obligation.

In a dynamic and increasingly complex world, to be in compliance involves identifying substances that are subject to reporting obligations. Industrial processes could involve elemental substances, compounds, alloys, and mixtures, all for legally permitted reasons. It is not always easy for the reporting enterprise to determine which category applies at any stage of source, release, or transfer, and differences in classification can occur.

Where reportable thresholds are above zero for each regulated category, it is conceivable that a facility may actually source, release, and transfer a large quantity of regulated substances and not be subject to reporting requirements if thresholds in each individual category are not exceeded. In this sense, setting out reporting thresholds can also operate to effectively permit or disregard releases or transfers up to the applicable thresholds. Over the long term, releases could continue below reporting threshold quantities and could bioaccumulate to toxic levels without triggering any requirements.

Lowering threshold quantities could result in greater transparency. However, the same transparency could compromise the confidentiality of business information. In extreme cases, full transparency could affect the existence of an enterprise. Disclosure of risks associated with business activities including GHG emissions and transfers into and out of the control of the business is now routinely required not only by regulators but also by service providers, employees, health and safety regimes, creditors, investors, and insurers. Even those parties who would typically have an interest in access to internal business information as part of their due diligence inquiries may not want the same information publicly accessible. Developing and maintaining an emission inventory therefore is complex task and involves balancing of interests.

See Also: Best Management Practices; Carbon Emissions Factors; Carbon Tax; Carbon Trading and Offsetting; Climate Change; Emissions Trading; Energy Audit; Energy Storage; Flex Fuel Vehicles; Food Miles; Global Warming; Green Energy Certification Schemes; Greenhouse Gases; Kyoto Protocol; Nonpoint Source; Volatile Organic Compound (VOC); Zero-Emission Vehicle.

Further Readings

Chow, Judith C., et al. "Will the Circle Be Unbroken: A History of the U.S. National Ambient Air Quality Standards." *Journal of the Air & Waste Management Association,* 57 (2007).
Environment Canada National Pollutant Release Inventory. http://www.ec.gc.ca/pdb/npri/npri_home_e.cfm (Accessed February 2009).
Jerry Bauer. "Conducting a Greenhouse Gas Emission Inventory." *Power Engineering* (November 2007).
"Mexico." http://www.epa.gov/ttn/chief/net/mexico.html (Accessed February 2009).
"North American Emissions Inventories—Canada." http://www.epa.gov/ttn/chief/net/canada.html (Accessed February 2009).
U.S. Environmental Protection Agency. http://www.epa.gov/ebtpages/airairpoemission inventory.html (Accessed February 2009).
"U.S. National Emission Inventory Database." http://www.epa.gov/ttn/chief/eiinformation.html or http://www.epa.gov/air/data/neidb.html (Accessed February 2009).

Lester de Souza
Independent Scholar

EMISSIONS TRADING

Emissions trading is an economic and market approach developed as a response to the perception that greenhouse gases are having an adverse effect on the climate. The approach is intended to control emissions into the environment from sources subject to human care and control that are identifiable and quantifiable and have consequences beyond the boundaries of their sources. An essential feature of this approach is the view that there are rights to emissions of identified substances that are organized as a product or property to which marketable title is available. This property right to emissions is generated by regulations or by agreement. Title to such property is described through transferable certificates, permits, or regulations or by voluntary registries. The overall principle in this approach is to make it profitable for those who control the sources of emissions to reduce the emissions.

Controls may be either by "command and control," as seen in regulatory regimes, or by the "market" or economic approach. Some regulatory regimes include economic motivation for compliance. The responsibility for the reductions can be allocated either to the persons controlling the emissions or the government on behalf of the public. For instance, regulations used to impose fines on polluters, issue permits to limit the amount of pollution, or tax pollution would allocate responsibility to the source of emissions for compliance. Supply chain subsidies or waste disposal facilities provided by the government are instances where the public accepts responsibility. Governments can also provide financial incentives through grants or tax reductions for corresponding pollution reductions. Emissions trading options can be seen in government-based, privately based, or hybrid mechanisms. Regulations can be used to establish rights for trading on governmental trading systems. Private registries in which subscribers establish rights by agreement offer an alternative trading forum. A hybrid form relies on government-established rights that are traded by agreement in private markets. The sale of rights may be either to individual purchasers or by auction.

In the United States, emissions trading began in 1975, when an Environmental Protection Agency (EPA) policy—eventually called the Emission Trading Program—was introduced. Trading was expected to lower compliance costs, provide firms with flexibility, and allow for new business activity and economic growth in places that had poor air quality without compromising environmental goals.

Those who control the sources of emissions have the option of either meeting the allocated amount of emissions or not. Where the allocation is not met, the options are to buy emission rights to cover pollution in excess of the allocated amount, or—if emissions are less than the allocated amount—the unused rights can be sold or banked for future use or trades. Participants in the Emission Trading Program system could net the emissions, offset emissions, create bubbles, or bank them.

Where the holder of emissions rights is sufficiently broadly defined, the rights holder can allocate emissions under its control to produce a result without a net change in emissions. Here the emphasis is on the net result for the emissions rights holder and not for the individual source or time of emission. In an offset strategy, the rights holder can structure the emissions to offset an increase in one location against a decrease in another. A bubble is created by aggregating emissions to maximize use of existing emission rights. Banking can be used to save rights either for subsequent use of the rights holder or for trade to other participants.

How Trading Works

Developing a trading system is a complex matter involving multiple design issues—transaction costs, market power, technology, uncertainty and risk, temporal issues, initial allocations, investment issues/risk, economic efficiency, political considerations, and pragmatic reasons including determinability and enforceability.

Marketable property requires a commonly recognized standard description of the intended property. Here, rights to emissions are identified in relation to limited amounts of specified substances that may be emitted into the environment. These rights and transfers of the rights are recorded in registries. Rights may be denominated in terms of quantity of emissions over a fixed time period.

The quantity or target of permitted emissions is defined at appropriate scales. It is the variation from the target levels that is tradable. Targets are also described as caps and are a prerequisite to trading. In the interest of reducing greenhouse gases, the target or cap to emissions is usually designed to decline over time. In practice, meeting the allocated target or cap exactly is unusual, and with a declining target, it is more difficult to achieve on an ongoing basis. More typically, emissions are either below or above the target—hence the opportunity to trade the amounts by which the target is not exactly matched. Whether improvements to emissions result in surplus rights that can be traded or banked depends on the design of the trading system. Emission levels greater than the target amount result in additional cost to the source of emissions.

In a government-regulated trading system, the target is stated in regulation and may be administered as a quota or a permit. These quotas or permits may be purchased in predefined quantities for use and/or trade either from the issuer or from another participant. In a nongovernmental trading system, the target may be set by agreement or by the exchange. In all cases, the targets may be designed for individual sources or standardized for a category of sources.

In a credit system, operators of sources can trade unused or banked credits for various management or commercial reasons in addition to compliance with the agreed target. For instance, if the cost of the credit is less than the cost of compliance, it would be possible for management to elect to purchase credits rather than modify production to achieve compliance. In addition, emission reduction increments can be implemented by providing for a tax on each trade.

A form of the trading system was used in the phasing-out of leaded fuel over 25 years and appears to have been successful in achieving the objective. From 1982 to 1987, the EPA awarded specific lead rights to each refinery. Where a refiner achieved lower lead content, the unused portion could be sold to another refiner or banked. The system worked for several reasons: Gasoline lead-content rights were well defined, its levels in the product could be monitored, and the goal was understood and agreed on and the rights readily tradable. In fact, trading in rights did occur, and the elimination of lead in gasoline in the United States was achieved by 1996.

An example of nongovernmental emissions trading systems is the Climate Exchange PLC, which owns exchanges in Chicago and Europe and affiliated exchanges in Montreal, Canada, and Tianjin, China. Using a cap-and-trade structure, the Chicago exchange provides emissions trading in all six greenhouse gases. The Chicago Climate Exchange is advertised as North America's only—and the world's first—legally binding multisectoral, rule-based, and integrated greenhouse gas emission registry, reduction, and trading system. Innovative tradable products now available include catastrophe event–linked futures and climate futures derivatives including standardized and cleared futures.

In Europe, in addition to the nongovernmental Climate Exchange PLC entity already noted, the European Union Emissions Trading Scheme commenced operation in January 2005 as the world's largest multicountry, multisector trading scheme.

What Trading Accomplishes

Trading systems alleviate a free-rider problem, which is where one entity incurs the cost of improvements and another obtains the benefit without cost. In the climate change context, several industrialized and industrializing countries were reluctant to take the initiative and bear the costs of reducing emissions, each alleging the other either had the benefit of past or future emission levels. As a result, several countries declined to subscribe to the Kyoto agreement and resisted regulating the greenhouse gases domestically. Government-sponsored emissions trading in these nonsignatory countries has been limited.

By graduating reductions and providing inherent flexibility in design, trading systems can also enable sources of emissions to be incrementally modified in a commercially reasonable manner. The overall objective of reduction of greenhouse gases in the environment can then be achieved by flexibly redistributing resources and costs among the trading participants.

Ideological issues with trading include questions about the efficiency of markets in reality and the accounting for market failures. In addition, the concept of a marketable property right to pollute is problematic to begin with and legitimized by authorizing tradable amounts of emissions. The question that survives the discussion is what is the source of the initial sovereignty over the global environment beyond national jurisdictions from which property rights—and particularly any right to pollute—can be based?

Internationally, as a result of U.S. efforts, the Kyoto agreement at Article 17 includes trading as an accepted device to achieve the objective of reducing greenhouse gas emissions. So, although some countries are not signatories to the Kyoto agreement, emissions reduction efforts have still proceeded at various domestic levels.

As noted above, the United States does have trading, even without being a signatory to Kyoto. Domestically, under U.S. law, sovereign jurisdiction to regulate emissions is constitutionally allocated. Thus, in its April 2, 2007, decision in the case of *Massachusetts et al. v. Environmental Protection Agency et al.* (No. 05-1120) 415 F. 3d 50, the U.S. Supreme Court noted that greenhouse gases, including criteria air pollutants caused by human activity, are a cause of climate change and the EPA is responsible for regulating these emissions. The EPA does permit trading in three greenhouse gases—carbon monoxide, sulfur dioxide, and nitrous oxides.

See Also: Carbon Trading and Offsetting; Emission Inventory; Green Energy Certification Schemes; Internal Energy Market; Kyoto Protocol; Metric Tons of Carbon Equivalent (MTCE).

Further Readings

Bogdonoff, Sondra and Jonathan Rubin. "The Regional Greenhouse Gas Initiative." *Environment.* v.49/2 (2007).

Emission Trading System (EU ETS). http://ec.europa.eu/environment/climat/emission/index_en.htm (Accessed February 2009).

Pinkse, Jonatan. "Corporate Intentions to Participate in Emission Trading." *Business Strategy & the Environment.* v.16/1 (2007).

Selin, Henrik and Stacy D. Vandeveer. "Canadian-U.S. Environmental Cooperation: Climate Change Networks and Regional Action." *The American Review of Canadian Studies.* (Summer 2005).

Woo, Jung-Hun, et al. "Development of North American Emission Inventories for Air Quality Modeling Under Climate Change." *Journal of the Air & Waste Management Association.* v.58/11 (2008).

Lester de Souza
Independent Scholar

ENERGY AUDIT

The rising costs and dwindling supplies of fossil fuels, as well as growing concerns about environmental externalities associated with fossil fuel use, have spurred growing interest in energy efficiency and conservation. The "energy audit" is one of many management tools that households, organizations, and businesses increasingly rely on to help them manage and reduce energy use. Audits typically include recommendations for reducing energy waste and optimizing energy, economic, and environmental performance. This article identifies some common characteristics of energy audits and illustrates practical application of two types of comprehensive energy audits: the household audit and the industrial firm audit.

The exact steps required to complete an energy audit vary depending on organizational needs, type, and size. However, most audits display the following characteristics:

- Physical inspection of facilities
- Evaluation of energy cost and consuming behaviors/processes of the household/organization
- Identification of high-energy-use mechanisms/activities
- Creation of a benchmark energy use profile
- Development of recommendations using energy efficiency "best practices" for reducing cost and energy use and for increasing environmental performance (through efficiency gains, purchase of new equipment, conservation, etc.).

Household Audit

A household energy audit typically begins with an analysis of the building envelope. In the United States, heating and cooling are responsible for more than half of household non-transportation energy use. The main goal is to determine whether or not the living space efficiently transfers and retains conditioned air. Accomplishing this task involves visual inspection of attic, floor, and wall insulation and identifying gaps in and around walls, doors, light fixtures, ceilings, ductwork, windows, and so on. Alternatively, some professional auditors use blower doors to fill the structure with pressurized air and then deploy infrared cameras to measure air infiltration and leakage. The audit may include conducting a room-by-room inventory of energy-using devices (e.g., furnace, air conditioner, refrigerator, oven, electronic appliances, lighting, etc.). Audits typically consider household energy use data such as monthly electricity bills. It is also important to develop an understanding of the energy consumption habits and beliefs (environmental concern, cost sensitivity, etc.) of household members. This information is then used to create an overall energy performance baseline profile. The household may then choose from a menu of "best practices" action items that can be implemented to reduce energy consumption (e.g., making more efficient use of energy-using devices, sealing gaps in thermal barriers, purchasing more energy-efficient products, behavior modification, etc.).

Industrial Firm Audit

The goals and methodologies used for industrial firm audits differ considerably from those for household audits. Private firms must consider how energy use affects market competitiveness, business risk, and shareholder value. First, energy inputs make up an increasingly large portion of business costs, so reducing energy use can have a significant effect on economic performance. Second, businesses face numerous regulatory requirements, including potential risks associated with managing criteria and greenhouse gas emissions associated with fossil fuel use. Third, consumers increasingly expect manufacturers to be sensitive to potential negative social and environmental externalities generated by business operations. In short, efficiently managing energy use just makes good business sense for most industrial firms.

Unlike in most households, the operation of industrial motors and machines is typically the highest energy-consuming activity of industrial firms. Processes, behaviors, and management integration can have as big an effect on energy use as the rated spec characteristics of specific machines. Therefore, many industrial firms assemble one or more cross-organizational teams to carry out the audit, usually in consultation with an external consultant. The audit team then engages in a physical inspection of the entire facility. This inspection may include:

- Machines and motors (turned off when not in use or operated in accordance with specified loads?)
- Lighting (lighting technology used, role of natural light, lighting provided only when and where needed?)
- Compressed air (leaks in compressors, is more pressure used than necessary?)
- Process heating (efficiency of heating technology, are heat temperatures excessive?)
- Heating/cooling and thermal envelope of buildings (integrity of envelope, efficiency of ventilation and conditioning systems, temperature levels, areas heated that are rarely used?)
- Review of historical energy use records by major categories and/or machine types.

Industrial firms use this information to develop a menu of options that may satisfy and balance the social and economic needs of the organization. This usually involves comparing firm benchmarks performance indicators with industry norms. Some common indicators include carbon dioxide emissions per dollar output, energy used per dollar output, or energy dollar cost per dollar output.

Some common strategies that industrial firms employ to reduce energy use include:

- Developing continuous productive maintenance programs that focus on maximizing production system efficiency.
- Replacing oversized and old equipment.
- Redesigning plant layout to improve production flow and reduce space required for productive activities.
- Training and encouraging workers to take an active role in increasing the efficiency of processes, machine operation, and error reduction.
- Any implementation plan should also include mechanisms for continuous evaluation and process improvement through use of a holistic management strategy (e.g., Six Sigma).

It is clear that the days of cheap energy are largely behind us. The main human-produced contributor to global warming is fossil fuel use, and emissions continue to rise worldwide. Many scientists predict that if carbon dioxide and other greenhouse gas emissions continue to grow at current rates, in a few short years humans will largely forfeit their ability to slow the process of planetary warming. Meanwhile, proven supplies of oil and many other raw materials are either peaking or declining. Households increasingly worry about paying next month's energy bill and how the climate crisis may affect their children's futures. Firms are driven by market forces to find new and innovative ways to reduce energy costs to stay competitive. Energy audits provide a powerful tool to individuals and organizations to help them potentially adapt and thrive in a carbon-constrained world: The best kind of kilowatt is one that never needs to be produced.

See Also: Appliance Standards; Best Management Practices; Energy Payback; Green Power; Home Energy Rating Systems.

Further Readings

Energy Information Administration. *Annual Energy Review 2007*. Washington, D.C.: Energy Information Administration (DOE/EIA-0384) (2007).

Energy Information Administration. "Energy Basics 101." http://www.eia.doe.gov/basics/ energybasics101.html (Accessed January 2009).

Environmental Protection Agency. "The Lean and Energy Toolkit." http://www.epa.gov/lean/ energytoolkit/index.htm (Accessed January 2009).

Harrington, Jonathan. *The Climate Diet: How You Can Cut Carbon, Cut Costs and Save the Planet*. London: Earthscan, 2008.

Johnston, David, et al. *Green Remodeling: Changing the World One Room at a Time*. Gabriola Island, Canada: New Society, 2004.

Kemp, William. *Smart Power: An Urban Guide to Renewable Energy and Efficiency*. Toronto: AZtext, 2004.

Jonathan Harrington
Troy University

ENERGY PAYBACK

The term *energy payback* or *energy payback time* (t_{EP}) is the amount of time it takes for a given device, system, technology, or process to produce (or conserve) as much energy as it took to construct. The concept of energy payback time has become increasingly important as policy makers attempt to transition society away from polluting fossil fuels to sources of green energy. The value of t_{EP} is given by embodied energy invested over the entire life cycle, E_E, divided by energy produced (or fossil fuel energy saved), $E_{P/S}$. Thus, t_{EP}, as measured in years, is:

$$t_{EP} = \frac{E_E}{E_{P/S}} \tag{1}$$

Similar terms are *net energy gain* and *EROEI*. Net energy gain refers to a surplus condition in the difference between the energy required to harvest an energy source and the energy provided by that same source. EROEI stands for energy returned on energy invested and is also sometimes referred to as EROI (energy return on investment). When the EROEI of a resource is equal to or lower than 1, that energy source becomes an energy sink and can no longer be used as a primary source of energy. This has happened in some of the worst cases of producing ethanol in an inefficient manner for use as a fuel.

The energy payback time is determined by first doing a simplified life cycle analysis (LCA) of the given device, system, technology, or process. An LCA is a means of quantifying how much energy and raw materials are used and how much (solid, liquid, and gaseous) waste is generated at each stage of a product, process, service, or system's life time. For determining the t_{EP}, only the embodied energy is necessary, so only an energy LCA needs to be performed. Ideally, an energy LCA would include the following: raw material production energy, manufacturing energy of all components, energy use requirements, energy generation (if any), end-of-use (disposal) energy, and the distribution/ transportation energies in between each stage. Complete energy LCAs are difficult but are possible to perform, especially on emerging green technologies, such as solar photovoltaic cells, the fabrication of which is constantly undergoing improvements and which have not been in mass production long enough for recycling or disposal to become established. Determining the $E_{P/S}$ is relatively straightforward, as renewable energy sources such as wind turbines can be metered, and the energy conserved from technologies such as light-emitting diodes can be readily measured.

Energy payback times are normally calculated for the purpose of guiding policy to drive down greenhouse gas emissions. Thus, care must be taken when comparing technologies such as "clean" (or dirty) coal, which produce greenhouse gas emissions during use, and renewable energy resources such as microhydro, which do not. For example, the energy payback time for extracting oil is fast, but the greenhouse gas emissions and climate forcing are relatively large. Energy payback time is most often calculated and relevant for renewable energy technologies such as solar photovoltaic cells. In the 1960s, there was some concern that solar cells did not produce as much energy over their lifetime as it took to fabricate them. This is no longer the case for all of the commercialized solar cells, as has been proven by numerous studies of the varying photovoltaic technologies and materials systems, which all find energy payback times of a few years. In general, solar cells are warranted for 20–30 years, and they thus produce many times as much energy as went into their construction. Photovoltaics are thus an energy-breeding technology.

Because of the current energy mix, all technologies that may produce or conserve energy are dependent to some degree on fossil fuel energy, and thus also contribute to greenhouse gas emissions and the concomitant climate destabilization. For a device, system, technology, or process to have a net negative effect on greenhouse emissions of the energy supply, first it must produce enough emission-free energy or conserve enough energy to offset the emissions that it is responsible for, and then it must continue to produce energy or conserve energy to offset emissions from existing or potential fossil fuel plants. Modern renewable energy technologies such as wind and solar (both photovoltaic and thermal) have energy payback times ranging from a few months to a few years, depending on local conditions and types of installations. These green energy sources thus easily pay for themselves in terms of energy over their lifetimes. This is also true of energy-conserving technologies such as high-performance windows and insulation.

Although each individual device may have a short energy payback time, producing a net negative effect on greenhouse gas emissions can be challenging in view of the rapid growth of a particular industry. This is because the construction of additional energy production plants (or the fabrication of many energy-conservation devices) to enable the rapid growth rate creates emissions that cannibalize the greenhouse mitigation potential of all the power plants viewed as a group. To illustrate this point, it is helpful to view all the energy plants of a given type as a single aggregate plant or ensemble and to look at its ability to mitigate emissions as it grows. This ability is first dependent on the energy payback time. For a generic energy-producing technology (or energy-conserving technology), an installed total capacity, C_T (in GW), produces tC_T of energy per year, where t is the time the plant is running at capacity in hours in a year. If we assume that in the same year, the industry of that technology grows at rate r, it will produce an additional capacity of rC_T. The amount of energy that the industry produces (or conserves) is obtained by multiplying by the time, and is thus $rC_T t$. For simplicity, assume that the additional capacity does not produce (or conserve) its energy, $rC_T t$, in that year, but only in subsequent years. The energy needed for the growth of the entire technology ensemble is given by the cannibalistic energy E_{Can}:

$$E_{Can} = \frac{E_E}{E_{P/S}} rC_T t \qquad (2)$$

Regardless of whether the technology is an energy producer or conserver, the technology ensemble will not produce any net energy if the cannibalistic energy is equivalent to the total energy produced. So by setting equation (1) equal to (2) and simplifying, it can be shown that the energy-neutral payback time is equal to the inverse of the growth rate. Thus, the growth rate of either an energy-conserving or energy-generating industry may not exceed the reciprocal of the t_{EP} to have a positive net energy. For example, if the energy payback time is five years and the capacity growth of a particular energy technology ensemble is 20 percent, no net energy is produced and no greenhouse gas emissions are offset. This is because the same analysis is again true for greenhouse gas emissions, as the embodied greenhouse gas emissions emitted to provide for the technology divided by the emissions offset every year must be equal to 1 over the growth rate of the technology simply to break even. This cannibalism of energy for both new energy-conserving technologies and energy-producing technologies has a profound effect on their ability to assist in the mitigation of greenhouse emissions. Recent research has shown that the limitations of even modest energy payback times demand extremely rapid transitions away from fossil fuels to non-greenhouse-gas-emitting renewable energy sources to stabilize the global climate. This is because the fewer fossil fuels in the energy mix, the faster the greenhouse gas emission payback times for a given energy payback time.

See Also: Embodied Energy; Energy Audit; Energy Policy; Life Cycle Analysis.

Further Readings

Fthenakis, Vasilis and Erik Alsema. "Photovoltaics Energy Payback Times, Greenhouse Gas Emissions and External Costs: 2004–Early 2005 Status." *Progress in Photovoltaics: Research and Applications.* v.14/ 3 (2006).
Gagnon, Luc. "Civilisation and Energy Payback." *Energy Policy.* v.36/9 (2008).
Pearce, J. M. "Limitations of Greenhouse Gas Mitigation Technologies Set by Rapid Growth and Energy Cannibalism." *Klima 2008/Climate 2008.* http://www.climate2008 .net/?a1=pap&cat=1&e=61 (Accessed November 9, 2009).
Pearce, Joshua and Andrew Lau. "Net Energy Analysis for Sustainable Energy Production From Silicon Based Solar Cells." *Proceedings of American Society of Mechanical Engineers Solar 2002: Sunrise on the Reliable Energy Economy*, R. Campbell-Howe, ed. (2002).

Pearce, Joshua M. "Thermodynamic Limitations to Nuclear Energy Deployment as a Greenhouse Gas Mitigation Technology." *International Journal of Nuclear Governance, Economy and Ecology.* v.2/1 (2008).

White, Scott W. and Gerald L. Kulcinski. "Birth to Death Analysis of the Energy Payback Ratio and CO_2 Gas Emission Rates From Coal, Fission, Wind, and DT-fusion Electrical Power Plants." *Fusion Engineering and Design.* v.48/3-4 (2000).

Joshua M. Pearce
Queen's University

ENERGY POLICY

A policy is a course of action taken to reach a desired goal. Polices can be the exercise of simple principles or complex programs of goals. The goals are determined to be desirable on the basis of values that are often held subjectively. Policies may be articulated clearly or be concealed.

Policies may be private—as in the courses of actions followed by private individuals, businesses, or corporations—or they may be public policies followed by local, regional, national, or international governments or organizations.

The principles that guide the course(s) of action taken to reach policy goal(s) may be ethical or unethical, although ethics often have to be defined before there can be agreement on whether a policy is ethical or not. However, in many cases, such as those of criminal conduct, the policy is to make money from enterprises such as selling addictive drugs. Most people would find the business policies (including murder) of such black market enterprises to be unethical regardless of their profitability.

Policies can be wise or unwise, practical or impractical, helpful to some but harmful to others. Government policies are adopted to serve some goal that will usually create winners and losers, regardless of the wisdom of the policy enacted.

The subject matter of policies can be virtually anything. Energy policy is about a basic part of the universe. It is composed of matter and energy that are, according to Albert Einstein's famous formula ($E = mc^2$), convertible. However, the amount of energy and matter are fixed and, according to the laws of conservation of matter and energy, can neither be created nor destroyed. However, energy can either be in a useful form or dissipated and not useful for human purposes.

Useful energy is potential energy. When put into action, it is kinetic energy. When used, it creates heat. When the heat is dissipated, the energy is no longer available for use. For example, a log next to a burning fire in a fireplace on a cold winter's day is potential energy. When used it is kinetic, and when it is fully burned and turned to ash it is dissipated heat.

Energy policies that are directed at increasing national supplies of energy sources usually address fuels, which are stored energy. Gasoline in the automobile tank is stored energy ready for use. So is water impounded behind a dam—a "fuel" that is available as potential energy to become kinetic energy as it drops down a 1,000-foot shaft to drive the blades of a turbine, which will create electricity.

Energy policy has grown in importance since the Industrial Revolution. The advent of the Industrial Revolution was accompanied by advances in steam engines that could drain water from deep shaft and deep pit mines. The technological advance of a steam engine that could be fired with coal became available about the same time that the forests of England and Europe were being exhausted as sources of charcoal for fuel for heating and other uses.

During the 19th century, the use of coal expanded globally as did the use of firewood in those places where it was still available in abundance. However, by the beginning of the 20th century, other kinds of energy became available. These included petroleum, natural gas, and hydroelectric power. After 1945, engineers developed ways to harness nuclear power for electrical production.

Energy policies involving these types of energy in many ways followed the politics in place at the time that type of energy came into use. Coal was in great use in the days of laissez-faire capitalism. Petroleum also was a fuel without much in the way of environment or market regulation until the abolition of the Standard Oil Company monopoly. Gas, with its need for pipelines and its explosive potential, required more safety regulation than either coal or oil. Nuclear power, born in secret in wartime, has been the most closely regulated of all energy sources for reasons of national security.

Traditional Policies

Energy policies have focused on a number of aspects of energy production, distribution, and consumption. In the case of coal production, numerous policies were instituted to encourage mining for the money it would bring to communities and to governments. However, coal mining and distribution eventually involved struggles with unions as miners and railroad workers fought for better wages and working conditions. Among their issues were concerns about wages, but safety to prevent mine disasters was also of great concern. As a consequence, mining policies were aimed at mine safety, which increased the costs of production in the short run.

Another important policy concern that has also increased over the decades has been the environmental costs of coal mines. The wastes from mines have often included very low-grade coal that is not commercially marketable. Wastes have been of policy concerns whether from the older eastern anthracite and bituminous mines or from the newer strip mines that include lignite.

Energy policies regarding petroleum have sought to regulate its marketing and environmental impact, as well as manage strategic concerns about supplies. Worries about sufficient supplies of fuel oil for heating and aviation gasoline have also evoked policy concerns at times.

Natural gas is closer to a natural monopoly than are coal and oil. Its transportation—originally in pipelines and later via overseas liquefied natural gas tanker ships—has been the subject of numerous laws and regulations regarding safety and monopoly regulation. State and local regulations of domestic gas service for heating and cooking are policy enactments that regulate pricing and safety.

Hydroelectric power generation has been regulated as much as a matter of dam safety as a fuel source. Until the terrorist attack on the United States on September 11, 2001, security at dams was mainly an issue involving the structural integrity of dams. After 9/11, policing security issues moved to the forefront of policy concerns. Geothermal power regulation was similar to electrical power regulation.

Energy is the source of power that enables anything in the world to take action. Basically, energy is derived from two sources: the heat in the Earth's core and sunlight.

Earthquakes, volcanoes, and continental movements resulting from plate tectonics are all products of energy released from the heat in the Earth's core. In contrast, all wind and wave action is derived from solar heating of the Earth's surface. The remainder of the energy supplies on Earth are derived from the sun via the biological process of photosynthesis—the basis of the biosphere and the food chain.

The traditional energy supplies have been fossil fuels: wood, peat, coal, oil, and gas. Nuclear power is a nonrenewable mineral source derived from uranium. Hydroelectric power depends on the flow of water over turbines. However, there are a limited number of rivers in the world, and damming them has ecological consequences. In addition, the life span of dams is eventually shortened by silt building up behind the dam, rendering it inoperable. Although this build-up is a long-term process, it is the inevitable end for the dam, after which either the silt has to be removed or the stream or river has to be rerouted. Fossil fuels, although currently abundant, are also nonrenewable and are the basis, especially in the case of oil, for a large number of the products consumed in modern industrial societies.

Energy policies adopted by fossil fuel suppliers, especially those of the oil-producing countries, have been nationalistic, particularly in the case of the Organization of Petroleum Exporting Countries, which is a cartel (monopoly) with 12 members that regulates their production of oil to stabilize world supplies and gain as much profit as they can. Their policies have often been more politically motivated than economically driven. At times, their policies have been focused on oil revenues that can be used for massive armaments purchases. Other smaller oil producers have taken a long-term view and have sought to build infrastructure and to make investments for the day when their oil resources are depleted.

By the end of 2000, several concerns were affecting the energy policies of the previous century. These were the effect of fossil fuels on the environment, the sustainability of petroleum supplies, the political and economic implications affecting consumer nations using the bulk of the world's oil, and the desirability of increasing global energy supplies that are renewable.

Governments have been adopting policies that focus on increasing energy consumption efficiency and on developing renewable sources of energy. Not only governments are concerned with energy costs—so are consumers, whether large or small, individual or corporate. As a consequence, more explicit and refined energy policies are going beyond "turning off the light."

New Policies

New government energy policy is focused on energy development, distribution, and consumption (including energy conservation) on a national and international basis. Policies are being developed to tax energy consumption, and thereby to force energy conservation through reduced usage. Other policy techniques include investing in alternative energy sources that range from practical to very experimental. In the case of wind power, the development of more efficient wind turbines is promoted with policy incentives such as subsidies. In the case of experimental sources such as wave energy, the use of research grants to cover the cost of experimentation is a policy action that is being used. The cost for some experimentation is such that virtually every idea is subsidized to enable any possibility to be examined and reviewed.

In the United States, nuclear fission as an electrical source has been blocked by environmental groups since the 1970s. However, in France and other countries, much of their electrical grid is based on fission reactors. There are four major problems with nuclear reactors. First, mining and processing uranium is dangerous work because exposure to radiation is possible. The risk in proper handling of uranium mine tailings is similar to the danger that occurs when the reactor's core rods are depleted. Also, where is the radioactive waste to be stored? Political opposition has to be subjected to rational policies, whether agreeable to some opponents or not. The other two problems are security from terrorism and security from nuclear plant accidents, such as the meltdown at the Chernobyl nuclear power plant in the Ukraine in 1986. Exposure of the nuclear core to the outside world creates extremely toxic pollution that has widespread health and environmental consequences for generations.

An alternative to nuclear fission is nuclear fusion, which is the same process by which the sun creates light. It is the fusion of hydrogen atoms into helium atoms with the emission of a quantity of energy that is received on Earth as sunlight. Research into this form of energy has and continues to be conducted. As nuclear fusion takes place, the temperature that has to be contained approaches the temperatures of the sun's surface—about 10,000 degrees Fahrenheit. Considering that steel is molten white hot at a mere 2,000 degrees Fahrenheit, the engineering challenges faced by trying to imitate the sun appear staggering.

Other alternative fuel sources include biofuels. These are fuels that are resource renewable and that can be used to manufacture methane or ethanol. Both of these fuels can be made from biomasses; that is, green plant material. Some parts of the United States, such as its southern states, are like southeastern China and other areas of the world in which masses of vegetation grow in abundance. In the hot, rainy season, millions of tons of vegetation could easily be harvested and then converted into methane for use as a renewable fuel for electric power plants or for home heating and cooking.

Solar power is also hailed as an important renewable energy source. It has the potential for supplying much of the electrical grid of the world. It is in abundance in tropical and desert areas of the world. Solar power could also be easily produced, given sufficiently efficient solar panels on the roofs of houses and other buildings.

Energy policy will have a set of goals to be achieved. Among these are planning, energy generation, transmission, usage, and national security. Generation has two major areas of concern in terms of energy usage: the electrical grid and the transportation grid. In terms of national security, energy policy includes planning for long-term generation supplies and for defense against sabotage or attacks. Implementation of plans—including testing—is also important.

Supplies are usually delivered by commercial concerns. Laws that regulate their activities are important to market stability, consumer protection, and energy security. In support of energy security, some nations have policy goals that include engaging in international agreements on energy.

The energy policy of the United States reflects its organization as a federal republic. Policies exist at the federal, state, and local levels. The policies increasingly address all issues concerned with energy production, distribution, and consumption. The policy goals are set at the federal level by federal legislation, by administrative rule-making to implement the legislation, by pertinent executive orders of the president, and by decisions of the federal courts as they adjudicate cases.

In the United States, the states may—and in many cases do—have stricter standards in their energy policies than does the federal government. Automobile fuel efficiency standards are set at the federal level, but states use gasoline taxes to encourage the purchase of more fuel-efficient cars.

Local standards can also be set; in many cases, programs for aiding the poor and elderly who live on fixed incomes with the cost of home insulation or other conservation measures are part of local energy programs. Home heating costs are also regulated at the state and local levels, and in some places, subsidies for the poor are available.

The administration of President Barack Obama entered office in 2009 and began to aggressively press for major changes in energy policy. Changes included using taxes—a system of "cap and trade"—to reduce greenhouse gas emission. This would affect coal-fired steam generation plants—in some cases, eventually driving them out of business. The policy goal would drive up the cost of electricity and other energy supplies to force the country away from using petroleum and coal for the major portion of its energy supplies. The administration was, with the aid of a Democrat-controlled Congress, spending large sums of money on alternative fuel research and development. Obama's stated policy goal was to create a new era of energy exploration.

See Also: California; Coal, Clean Technology; Department of Energy, U.S.; Nonrenewable Energy Resources; Nuclear Power; Organization of Petroleum Exporting Countries; Public Utilities; Renewable Energies; Smart Grid.

Further Readings

Allison, Lincoln. *Ecology and Utility: The Philosophical Dilemmas of Planetary Management.* New York: Continuum International Publishing, 1992.

Bunn, Derek W., ed. *Systems Modeling for Energy Policy.* New York: John Wiley & Sons, 1997.

Cook, Earl. *Man, Energy, Society.* San Francisco: W. H. Freeman, 1976.

Doty, Steve. *Commercial Energy Auditing Reference Handbook.* London: Taylor & Francis, 2008.

Dukert, Joseph M. *Energy.* Westport, CT: Greenwood, 2008.

Hinrichs, Roger A. *Energy: Its Use and the Environment.* Florence, KY: Thompson Learning, 2002.

Horlock, J. H. *Energy: Resources, Utilisation, and Policies.* Malabar, FL: Krieger, 2009.

Jacobs, Noah B. *Energy Policy: Economic Effects, Security Aspects and Environmental Issues.* Hauppauge, NY: Nova Science, 2009.

Kalicki, Jan H. and David L. Goldwyn, eds. *Energy and Security: Toward a New Foreign Policy Strategy.* Baltimore, MD: Johns Hopkins University Press, 2005.

Orr, Lloyd, ed. *Energy: Science, Policy, and the Pursuit of Sustainability.* Washington, D.C.: Island Press, 2002.

Peters, B. Guy. *American Public Policy: Promise and Performance.* Washington, D.C.: Congressional Quarterly, 2006.

Ruschmann, Paul. *Energy Policy.* New York: Facts on File, 2009.

Andrew J. Waskey
Dalton State College

ENERGY STORAGE

Energy storage refers to any means by which excess forms of energy are accumulated and retained until further required. An imbalance between energy supply and demand necessitates energy storage, occurring commonly in nature and by human design. In nature, the sun, fossil fuels, and food are in fact sources of stored energy. In human design, a battery

In early 2009, this engineer at the U.S. Department of Energy's Sandia National Laboratories was working on a hydrogen energy storage system for General Motors that uses sodium alanate to store hydrogen.

Source: U.S. Department of Energy, Sandia National Laboratories

is an example of a familiar form of storage for electricity. Energy in the form of firewood is a source of heat and light that may be used intermittently or fed in units to maintain a continuous supply. Energy must be collected before it is transformed into a usable form. Generators collect and transform one form of energy into another, such as energy from flowing water that is changed to electricity. Energy harvesters such as nuclear reactors provide a means of accessing and converting the energy from atoms into usable energy transmitted to users on the electricity grid. In this sense, a generating station harvests energy that may then be used immediately or stored for subsequent use. The parameters for energy storage are determined by supply, demand, and storage technologies. In a complex, dynamic human society, with increasingly multiple and diverse supply and demand points, energy storage in bulk and in retail is proportionate to demand. As a consequence, emergent energy storage technology and its management are appearing in a comparable variety of forms.

Energy storage requires the use of harvesters, which are systems used to collect, store, and manage energy until it is required for use. Several different categories of energy storage exist depending on the primary energy source and materials employed. Electrical storage systems include unique internal resistance and charge and discharge features. In some nanotechnology systems, efficiency is achieved by coupling supercapacitors and batteries to access and provide variable amounts of energy. Where batteries are of the thin-film variety, equivalent series resistance or internal resistance in the storage unit can be reduced to low or none, which results in proportionately higher power efficiency. More complex systems can include piezos (electricity generation from physical deformation using specialized materials, such as crystals) and magnetorestrictive layers (a material that stretches and distorts in response to a magnetic field resulting in power generation). Electrochemical energy storage systems include photovoltaic cells or solar batteries and can be coupled with a thin-film battery to make a self-charging energy storage unit with low or intermittent energy discharge capability.

Thermal energy storage systems can be used for heating and cooling either in closed or open systems. The storage medium is usually water that is transported through systems in which heat is redistributed. The medium transporting the heat is pumped through the system to locations where heating-cooling exchangers are placed. Where the medium remains in a closed loop, it is a closed storage system. In a closed system the medium may or may not remain in the original single phase through the heating and cooling cycle. Conversely, an open system harvests and disposes the energy from outside the system and can be seen where the medium is accessed from and returned to aquifers or underground caverns. An example of an open system is the use of water from the Great Lakes to cool commercial buildings in cities on the banks of the lakes. Cold water is drawn from the lake, cools the buildings, and is piped back to the lake. In a closed system, the segment of the system in the lake would be continuous and energy would be accessed through a heat exchanger in the lake. In both the closed and open systems cases for this example the lake provides the energy storage. Geothermal energy can also operate as a thermal energy storage system.

Mechanical energy storage results from an object acquiring energy due its position (potential energy) or its motion (kinetic energy). Pumped hydroelectric storage (PHS) plants are one of the largest mechanical energy storage systems currently available, capable of producing 100–2000 MW of power with a typical energy rating of several hours. PHS systems pump water from lower elevation reservoirs to higher elevation reservoirs (increasing the gravitational potential energy of the water) during non-peak hours (usually nighttime). During periods of higher electricity demand, the water from the higher reservoir is released (increasing the kinetic energy of the water) through turbines to the lower level reservoir and electricity is generated. Other examples of mechanical energy storage systems include compressed air energy storage, spinning flywheels, and hydraulic accumulators.

New technologies contribute to the emergence of energy storage systems with improved economic and technological efficiencies and capabilities. For instance, internal resistance in electricity storage systems can be reduced without loss of physical strength with the use of a type of carbon molecule known as a Buckminsterfullerene. Using these molecules, a material known as Buckypaper was invented in 2005. Buckypaper is twice as hard as diamonds and one of the most conductive or low-resistance materials for heat and electricity. Such materials can result in technological improvements that reduce inefficiencies in systems, resulting in improved supply of energy and reduced energy requirements or demands. Reduced resistance and improved conductivity can also have an effect on improving the quality of the energy being delivered on demand.

In the move to renewable energy, the availability of energy storage appears to play a critical role in advancing the stability of energy supplies. Developers and researchers are considering energy storage in the form of compressed air and compressed gas, which are conceptually not unlike the hydroelectric systems that generate electricity from the flow of water to power turbines. With further development, sequestered carbon can conceivably be configured to form energy storage units.

Stability of energy supplies is secured using conventional technologies, which include back-up diesel generators, uninterruptible power supplies, and valve-regulated and wet cell lead acid batteries, as well as mechanical devices such as rotary flywheels. Natural gas– and propane-fueled generators are an alternative to diesel generators. These fuel supplies as well as steam can be used to drive turbine generators, often in a cogeneration configuration. Start-up times for such generators render them generally unsuited to emergency requirements where instant demand is required. Examples of these situations are in life-support systems and in computer power systems.

New chemical batteries such as sodium-sulfur units and vanadium batteries with proton exchange membranes are also being developed, as are the use of new materials like one-atom-thick graphene sheets to produce ultracapacitors. Capacitors and ultracapacitors act like batteries, in that they can be charged and discharged. Typically, they have lower storage capacity than chemical batteries but charge and discharge faster and last longer, as they do not use chemicals. Capital intensive options include the use of superconducting magnetic energy storage systems that can be coupled to wind or solar energy farms.

Inefficiencies in harvest and storage systems arise out of energy dissipation in transmission through radiant energy that may be in the form of heat, sound, or light. Energy leakage in storage units also results in the loss of efficiencies. Cogeneration can be used to improve efficiencies in conventional energy harvesting and storage. Cogeneration, or "cogen," is also known more descriptively as combined cycle, or combined cooling, heating, and power. These designs harvest radiant energy in conventional operations and make it available to the facility either for its own use or for distribution externally.

The variations between supply and demand can be reduced if the suppliers and the users can cooperate with each other. For instance, industrial users can be offered special energy pricing if they reduce demand on the grid at peak or specified times. This form of cooperation relies on distributed storage and local generation capabilities in which the storage and generation capacity available at the user's facility is integrated into the grid. In doing so, the supplier reduces the costs of installing storage and the user deploys locally installed generation and storage capacity to reduce energy demand on the grid and costs at the facility.

The evolution of the energy grid into a localized or distributed system provides an opportunity for innovative solutions. Energy storage systems are essential in designing redundancy into operations. Flexibility to operate on multiple energy sources is featured in consumer products such as hybrid and ecofriendly automotive vehicles. Including vehicle batteries in the distributed storage grid is also currently feasible without extensive capital investment: Plugging a vehicle battery into an outlet can immediately bring its energy storage online to the grid.

A promising option currently in development relies on hydrogen fuel cells to store energy harvested from solar radiation. In one model, photovoltaic cells convert sunlight into electricity that is used to electrolyze water into hydrogen and oxygen, which can then be stored in cells and converted back to electrical energy when required. In another alternative, mimicking chlorophyll in plant leaves, the sunlight can be captured by dyes and directly used in the electrolytic reaction to produce hydrogen for storage in cells.

Energy storage capabilities need to be managed much like the levels of fat in a human body. At the risk of oversimplifying the analogy, in a human being, fat cells harvest and store energy when there is a surplus and release it when required. A certain amount of fat is necessary for energy storage critical to survival and optimal operation of the body. Where the energy storage system is inadequate or disproportionately large or inappropriately distributed or dysfunctional, the energy storage capabilities can be counterproductive, causing damage and reducing the longevity of the systems it serves.

See Also: Alternative Energy; Alternative Fuels; Batteries; Biomass Energy; Carbon Sequestration; Combined Heat and Power; Electric Vehicle; Electricity; Exergy; Flex Fuel Vehicles; Fuel Cells; Fusion; Green Power; Grid-Connected System; Hydroelectric Power; Hydrogen; Landfill Methane; Nuclear Power; On-Site Renewable Energy Generation; Photovoltaics (PV); Plug-In Hybrid; Power and Power Plants; Renewable Energies; Smart Grid.

Further Readings

Briat, O., J. M. Vinassa, W. Lajnef, S. Azzopardi, and E. Woirgard. "Principle, Design and Experimental Validation of Flywheel-battery Hybrid Source for Heavy-duty Electric Vehicles." *IET Electric Power Applications.* v.1/5 (2007).

Buckypaper. www.fsu.edu/news/2008/07/21/buckypaper.honored/ (Accessed March 2009).

Bullis, Kevin. "Sun + Water = Fuel." *Technology Review* (November /December 2008). www.technologyreview.com/energy/21536 (Accessed March 2009).

Electronic Design. Available online at: http://electronicdesign.com/Articles/Index .cfm?ArticleID=20222 (Accessed March 2009).

Garimella, Srinivas. "Sustainable Energy: Of Course; But How?" *HVAC and R Research*. v.14/3 (2008).

Hu, Yuantai, Ting Hu, and Quing Jiang. "On the Interaction Between the Harvesting Structure and the Storage Circuit of a Piezoelectric Energy Harvester." *International Journal of Applied Electromagnetics and Mechanics*. v.27 (2008).

Murray, Charles J. "Engineers Search for Electrical Storage Solution." *Design News* 24 October 2008). www.designnews.com/article/48883-Engineers_Search_for_Electrical_ Storage_Solution.php (Accessed March 2009).

Lester de Souza
Independent Scholar

ENTROPY

Derived from the Greek word for "transformation," entropy is the part of energy not converted to work but is instead dissipated to its surroundings. The concept of entropy stems from the second law of thermodynamics that states isolated systems tend toward disorder and entropy is a measure of that disorder. However, in the applied science and agricultural systems, entropy is more commonly defined as a measure of the degree of "spreading out" rather than of disorder. Measuring the degree of entropy overproduction has been proposed as a way to quantify the "sustainability" of agroecosystems. According to the laws of thermodynamics, all ecosystems, including agroecosystems, are open thermodynamic systems. Although climax or natural ecosystems maintain a state of equilibrium, in which the amount of entropy produced by the system equals the amount exported, many agroecosystems accumulate excess entropy.

A natural ecosystem in its climax state has achieved thermodynamic equilibrium with the environment, meaning there has been no net change in entropy between the system and the environment. Beginning with an initial thermodynamic equilibrium, work done on an ecosystem by the environment (e.g., solar energy) decreases entropy by creating new biomass (e.g., through photosynthesis). This decrease in entropy is followed by a spontaneous and reversible process in which the system returns to an initial state through respiration and decomposition—processes that increase the system's entropy. Both increasing and decreasing entropy occur simultaneously: Entropy accumulates and leaves a system at the same time. The cycle of increasing and decreasing entropy can be repeated infinitely, maintaining a thermodynamic equilibrium indefinitely.

Unlike human-dominated systems, natural ecosystems do not accumulate entropy over time, as excess entropy is constantly exported to the surrounding environment. For the (solar) energy "pump" to work, only heat can be transferred from the ecosystem to the environment. The entropy pump is quantitatively defined as the following: entropy export = $\Delta S_i = \Delta G/T$ (ΔG = gross primary production, T = time; assumption made that gross primary production is transformed into heat).

When the human-dominated ecosystem, or technosphere, invests energy to compete for natural resources with the natural ecosystem, disequilibrium can occur. Entropy excesses from the technosphere can be compensated through three processes: (1) biosphere degradation, (2) climatic changes, or (3) artificial spatial redistribution. This latter compensation is sometimes considered a "sustainable development" strategy locally but is made possible only through "entropy dumps" elsewhere. An example of this is the removal of waste from cities to remote rural areas or even to overseas destinations.

In agroecosystems, the natural entropy cycle is disrupted through entropy overproduction caused by artificial energy inputs. Artificial energy inputs used to increase crop production (at least in the short term) include fertilizers, herbicides, and combustibles used to drive complex machinery. To the extent increased entropy associated with artificially intensified crop production fails to be exported to the environment, ecosystem degradation (e.g., soil erosion) occurs as the system strives to retain equilibrium. Entropy can therefore be usefully employed as an ultimate quantitative measure of agroecosystem sustainability. The entropy balance equation for agroecosytems is described as follows:

$\sigma = 1/T \left[y(1/n + 1/s) - 1 \right] - P_0$

σ = entropy

T = time

$s = k(1 - r)$

r = mean respiration coefficient

P_0 = gross primary production of ecosystem in "climax state" $\rightarrow$ "natural ecosystem"

y = crop yield = $k(1 - r) \times P_1$

k = fraction of net production that is being extracted from system

P_1 = gross agroecosystem production

According to David Pimental, although agroecosystems have varied energy efficiency coefficients, they all share a threshold past which sustainability is lost. Therefore, if $\sigma > 0$, the system "accumulates" entropy and the ecosystem becomes degraded, leading to system death and the formation of a new equilibrium. This process can be reversed if an energy source can export excess entropy out of the system. In sum, excess entropy can either be (1) accumulated in a system, causing degradation and eventually death, or (2) removed from the system and equilibrium reestablished. From the vantage of coupled human–environment systems, environmental degradation can be conceived of as a type of entropy tax levied on modern industrial agriculture. Whether this tax is paid by farmers, taxpayers, consumers, or all of the above is predicated on political–economic subsidies, laws, and production and distribution incentives and norms.

Entropy has been used as a sustainability measure for agroecosystems in recent years. Monocultures are efficient in the short term but lack resilience and remove mass from the ecosystem. For example, in northern Germany, Wolf Steinborn and Yuri Svirezhev demonstrate how entropy changes with varying land management strategies: Land uses that reduced artificial energy inputs moved the system closer to sustainability. However, the system failed to achieve suitability, as it still created more energy than could be exported.

The unsustainability of the maize agroecosystem, recently in high demand as a result of ethanol incentives, has been well documented by several authors. Yuri Svirezhev and Anastasia Svirejeva-Hopkins, for example, predicted that Hungarian maize agriculture will lead to "agricultural disaster within 30–40 years" as a result of excess entropy, taking the form of physical and chemical soil degradation and toxic effluent runoff. Maize agriculture in the United States may be similarly threatened by entropic outcomes. Erosion from U.S. maize agriculture reaches an average of 23–45 tons per hectare per year, 5–10 times greater than the erosion produced from conventional agriculture, and up to 30 times the erosion rate of no-till agriculture. In addition to causing erosion, agroecosystem entropy can also manifest itself in soil acidification, build up of metal compounds and toxic residues, and air pollution (ammonia, nitric oxide, carbon dioxide, monoxide, methane). Conversely, organic maize agriculture may be considered sustainable if it generates only as much entropy as the displaced native prairie ecosystem. This would require that the sources of generated entropy are derived not from chemical and fossil fuels (that the excess can be radiated into space) but only from solar energy and recycled local biomass.

Although each crop yields different values of overproduction of entropy, much of the difference in entropy among crops is the result of high artificial energy inputs more than inherent characteristics of the crop itself. What remains clear is that a decrease in artificial energy inputs would improve agroecosystem sustainability. It is also clear that current farming techniques remain far from achieving a sustainable thermodynamic state. Minimum entropy production will remain the ultimate sustainability criterion. On the basis of this definition, farming systems that approach steady states will be increasingly necessary in a world of continued population growth and even more rapid growth in energy and food energy consumption.

See Also: Ethanol, Corn; Sustainability.

Further Readings

Addiscott, T. M. "Entropy and Sustainability." *European Journal of Soil Science.* v.46/2 (2005).

Eulenstein, F., W. Haberstock, W. Steinborn, Y. Svirezhev, J. Olejnik, S. L. Schlindwein, and V. Pomaz. "Perspectives from Energetic-Thermodynamic Analysis of Land Use Systems." *Archives of Agronomy and Soil Sciences.* v. 49 (2003).

Georgescu-Roegen, Nicholas. *The Entropy Law and the Economic Process* Boston, MA: Harvard University Press, 1971.

Leff, Harvey S. "Entropy, Its Language and Interpretation." *Foundations of Physics.* v. 37 (2007).

Patzek, Tad W. "Thermodynamics of Agricultural Sustainability: The Case of U.S. Maize Agriculture." *Critical Reviews in Plant Sciences.* v. 27 (2008).

Pimental, David, ed. *Handbook of Energy Utilization in Agriculture.* Boca Raton, FL: CRC Press, 1980.

Steinborn, Wolf and Yuri Svirezhev. "Entropy as an Indicator of Sustainability in Agro-ecosystems: North Germany Case Study." *Ecological Modeling.* v.133 (2000).

Svirezhev, Yuri M. and Anastasia Svirejeva-Hopkins. "Sustainable Biosphere: Critical Overview of Basic Concept of Sustainability." *Ecological Modeling*. v.106 (1998).

David L. Carr
University of California, Santa Barbara

Leah Bremer
San Diego State University

ENVIRONMENTALLY PREFERABLE PURCHASING

Environmentally preferable purchasing, or EPP, is the practice of specifying products that have a lesser or reduced effect on the environment or human health when compared with conventional products used for the same purpose. These may include those products that are made with reduced packaging; are reusable; generate few emissions in their production; are energy efficient; or contain recycled, rebuilt, or remanufactured materials. EPP guidelines take into account not only the raw materials that make up the product but also the environmental impacts associated with the manufacturing, packaging, distribution, use, reuse, operation, maintenance, and disposal of the product. Because of the environmental and economic benefits of EPP, it is becoming more common for businesses and government agencies to develop policies for EPP.

Although many organizations and agencies previously considered only product needs and price, many are now considering the environmental impacts of the products they purchase by adopting EPP guidelines. The products covered under EPP encompass those used in all operations of an organization or agency, including products used in manufacturing their products and services, in daily operations (e.g., paper, print cartridges), in building construction and operation (building materials, lighting), in landscaping, in cleaning, and in vehicles. Recognized benefits of using EPP include:

- Improved ability for an organization or agency to meet environmental goals
- Improved worker safety and health through the elimination or reduction of toxics that create hazards to workers and the community
- Reduced worker- or consumer-related liabilities
- Reduced health and disposal costs
- Increased availability of environmentally preferable products in the marketplace and encouragement for other purchasers in the community to adopt similar guidelines
- Reduced environmental and human health impacts throughout the product supply chain (raw material extraction, manufacturing, distribution, use, and disposal)
- Support for strong recycling markets (or "closing the loop" for recycled products)
- Rewards for manufacturers and vendors that reduce environmental impacts in their production of goods
- Minimized impacts such as pollution of land and water and reduced use of natural resources and energy

To develop guidelines and implement EPP, organizations often develop internal guidelines or adopt existing purchasing guidelines that consider the environmental benefits or consequences before the bidding or purchasing of products. Typically, existing vendors are consulted (either in person or through surveys) to identify environmentally preferable products that they currently offer, and new vendors are sought out when existing ones cannot meet EPP needs. Reporting on purchasing and monitoring of progress is often employed to assist in implementation of EPP. Designing an environmentally preferable purchasing guideline and strategy may have up-front time and costs associated with planning and adjusting to new products; however, these costs can often be recovered by reducing the need for product replacement (as many environmentally preferable products are more durable than their counterparts) and by lowering disposal fees.

Products that would qualify under EPP may include some or all of the following elements:

- Raw materials sourced from widely available recycled or remanufactured elements (postconsumer or postindustrial wastes) or rapidly renewable and organically raised plant-based raw materials (e.g., agricultural residues, bamboo, sustainably harvested wood, etc.)
- Manufactured using 100 percent renewable energy and/or grown organically
- Little to no waste generation during production
- No products containing materials that are highly toxic, are carcinogenic, are flammable, cause skin irritation, cause respiratory problems, or cause allergic reactions in production or generated during product use
- No products containing chemicals that might end up in local wastewater systems or contribute to ozone depletion
- No packaging/use of reusable, recyclable, or compostable packaging when available
- Quality and durability appropriate for intended use
- Designed for reuse and/or easy recycling or composting
- Minimal transportation energy effects

When products that are necessary for continued operation of a business cannot be found to meet the above guidelines, preference can be given to products with recycled and recyclable content, those that meet Energy Star requirements, and products that are third-party certified as "eco-friendly" (e.g., those that bear the Green Seal label).

The U.S. Environmental Protection Agency (EPA) developed an EPP Program in 1993 to conserve resources, prevent waste, promote recycling, and to develop guidelines for EPP (Executive Orders 12873, 13423, 13101). As the single largest consumer of goods and services within the United States, and one of the largest consumers in the world, the U.S. government spends an estimated $350 billion per year on goods and services. With this purchasing power and the adoption of EPP guidelines, the federal government has the potential to influence further production of products that have minimal environmental impact. According to the 1999 EPA guide on Preferable Purchasing, the EPA adopted the following five guiding principles: (1) environmental considerations should become part of normal purchasing practice, consistent with such traditional factors as product safety, price, performance, and availability; (2) consideration of environmental preferability should begin early in the acquisition process and be rooted in the ethic of pollution prevention, which strives to eliminate or reduce up-front, potential risks to human health and the environment; (3) a product or service's environmental preferability is a function of multiple attributes from a life cycle perspective; (4) determining environmental preferability might involve comparing environmental impacts such as the reversibility and geographic scale of the environmental impacts, the degree of difference among competing products or services, and the overriding importance of protecting human health; and (5) comprehensive, accurate, and meaningful information about the environmental performance of products or services is necessary to determine environmental preferability.

Requirements under the EPA's EPP guidelines include preferred purchasing for the following products: bio-based products, which, as defined by Executive Order 13101 Section 201, are products made from renewable products such as biological products or renewable domestic agricultural (plant, animal, and marine) or forestry materials; energy efficient products; recycled-content products; and water-efficient products.

Organizations and agencies have developed EPP databases to ease the selection and purchasing of products that have a decreased environmental impact. Examples of databases include those published by county solid waste departments, national government databases published by the EPA, for-profit product guides, and product databases developed by nonprofit organizations.

See Also: Best Management Practices; Carbon Footprint and Neutrality; Environmental Stewardship; Life Cycle Analysis; Recycling.

Further Readings

Science Applications International Corporation. "Environmentally Preferable Purchasing." Prepared for the New York City Department of Waste Prevention, Reuse and Recycling. April 2001.

U.S. Environmental Protection Agency. "Environmentally Preferable Purchasing." http://www.epa.gov/epp (Accessed November 2008).

U.S. Environmental Protection Agency. "Final Guidance on Environmentally Preferable Purchasing." 1999.

Stacy Vynne
University of Oregon

ENVIRONMENTAL MEASURES

The World Energy Council (WEC) has defined the three main objectives in the development of the energy sector as (1) affordability, (2) availability, and (3) acceptability. Affordability is needed to provide energy to the entire population of the world—poor and rich alike—by making it cheaper. This could be achieved by creating more nonconventional (renewable) sources of energy than presently available. Availability of energy can be achieved through the generation of more power through more sources including renewable energy sources (RES) to ensure long-term security of the energy supply. Acceptability means making energy development compatible with environmental protection and creating a perfect public attitude toward the energy sources. RES are unanimously accepted as clean sources.

The WEC has forecast, however, that fossil fuels will remain the largest source of energy during the first half of the 21st century, accounting for 85 percent of global demand. As such, carbon dioxide emissions will also double, to 46 billion tons a year in 2050. With such striking figures, many countries are taking note of the harmful effects of fossil fuels. As the WEC recognized, governments must remain open-minded toward "greener" fuel sources and technologies that may aid us in lessening global warming. Nuclear power for electricity, large-scale hydroelectric, affordable RES, and simultaneous carbon capture and storage are options to meet the soaring energy demand while working to conserve the environment.

We all know that energy production in the world is weighed in terms of profitability. No energy producers in the world, including governments, see environmental protection and preservation as the topmost priorities for the production and distribution setting. Although there is an "environmental protection" string attached to any new or old energy production source setup or functioning, by means of law or public protest/demonstration, respectively, this string is never followed strictly. In poor countries,

Environmental measures include constructing "green" buildings such as this Leadership in Energy and Environmental Design (LEED)-certified laboratory in Albuquerque, New Mexico, which has a built-in water recycling loop and whose landscaping can survive with little water.

Source: U.S. Department of Energy Sandia, National Laboratories

where livelihood is measured by food security only, governments turn a blind eye to the environmental damage that is caused by energy production; for example, polluting the air by setting up coal-fired thermal power plants or inundating vast tracts of natural resource–rich land by constructing huge hydropower plants. They do this because their main goal is to produce energy or supply irrigation water to produce food. Even in rich countries, governments do not forcefully apply environmental protection or carbon emission laws to safeguard industries' profitability; for example, allowing nuclear power plants to dump nuclear waste in unsafe ways, not forcing automobile companies to produce higher-mileage automobiles, or allowing coal-fired power plants to flourish, as coal is abundantly available. Therefore the time is now ripe to enforce these environmental protection measures forcefully and with sincerity.

Now is also the time to find and implement new measures so that environment protection would be considered a prime goal. Thus we would be able to avoid or at least reduce serious consequences of global warming and climate change. As poor and underdeveloped countries will contribute more toward environmental damage through their pursuit of high-energy production, these new environmental measures—in the form of enlightened legal policies—should be relayed to them.

According to a WEC estimation, "with the right policies and technologies in place, the rise in greenhouse gas emissions can be tempered in the short-term, their absolute level can be stabilized in the medium-term, and reduced in the longer-term." Following are a few policies and technologies that can be put in place as environmental measures so that the world can meet its soaring energy demand without hampering the environment.

- Strong renewable energy standards: This would allow boosting production of energy from renewable and somewhat carbon-neutral sources such as wind, biomass, geothermal, and solar energy through already available technologies. Establish a streamlined approval process for federal or private renewable energy projects so that renewable energy investors would not be discouraged from pursuing the unchartered and untested area of investment. Governments may create a pricing system for renewable energy so that rates can be locked to provide confidence to investors.

- Promote carbon-neutral technologies: Rich countries should promote research and development (R&D) to develop and advance carbon-neutral technologies to make energy production through RES cheaper and affordable. These technologies should be transferred to developing and poor countries with intellectual property rights attached so that they can replicate the process. Developing countries that solely depend on coal-fired thermal power–generation technology should be provided with carbon-neutral technologies for energy production; for example, nuclear power generation. Thus they can meet their soaring energy demand without harming the environment.
- Strong energy-efficiency incentives and standards: This would help achieve the easiest and most cost-effective reductions in greenhouse gas emissions and consequently preserve the environment. The American Council for an Energy Efficient Economy estimates that lighting standards alone could reduce global warming pollution by 100 million metric tons by 2030, and at the same time consumers and businesses would be able to save billions of dollars. Government should promote a culture of energy conservation by assisting homeowners, government, schools, and industrial employers in transitioning to lower energy use.
- Increased fuel economy standards: Corporate Average Fuel Economy regulations have not been significantly raised since the mid-1970s. A higher per gallon—say, 35 miles per gallon—automobile standard should be imposed so that automakers would pursue high-end technologies like hybrid, electric, or hydrogen cell (using compressed or liquid hydrogen) to produce automobiles with high mileage. This would not only reduce the consumption of fossil fuels but also enrich the customers and the car manufacturers. With government intervention, autos need to be developed that would use ethanol and methanol, liquidated coal and biomass, and dimethyl ether and fatty-acid methyl esters instead of conventional fossil fuel that produces greenhouse gases.
- Energy-efficient building standards: In developed countries like the United States, buildings waste a huge amount of energy by not being energy efficient. Government-fixed standards for environmental measures can conserve huge amounts of energy. They can be achieved through vacuum insulation, light-emitting diode installation, advanced energy management system inclusion, and energy production through rooftop solar panels, and so on, in the buildings.
- A strong/high carbon cap: A strong carbon cap on high-polluting industries like coal-fired thermal power plants would discourage production of a high rate of carbon emissions to the atmosphere. The industries would be encouraged to find new clean-coal technologies to reduce the carbon exhaust from coal burning. They also may find resources to sequester carbon underground or separate carbon from exhaust to use for other processes.
- R&D initiative to reduce cost of carbon capture and storage: With the goal of making almost all industries of the world zero-emission industries, government should provide funding and encourage investors to fund research to develop technologies for reduced-cost carbon capture and storage.
- R&D initiative for second-generation biofuel generation: Recently, it was found that biofuel from corn or other crops disrupted the food chain in the world market. The price rise in food products was one of the main outcomes of the biofuel initiative. Therefore, government should facilitate R&D in biofuel development from other sources like algae; human, animal, or bird excreta; grass; and others so that the food chain should not be disrupted.
- Transfer of energy storage and transmission technology to developing countries: Developing countries like India lose almost 30–40 percent of their energy as a result of faulty storage and transmission systems. Policies favoring the transfer of better technology on this front

should be developed so that transfer of advanced technologies could take place between developed and developing countries. Promising energy storage techniques that help balance the flow of electricity between high and low periods of demand include use of hydrogen compressed air, pumped water for generating hydroelectricity, use of next-generation batteries, flywheels, and ultra capacitors. These technologies need to be transferred to wanting countries. Saving energy through these useful technologies will reduce the environmental damages at a cheaper cost.

Along with these environmental measures for increases in energy efficiency, governments and the public alike should learn and follow other emissions-stabilizing strategies like using energy-efficient equipment and reducing waste not only in energy by consuming less but also in food, clothing, and shelter requirements. The possible energy production from newer sources like nuclear fusion technology (third-generation nuclear power generation), sewer discharge, and waves and ocean tides can meet the demand of energy with no harm to the environment. Micro-combined-heat-and-power plants, advanced heat pumps, and active and passive solar heating are the next-generation green energy technologies that would help in environmental preservation.

See Also: Alternative Energy; Alternative Fuels; Biomass Energy; CAFE Standards; Carbon Sequestration; Coal, Clean Technology; Energy Policy; Ethanol, Corn; Fossil Fuels; Geothermal Energy; Greenhouse Gases; Hydroelectric Power; Nuclear Power; Renewable Energies; Solar Energy; Tidal Power; Wind Power.

Further Readings

Calvert Group, Ltd. "Investors Seek Strong Environmental Measures in Energy Bill." http://www.csrwire.com/News/10159.html (Accessed February 2009).
"Ontario's Green Energy Act Promises Jobs, Infrastructural Change." http://www.environmentalleader.com/2009/02/24/ontarios-green-energy-act-promises-jobs-infrastructural-change/ (Accessed February 2009).
World Energy Council. *The Energy Industry Unveils its Blueprint for Tackling Climate Change: WEC Statement 2007.* London: WEC, 2007.

Sudhanshu Sekhar Panda
Gainesville State College

ENVIRONMENTAL STEWARDSHIP

The concept of environmental stewardship encompasses a diverse group of motivating factors and behavioral outcomes centered on the idea that human beings have a responsibility to live sustainably and to promote an ecologically healthy environment by actively managing their behavior through informed and conscientious use of the world's environmental resources. The concept of environmental stewardship charges humans with the responsibility to act as stewards of, or caretakers for, the Earth by minimizing the potentially negative environmental impacts of human behaviors and actively engaging in the protection of ecosystems. Environmental stewardship can guide decisions at a wide array of scales, from individual behaviors to global institutions and international policy. Environmental concerns often considered from this perspective include protection of rare and threatened species and habitats, ecosystem health, and the environmental consequences of human behaviors such as food production, transportation, and energy generation, distribution, and consumption.

After introducing the meaning of environmental stewardship and the historical foundation of the concept, different motivations for and interpretations of environmental stewardship are reviewed. Environmental stewardship is distinct from the popular environmental concepts of preservation, conservation, and sustainability, although it is similar in many ways to conservation and sustainability. Yet stewardship has unique contributions to and implications for consideration of the world's green energy choices. Because of its wide appeal to diverse audiences and their many different motivating factors, environmental stewardship is an important concept for promoting green energy options.

The word *stewardship* incorporates a personal responsibility for care and maintenance; in an environmental context, it means the human responsibility to consider and care for the Earth. This sense of responsibility for the health and sustainability of the Earth can be applied at many levels, from individuals and communities to companies and governmental and nongovernmental organizations. According to the U.S. Environmental Protection Agency, environmental stewardship is manifested by "a commitment to efficient use of natural resources, protection of ecosystems, and, where applicable, ensuring a baseline of compliance with environmental requirements."

Origins of Environmental Stewardship

The origins of environmental stewardship in the United States can be traced back to the analogous idea of the "land ethic" proposed by Aldo Leopold in his 1949 collection of essays, *A Sand County Almanac*. In Leopold's words, "The land ethic simply enlarges the boundaries of the community to include soils, waters, plants, and animals, or collectively: the land. . . . [A] land ethic changes the role of *Homo sapiens* from conqueror of the land-community to plain member and citizen of it. It implies respect for his fellow-members, and also respect for the community as such."

Informing Leopold's ideas of the land ethic and today's concept of environmental stewardship is the ecological "systems view" of nature. Indeed, an "ecosystem" comprises many interdependent components, both biotic and abiotic, on which the functioning of the system and the provisioning of ecosystem goods and services (the components of nature that humans use directly) depend. Leopold recognized that although some of these components are useful from an anthropogenic perspective (such as timber, oil, and coal), and some are not (such as leopards), the presence of all of these various components is essential to the functioning of the ecosystem as a whole. Environmental stewardship requires that humanity pay attention to these ecological connections to ensure that human behavior does not adversely affect the health of the ecosystem. Environmental stewardship is a growing component of the broader environmental conservation movement and empowers individuals to contribute to sustainability.

The active and mutually beneficial relationship between humans and the environment promoted by environmental conservation and stewardship differs from the preservationist environmental agenda. In the United States, the idea of preservation is embodied by traditional nature preserves focused on the persistence of single species and grew out of the broader romantic-transcendentalist cultural movement of the 19th century. For foundational readings on preservation, see the works of Ralph Waldo Emerson, Henry David Thoreau, and John Muir. This perspective placed humans outside "nature," meaning that use of and intervention in nature by humans was deemed unnatural and destructive.

In contrast, the conservation movement and the concept of environmental stewardship acknowledge that humans are one component of the ecosystem and promote a mutually beneficial relationship between humans and the environment in which humans actively manage their interactions with and effects on the environment to promote ecological sustainability. Environmental stewardship prioritizes the responsibility of humans to protect and care for the natural world by promoting and protecting prosperous ecosystems, environmental health, and sustainable human-nature relationships. Humans benefit from this relationship through the continued provision of natural resources and ecosystem services for human consumption and enjoyment. Yet relative to ideas such as conservation and sustainability, environmental stewardship can motivate behavior for a myriad of reasons, from religion to economics, and does not rely exclusively on an intrinsic valuation of nature or on the exclusive consideration of environmental protection.

There are many different motivations for, interpretations of, and practical applications characterized by environmental stewardship, from economic arguments related to resource sustainability to religious and moral arguments. Economic factors motivating environmental stewardship are often linked to the concern for continued profitability and to the expectations of consumers and investors that firms will make environmentally sound business decisions. For example, product stewardship aims to address the environmental concerns of both the firm and the consumer by minimizing the negative environmental impacts incurred not only during manufacturing but also during the full life cycle of a product. Embodiments of product stewardship include "design for environment" and the life cycle approach, both of which attempt to use natural, recycled, and recyclable products; to minimize production waste; and to consider the entire life cycle of a product from creation to obsolescence to minimize negative environmental impacts.

Motivations for environmental stewardship can also be driven by values founded on religious morals or spiritual beliefs. Spiritual motivations for environmental stewardship have been posited from numerous theological perspectives, including those of Christianity, Judaism, Buddhism, Confucianism, Hinduism, Jainism, and Islam. Common to all of these faiths is conviction in the responsibility of humanity to care for that which was divinely created and to serve as stewards of this creation. Moral arguments regarding environmental stewardship can also be found in secular philosophical traditions.

A variety of pragmatic, nonspiritual values, including claims to ownership, can also motivate environmental stewardship. According to the ownership argument, the planet belongs not to any individual but to all inhabitants across the globe, as well as to future generations. Therefore, each human being has a duty to live in an environmentally responsible way, so as to not violate the ownership rights of other human beings now or in the future.

Environmental Stewardship and Global Energy Issues

With regard to global energy issues, environmental stewardship can inform perspectives and decisions in several ways. First, responsibility to protect the Earth means resourceful and efficient energy use, pursuing energy conservation and efficiency whenever possible as a cost-free or inexpensive mechanism of environmental stewardship. From walking and biking, purchasing energy-efficient appliances and automobiles, and promoting efficiency standards for electric utilities, energy efficiency and conservation in homes and commercial buildings are ideal means of acting as an environmental steward. Energy conservation and efficiency can be pursued at many levels, from individual decisions to national and international energy policy, and an environmental stewardship approach would support energy efficiency and conservation at all scales.

Second, renewable energy options are a means of actively engaging in environmental stewardship. Exploiting nonrenewable resources such as fossil fuels is not an approach to energy use founded in environmental stewardship. Specifically, both the extraction and use of nonrenewable resources is energy intensive and environmentally detrimental across the life cycle stages of these resources. Immense amounts of energy and other resources, such as water, are used in the initial extraction of fossil fuels, the transportation of fossil fuels globally, the production and generation of electricity from such resources, and the end products made from such sources (such as vehicle fuels, plastics, and a broad host of synthetic materials). A variety of toxic waste products are created and ecological degradation occurs during each of these stages.

Stewardship is, at its core, about responsibility to care and protect for the Earth; renewable energy sources are an ideal means of pursuing environmental stewardship. Specifically, solar, wind, and biomass electricity generation offer solutions to energy demand with minimal environmental impacts. Distributive electricity generation is another possible means of acting as a steward of the Earth. The development of alternative fuel options such as biofuels and electric vehicles is being pursued as a means of creating choices that reflect environmental stewardship—or care for and responsibility toward the Earth.

Environmental stewardship offers a third perspective on energy issues, a unique contribution to debates on global energy choices. The life cycle approach to production, with its foundation in an environmental stewardship perspective, suggests that it is not enough to question energy sources and uses. When considering energy choices, it is important to also consider and compare how the extraction, production, and waste from different forms of energy affect the natural world. This could mean considering how photovoltaic cells are produced, the environmental consequences of biofuel production, the effects of disposing of waste from nuclear power plants, or the potential effects of an immediate transition to electric vehicles.

When exploring the world's energy choices, the framework offered by environmental stewardship offers a unique lens for considering more environmentally sustainable options for future energy development and growth. Environmental stewardship is a widely appealing concept because it can be justified employing moral, religious, rational-legal, or anthropocentric-utilitarian arguments. Regardless of the motivation, environmental stewardship calls for responsibility and care toward the Earth as well as management of our individual behaviors, national policies, and global energy decisions to act as stewards of the Earth.

See Also: Alternative Fuels; Appliance Standards; Bicycles; Carbon Footprint and Neutrality; Renewable Energies; Sustainability.

Further Readings

Carr, Anna. *Grass Roots and Green Tape: Principles and Practices of Environmental Stewardship*. Australia: Federation Press, 2002.

DeWitt, Calvin B. *Caring for Creation: Responsible Stewardship of God's Handiwork.* Grand Rapids, MI: Baker, 1998.

Emerson, Ralph Waldo. *Nature and Selected Essays.* New York: Penguin Books, 2003.

Esty, Daniel C., et al. *2005 Environmental Sustainability Index: Benchmarking National Environmental Stewardship.* New Haven, CT: Yale Center for Environmental Law & Policy, 2005. http://www.yale.edu/esi/ (Accessed March 2009).

Hart, Stuart L. "Beyond Greening: Strategies for a Sustainable World." *Harvard Business Review* (January/February 1997).

Leopold, Aldo. *A Sand County Almanac: With Essays on Conservation From Round River.* New York: Ballatine, 1986.

McDonough, William and Michael Braungart. *Cradle to Cradle: Remaking the Way We Make Things.* New York: North Point, 2002.

Muir, John. *John Muir: Nature Writings.* William Cronon, ed. New York: Library of America, 1997.

Thoreau, Henry David. *Walden and Civil Disobedience.* New York: Barnes & Noble, 2003.

U.S. Environmental Protection Agency. "Everyday Choices: Opportunities for Environmental Stewardship." http://www.epa.gov/stewardship/ (Accessed March 2009).

Van Dyke, Fred, et al. *Redeeming Creation: The Biblical Basis for Environmental Stewardship.* Downers Grove, IL: Intervarsity Press, 1996.

Yale School of Forestry and Environmental Studies, "Forum on Religion and Ecology." http://fore.research.yale.edu/main.html (Accessed March 2009).

Chelsea Schelly
Jessica Price
University of Wisconsin–Madison

ETHANOL, CORN

Ethanol from corn is at the center of debate about the sustainability of biofuels. Whether or not it can contribute significantly to the fuel supply depends upon if the amount of energy that is gained by burning it exceeds the amount of energy that goes into producing it. Benefits of ethanol are weighed against its costs, such as higher corn prices and reduced supplies of grain for food and animal feed.

Need for Gasoline Substitutes

An alternative to gasoline is needed: humans are pumping about 1,000 barrels of oil out of the ground every second, a rate that cannot be sustained. In just over a century, about half of the recoverable oil found in the ground has been extracted, an energy resource that has taken millions of years to accumulate. But the earth has been nearly depleted of oil, and oil that is easy to extract from the earth has already been extracted. Society is reliant on oil because mechanized modern farming, transportation, and industry depend upon it.

Global warming demands reduced fossil fuel use in any case. Because there is already a large infrastructure based on internal combustion engines, a fuel that could substitute for gasoline from oil in those engines is urgently needed, and ethanol could potentially meet that need because it can be burned in existing vehicles. Fuel containing 5 percent ethanol can be

These U.S. Department of Agriculture (USDA) Agricultural Research Service scientists are comparing corn varieties for ethanol yield and are shown introducing yeast to a bioreactor to initiate fermentation.

Source: USDA Agricultural Research Service/ Scott Bauer

used without any engine modifications, while E85 'gasohol' with 85 percent ethanol and 15 percent gasoline requires minor changes.

Ethanol has long been the stepchild of vehicle fuels. When Henry Ford developed his Model T, he foresaw farmers distilling their own ethanol from crop waste. This idea was defeated by John D. Rockefeller's Standard Oil Trust offering inexpensive gasoline. During the 1920s prohibition discouraged any production of ethanol, even as an additive. The alternative was tetraethyl lead, which was inexpensive and easier for refiners to handle, even though it was known to be toxic decades before it was outlawed. Only recently has ethanol reemerged as a substitute for dwindling oil.

The United States presently devotes 14 percent of its corn crop to replace 3.5 percent of its fuel, so there will not be nearly enough available. The entire corn crop would yield about one-fourth of the fuel that Americans now consume. Because the production technology is fairly mature, major gains in efficiency are unlikely, though efficiency is still improving. Corn ethanol is viable because of a 51 cents/gallon subsidy, but since ethanol provides about two-thirds as much energy per gallon as gasoline, the subsidy is equivalent to 75 cents per gallon of gasoline displaced, or $31 per barrel. Nonetheless, the U.S. Department of Agriculture projects that ethanol will consume 31 percent of the corn crop by 2017, providing 7.5 percent of vehicle fuel. Both exports and domestic grain supply would decline as a fraction of total corn production.

Producing Ethanol

Ethanol is not an oil-based fuel. It is an alcohol, the same chemical molecule that is found in alcoholic beverages. The molecular structure of ethanol is similar to small hydrocarbon (oil and gas) molecules, with an added oxygen atom. Unlike hydrocarbons, it is soluble in water.

Corn is converted into ethanol, also called ethyl alcohol, in several steps. First the corn is separated from the husks and ground to a rough flour. In dry milling operations, the corn meal is heated with water and enzymes to produce liquefied corn starch. A second enzyme breaks apart the starch to sugars that are combined with water and yeast. The product is warmed to aid fermentation—the yeast metabolizes the sugar into ethanol and carbon dioxide. The chemistry is similar to the fermentation of grapes into wine. When the alcohol concentration reaches about 15 percent, the mixture becomes toxic to the yeast cells. After they die the liquid is separated from the remaining mash, which can be used for cattle feed, but must be returned to the soil if fertility is to be sustained. The alcohol is separated from the water by boiling off the alcohol, which has a lower boiling point than the water, a process analogous to distilling wine into brandy. Not all of the water can be removed in this way, however, because some of it evaporates and contributes to the distillate. The remaining water is removed by chemical processes, and a small amount of gasoline is added to make the final product undrinkable.

All of these steps require energy, most of it coming from fossil fuels, though in theory renewable energy sources could be used. The heat needed for fermentation and distillation comes mostly from oil and natural gas. If coal is used, the process releases more of the greenhouse gas carbon dioxide than gasoline to provide the same energy. The most readily available fuel is the ethanol itself, though burning it in the production process would reduce yields to unviable levels. In the Midwest, where most ethanol plants are located, most of the electricity that runs pumps and other machinery comes from coal. The fuel must then be shipped to distant markets because most production is near cornfields, far from population concentrations on the coasts. It cannot go into conventional pipelines, normally the most efficient delivery method, because it mixes readily with water that would have to be removed at the consumer end of the pipeline to reconstitute a usable fuel. Ethanol is also corrosive to pipelines, pumps, and valves, so delivery is made by truck or railroad tank car over thousands of miles, increasing both its cost and the fossil fuel requirements for the total production cycle.

Measuring the Net Energy Benefit

Still, the promise of an unending supply of renewable energy is welcomed by most politicians and by all corn farmers. Yet expert opinion is divided on whether producing this alternative fuel will increase the total fuel supply. The U.S. Department of Agriculture calculates that one unit of energy input yields about 1.3 units of output.

Now the data are available to decide the issue. As production ramps up, gasoline use should ramp down if ethanol is a net benefit. The most complete figures come from 2006. Since ethanol made up about 3 percent of vehicle fuel at the time, there should have been a 3 percent decrease in gasoline use. But U.S. gasoline use increased by 1.4 percent annually from 2003 to 2007. If ethanol had replaced gasoline, there would have been a 1.6 percent decrease in gasoline use. The lack of savings in gasoline might be explained if fuel demand had significantly increased, but according to the Federal Highway Administration total miles driven increased only 1.2 percent from 2005 to 2006. The only viable theory is that growing corn and distilling it into ethanol uses as much energy as it offers, and that the benefit cancels itself out. Intuitively, it seems that converting corn to biofuel should substitute sunlight for fossil fuel, reducing fossil fuel demand. But the appearance of corn grown solely by using solar energy is an illusion. About nine-tenths of the energy currently used in growing American corn comes not from sunlight, but from oil and gas. Tractors run on oil, fertilizer is made mostly from natural gas, combines harvest the grain using oil, and the grain must be delivered to a distillery, converted into ethanol in the steps outlined above, and delivered. The fossil fuel that dominates corn calories makes it difficult for ethanol to achieve positive energy balance, although implementing green energy sources to operate farming equipment could change that.

Small variations in statistics do little to change the perception that ethanol production does not contribute significantly to U.S. fuel supplies, and that the side effects of its production will increase as new distilleries come on line. As more and more ethanol is distilled, less corn is left for food and feed. The United States is the world's largest consumer of fossil fuels per person, but it is also the world's largest grain exporter. The United States, in short, converts oil into grain to feed the world. Ethanol production reduces the amount of exportable grain without reducing oil imports. In statistical summaries, the ethanol is counted twice— once as the fuel that went into producing it, and again as vehicle fuel. The fact that ethanol production yields no net equal energy gain does not enter into the total consumption tallies.

Popularity of Ethanol

Corn-based ethanol does not increase the net vehicle fuel available to the public, and is viable only with heavy subsidies, yet its production remains popular. There are several reasons for this. The first reason is political; ethanol increases the demand for corn, keeping prices high, and boosting the incomes of corn farmers and the economy surrounding them. The benefits are concentrated and obvious. The resulting price increases for corn products are more dispersed and their cause is less obvious. The neutral energy balance is invisible.

The second reason is psychological; standing in front of a giant ethanol distillery in Iowa, one sees corn going in one end and fuel coming out the other. The fossil fuel inputs are not as noticeable. Burning sun-dried corn as a fuel for electrical power plants, partially replacing coal and natural gas, would be more efficient than converting it to biofuel, in terms of the energy return on investment, because the energy demands of distillation would be eliminated.

There is also a phenomenon of optimism; when there is a problem, such as the looming shortages and price spikes of fossil fuels, people naturally gravitate to solutions, even substandard ones. In the long run, solutions like vehicles powered by electricity, wind, or nuclear power, cities that are more accessible to pedestrians and biking, and electric mass transit would likely provide more sustainable alternatives to fossil fuels. These alternatives have not yet been developed because people naturally use the easiest resource first—in this case, inexpensive oil. Ethanol will continue to be added to gasoline in small amounts as a pollution-reducing oxygenator, safer than the methyl tertiary butyl ether (MTBE) it replaces, but corn-based ethanol as a substitute for gasoline is likely not the answer to our fossil-fuel problems.

See Also: Alternative Fuels; Biomass Energy; Ethanol, Sugarcane; Gasohol; Sustainability.

Further Readings

Cardona, C., O. Sanchez, and L.Gutierrez. *Process Synthesis for Fuel Ethanol Production (Biotechnology and Bioprocessing)*. Boca Raton, FL: CRC, 2009.

Dewsbury, Suzanne, and Ian M. Dewsbury. *Ethanol*. Farmington Hills, MI: Gale Group, 2009.

Ferris, John. "Agriculture as Energy? The Wisdom of Biofuels." *Harvard International Review*, v.31/2 (June 2009).

Goettemoeller, Jeffrey, and Adrian Goettemoeller. *Sustainable Ethanol: Biofuels, Biorefineries, Cellulosic Biomass, Flex-fuel Vehicles, and Sustainable Farming for Energy Independence*. Maryville, MO: Prairie Oak Publishing, 2007.

Icon Group International. *The 2009 Report on Fuel Ethanol Manufactured by Dry Mill Distillation: World Market Segmentation by City.* San Diego, CA: Icon Group International, 2009.

Patzek, Tad W. "Thermodynamics of the Corn-Ethanol Biofuel Cycle." *Critical Reviews in Plant Sciences.* v.23 (2004).

Pahl, Greg, and Bill McKibben. *Biodiesel: Growing a New Energy Economy.* White River Junction, VT: Chelsea Green Publishing, 2008.

Pimentel, David. "Ethanol Fuels: Energy Balance, Economics, and Environmental Impacts are Negative." *Natural Resources and Research.* v.12 (2003).

Westcott, Paul C. "Ethanol Expansion in the United States: How Will the Agricultural Sector Adjust?" *USDA Report* (2007). http://www.ers.usda.gov/Publications/FDS/2007/05May/FDS07D01/fds07D01.pdf (Accessed March 2009).

Bruce Bridgeman
University of California, Santa Cruz

ETHANOL, SUGARCANE

According to the World Energy Council (WEC) forecast, to meet a growing demand for energy, fossil fuels will soon account for up to 85 percent of total energy sources. As such, carbon dioxide emissions will double, to 46 billion tons a year in 2050. Increases in carbon dioxide emissions will accelerate some of the forecasted impacts from global warming and climate change. Not just the WEC but also governments and other entrepreneurs are recognizing this crisis and applying resources to find greener energy sources. Renewable energy sources (RES) are green energy sources that help reduce greenhouse gas production.

Sugarcane is fermented with yeast, as shown here, to produce ethanol. The process takes half as much energy as converting corn to ethanol.

Source: iStockphoto.com

Biofuel

Biofuel is one such RES that is produced from biomass. Biofuel is defined as solid, liquid, or gaseous fuel that is derived from biological material available in its present stage, unlike the fossil fuels that are developed from long-dead biological materials decomposed over the years. The most common sources of biofuels are carbon sources like photosynthetic plants. Biofuel is not as clean as the other RES like solar, wind, geothermal, tidal, hydrogen, or nuclear power, but it is better than the energy generated from fossil fuel or coal. Biomass itself is carbon neutral, but when used for biofuel production, it blocks methane emission, thus becoming carbon negative. If the biomass were allowed to decay naturally, it would release stored carbon into the atmosphere. Thus, the use of biomass in production of biofuels is perhaps the best carbon sequestration process.

Ethanol From Biosources

Electricity is generated via the direct burning of biomass to produce heat energy or steam to run turbines. In an indirect form, liquid or gaseous biofuel is produced from biomass of various forms such as wood, grasses, crops, and crop residues. These biomasses are converted to energy through thermal or biological conversion process to produce liquid or gaseous biofuels, which are used for transportation vehicles. Thus, liquid biofuels are an alternative to the greenhouse gas–producing fossil fuels.

Agrofuels are biofuels produced from agricultural products such as crops, seeds, crop residues, and so on. Crops with higher sugar, starch, and oil contents are perfect for agrofuel production. Soybean, sunflower, palm, jatropha, *pongamia pinnata*, and algae are crops, plants, or other living biomasses that are rich in oil. These oils are heated to reduce their viscosity and are then directly used in diesel engines as fuels. Sugarcane, sugar beet, and sweet sorghum—crops with high sugar content—and corn—with its high starch content—are fermented with yeast to produce ethyl alcohol, or ethanol.

Ethanol, a straight chain alcohol with the molecular formula C_2H_5OH, is also known as ethyl alcohol, pure alcohol, grain alcohol, or drinking alcohol. It is a volatile, flammable, colorless liquid. The flammability of the ethanol is perfect for its use as motor fuel. In addition to automobiles, ethanol can also be used to power farm tractors and airplanes.

Producing Ethanol From Sugarcane

Ethanol produced through fermentation of sugar was one of the earliest discoveries by man. Ethanol is produced from sugar/glucose ($C_6H_{12}O_6$) through the reaction process $C_6H_{12}O_6 \rightarrow 2C_2H_6O$ (ethanol) $+ 2CO_2 +$ heat.

Four basic steps are used in the industrial production of ethanol from sugarcane: microbial (yeast) fermentation of sugar/glucose, distillation, dehydration, and denaturing. In the distillation process, water is removed from the fermented glucose or ethanol. However, the water removal is limited to 95–96 percent because of the formation of low-boiling water ethanol azeotrope—a mixture of two or more chemicals in such a ratio that its composition cannot be changed by simple distillation. The dehydration process is used to remove the water from an azeotropic ethanol/water mixture. Denaturing is an optional process that is used to detoxify the produced ethanol.

Ethanol can be produced from other by-products of sugarcane such as molasses, raw sugar, refined sugar, bagasse, and sugarcane trash. Sugarcane bagasse is the material left over after sugar juice is squeezed from cane stalk, and sugarcane trash is stalks and leaves that are completely left behind in the field when sugarcane stalks are transported to the factories. Ethanol is produced from molasses and raw or fine sugar using the same fermentation process as sugarcane, but the bagasse and sugarcane trash do not have glucose present in them. They are processed to produce ethanol using a cellulose-processing technique similar to that used in ethanol production from corn.

Ethanol Versus Gasoline in Auto Engines

Energy per unit volume of ethanol (E100, no mixture) is 34 percent lower than gasoline. Therefore, consumption of E100 is approximately 51 percent higher than for gasoline in automobiles. However, better fuel economy can be achieved with higher compression ratios in an ethanol-only engine, and in general, ethanol-only engines are tuned to give slightly better power and torque output than gasoline-powered engines, so that the discrepancies of energy production are eliminated. Researchers have also argued that if and when high-compression ethanol-only vehicles are produced as a result of a future sufficiency of ethanol, the fuel efficiency of such engines should be equal to or greater than that of current gasoline engines. However, a mixture of E10 or E20 with gasoline does not provide any significant energy loss for automobiles.

Ethanol use in automobiles may stall the engine as a result of the slugs of water (from the ethanol) in fuel lines because it is not possible to have 100 percent water-free ethanol production. To counter this, the fuel must exist as a single phase. The fuel mileage decreases because of the presence of more water in fuel lines. Again, the amount of water in the fuel line increases with the increase of ethanol percentage, say from E10/E20 to E70/E85. It is also argued that with sufficient availability of ethanol, auto engines would be developed to be compatible with even E100.

Ethanol Production in the World

More than half the world's ethanol is produced from sugar or sugar by-products, and the majority of the world uses sugarcane as the main source for these products. Brazil has the largest ethanol fuel production industry in the world that uses mainly sugarcane. Gasoline sold in Brazil contains more than 25 percent ethanol, and more than 90 percent of new cars sold in Brazil are run with an ethanol-mix fuel. Thus the use of sugarcane in ethanol production in Brazil is praised for its high carbon sequestration capabilities and its positive effect on combating global warming and climate change.

By 2006, 16.3 billion liters—33.3 percent of the world's total ethanol or 42 percent of the world's ethanol used as fuel—of ethanol was produced by Brazil from sugarcane. Over 3.6 million hectares of land in Brazil (only 1 percent of total arable land in the country) was cultivated with sugarcane for ethanol production. At this time, in the United States there is no ethanol produced from sugarcane or sugar beets. Ninety-seven percent of ethanol produced in the United States is from corn.

Which Source Is Better for Ethanol Production?

Because of the low price of corn in most parts of the world, it might seem more prudent to produce ethanol from corn than from sugarcane. One gallon of ethanol can be produced from cornstarch at a cost of $1.03–$1.05 per gallon compared with $2.40/gallon for raw sugarcane. However, sugarcane cultivated for ethanol production is more fully used. The sugarcane stalk, the bagasse, and the sugarcane trash are used to produce ethanol. However, the starch contained in corn seeds is only used for ethanol production.

According to J. Jacobs, research shows that one dry ton of sugarcane bagasse can produce only 80 gallons of ethanol compared with 98 gallons of ethanol from a ton of corn. In contrast, when the ethanol produced from sugarcane's glucose and from other cellulose sources are added together, it provides a much larger production volume than from only cellulose source biomaterials like corn. Again, unlike corn, other forms of sugarcane can be processed to produce ethanol, and with a much lower processing cost than corn. Molasses and raw/refined sugar are cost competitive with corn in the United States, as ethanol can be produced with 50 percent less investment than corn. The processing cost of ethanol production from molasses in the United States is close to half ($0.36) the cost per gallon compared with the cost of producing ethanol from corn ($0.63). Researchers are now also trying to enhance the amount of ethanol production from sugarcane bagasse and trash, which would make sugarcane the most economic source to produce ethanol.

Above all, corn is a food grain for a majority of world's population. Because of the use of corn in ethanol production in the United States, the price of corn has gone up drastically in the last few years, in concert with a food shortage problem worldwide. Therefore, it is wise to use other biosources like sugarcane rather than the staple food grain corn.

Brazil produced 7,500 liters of ethanol per hectare in 2006. However, the ethanol productivity from corn in the United States was only 3,000 liters per hectare in 2006. This is one more reason why sugarcane is a better biomaterial for ethanol production than corn. Producing ethanol from sugarcane is easier than from corn because converting corn into ethanol requires additional cooking and enzyme application, whereas ethanol from sugarcane is produced by a simple yeast fermentation process. Therefore, the energy required to obtain ethanol from sugarcane is half that of corn.

Sugar beets are another alternative to sugarcane for ethanol production, but global output of sugar beets is very low compared with sugarcane, and production of ethanol from sugarcane is more expensive than that from sugar beets. One cautionary note is that if sugarcane or sugar beets are not processed quickly after harvest, ethanol production decreases because of sucrose deterioration. Therefore, sugarcane is initially processed into raw sugar at mills near the cane fields and later used for ethanol production.

Viability of Ethanol

In recent years, high global oil prices have spurred interest in ethanol. When the price of gasoline was above $3.50 per gallon, ethanol was considered to be a cost-competitive fossil fuel alternative, but with the depletion of global gas prices (less than $2.00 per gallon), it seems not sensible to produce ethanol from biosources including sugarcane. According to a U.S. Department of Agriculture report, the use of raw or refined sugar as a feedstock would be more profitable, and it would not be profitable to produce ethanol from sugar and sugar by-products—to see a profit, ethanol prices cannot drop below $2.35 a gallon. But we cannot put a price on the present catastrophe—global warming and climate change. It is now practical to use somewhat high-priced ethanol in automobiles instead of gasoline to save the planet.

See Also: Alternative Fuels; Biomass Energy; Ethanol, Corn; Gasohol; Sustainability.

Further Readings

"Biofuels: The Promise and the Risks, in World Development Report 2008." The World Bank, 2008. http://siteresources.worldbank.org/INTWDR2008/Resources/2795087-1192112387976/WDR08_05_Focus_B.pdf (Accessed February 2009).

Energy Information Administration (EIA). "Alternative Fuel Efficiencies in Miles per Gallon." http://www.eia.doe.gov/cneaf/solar.renewables/alt_trans_fuel/attf.pdf (Accessed February 2009).

Jacobs, J. "Ethanol From Sugar: What Are the Prospects for US Sugar Co-Ops?" http://www.rurdev.usda.gov/rbs/pub/sep06/ethanol.htm (Accessed February 2009).

Myers, Richard L. and Rusty L. Myers. *The 100 Most Important Chemical Compounds: A Reference Guide*. Westport, CT: Greenwood Press, 2007.

Panda, S. S. "Biomass." In *Encyclopedia of Global Warming and Climate Change*, S. G. Philander, ed. Thousand Oaks, CA: Sage, 2008.

Panda, S. S. "Energy, Renewable." In *Encyclopedia of Global Warming and Climate Change*, S. G. Philander, ed. Thousand Oaks, CA: Sage, 2008.

Reel, M. "Brazil's Road to Energy Independence." *Washington Post* (August 19, 2006).

Stauffer, N. "MIT's Pint-Sized Car Engine Promises High Efficiency, Low Cost." Massachusetts Institute of Technology. http://web.mit.edu/newsoffice/2006/engine.html (Accessed February 2009).

World Energy Council. "The Energy Industry Unveils Its Blueprint for Tackling Climate Change: WEC Statement 2007." London: World Energy Council, 2007.

Sudhanshu Sekhar Panda
Gainesville State College

EXERGY

The first law of thermodynamics tells us that energy can be neither created nor destroyed. The second law places additional limits on energy transformations and reflects qualitative characteristics. It states that energy can only be transformed by the consumption of quality. Locally, the quality can be improved, but this can only occur at the expense of a greater deterioration of the quality elsewhere. The level of quality deterioration or disorder is measured through the property of entropy. Hence, the second law of thermodynamics can be formulated as follows: In all real processes of energy transformation, the total entropy of all involved bodies can only be increased or, in an ideal case, unchanged. Beyond these conditions, the process is impossible even if the first law is fulfilled. The combination of both laws indicates that it is not a question of the existent amount of mass or energy but of the quality of that mass or energy—in other words, on its exergy content.

Technically, exergy is defined as the maximum amount of work that may theoretically be performed by bringing a resource into equilibrium with its surrounding environment by a sequence of reversible processes. The surrounding environment (or reference environment) must be characterized by a set of intensive properties such as temperature (T_0), pressure (p_0), or chemical potential of the substances in the environment (μ_{0i}), or extensive ones such as specific volume (v_0), entropy (s_0), and the number of moles of substance i in the environment (N_{0i}) plus height (z_0) and velocity (c_0). The exergy of a system gives an idea of its evolution potential for not being in thermodynamic equilibrium with the environment. Unlike mass or energy, exergy is not conservative. It is an extensive property with the same units as energy. In all physical transformations of matter or energy, it is always exergy that is lost.

The specific exergy of a system per unit of mass (b_T) is generally defined as:

$$b_T(kJ/kg) = (u-u_0) + p_0(v-v_0) - T_0(s-s_0)$$
$$+ \frac{1}{2}(C^2 - C_0^2) + g(z-z_0) - \sum_i^n \mu_{0i}(N_i - N_{0i}) \tag{1}$$

Exergy analysis is a powerful tool for improving the efficiency of processes and systems. This leads to fewer resources being used and the emission of fewer wastes to the environment. Exergy analysis can also be applied to natural resources accounting.

All materials have a definable and calculable exergy content with respect to a defined external environment. The consumption of natural resources implies destruction of organized systems and pollution dispersion, which is in fact generation of entropy or exergy destruction. Furthermore, exergy has the capability to aggregate heterogeneous energy and material assets, and being a physical property, it provides objective information about resources, as opposed to economic valuation methods, which usually depend on monetary policy and currency speculation. This is why exergy analysis can describe perfectly the degradation of natural capital. For that reason, an increasing number of scientists believe that exergy provides useful information within resource accounting and can adequately address certain environmental concerns.

The exergy of a substance deriving only from its concentration (b_c) and chemical components (b_{ch}) can be calculated as follows:

$$b_T(kJ/kg) = \underbrace{RT_0 \Sigma x_i \ln \frac{c_i}{c_{i,0}}}_{b_c} + \underbrace{\left[\sum_i y_i \left(\Delta G_f + \sum_e n_e b_{chne} \right)_i \right]}_{b_{ch}} \tag{2}$$

In the concentration exergy term (b_c), R is the universal gas constant (kJ/kgK); x_i the molar concentration of substance i in the mixture (kmol/kmol); c_i and $c_{i,0}$ the concentration of substance i in the system and in the reference environment, respectively (mol/kg). In the case of liquid solutions, activities rather than concentrations are used.

The chemical exergy term (b_{ch}) is equivalent to the chemical potential term of equation 1. This expression is usually more convenient to use, as the Gibbs free energy (ΔG_f, kJ/kmol) and the chemical exergy of the elements that compose substance i (b_{chne}, kJ/kmol) are easy to find in tables. Variables y_i and n_e represent, respectively, the relative molality of substance i (mol/kg) and the amount of each element e in 1 kmol of the substance.

When assessing natural resources, neither their mass nor their energy contents can set off intrinsic qualities. To overcome that problem, thermodynamics proposes to measure things in units of actual capability to perform work; that is, exergy. For instance, exergy is sensible to the ore grade of a mineral in a mine; meanwhile, its mass or its energy are the same whether the mineral is dispersed or concentrated. The same thing happens with pollution processes. Exergy provides meaningful, homogenous, and universal units.

Thermodynamics provides limits of behavior, so exergy is a minimum value that is far from actual numbers. The real exergy expenditure in a given process is in many cases far greater than its exergy. We name this actual exergy expenditure the exergy cost of the production process once the limits of the analysis, the process itself, and the efficiencies of each process component have been defined. This also implies the definition of what the raw materials, fuels, products, co-products, by-products, and wastes are.

The unit exergy cost is the ratio between the real exergy cost of a product and its exergy. Its value is dimensionless and measures the number of units required for obtaining a unit of product. For instance, the unit cost of producing electrical energy from a thermoelectric source is around 3, as the efficiency of that process is 0.33.

The exergy cost provides actual numbers that sound familiar for practitioners. In contrast, exergy is the reference point of the minimum physical cost of things. The difference between the exergy costs and the exergy measures our thermodynamic ignorance.

The exergy and exergy cost concept can be applied to any natural resource such as a mineral deposit. The extraction of materials from the Earth's crust implies a net reduction of the natural's exergy stock.

The exergy of a mineral deposit with a given composition and concentration is defined as the minimum work needed to produce it from the substances included in the reference environment (RE) and is calculated using equation 2.

The exergy replacement cost of a mineral deposit is defined as the exergy that would be expended in recovering the deposit from the RE with the best available technology plus the environmental exergy costs for restoring the mining zone. The latter can be approximately obtained from the best available practices provided from the mining industry in soil and landscape restoring techniques.

The depletion of natural resources is one of the key issues man is facing in this century. However, not much attention has been paid to nonfuel mineral depletion. Throughout the 20th century, the best mines with the best ore grades have been mined. This implies that the remaining deposits need much more extraction exergy—more land degradation and water use than the previous ones—and this tendency becomes exponential. As a consequence, the Earth is gradually becoming a degraded planet of minimum exergy, with the absence of fuel and nonfuel mineral deposits. This is a consequence of the second law of thermodynamics. Nature cannot be compensated with money but with counteractions like recovering, restoring, and replacing techniques. For an appropriate management of resources, society—and especially decision makers—should know not only the yearly depletion rates of fossil fuels but also the rates of all kinds of minerals. Using the replacement exergy costs of minerals as an indicator of the mineral capital on Earth would allow a careful husbandry of materials and energy resources, helping to slow down the Earth's degradation rate.

See Also: Best Management Practices; Entropy; Fossil Fuels.

Further Readings

Ayres, R., et al. *Sustainable Metals Management: An Application of Exergy Accounting to Five Basic Metal Industries*. New York: Springer, 2006.

CIRCE. "The Exergoecology Portal." 2007. http://www.exergoecology.com (Accessed January 2009).

Dincer, I. "Thermodynamics, Exergy and Environmental Impact." *Energy Sources*, 22:723–32 (2000).

Dincer, I. and Y. A. Cengel. "Energy, Entropy, and Exergy Concepts and Their Roles in Thermal Engineering." *Entropy*, 3:116–149 (2001).

Jorgensen, S. and E. Tiezzi, eds. *Eco-Exergy as Sustainability*. Ashurst Lodge, UK: WIT, 2006.

Rant, I. "Exergie, ein neues Wort fur 'Technische Arbeitsfahigkeit'" [Exergy, A New Word for 'Technical Available Work.'] *Forschung auf dem Gebiete des Ingenieurwesens*, 22:36–37 (1956).

Rosen, M. "Can Exergy Help Us Understand and Address Environmental Concerns?" *Exergy*, 2:214-217 (2002).

Szargut, J. *Exergy Method: Technical and Ecological Applications*. Ashurst Lodge, UK: WIT, 2005.

Szargut, J., et al. "Depletion of the Non-renewable Natural Exergy Resources as a Measure of the Ecological Cost." *Energy Conversion and Management* (2002).

Valero, A., et al. "Exergy Accounting of Natural Resources Exergy, Energy System Analysis, and Optimization." In *Encyclopedia of Life Support Systems*. Oxford: UNESCO EOLSS. http://www.eolss.net (Accessed May 2005).

Valero, A. and S. Ulgiati, eds. "Thermoeconomics as a Conceptual Basis for Energy-ecological Analysis Advances in Energy Studies." *Energy Flows in Ecology and Economy* (1998).

Valero D., A. Valero, and I. Arauzo. "Evolution of the Decrease in Mineral Exergy Throughout the 20th Century. The Case of Copper in the US." *Energy*, 33:107–115 (2008).

Wall, G. *National Exergy Accounting of Natural Resources*. Oxford: UNESCO EOLSS. http://www.eolss.net (Accessed May 2005).

Wall, G. and M. Gong. "On Exergy and Sustainable Development, Part 1: Conditions and Concepts." *Exergy*, 1:128–145 (2001).

Antonio Valero
Alicia Valero
Independent Scholars

Exxon Valdez

On March 24, 1989, the *Exxon Valdez*, an oil tanker owned by Exxon Corporation, went aground on Bligh Reef in Prince William Sound, Alaska. The oil tanker had just departed the Valdez terminal with over 53 million gallons of crude oil, bound for Exxon's West Coast refineries. The vessel spilled 10.8 million gallons of crude oil into Prince William Sound, and the oil eventually covered 11,000 square miles of ocean and 1,300 miles of shoreline. The oil spill immediately killed between 250,000 and 500,000 seabirds, more than 1,000 sea otters, 300 harbor seals, 250 bald eagles, 22 orca whales, and billions of herring and salmon eggs. Today, 20 years after the spill, 26,000 gallons of oil remain contaminating roughly 6 kilometers of shoreline.

The *Exxon Valdez* oil spill is still considered the most environmentally damaging oil spill to date. As the *Exxon Valdez* Oil Spill Trustee Council has indicated, "[t]he timing of the spill, the remote and spectacular location, the thousands of miles of rugged and wild shoreline, and the abundance of wildlife in the region combined to make it an environmental disaster well beyond the scope of other spills."

The Accident

After a harbor pilot successfully navigated the *Exxon Valdez* through the Valdez Narrows, he returned control of the ship to Captain Joseph Hazelwood. To avoid icebergs in the outbound shipping lane, Hazelwood maneuvered the ship into the inbound shipping lane. Hazelwood then put the ship on autopilot and left a third mate in charge of the wheelhouse and an able seaman at the helm. The crew failed to reenter the outbound shipping lane. While Hazelwood was relaxing in his stateroom, the *Exxon Valdez* went aground on Bligh Reef, rupturing eight of her 11 cargo holds.

Hazelwood, whom Exxon knew was an alcohol abuser who had not completed treatment and had stopped attending Alcoholics Anonymous meetings, had drunk 15 ounces of 80-proof alcohol shortly before leaving Valdez. The National Transportation Safety Board's investigation of the accident identified five factors that contributed to the grounding of the *Exxon Valdez*: the third mate failed to properly maneuver the vessel, possibly as result of fatigue and an excessive workload; the captain failed to provide navigation watch, possibly because of impairment from alcohol; Exxon failed to supervise the captain and provide a rested and sufficient crew for the vessel; the U.S. Coast Guard failed to provide an effective vessel traffic system; and there was a lack of effective pilot and escort services from the Valdez terminal through Prince William Sound.

This worker was attempting to absorb oil from the *Exxon Valdez* oil spill that had reached the shore of Prince William Sound, Alaska. The spill affected 1,300 miles of shoreline.

Source: National Oceanic and Atmospheric Administration/Exxon Valdez Oil Spill Trustee Council

Litigation

Five separate sets of lawsuits arose out of the *Exxon Valdez* oil spill. First, Exxon Shipping pled guilty to negligent discharge of pollutants under Clean Water Act section 309 as well as criminal violations of the Refuse Act and the Migratory Bird Treaty Act. Exxon was fined $150 million, the largest fine ever imposed for an environmental crime. The court forgave $125 million of that fine in recognition of Exxon's cooperation in cleaning up the spill and paying certain private claims. As criminal restitution for the injuries caused to the fish, wildlife, and lands of the spill region, Exxon agreed to pay $100 million, equally divided between the federal and state governments.

Second, the federal and state governments sued Exxon Shipping and Exxon under Clean Water Act section 311 and the Comprehensive Environmental Response Compensation and Liability Act section 107, to recover damages to natural resources for which the governments are trustees. In settlement of those civil claims, Exxon agreed to pay $900 million, with annual payments stretched over a 10-year period. The settlement also contained a $100 million reopener for funds to restore resources that suffered a substantial loss or decline as a result of the oil spill, the injuries to which could not have been known or anticipated by the trustees at the time of the settlement. The United States demanded the full $100 million under the reopener provision in 2006.

Third, within two or three years of the accident, Exxon settled the claims of various fishermen and property owners for $303 million.

Fourth, a class action lawsuit involving tort claims against Exxon, Hazelwood, and others by commercial fishermen, Native Americans, and property owners resulted in a $5 billion jury verdict against Exxon. That verdict was reduced to $2.5 billion, and in *Exxon Shipping Co. v. Baker* the U.S. Supreme Court vacated the 9th Circuit award, limiting punitive damages against Exxon to $507.5 million—the same amount of compensatory damages—in addition to the compensatory damages due to plaintiffs.

Finally, Captain Hazelwood was prosecuted by the state of Alaska for operating a vessel while under the influence of alcohol and negligent discharge of oil. Despite evidence that Hazelwood had consumed numerous alcoholic beverages before departing Valdez and still had alcohol in his blood many hours after the accident, an Alaskan jury found him not guilty of the operating-under-the-influence charge. The jury did find him guilty of negligent discharge of oil. Hazelwood was fined $50,000 and sentenced to 1,000 hours of community service in Alaska.

Resulting Legislation

Frequently, environmental legislation is the result of a dramatic event or environmental accident. In the case of the *Exxon Valdez* oil spill, Congress reacted by enacting the Oil Pollution Act of 1990, creating a fund to finance oil-spill cleanup when parties do not voluntarily clean up oil spills for which they are responsible, set standards for oil tankers and oil storage facilities to avoid future spills and improve spill response, and sought to improve emergency responses to oil spills through regional contingency planning.

See Also: Environmental Measures; Environmental Stewardship; Oil; Pipelines.

Further Readings

Exxon Shipping Co. v. Baker. No. 07-219 (June 25, 2007) http://www.supremecourtus.gov/opinions/07pdf/07-219.pdf (Accessed February 2009).

Exxon Valdez Oil Spill Trustee Council. http://www.evostc.state.ak.us (Accessed February 2009).

State of Alaska v. Hazelwood. Supreme Court No. S-5311 (AK Sept. 3, 1993) http://touchngo.com/sp/html/sp-4034.htm (Accessed February 2009).

Susan L. Smith
Willamette University College of Law

FEDERAL ENERGY REGULATORY COMMISSION

The Federal Energy Regulatory Commission (FERC) is an independent government agency that was created to regulate energy companies, to protect the public, and to protect energy customers in a nonpartisan political manner. It is an independent agency that is officially organized as a part of the Department of Energy. Its headquarters are in Washington, D.C., and it has regional offices in Atlanta, Chicago, New York City, Portland, and San Francisco.

The agency is led by five members who are nominated to five-year terms by the president. Their appointments are then approved with the advice and consent of the Senate. However, to protect the FERC from undue political influence, its membership must have no more than three members of the same political party. In practice, in the two-party system of the United States, this means either three Republicans and two Democrats or three Democrats and two Republicans. The head of the FERC is the chairman, who is appointed by the president and approved by the Senate for this leadership designation.

Every year Congress appropriates money for the operations of the FERC; however, it is able to recover the full cost of its regulatory and other operations through the annual charges and filing fees it assesses on the industries it regulates. The authority to collect these moneys from industry was authorized in the Federal Power Act and Omnibus Budget Reconciliation Act of 1986. Among the nonlegislative activities of Congress is its power to exercise oversight of its agencies. Congress regularly exercises oversight of the FERC when it hears reports and testimony from the commission on energy-related topics.

Goals and Responsibilities

The FERC has been assigned regulatory responsibility for interstate transmission of natural gas, oil, and electricity. Its regulatory authority applies to the construction of interstate natural gas pipelines, storage facilities, and liquefied natural gas terminals. It also approves the routing and abandonment of pipelines so that they can operate efficiently and safely. Regulation of both natural gas pipelines and electrical transmission lines is authorized under the constitutional power of Congress to regulate interstate commerce (Article I, Section 8, Clause 3).

Regulation of wholesale electricity sales is another FERC responsibility. The states regulate retail sales of electricity. Because electricity is generated by hydroelectric plants that are built at river dam sites, the FERC has been assigned the duty of licensing and inspecting hydroelectric projects. This includes private, state, and municipal hydroelectric projects. It also supervises the reliability of high-voltage interstate electrical transmission lines as part of different grids.

A third goal is to fulfill its charge to monitor and investigate energy markets. The FERC regulates and oversees energy industries in the U.S. economy. As it exercises this authority, it has several goals to balance. One goal is to promote abundant and reliable supplies of energy that are provided to the public. A second goal is to see that the private investor companies that produce and deliver energy to customers do so in fair and competitive markets. To ensure that the public interest is well served, it monitors the energy industries under its regulatory authority. In addition, it can investigate energy markets either for criminal misconduct or for business practices that may not promote the best interests of either energy customers or the general public.

FERC regulatory authority extends to examining the accounting and financial reporting practices of energy companies. Similar to other regulatory agencies, the FERC has both quasi-legislative and quasi-judicial functions. It develops regulatory rules and can enforce these through administrative law procedures. It can therefore use civil penalties and other means to stop energy organizations or individuals from violating FERC rules of conduct in energy markets.

A fourth FERC goal is environmental protection. In the course of its regulatory supervision it, like other federal agencies, is charged with environmental protection. This duty is exercised when pipelines, transmission lines, liquefied natural gas terminals, dams, and other electrical projects are proposed and planned. Environmental protection also extends to cultural and historic site protection. Sites in these categories include structures and objects of a cultural or community life that have scientific, traditional, religious, or other interests, the destruction of which would deny understanding of past life ways or of the course of future cultural development.

Limits on Authority

The authority of the FERC is limited. It is not authorized to go beyond its jurisdiction even if its members believe that the practices of energy companies or individuals are not in the public's interests. Among the many things the FERC does not do is regulate retail electricity and natural gas sales to consumers. It also does not inspect retail gas facilities such as railroad-tank-car-sized tanks or smaller tanks that may be placed in locations that are dangerous to the public. When a "bleve" (boiling liquid expanding vapor explosion) occurs from a propane tank, it is equivalent to a bomb exploding. Bleve explosions occur on a regular basis when propane tanks are allowed to leak or are ruptured in an accident. However, these dangerous materials are outside of FERC regulatory authority.

The authority of the FERC to regulate electrical generation and transmission lines is limited. It regulates hydroelectric generation and transmission lines, as well as the main transmission lines in transmission corridors, but the remaining portions of the electrical grids are outside its jurisdiction. This includes federal electricity-generating agencies such as the Tennessee Valley Authority, most rural electric cooperatives, municipal power systems, and other systems including nuclear power plants. The latter are regulated by the Nuclear Regulatory Commission.

Although the FERC supervises gas and oil pipelines, it does not exercise oversight of the construction of pipelines once it has issued certificate to authorize building pipelines. Responsibility for pipeline safety is handled by other agencies. It supervises the abandonment of pipelines but not the related oil facilities or oil (and gas) fields. It also does not have any responsibility for pipelines crossing or operating on the continental shelf.

The FERC regulates the operations of energy companies, but it does not supervise their acquisitions and mergers. Nor does it have any responsibility for the development of automobiles or other vehicles that use natural gas, electricity, or oil transmitted in the transmission lines or pipelines it regulates.

The duties that Congress has assigned to the FERC require it to balance the interests of energy producers, energy customers, and the public, especially in the case of the latter in regard to the environment. For energy companies to get licenses for new energy projects, they must demonstrate the environmental impacts that their projects will have. This means that licensing requests must be submitted with an environmental impact statement (EIS). Impacts are consequences that result from activities. They can be positive or negative or both. However, environmental impacts that are negative are not the only consideration. The FERC is required to consider the energy availability and the current cost of energy in the project's proposed locality.

The National Environmental Policy Act of 1969 requires the FERC to analyze the environmental impacts of proposed projects. It also mandates considering alternatives and appropriate mitigation measures that will safeguard the environment and allow the energy project to proceed. The EIS uses an environmental assessment, which is a comprehensive and systematic evaluation of the effects the project will have on both the natural and the cultural environment. The process for developing the EIS usually includes public hearings at which citizens may appear and testify. Among the considerations are the impact on recreation, wetlands, water quality, fishing, wildlife; cultural impact; aesthetic impact; and economic impact.

To facilitate its work, the FERC uses an environmental measures effectiveness database. The database tracks the success of environmental requirements in licenses so that it accumulates an electronic history of the effects that different impacts have had in different projects. The tracking informs administrative decision making with a body of knowledge that measures experience with impacts that are thought to be positive or negative. This capability is a growing edge in the greening of the regulatory work of the FERC.

See Also: Department of Energy, U.S.; Hydroelectric Power; Natural Gas; Nuclear Power; Pipelines; Public Utilities.

Further Readings

Diane Publishing, ed. *Electricity Regulation: Factors Affecting the Processing of Electric Power Applications.* Darby, PA: Diane Publishing, 1994.

Diane Publishing, ed. *Electricity Regulation: FERC's Efforts to Monitor and Enforce Hydroelectric Requirements.* Darby, PA: Diane Publishing, 1996.

Diane Publishing, ed. *Natural Gas: Factors Affecting Approval Times for Construction of Natural Gas Pipelines.* Darby, PA: Diane Publishing, 1993.

Echeverria, John D., et al. *Rivers at Risk: Concerned Citizen's Guide to Hydropower.* Washington, D.C.: Island Press, 1989.

Enholm, Gregory B. and J. Robert Malko, eds. *Reinventing Electric Utility Regulation.*
    Vienna, VA: Public Utilities Reports, Inc., July 1995.
Segers, Marc, et al., eds. *Federal Regulatory Directory.* Washington, D.C.: CQ Press,
    2007.
U.S. Department of Energy, Federal Energy Regulatory Commission. *Federal Energy
    Regulatory Commission Annual Report 2008.* Washington, D.C.: Government Printing
    Office, 2009.

Andrew J. Waskey
Dalton State College

FEED-IN TARIFF

A feed-in tariff can be described as a policy structure operating as an incentive to promote the use of specific products and/or services. Often these policy structures are needed to promote products and/or services not mature enough to be exposed to a competitive market. Feed-in tariffs are widely used in the energy markets, particularly electricity, as an incentive mechanism to adopt electricity generation from renewable energy sources (e.g., solar) and encourage development of renewable energy technologies (e.g., photovoltaic panels). Policy supporters argue feed-in tariffs encourage new job development, reduce imported energy supplies, improve energy security, and promote renewable energy implementation. However, critics contend feed-in tariffs are incompatible with the liberalization of energy markets, distorting the competition with the introduction of costly and inefficient methods for power generation. Feed-in tariff policies for the renewable energy market are pervasive in Europe but very limited in other regions, including the United States.

The main principle of the feed-in tariff is that the electricity network utilities are obliged to purchase electricity generated from renewable energy sources at a premium rate that is significantly above market rate for electricity. In this way, renewable energy investments are encouraged and investors are compensated for the higher production costs they incur compared with generators based on fossil fuels. The funds used to pay for the premium rate may be sourced from carbon taxes or other governmental budgets or shared among consumers.

The European Environment Agency defines a feed-in tariff for energy markets as a government regulated rate for the price of electricity that a utility supplier has to pay for renewable electricity from private generators. In other words, those that generate electricity using renewable energy technologies (e.g., wind turbines) will receive payment for producing rather than consuming electricity. This is different than the concept of net metering, an electricity policy that enables customers to offset their energy consumption with energy production using specialized utility meters that record both energy output and input. In Gainesville, Florida (U.S.), a solar feed-in tariff bill passed in 2009 that enables residents with home solar panels to receive 32 cents per kilowatt-hour (kWh) when electricity is produced whereas residents acquiring electricity from the utility company pay an average of 12 cents per kWh.

The Public Utility Regulatory Policies Act of 1978 ignited the first feed-in tariff application in California. In particular, the state utility commissions specified prices for electricity generated by Qualified Facilities that mostly included small-scale renewable energy installations. Electricity utilities were then obliged to purchase electricity from these facilities—a measure that was considered a significant success. Because the act directed that the rates should be equal to the marginal cost of generating electricity with alternative fuels, the commissions often linked the prices to the (by that time high) oil prices, which provided for a considerable incentive.

The feed-in tariff is considered a successful policy measure for the promotion of renewable energy sources, having provided and still providing for excellent results in various countries where it has been adopted. Germany, the world leader in renewable energy installations, introduced the feed-in tariff in 1991 and revised the policy several times in 1998, 2000, and 2004 to account for changes in technology and pricing. The latest German feed-in tariff specifically addresses electricity generated by wind turbines; the fixed rate was set in 2000 at €91 per megawatt-hour (MWh) (or approximately $130/MWh) for the first five years of operation and €61.9/MWh (or approximately $88/MWh) for the following 15 years. After Germany, Denmark has the highest installed capacity of wind energy, peaking in the mid-1980s and 1990s. This has been a result of the generous feed-in tariff system that has provided premium rates for a variety of renewable energy sources (e.g., biomass, wind, solar). In particular, wind turbine installation was particularly encouraged in Denmark; however, a recent law introduced in 2000 enables the feed-in tariff only for existing wind turbine installations as part of a phase-out process that started in 2003.

Southern Europe has also seen the implementation of renewable energy feed-tariffs, particularly related to solar power in one of the sunniest regions on the continent. In Greece, the introduction of Law 3468 in June 2006 set ambitious targets for the promotion of electricity generation with solar photovoltaic systems. The Greek government supports solar feed-in rates for new installations of €0.45/kWh (approximately $0.64/kWh) for inland applications and €0.50/kWh (approximately $0.71/kWh) for applications on islands off the interconnected grid for the first 100 kW with a slightly reduced rate for electricity generation over 100 kW. Spain also has a lengthy tradition—since 1980—of supporting renewable energy sources; however, a feed-in tariff scheme was adopted only in 1997. The most recent version of this law (2004) provides for a complete renewable energy framework with premium tariffs for energy coming from solar (photovoltaic and thermal), wind (onshore and offshore), geothermal energy, ocean power, hydroelectric and biomass.

In Asia, the feed-in tariff policies for renewable energy adoption has had limited success in the two most populous countries in the world. The Chinese Electricity Regulatory Commission (SERC) has introduced legislation supporting renewable energy sources since 2005, but without achieving remarkable results due to: 1) the heavy reliance on coal for electricity generation; and 2) the high cost of electricity production from renewable energy sources. The most recent version of the Chinese policy on renewable energy attempts to change this situation. The Chinese National Development and Reform Committee now demands that the electricity distributors pay a price fixed by the government when they purchase electricity generated by renewable energy sources. Renewable energy feed-in tariffs are also a relatively new concept in India, first initiated nationally in 2003 by the Indian Electricity Act; however, in the states of Gujarat and Tamil Nadu attractive rates for wind power electricity generation were first offered a decade earlier. The Indian governmental plan also supports photovoltaic installation, stipulating rates of up to Rs15/kWh (approximately $0.31/kWh).

Although feed-in tariff is considered to be a successful policy mechanism for supporting renewable energy generation, it is not the only one. One of the major policy alternatives to feed-in tariffs are the renewable energy certificates, according to which all distributors are obliged to purchase and distribute a certain percentage of their electricity by renewable generators. The generators classified as renewable ones are awarded with the renewable energy certificates. This system has been adopted successfully in a number of countries such as Poland, Italy, Belgium, United States, and others; the renewable obligation scheme in the United Kingdom displays similar characteristics. Another alternative policy to incentivize the use of renewable energy sources is carbon taxing, according to which energy from fossil fuels is taxed and therefore less attractive than renewable electricity. Finally, the emissions trading scheme that is already applied in the European Union is another, if indirect, way of promoting renewable energy by increasing the costs of fossil fuel–fired electricity generation.

See Also: Carbon Tax; Carbon Trading and Offsetting; Emissions Trading; Green Energy Certification Schemes; Green Power; Renewable Energies.

Further Readings

Butler, Lucy and Karsten Neuhoff. "Comparison of Feed-In Tariff, Quota and Auction Mechanisms to Support Wind Power Development." *Renewable Energy.* v.33/8 (2006).
Galbraith, Kate. "Feed-In Tariffs Contemplated in the U.S." *The New York Times* (February 9, 2009).
Sijm, J. P. M. "The Performance of Feed-In Tariffs to Promote Renewable Electricity in European Countries." ECN-C-02-083 (2002).

Konstantinos Chalvatzis
University of East Anglia

FLARING

Flaring is the combustion of excess gas typically at sites of fossil fuel extraction, or in industrial manufacturing facilities. It involves the burning off the natural gas associated with the deposit or using the gas stream to burn off waste. These associated gases, sometimes called solution gases, are mixtures of collections of different hydrocarbons, primarily methane, given off in the process of crude oil mining or extraction; they can be seen as accompanying gases that follow crude oil to the surface during mining. Flaring has always been seen as an important operation carried out with safety of both facility and personnel in mind, it is used in such events as unplanned maintenance, emergency shutdowns, or disruptions in the processing systems during pressure build ups in the extraction of crude oil from the earth. These gases are automatically diverted through pressure relief valves to flare stacks for burning off, which in turn reduce the methane contents by direct reaction with the atmosphere to produce carbon dioxide, water vapor and enormous heat. Some argue that the flaring of natural gas wastes potential fossil fuel energy.

Flaring has been extensively employed around oil production rigs, in refineries, and in chemical plants. Its primary purpose has been to act as a safety measure to protect vessels or pipes from excessive pressuring as a result of unplanned upsets; thus, when plant equipment is overpressured, the pressure relief valves on it opens and releases gases (sometimes with some liquids) into flare stacks through the flare headers, where the gases are burned as they exit the stacks. Flaring is also employed in landfills with the aim of burning waste gases that result from the decomposition of materials in the dump. This process prevents methane, a very strong greenhouse gas (GHG; 23 times more powerful than carbon dioxide [CO_2]), from reaching the atmosphere by converting it through combustion process to water vapor, CO_2, and heat.

Gas burning in the oil mining industries has been found to contribute an enormous amount of CO_2 to the Earth's atmosphere; this has invariably been found to greatly affect the structure of the planet's greenhouse. Recently, global warming has become a widespread concern because of the increasing trend of atmospheric CO_2 level, and flaring is a major

The photo shows multiple flare stacks in operation at a large refinery. According to 2004 estimates, more than 100 billion cubic meters of natural gas are flared or vented every year.

Source: iStockphoto.com

culprit. It contributes huge amounts of CO_2, water vapor, and other GHGs such as sulfur dioxide, nitrous oxide, and possibly methane to the atmosphere. Moreover, gas flaring can have potentially harmful effects on human health because of regular exposure to some by-products of combustion that include sulfur dioxide, nitrogen dioxide, and some volatile organic compounds like benzene, toluene, hydrogen sulfide, and xylene. Humans exposed to such compounds can suffer from a variety of respiratory problems, which have been reported among many children in environments where gas flaring has been rampant. These substances can also aggravate asthma, causing breathing difficulties and pain, as well as chronic bronchitis. More so, benzene has been discovered to be a causative agent for leukemia, some other blood-related diseases, and cancer.

Environmentally, flaring has been found to have tremendous effects. Communities around flaring sites have been found to experience acid rains on a regular basis, as well as air pollution, heat, and at times soot cover where the flare stack is old and inefficient. Globally, the major causes of the climate change phenomenon have been attributed to the activities of anthropogenic emissions, of which flaring has been recognized has a major contributor. It was reported in 2004 that the presence of a large deposit of greenhouse gases in the atmosphere is a result of the activities of the big eight gas-flaring nations of the world, which produce about 94 percent of the 75 percent emissions produced by the first 10 gas-flaring countries; Nigeria is the leading nation, flaring over 24 billion cubic meters of associated gases annually. Other countries (with the approximate amount of gas in billions of cubic meters) include in descending order Russia (14.9), Iran (13.3), Iraq (8.6), Angola (6.8), Venezuela (5.4), Qatar (4.5), Algeria (4.3), Indonesia (3.7), Equatorial Guinea (3.6), United States (2.8), Kuwait (2.7), Kazakhstan (2.7), Libya (2.5), Azerbaijan (2.5), Mexico (1.5), United Kingdom (1.6), Brazil (1.5), Gabon (1.4), and Congo (1.2).

Global gas flaring activities, though, differ from nation to nation, and its contribution to global warming and climate change is on the increase—in 1997, it was reported that Nigeria alone flared about 21 billion standard cubic meters (compared with 24.1 billion cubic meters in 2004) of associated gas, representing about 19 percent of the total gas flared worldwide in that year alone. Regionally, Africa flared about 38 percent, North

America 17 percent, Middle East 15 percent, Central and South America 12 percent, Eastern Europe together with the former Soviet Union 7 percent, and Western Europe 3 percent, all within 1997. Thus, unless efforts are focused toward gas flaring reduction, the fight against global warming and climate change may not be very effective.

Incidentally, the national governments of countries and international communities all agree that gas flaring has a negative effect on the global environment and the economy of states. The World Bank reported in 2004 that over 100 billion cubic meters of natural gas are flared or vented annually—an amount worth approximately $30.6 billion, and equivalent to the combined annual gas consumption of Germany and France, twice the annual gas consumption of Africa, three-quarters of Russian gas exports, or enough to supply the entire world with gas for 20 days, making it necessary for states to look into ways of harnessing the natural and associated gas for productive use rather than just flaring, which has been found to be cheap. However, efforts at stopping gas flaring have been slow in implementation. This may be because of the high cost of separating and recycling the gases. Moreover, the ideology behind gas flaring initially involves the fact that the gases flared are of no use, although research has shown that many of these gases can have numerous uses and can be good alternative sources of energy production using gas turbines—domestically as cooking gas and jet fuels, and in the gas-to-liquid technology, converted to form lubricants, olefins, and methanol. It can also be reinjected into oil wells to maintain reservoir pressure during production and to improve oil recovery.

See Also: Climate Change; Global Warming; Nitrogen Oxides; Sulfur Oxides (SO$_x$).

Further Readings

Akinrele, F.O. & Co. "Gas Flaring—Is There an End in Sight?" http://www.foakinrele.com/pdfs/oil&gas/gasflaring.pdf (Accessed January 2009).

Falola, Toyin and Ann Genova. *The Politics of the Global Oil Industry: An Introduction.* Westport, CT: Praeger Publishers, 2005.

Odu, C. T. I. "Gas Flare Emissions and Their Effects on the Acidity of Rain Water in the Ebocha Area." http://www.elaw.org/system/files/Ng.GasFlares.AcidRain.Ebocha.doc (Accessed January 2009).

Patin, Stanislav. *Environmental Impact of the Offshore Oil and Gas Industry.* East Northport, NY: EcoMonitor Publishers, 1999.

Ajayi Oluseyi Olanrewaju
Covenant University, Nigeria

FLEX FUEL VEHICLES

Flex fuel vehicles (FFVs) are motor vehicles that do not require traditional energy sources such as gasoline or diesel fuel. They can run efficiently on gasoline but can also be fueled by up to 85 percent ethanol (E85) fuel. At this time, automobile manufacturers are developing engines that can run entirely on ethanol. In the United States, FFVs bring the benefits of less expensive fuel that is developed from resources within the country, thus reducing dependence on foreign oil sources. In addition, it is a cleaner fuel and therefore better for the environment, and it is also relatively renewable because corn is a major crop in the United States.

FFVs are not a new technology. Although they have been in mass production only since the 1980s, they have been around as long as cars. Today, many cars can be run on E85, unbeknownst to the drivers. To designate an FFV vehicle, typically a label will be in the owner's manual or near the fuel repository. At this time, the nations with the highest proportion of FFVs are the United States, Canada, Brazil, and Sweden, although FFVs can be found throughout the European Union.

The first flex fuel car available in the United States was the Ford Model T, made between the years 1908 and 1927. The Model T could run on gasoline, ethanol, or any combination blend. During the years of Prohibition (1919–33), when manufacturing, trading, or selling alcohol was outlawed in the United States, Henry Ford of Ford Motor Company was a chief advocate for ethanol as fuel.

Technology then developed with a chief focus on gasoline as fuel until political and economic reasons forced nations to look for supplemental fuel sources. The flex fuel technology was revolutionized by Brazil in the 1970s and 1980s in response to the oil crises of the 1970s. Oil prices soared, and it became prohibitively expensive, as well as socially unfavorable, to use foreign oil to fuel vehicles. Brazilian automakers developed cars that could run on ethanol made from sugarcane, a copious crop in the southern hemisphere nation. In fact, in 1976 a law was enacted in Brazil that required gasoline to be at least 10 percent anhydrous ethanol. These flex cars dropped in popularity, however, in the late 1980s, when sugar prices rose dramatically. Nevertheless, the momentum behind FFVs in Brazil has remained. In contemporary times, the laws that quantify fuels have been edited several times, and as of 2007, fuel is required to be one-quarter anhydrous ethanol blended into gasoline.

In the last quarter of the 20th century, public concern for global warming grew and new technologies were sought to attenuate the problem. The United States revisited flex fuel engineering in the 1980s in response to the state of California's request for a car that ran entirely on methanol, in hope of preserving air quality. As before, the Ford Motor Company was the first to make an effort. By the end of the 20th century, corn-produced ethanol was the preferred fuel. It had the necessary support from both the government and the influential farming community. Because of E85's source in corn, the Midwest region of the United States is a large market for FFVs.

Car companies that currently offer FFVs are DaimlerChrysler, Ford, General Motors, Isuzu, Mazda, Mercedes, Mercury, and Nissan. Models that can be FFVs come in all sizes including the Toyota Tundra and Sequoia, Chrysler Sebring, Jeep Grand Cherokee, Chevrolet Impala, Hummer H2, and Ford Explorer and Taurus.

Benefits of FFVs and E85 include it being a cleaner fuel source than gasoline, a renewable fuel source (E85 is made from corn), and a fuel source that is generated in the United States, thus reducing the need for foreign oil. In addition, scientists are developing methods of using bacteria to metabolize what would otherwise be waste products from corn, such as the husks, as well as plant-based garbage, into ethanol. This technology is being developed by companies such as Illinois-based Coskata. The company, supported by General Motors, predicts its fuel will be produced for less than $1 per gallon. FFVs can run on mixtures of fuels as well, so a car with half a tank of E85 can be filled with gasoline.

Drawbacks to FFVs include a reduced ability to start the car in colder weather. To avoid problems, a winter E70 blend of 70 percent anhydrous ethanol and 30 percent gasoline is sold in the United States (E85 is 15 percent gasoline and 85 percent anhydrous ethanol). Furthermore, one gallon of E85 fuel has less energy than one gallon of traditional gasoline; therefore, a car running on E85 will go for fewer miles per gallon. The difference is generally 20–30 percent. Another consideration for crop-based fuels around the globe is the food versus fuel debate. Some have expressed concern over the use of land to grow crops that will be used to fuel vehicles for the upper economic classes when the land could be used to grow crops that will feed people in the lower economic classes.

In addition, there is growing concern, especially in Brazil, over just how environmentally friendly sugar ethanol is. The deforestation to clear more land for sugarcane and increased use of pesticides and fertilizers, as well as the traditional practice of burning sugarcane fields before harvest, and now machine-harvesting the sugar, all dampen the environmental benefits of ethanol fuel. Scientists are working to determine which fuel source has the greatest net benefit when all these factors are taken into account.

FFVs are not to be confused with bi-fuel vehicles that have separate fuel tanks for different fuel types, such as hydrogen or compressed natural gas.

See Also: Alternative Fuels; Automobiles; Biodiesel; California; Ethanol, Corn; Ethanol, Sugarcane; Gasification; Gasohol; Greenhouse Gases; Hydrogen; Natural Gas; Oil; Plug-In Hybrid.

Further Readings

Blume, D. *Alcohol Can Be a Gas!: Fueling an Ethanol Revolution for the 21st Century.* International Institute for Ecological Agriculture, 2007.

Goettemoeller, J. and A. Goettemoeller. *Sustainable Ethanol: Biofuels, Biorefineries, Cellulosic Biomass, Flex-Fuel Vehicles, and Sustainable Farming for Energy Independence.* Marysville, MO: Prairie Oak, 2007.

McGuire, B. "Ethanol Alternative Fuel—E85 Secrets." *Hot Rod Magazine.* http://www .hotrod.com (Accessed May 2009).

Squatriglia, C. "Startup Says It Can Make Ethanol for $1 a Gallon, and Without Corn." *Wired Magazine.* (January 24, 2008).

Claudia Winograd
University of Illinois at Urbana-Champaign

FOOD MILES

Consumers are paying more attention to where their food comes from and to the environmental and social impacts of their diet. Food miles have emerged as an important tool to enable consumers, industry, and governments to visualize more effectively the environmental impacts of food. Food miles are typically defined as the distance food travels from where it is grown or raised to where it is sold to consumers. Higher food miles are associated with higher air pollution, greenhouse gas emissions, and energy use. Food miles have been used to promote locally sourced foods on the assumption that lower food miles are beneficial. However, food miles only capture some transportation impacts, whereas food production may be far more environmentally damaging.

The concept of food miles originated in the United Kingdom in the early 1990s in a nongovernmental organization report from Sustainable Agriculture, Food, and the Environment. The industrial food system is characterized by massive food miles and high inefficiency in energy use. In the United States, foods are transported a distance of 1,640 kilometers. In the United Kingdom, government researchers estimate that food transport was responsible for 30 billion kilometers in 2002, or 25 percent of all truck kilometers, emitting 19 million tons of carbon dioxide (or 12 percent more than in 1992).

Some of the reasons for the long distances include changes in the food industry, which is increasingly centralized and concentrated. In particular, supermarkets have built complex distribution systems that function in a hub-and-spoke way. Foods are shipped from producers or processors to a central warehouse and then to stores, increasing the distances traveled. Just-in-time deliveries also have come to prevail, so that supermarkets can reduce their inventories. In turn, international trade in food has expanded greatly since the 1970s, aided by improved packaging and refrigeration technologies, as well as falling trade barriers and cheaper transportation costs. Retailers increasingly source their foods globally; their sourcing decisions are probably far more important than consumer choices. Supermarkets now prefer to source foods from a few large-scale suppliers that can produce to uniform, predictable standards. Consumers are now accustomed to year-round availability, regardless of seasonality, as retailers switch between suppliers in different parts of the world.

The concept of food miles is very useful in illuminating some costs of producing food. Transporting food over long distances can cause excessive energy use, traffic congestion, noise, accidents, road infrastructure spending, air pollution, and climate change impacts. For example, Hawaiian pineapples flown to California and trucked to Iowa (a total of 4,234 miles) can use 250 times more energy and emit 260 times more carbon dioxide per pound of produce than apples grown in Iowa. Moreover, the mode of transportation is important because different modes vary greatly in their environmental impacts. Food transported by airplanes and trucks has significantly higher carbon emissions and energy use compared with cargo ships and railways. Thus, even if food is transported long distances, it can still have low impacts if it uses an efficient, low-carbon mode.

To calculate food miles, one approach is to use the "Weighted Average Source Distance" method. In other words, the distances for the various sources of a specific food are estimated geographically and averaged according to the market share of each source. For example, researchers at Iowa State University calculated the food miles for a typical Iowa dinner. The average distance was 2,577 kilometers, with purple cabbage moving 2,720 kilometers from California, chuck roast traveling 1,080 kilometers from Colorado, and potatoes 2,800 kilometers from Idaho. In contrast, a meal with ingredients wholly produced in Iowa had an average distance of 74 kilometers. Food miles calculations will vary based on the location of consumers. Californians are likely to have relatively lower food miles compared to most Americans because they live nearer major vegetable- and fruit-producing regions. Most studies have focused on specific foods such as fruit, meat, milk, or vegetables. Relatively few studies have examined baskets of foods or processed foods.

Some policy solutions include sourcing foods more locally, making vehicles more efficient, increasing the use of rail over trucking, and decreasing the use of cars in shopping. Supermarkets, especially in Europe, are changing their purchasing policies to reduce the food miles of their offerings and to preferentially buy from local suppliers. Food miles have been used widely to emphasize the benefits of locally sourced foods, including fewer transportation impacts, greater freshness and nutritive value, potentially better-treated agricultural workers, and income retained within local areas.

Analysts, however, have pointed out that food miles have significant limitations. First, food miles measure only transportation impacts, whereas other parts of the food production chain may be far more environmentally damaging. Recent research suggests that 83 percent of greenhouse gas emissions associated with food occurs in the production stages, compared with only 11 percent for all transportation in the supply chain and 4 percent for what food miles usually measure (i.e., transportation from producers to retailers). This research highlights the fact that total transportation within the full supply chain (such as transporting coal to power plants or fertilizers to farms) may be four times greater than "food miles" per se. The materials, chemicals, energy, and equipment used to produce food may make a food unsustainable, whereas considering only food transportation may make it sustainable.

One classic case often mentioned is tomatoes. Tomatoes grown in open air fields in Spain and imported into Sweden have markedly greater food miles compared with tomatoes grown in Sweden and Denmark. Yet the Scandinavian-grown tomatoes may have greater overall energy and greenhouse gas impacts if they are produced in heated greenhouses. In other words, food miles need to be contextualized.

Second, the complexity of processed foods makes calculating food miles a real challenge. Most foods—up to 70 percent of the U.S. food supply—are now processed. Processed food ingredients can be sourced from many locations around the world. To some extent, the concept of the Weighted Total Source Distance can deal with this. Food manufacturers, though, are reluctant to disclose where they obtain their ingredients.

Third, food miles have largely ignored distributive and social justice issues. Developing countries often complain that food miles may be used to discriminate against their products, threatening the livelihoods of poorer producers. In addition, New Zealand is lobbying British retailers to continue imports of fruits that are highly valuable to the country's economy. Moreover, requiring consumers to buy locally sourced foods may have a regressive effect: poorer consumers may be less able to afford such foods, particularly if they are organic. The notion of the "local trap" is becoming more prevalent: the assumption that local is intrinsically positive may not be realistic. Local food systems can still be polluting and abusive of worker welfare.

As a result, the U.K. government has decided to use multiple indicators, including urban food kilometers, life cycle greenhouse gas emissions, and measures that explicitly identify transport mode. Many researchers also say that local sourcing of food may not be sufficient in itself. Consumers may reduce their environmental impacts much more if they change their diets, such as switching from red meat to fish. Livestock production is known to be a major source of greenhouse gas emissions. In conclusion, food miles have highlighted what had been hitherto a neglected issue: the environmental impacts of food. Food miles, though, are increasingly being used as part of a suite of metrics that evaluate the environmental impacts of food more holistically.

See Also: Carbon Footprint and Neutrality; Climate Change; Sustainability.

Further Readings

Halweil, B. *Eat Here: Reclaiming Homegrown Pleasures in a Global Supermarket.* New York: W. W. Norton, 2004.

Jones, A. "An Environmental Assessment of Food Supply Chains: A Case Study on Dessert Apples." *Environmental Management,* v. 30/4 (2002).

Pirog, R., et al. *Food, Fuel, and Freeways: An Iowa Perspective on How Far Food Travels, Fuel Usage, and Greenhouse Gas Emissions.* Ames, IA: Leopold Center for Sustainable Agriculture, 2001.

Smith, A. *The Validity of Food Miles as an Indicator of Sustainable Development.* London: UK DEFRA, 2005.

Alastair Iles
University of California, Berkeley

FORECASTING

A broad range of quantitative and qualitative techniques and methodological approaches have been applied in the area of future studies, one of which is forecasting. Forecasting is often defined as the estimation of the value of a variable (or set of variables) at some future point in time. The terms *forecast, prediction, projection,* and *prognosis* are typically used interchangeably to describe forecasting. Risk and uncertainty are also central to forecasting and prediction.

The motivation for conducting future studies is to think about how firms and other actors might respond to a range of potential changes in the future. Preparedness (which is likened to agility) is therefore central to the value of futuring and forecasting as a methodological tool.

Forecasts may also be conditional, that is, if policy A is adopted, then outcome C is likely, but if policy B is adopted, then outcome D is more likely to take place. Forecasts are sometimes future values of a time-series; for example, the number of houses sold in a year, or the likely demand for electric cars. Forecasts can also be one-off events such as the opening of a new power station or a new energy policy. Forecasts can also be distributions, such as the locations of wind farms or the installation of cavity insulation among different age groups. Forecasting also includes the study and application of judgment as well as of quantitative (statistical) methods.

Forecasting exercises are often carried out as an aid to decision making and in planning the future. They typically work on the premise that if we can predict what the future will be like, we can modify our behavior now to be better positioned for the future than we otherwise would have been. Examples of applications for forecasting include inventory control/production planning, investment policy, and economic policy.

It is useful to consider the timescale involved when classifying forecasting problems—that is, how far forward into the future are we are trying to forecast? The usual categories are short, medium, and long term, but the actual meaning of each category will vary according to the situation that is being studied. For example, when forecasting energy demand to construct power stations, 5–10 years would be considered short term and 50 years long term.

Forecasting methods can be classified roughly into four methods:

- Qualitative methods: no formal quantitative or mathematical model or numbers
- Regression methods: an extension of linear regression where a variable is thought to be linearly related to a number of other independent variables
- Multiple equation methods: where there are a number of dependent variables that interact with each other through a series of equations (e.g., economic models)
- Time series methods: where a single variable changes with time and whose future values are related in some way to its past values

It is necessary to understand the different forecasting techniques and their relative qualities and thus be able to choose which method to apply in a particular circumstance. The accuracy of forecasting is compared by measuring errors. This error measure should be the one that relates most closely to the decision being made. Ideally, it should also allow you to compare the benefits from improved accuracy with the costs of obtaining the improvement. This is seldom possible to assess, however, so you might simply use the method or methods that provide the most accurate forecasts. The selection of which measure to choose depends on the purpose of the analysis. When making comparisons of accuracy across a set of time series, for example, it is important to control the scale, the relative difficulty of forecasting each series, and the number of forecasts being examined.

Forecasting the future of technology can be a hazardous enterprise. There is often a shortsightedness, even among experts, that causes them to focus on the future in terms of present conditions. One famous example is the call by the U.S. Commissioner of Patents in 1899 to abolish the Patent Office on the grounds that there was nothing left to invent. Futures studies also sometimes suffer from being perceived as an attempt to forecast or foresee the future. Prediction is not their purpose, however—their usefulness is in helping people and firms prepare for an uncertain future by producing a range of possible futures and identifying potential risks and opportunities to inform current decision making.

It is a general weakness of future-oriented methodologies in general that it is much more difficult to imagine a radically transformed future than to extrapolate current trends forward through time. This also partly explains the popularity of information technology growth, education, sustainability, and increased global competition within forecasting reports, as they are long-standing current preoccupations within government, firms, and society more generally. This also highlights a paradox within many forecasting studies—they are often intended to help understanding and to deal with a rapidly changing world, but they usually do so with reference to past and current trends and ideas. There is also debate as to what are "good" and "bad" forecasts, as well as for whom they might be good and bad. A highly regulated and standardized future energy sector may be able to produce environmentally efficient and functionally adequate energy, but what might be the effect on cost and functionality for consumers? How can potentially expensive energy-efficiency measures for buildings be reconciled with a perceived need to increase the productivity and profitability of building and construction work?

It is a common assumption that progression toward an envisioned future moves along an incremental path, and that incremental adaptation is required to keep pace with what changes are occurring. However, given both the variety of factors and the wide-sweeping nature of factors such as climate change, as well as the possibilities for the same factors to produce significantly different outcomes, forecasting on its own may be insufficient. Multiple, irregular, and potentially discontinuous paths might lead toward numerous futures, and more radical steps may be required to merely survive, let alone perform more effectively and productively.

Examples of recent forecasts include the United Nations, which reports that millions of new jobs will be created worldwide over the next few decades by the development of alternative energy technologies, with those working in biofuels rising from one million today to 12 million by 2030. Another report from the American Solar Energy Society shows that as many as one of four workers in the United States will be working in the renewable-energy or energy-efficiency industries by 2030. These industries already generate 8.5 million jobs in the United States, and with appropriate public policy, it is forecast that this could grow to even 40 million jobs by 2030.

See Also: Energy Policy; Innovation; Risk Assessment.

Further Readings

Armstrong, J. Scott, ed. *Principles of Forecasting: A Handbook for Researchers and Practitioners.* Norwell, MA: Kluwer Academic, 2001.
International Institute of Forecasters. http://www.forecasters.org (Accessed February 2009).
International Institute of Forecasters. *International Journal of Forecasting.* http://www.elsevier.com/locate/ijforecast (Accessed February 2009).
Journal of Forecasting. http://eu.wiley.com/WileyCDA/WileyTitle/productCd-FOR.html (Accessed February 2009).
Makridakis, Spyros, et al. *Forecasting: Methods and Applications.* New York: John Wiley & Sons, 1998.
Rescher, Nicholas. *Predicting the Future: An Introduction to the Theory of Forecasting.* Albany, NY: State University of New York Press, 1998.

Chris Goodier
Loughborough University

FOSSIL FUELS

Fossil fuels are nonrenewable energy resources formed from dead remains of plants and lower animals including phytoplankton and zooplankton that have settled to the sea or lake bottom in large quantities under anoxic conditions. They are nonrenewable resources because they take millions of years to form, and reserves are being depleted much faster than new ones are being formed. They formed from ancient organisms that died and were buried under layers of accumulating sediment. As additional sediment layers built up over these organic deposits, the materials were subjected to increasing temperatures and pressures, leading to the formation of a waxy material known as kerogen, and subsequently hydrocarbons. The processes involved are diagenesis and catagenesis.

The three types of fossil fuel are coal, natural gas, and petroleum (crude oil). They are widely different in their physical properties and molecular composition. Carbon is the predominant element present in them. They differ in the amounts of hydrogen and carbon they contain, and this accounts for the differences in fuel energy content and physical state.

The ratio of hydrogen atoms to carbon atoms is approximately 1:1, 2:1, and 4:1 in coal, crude oil, and natural gas, respectively. A variety of sophisticated instruments are used to locate underground petroleum, natural gas, and coal deposits. These instruments allow scientists to interpret the geologic composition, history, and structure of sedimentary basins in the Earth's crust.

Coal

Coal is a heterogeneous solid fuel with no definite molecular structure. Various chemical reactions and instrumental analyses of coal indicate that it has a three-dimensional structure and consists of aromatic ring systems cross-linked by aliphatic carbon bridges or ether oxygen bridges. Coal contains between 65 and 95 percent carbon. The successive stages of coal formation are peat, lignite, subbituminous coal, bituminous coal, and anthracite. Anthracite is the most carbon rich, and hence, it has the highest heating value. The process that transformed the plant remains into coal by chemical and geological alterations is called coalification.

The components of coal are volatile matter, fixed carbon, moisture, and ash. The percentages of these components in a given sample of coal vary depending on the source and type of coal. Volatile matter and fixed carbon are the fuel components of coal, whereas moisture and ash are not combustible. The percentage elemental composition of coal varies depending on the type and source of coal. Coal contains trace amounts of every known element except the noble gases and the highly unstable elements made in nuclear reactors. Important physical properties of coal that determine its behavior during handling and processing are porosity, density, hardness, and grindability.

There are three main areas where coal is used, namely carbonization, conversion, and combustion. Heating coal in the absence of air is called carbonization. High- and low-temperature coal carbonization processes give three products each; namely, coal gas, coal tar, and coke. About 90 percent of all coke produced by high-temperature carbonization is used in blast furnaces for making pig iron. The rest is used in foundries for making cast iron, for reaction with calcium carbide, for reaction with steam and air to give water gas and producer gas, for domestic cooking, for home heating in cold countries, for production of acetylene and synthesis gas, as boiler fuel for generating steam, and for sintering iron ore into pellets meant for blast furnaces. Coal gas is used as fuel; the soluble components of the coal gas include ammonia, hydrogen sulfide, hydrogen cyanide, hydrogen chloride, and carbon dioxide (CO_2). Coal tar is refined to create lubricants, heating oils, and gasoline. Hundreds of chemical components can be isolated from coal tar by careful separation and analyses; they include benzene, toluene, xylenes, ammonia, naphthalene, anthracene, phenol, aniline, and pyridine. The three general methods used for the conversion of coal to synthetic fuels are gasification, indirect liquefaction, and direct liquefaction. Synthetic fuels from coal are easier and more convenient to transport and distribute than coal.

Coal combustion for the production of steam for the generation of electricity is the most important application of coal in industrialized countries. Coal is burned in power plants to produce electricity and in steel mills to make coke for the production of steel. About 50 percent of worldwide coal consumption is used for generation of electricity. Traditionally, coal has been burned to provide heat and power for residential and manufacturing needs; for example, for firing kilns in manufacturing cement. However, some of the organic chemicals derivable from coke and coal tar, for example, acetylene and synthetic gas, are now derived from natural gas and crude oil. Thus, the modern organic chemical industry is largely a petrochemical industry.

Natural Gas

Natural gas is a mixture of hydrocarbons with a small proportion of nonhydrocarbons. Methane is the principal hydrocarbon component of natural gas. Natural gas contains over 90 percent methane. The nonhydrocarbon components of natural gas include CO_2, hydrogen sulfide, nitrogen, inert gases, and water vapor. The actual composition of natural gas varies from place to place. A greater proportion of natural gas is used as fuel. The principal nonfuel use of natural gas is the manufacture of synthesis gas for the production of ammonia, methanol, hydrogen, aldehydes, alcohols, and acids. The main challenges of natural gas as fuel are encountered in its storage, long-distance transportation, and incomplete combustion of methane. Because of its very low density at room temperature and atmospheric pressure, storage of natural gas requires its liquefaction at very low temperatures. Storage of natural gas is, therefore, expensive in comparison with liquid petroleum products and solid coal. Also, as gas leakages may not immediately be noticed, a potentially hazardous accumulation of gas may build up from the storage. Similarly, intercontinental transportation of natural gas is very expensive and potentially hazardous, as it involves liquefaction of large volumes of liquefied natural gas by cooling it to minus 162 degrees C. The manufacture of liquefied natural gas is expensive in terms of both the process plants and the sophisticated ships that carry the cargo. The fuel uses of natural gas involve combustion of methane. However, in a limited supply of oxygen, incomplete combustion of methane takes place, giving off carbon monoxide and water. Incomplete combustion is undesirable because it produces less heat and more carbon monoxide, which is an extremely poisonous gas. It also produces soot, which is undesirable in a gas combustion system.

Petroleum

Petroleum is a homogeneous liquid containing several hundred to several thousand individual chemical compounds. Analyses of samples of petroleum from different parts of the world show that the elemental composition of petroleum varies over a narrow range: 82–87 percent carbon, 11–16 percent hydrogen, 0–1 percent sulfur, 0–7 percent oxygen plus nitrogen, and a few parts per million metals. All the compounds in petroleum may be grouped into two broad categories: hydrocarbons and nonhydrocarbons. About 90 percent by weight of the compounds present in petroleum are hydrocarbons. The hydrocarbons present in petroleum may be further grouped into alkanes, cycloalkanes, and aromatic hydrocarbons. The proportions of these three types of hydrocarbons vary with the source of the petroleum. In addition to hydrocarbons, petroleum generally contains relatively small proportions of sulfur, nitrogen, oxygen-containing compounds, and organo-metallic compounds. Petroleum could be classified on the basis of its composition, geochemical considerations, and American Petroleum Institute gravity.

Petroleum refining for production of hydrocarbon fuels involves a number of unit operations that are designed to achieve four main objectives:

- Separation of crude oil into fractions with different boiling ranges that are suitable for specific uses
- Purification of the petroleum products

- Conversion of those products with a relatively low market demand to the products with higher market demand
- Improvement of the quality of the refined products

Petroleum is a major energy source for domestic and industrial activities. Fuel gases, liquefied petroleum gas, gasoline, aviation turbine kerosene, dual-purpose kerosene, diesel fuel, fuel oil, lubricants, and bitumen are important fractions from petroleum whose applications cut across land and air transportation, cooking, heating, lubrication, and road construction, among others. A major challenge in the use of petroleum fractions is the high cost of petroleum refining and the maintenance of the refineries. Pipeline leakages lead to oil spillage, and consequently environmental pollution. Petroleum combustion also pollutes the environment, as undesirable gases are evolved. In addition, many of the fractions could easily ignite, leading to fire explosions.

Deposits of other hydrocarbons, including gas hydrates (methane and water), tar sands, and oil shale, have been identified. Vast deposits of gas hydrates are contained in ocean sediments and in shallow polar soils. In these marine and polar environments, methane molecules are encased in a crystalline structure with water molecules. Because technology for the commercial extraction of gas hydrates has not yet been developed, this type of fossil fuel is not included in most world energy resource estimates.

Tar sands are heavy, asphalt-like hydrocarbons found in sandstone. Tar sands form where petroleum migrates upward into deposits of sand. When the petroleum is exposed to water and bacteria present in the sandstone, the hydrocarbons often degrade over time into heavier, asphalt-like bitumen. Oil shale is a fine-grained rock containing high concentrations of kerogen. Oil shale forms on lake and ocean bottoms where dead algae, spores, and other microorganisms died several years ago and accumulated in mud and silt. The increasing pressure and temperature from the buildup of overlying sediments transformed the organic material into kerogen and compacted the mud and silt into oil shale. However, the insufficiency of the pressure and heat chemically break down the kerogen into petroleum. Because the hydrocarbons contained in tar sand and oil shale are not fluids, these hydrocarbons are more difficult and costly to recover than liquid petroleum.

When a fossil fuel is burned, the atomic bonds between carbon and hydrogen atoms are broken, releasing chemical energy. This energy provides heat, which is converted to electrical energy in the power plant. They provide about 95 percent of the world's total energy demands and about 66 percent of the world's electrical power. The amount of energy that can be obtained by burning a specified amount of fossil fuel is called its heating value. The production and use of fossil fuels raise environmental concerns. The quantity of fuel burned largely determines the amounts of environmental contaminants that are released. Acid rain, global warming, and ozone depletion are some of the most serious environmental issues related to large-scale fossil fuel combustion. Other environmental problems, such as land reclamation and oil spills, are also associated with the mining and transporting of fossil fuels.

When fossil fuels are burned, sulfur, nitrogen, and carbon combine with oxygen to form their oxides, and when these oxides are released into the air, they react chemically with atmospheric water vapor, forming acid rain that can enter the water cycle and subsequently harm the biological quality of forests, soils, lakes, and streams. Burning natural gas that contains hydrogen sulfide releases oxides of sulfur into the atmosphere. The presence of sulfur oxides in the atmosphere lowers the pH of rainwater and gives rise to the challenge of acid rain. Hydrogen sulfide and sulfur oxides, its combustion products, are injurious to health. In humans, acid rain can cause or aggravate lung disorders. It also damages many building materials and automobile coatings.

CO_2 is one of the greenhouse gases that lead to global warming. Global warming refers to the rise in atmospheric pressures resulting from the heating of the atmosphere from increasing amounts of CO_2 and higher concentration of greenhouse gases. Global warming engenders climate change. CO_2 and other greenhouse gases released into the atmosphere from human activities trap the infrared wavelengths of radiant energy so that the heat is not dispersed into space. Rapid industrialization has resulted in increasing fossil fuel emissions, raising the percentage of CO_2 in the atmosphere by about 28 percent. This increase in CO_2 level has led some scientists to predict a global warming scenario that could cause numerous environmental problems, including disrupted weather patterns and polar ice cap melting.

Ozone is a form of oxygen naturally created in the stratosphere when ultraviolet radiation breaks down the oxygen molecule into two individual oxygen atoms. Ozone is essential in the stratosphere but a pollutant in the troposphere. Ozone is broken down by human-made pollutants such as chlorofluorocarbons, halons, methyl chloroform, and carbon tetrachloride. A decrease in stratospheric ozone results in an increase in surface ultraviolet light levels. High levels of ultraviolet light cause skin cancer and cataracts and also weakened immune systems. Ultraviolet light destroys chlorophyll and could damage the plankton that forms the base of the food chain in Antarctica.

Another environmental problem associated with coal mining occurs when freshly excavated coal beds are exposed to air. Sulfur-bearing compounds in the coal oxidize in the presence of water to form hydrogen tetraoxosulfate acid. When this acid solution, known as acid mine drainage, enters surface water and groundwater, it can be detrimental to water quality and aquatic life. When coal is burned in combustion systems, three main forms of environmental pollutants (sulfur oxides, nitrogen oxides, and coal ash) are formed. CO_2 is also increasingly emitted during coal combustion.

Combustion of fossil fuels produces unburned fuel particles, known as ash. Although petroleum and natural gas generate less ash than coal, air pollution from fuel ash produced by automobiles may be a problem in cities where diesel and gasoline vehicles are concentrated. Petroleum is usually transported long distances by pipeline or tanker to reach a refinery. Transport of petroleum occasionally leads to accidental spills. Oil spills, especially in large volumes, can be detrimental to humans, wildlife, and habitat. Oil spills have been a challenge. Environmental problems are created by drilling oil wells and extracting fluids because the petroleum pumped up from deep reservoir rocks is often accompanied by large volumes of saltwater. This brine contains numerous impurities, so it must either be injected back into the reservoir rocks or treated for safe surface disposal.

Reserves Versus Resources

When estimating the world's fossil fuel supply, reserves are usually distinguished from resources. Reserves are fossil fuel deposits that have already been discovered and are immediately available, whereas resources are fossil fuel deposits that are believed to be located in certain sedimentary basins but have not yet been discovered. Fossil fuel reserves can be further divided into proved reserves and inferred reserves. Proved reserves are deposits that have been measured, sampled, and evaluated for production quality. Inferred reserves have been discovered but have not been measured or evaluated.

Fossil fuel resources can be narrowed to technically recoverable resources. Here, the consideration is whether the fossil fuel can be recovered using existing technology. Fossil fuels take several years to form and cannot be replaced in the same way. A profound concern is that someday these resources will run out. Thus, nation-states depending on the refining of these finite resources must find other alternatives that are renewable for their economic growth and development.

See Also: Coal, Clean Technology; Natural Gas; Oil; Oil Sands; Oil Shale.

Further Readings

Kiely, G. *Chemical and Petroleum Engineering Series.* Columbus, OH: McGraw-Hill International Editions, 1998.
Obuasi, P. *The Origin of Petroleum, Natural Gas and Coal.* Nsukka, Nigeria: University of Nigeria, 2000.
Rubin, E. S. *Introduction to Engineering & the Environment.* Columbus, OH: McGraw-Hill, 2000.

Akan Bassey Williams
Covenant University

FUEL CELLS

Fuel cells, like batteries, convert chemical energy to electricity. Unlike batteries, "fuel" can be fed continuously into a fuel cell from a reservoir that can be refilled, even during operation. Fuel cells have advantages over internal combustion engines (ICEs): (1) they can convert chemical energy with higher efficiency into electricity (and in conjunction with electric motors, into mechanical energy), (2) all waste products are recyclable, (3) they typically produce no toxic emissions, and (4) they operate with no moving parts to wear out. With no combustion, fuel cells provide the cleanest fuel-consuming energy technology, with near-zero smog-causing emissions. Many see these environmental advantages to be critical for a future "green" economy.

Fuel cells produce electricity quietly, with efficiencies of up to 60 percent (80–90 percent overall efficiency if heat energy is used and accounted for). For comparison, diesel power turbines covert only about 40–50 percent of the chemical energy into electricity, whereas gasoline internal combustion engines generators reach only 10–20 percent efficiency.

How Fuel Cells Work

Different materials (metals, hydrogen, and oxygen) have different affinities for electrons. When two dissimilar materials make contact, or connect through a conducting medium, electrons tend to pass from the material with the lesser affinity for electrons, which becomes positively charged, to the material with the greater affinity, which becomes negatively charged. The potential difference between the materials results from an electrochemical oxidation-reduction reaction that builds up to the "equilibrium potential," which balances the difference between the propensity of the two materials to gain or lose electrons. Batteries and fuel cells can deliver this electrical energy on demand. The fuel cell's electrochemical reactions can be reversed by electrolysis. Where "excess" renewable energy (e.g., excess wind or hydro power) is available for regenerating the fuel, this provides an environmentally friendly cycle.

Fuel cells directly convert chemical energy to electrical energy and consist of at least three components:

1. The anode (negative electrode) releases electrons to the external circuit during the electrochemical reaction. A metal alloy or hydrogen is commonly used. This process oxidizes the "fuel" to form positive ions.

2. The cathode (positive electrode) accepts electrons from the external circuit during the electrochemical reaction. A metallic oxide, a sulfide, or oxygen commonly serves this purpose. This process reduces the oxide to form negative ions.

3. An electrolyte (the ionic conductor) transfers a charge (as ions) between the anode and cathode; however, it does not conduct electrons.

Fuel cells produce electricity that can then be sent through power lines for typical electrical applications (or wires for potable uses), can power electric motors for transportation (automotive, locomotive, specialty vehicles), or can provide combined heat and power.

Fuels Cells Versus Batteries

Both batteries and fuel cells use electrochemical reactions based on electrochemical potentials. However, a battery has a fixed life based on its original constituents, and links power (Watts, W) to energy (Watt-hours, Wh). Rechargeable batteries can usually only be recharged with electricity. Some newer "refuelable batteries" and the variations called "flow batteries" may be considered a type of a fuel cell.

The total energy efficiency of fuel cells generally exceeds that of batteries. Fuel cell efficiencies for conversion of chemical energy to electricity vary by type, but today's most efficient fuel cells exceed 60 percent efficiency, whereas a nonrechargeable battery returns only about 2 percent of the energy used for its production. When fuel cells are configured to use their waste heat, the combined systems can exceed 80 percent efficiency.

Fuel Cells Versus Internal Combustion Engines

Fuel cells convert chemical energy to electricity, and if desired, that electricity can be converted into mechanical energy. ICEs convert chemical energy to heat, heat to mechanical energy, and if desired, mechanical energy to electrical energy. Each transformation entails losses, mostly as waste heat.

Fuel cells, similar to ICEs, allow refueling during operation, and preparation of "fuel" can take place elsewhere. Both draw fuel from an external fuel storage system, so they can be sized for the required power (W; horsepower, hp), and only the availability to store and feed in the fuel governs the total energy (Wh; horsepower hours, hp-h) that can be produced. Thus, they both separate power from energy, and both can be sized for optimum power (W or hp), not energy (Wh, hp-h). Both typically use hydrogen-rich fuels and oxygen from the air as the oxidant. Both give off heat (and may require cooling).

Fuels cells may also require direct current to alternating current conversion and power conditioning (~90 percent efficient), controller and heat management systems, air compression (~80 percent efficient), and a motor (~90 percent efficient). These losses slightly reduce the overall efficiency of fuel cells in providing either household electricity or motive power. Hydrogen fuel cells may also require a hydrogen handling system and hydrogen storage (typically ~85 percent efficient).

Types of Fuel Cells

When most people think of fuel cells, they most frequently think of hydrogen fuel cells, using hydrogen-oxygen electrochemical processes. Several other types of fuel cells have been developed or are in various stages of research and development. One of those making substantial progress is the zinc-air fuel cell (discussed here).

Other electrochemical processes that have been made into "fuel cells" (some called "refuelable batteries" or "flow batteries") use a reservoir of rechargeable electrolytes. These include the use of reactions with zinc/bromine, sodium/bromine, or a vanadium redox reaction. Additional work has also been done using consumable alcohols, dimethyl ether, and bacterial products.

All of the waste from fuel cells can be recycled, and the electrochemical processes can typically be reversed by electrolysis, which can be powered from renewable sources.

Hydrogen Fuel Cells

Hydrogen has the highest energy-to-weight ratio of any fuel. Its low density as either pressurized gas or a liquid means that large volumes are required to carry the energy equivalent of conventional fuels. This plus the containment problems requiring heavy tanks raise challenges for the mobile use of hydrogen fuel cells. The high efficiency of fuel cells does not come close to making up for the difference in storage volumes. Even at 350 bar pressure, it requires 11 times the volume of hydrogen than that of gasoline for the equivalent energy, and (including containment) eight times the weight. Even liquid hydrogen requires four times the volume and three times the weight, and liquefaction can cost as much as a 40 percent energy penalty.

Alternative storage methods, such as chemical binding in hydrides or sorption in nanotubes, continue to be researched, with no clear solutions yet fully demonstrated. The alternative to carrying a fossil fuel for onboard production of hydrogen solves none of the environmental issues. However, onboard hydrogen production from other chemical reactions or biofuels remains possible. For example, Millennium Cell's Hydrogen on Demand system generates pure hydrogen from sodium borohydride. Dissolved in water and passed through a proprietary catalyst chamber, sodium borohydride releases pure hydrogen on demand. (Borohydride comes from sodium borate, commonly known as borax, which is found in substantial natural reserves globally.)

Hydrogen fuel cell types can be classified by their operating temperatures as either moderate or high temperature. Moderate-temperature fuel cells include alkaline, proton exchange membrane (PEM), polymer electrolyte, direct methanol (related to PEM), and solid acid fuel cell. High-temperature fuel cells include phosphoric acid, molten carbonate (sometimes called direct fuel cells), and solid oxide. Each has certain advantages and disadvantages (e.g., PEM cells have been harnessed for transportation because of ease in rapid start-up and load following, whereas solid oxide fuel cells require long start-up times but are much more tolerant of fuel impurities).

Zinc-Air Fuel Cells

Of the many potential types of fuel cells, zinc-air (ZAFC) technologies show particular promise and are being rapidly commercialized. ZAFCs use the electro potential between zinc and oxygen, which provides theoretical efficiency limits higher than for hydrogen/oxygen and has electrochemical reversibility. Zinc carries a unique set of properties that provide advantages over hydrogen. These include high volumetric energy density, high specific energy, good conductivity, abundance, low cost, low toxicity, and ease of handling (especially compared with hydrogen).

Typical ZAFCs use a gas diffusion electrode that allows atmospheric oxygen to pass through a zinc anode separated by electrolyte, and some form of mechanical separators. Hydroxyl ions travel through the electrolyte and reach the zinc anode. There, they react with the zinc and form zinc oxide. ZAFCs contain a "fuel tank" that feeds zinc (e.g., as pellets) into the cell. The resultant zinc oxide can be regenerated quickly by electrochemical processes, with high efficiency.

One fueling system uses a unique cell construction that allows zinc pellets to flow through the system. As the zinc reacts, the pellets shrink and sink farther into the tapered electrode chamber, until the final waste product (a zinc hydroxide) flows out the bottom for collection and recycling. Electrolysis then reforms zinc pellets from the residue. Another variation on the zinc feed system uses replaceable fuel "charges" that can be replaced quickly and then reprocessed at another location.

Demonstrations have been conducted with zinc-air fuel cell vehicles, and full commercialization may follow soon. Infrastructure investment would not be trivial for transportation applications but could be small in comparison with that required for hydrogen. Perhaps the electrolyzer and fuel-forming apparatus could be located at a gas station to regenerate using night-rate electricity.

Other Promising Fuel Cells

The direct carbon conversion fuel cell also shows much promise. Electricity comes from the carbon-to–carbon dioxide electrochemical reaction. The total efficiency can potentially reach 80 percent, with all of the carbon consumed in a single pass. Carbon is readily available from coal. Although the reaction creates carbon dioxide from fossil carbon, the output is pure carbon dioxide, which can easily be captured for sequestration.

Research on the microbial fuel cell is another development worth following. The microbial fuel cell converts chemical energy, available in a bioconvertible substrate, directly into electricity. To achieve this, bacteria are used as a catalyst to convert a substrate into electrons. Very small (approximately 1 µm) bacteria can convert a large variety of organic compounds into carbon dioxide, water, and energy. The microorganisms use the produced energy to grow and to maintain their metabolism. However, the microbial fuel cell can harvest a part of this microbial energy in the form of electricity.

Potential Fuel Cell Applications

Many see fuel cells with electric motors as replacing ICEs in automobiles. Manufacturers have already made some inroads with specialty vehicles, such as mine or warehouse applications, where noxious emissions present a major concern. Several carmakers now have demonstration vehicles operating in test mode. In the process, current hybrid vehicles are helping refine many of the technologies required for fuel cell vehicles, such as efficient electric motors and regenerative braking for small cars. In terms of "well to wheel" efficiency, current internal combustion engine vehicles yield about 14 percent efficiency, and the Toyota Prius about 28 percent, whereas studies suggest that fuel cell vehicles using hydrocarbon fuels can yield 40 percent. Except for onboard reforming of gasoline to provide hydrogen, fuel cell fuels will require major changes to our fueling infrastructure.

Stationary power from fuel cells also has found niche markets in which they appear to be practical today. Stationary applications usually have ample space for storing large volumes, which can pose a problem for vehicular use. Village power based on intermittent renewables can benefit from excess electric capacity to electrolyze water (or other fuel cell fuels) to provide energy storage. Fuel cells can then provide clean, quiet power and heat at individual buildings. Similarly, small systems can provide back-up power for critical facilities, such as hospitals. As of early 2008, over 2,000 homes in Japan were using hydrogen reformed from natural gas to run fuel cells to provide their electricity and heat their water. Although the natural gas reforming gives off carbon dioxide, the entire process produces far less greenhouse gas per W than traditional generation. The Japanese government is so bullish on this technology, it has earmarked $309 million a year for fuel cell development and plans for 10 million homes (about one-fourth of all Japanese households) to be powered by fuel cells by 2020.

Portable fuel cells also show promise for a variety of small-scale applications and may eventually replace batteries, especially nonrechargeable ones. Manufacturers plan production of portable fuel cells for purposes ranging from military communications to laptop and cell phone operations. They will probably be cost competitive for niche applications before long.

Conclusion

Fuel cells provide the opportunity to efficiently use nonpolluting energy carriers based on renewable resources. The new fuels have the potential to substantially decrease emissions of carbon dioxide, sulfates, and nitrates, especially if the base energy sources are renewable. Fuel cells offer increased efficiency compared with internal combustion engines, but carrying hydrogen fuel still presents an issue—other fuels may become more popular in the long run.

Stationary applications (such as home power and heat) appear to be filling niche markets and will no doubt expand as technologies continue to improve and costs decline. Although cost competitive for certain applications now, considerable development and demonstration will be necessary to convince the public to fully embrace fuel cell vehicles with the accompanying infrastructure costs it will require.

See Also: Alternative Energy; Automobiles; Combustion Engine; Hydrogen.

Further Readings

Schlumberger Excellence in Educational Development, Inc. "Alternative Energy Sources: Fuel Cell Energy." http://www.seed.slb.com/en/scictr/watch/climate_change/alt_fuelcell.htm (Accessed January 2008).

William Isherwood
Chinese Academy of Sciences

FUSION

Fusion is the energy source that powers the sun and stars, thermonuclear weapons, and a range of experimental efforts to develop a sustainable source of commercial energy. The term also represents the class of nuclear reactions that produce these forms of energy. In those reactions, lighter nuclei combine to produce heavier ones, releasing energy in the process. Controlled fusion is the goal of researchers and advocates who regard it as a potentially safe, reliable, and virtually unlimited energy source. Although researchers have achieved important technical successes along the path to controlled fusion, a practical and economically viable technology has yet to be developed. Whether fusion energy would be truly "green" in terms of its environmental impact is a question debated by advocates and critics.

Fusion is distinguished from another class of nuclear reactions known as fission, in which heavier nuclei split to produce lighter nuclei. Fission reactions are the energy source for first-generation atomic weapons (as opposed to hydrogen, or thermonuclear, weapons) and for the nuclear power reactors that have been built to date.

In the most basic fusion reaction, two hydrogen nuclei, or protons, combine to produce a nucleus of deuterium, also known as heavy hydrogen, consisting of a proton and a neutron bound together by the strong nuclear force. A positron and a neutrino are also released in this process, along with energy. In the sun, this reaction links with others to ultimately convert hydrogen into helium with a net release of energy. The complete set of reactions is known as the proton-proton chain. Other processes, such as the carbon cycle, operate in more massive stars.

For such reactions to occur, the positively charged nuclei must overcome the repulsive electrostatic force, known as the Coulomb barrier, that would normally push them apart. In the stars, energy sufficient to overcome (or, more accurately, tunnel through) this barrier is provided by the high temperatures and densities that prevail in stellar interiors. In thermonuclear weapons, the energy is provided by a fission explosion, known as the primary, which produces conditions suitable for a secondary fusion explosion.

Laboratory experiments producing small-scale fusion reactions took place in the 1930s, before the invention of atomic and hydrogen weapons. Larger-scale efforts directed toward developing controlled and sustainable fusion began in the 1950s, at a time of great scientific and technological optimism. Since that time, it has been a standard joke among researchers and critics that fusion is a highly reliable energy source, always just 50 years away. In practice, the technological challenges associated with fusion are daunting.

In the sun and stars, thermonuclear weapons, and laboratory experiments, fusion reactions take place in a plasma, a state of matter in which atoms have been separated into their constituent nuclei and electrons. The most common approach to controlled fusion has been magnetic confinement, in which plasmas are heated to high temperatures and confined by strong magnetic fields. Successive generations of experiments have brought researchers closer to reaching the breakeven point at which the energy produced exceeds the energy used to heat and confine the plasma, and ignition or burning, a condition in which the fusion reaction is self-sustaining. Plasma physics is the science of matter in the plasma state and the foundational science of nuclear fusion research.

A class of fusion reactors known as "tokamaks" (derived from the Russian words for "toroidal chamber and magnetic coil") has been one of the more promising experimental technologies. The most powerful tokamak experiments have used the deuterium-tritium reaction, in which the nuclear fuel consists of two heavier isotopes of hydrogen. Deuterium is a harmless isotope available naturally as a small fraction of the world's natural hydrogen, and tritium is a radioactive isotope that must be produced artificially. Current experiments rely on tritium produced externally in fission reactors. Future devices would incorporate a supply of lithium that would be converted to tritium by means of additional reactions. Critics regard fusion technologies relying on tritium as problematic because of their use of radioactive material and their present connections to fission technologies. In addition, the deuterium-tritium reaction produces energetic neutrons that bombard the reactor chamber, producing radioactively contaminated materials that must be handled remotely and disposed of appropriately. Although these concerns are far less severe than those associated with fission reactors, they are still considered significant by critics.

Two more-advanced tokamak reactors are intended to take the fusion research and development program further. ITER (originally the International Thermonuclear Experimental Reactor), in Cadarache, France, may produce pulses of fusion power of up to 1,000-second duration using deuterium and tritium fuel. The power produced may be approximately 10 times the power used to heat and confine the plasma, although the experiment will not convert the power produced to a commercially usable form. The next-generation device, called DEMO (for demonstration power plant) would produce continuous fusion power at a level comparable to the output of a commercial nuclear fission plant and would convert that power to electricity. These experiments are of interest to social scientists and policy makers as case studies in "big science," involving large budgets, large-scale international and interdisciplinary collaboration, and long time frames for development and completion.

An alternative route to commercial power is known as inertial confinement fusion or inertial fusion energy. The National Ignition Facility at the U.S. Department of Energy's Lawrence Livermore Laboratory is the premier inertial confinement experiment, using 192 powerful lasers to ignite tiny pellets of deuterium-tritium fuel. Inertial confinement fusion research also supports the U.S. nuclear weapons "stockpile stewardship" program, intended to explore the design principles for new thermonuclear weapons without large-scale explosive testing. The U.S. budget for inertial confinement fusion research has exceeded the corresponding budget for magnetic fusion research since 1996.

Another approach that has received wide public attention but remains highly speculative is cold fusion. In 1989, two researchers announced that they had produced fusion under room-temperature conditions using an electrochemical apparatus. Their findings have been discredited by the majority of experts and are now widely viewed as a case study in scientific error. Nevertheless, some researchers continue to explore the possibilities for room-temperature fusion with modest amounts of private and government funding under a new and less-controversial name, condensed matter nuclear science.

Critics express a range of concerns about both magnetic and inertial confinement fusion, including their costs and uncertainties and their reliance on complex, centralized technologies. In the case of inertial confinement, additional concerns surround the technology's close connections to nuclear weapons research and development. At the same time, advocates point to fusion's potential for providing large amounts of energy without the atmospheric impacts of fossil fuels or the more severe problems of safety and waste disposal associated with nuclear fission. Meanwhile, most scientists remain highly skeptical of the prospects for cold fusion. It is likely to be some time before these debates are resolved and the technological and policy questions related to fusion are answered conclusively.

See Also: Department of Energy, U.S.; Nuclear Power.

Further Readings

Fowler, T. Kenneth. *The Fusion Quest*. Baltimore, MD: Johns Hopkins University Press, 1997.

International Atomic Energy Agency. *World Survey of Activities in Controlled Fusion Research*, 11th ed. Vienna: International Atomic Energy Agency, 2007. http://nds121.iaea.org/physics (Accessed February 2009).

ITER International Fusion Energy Organization. "The ITER Project." http://www.iter.org/ (Accessed February 2009).

Lawrence Livermore National Laboratory. "The National Ignition Facility and Photon Science." https://lasers.llnl.gov (Accessed February 2009).

McCracken, Garry M. and Peter Stott. *Fusion: The Energy of the Universe*. Boston, MA: Elsevier Academic, 2005.

William J. Kinsella
North Carolina State University

GASIFICATION

Gasification is the thermochemical production of gaseous fuels from carbonaceous material such as coal, industrial and municipal waste, and biomass. Whereas simple combustion of this material releases thermal energy, gasification produces gaseous fuel that can be combusted at higher efficiency than solid fuels or employed as a precursor for higher-value fuels and chemicals. Removing pollutant-forming compounds from the output gas stream can significantly reduce the level of air pollution resulting from use of the fuel when compared with pollution formed when using conventional solid fuels. Carbon dioxide can be separated from product gas into a concentrated stream of carbon dioxide, facilitating the integration of carbon capture and sequestration within power systems and reducing greenhouse gas emissions associated with fossil fuel use. Using sustainably produced biomass feedstock as an input to gasification can achieve further reductions in greenhouse gas emissions.

Process Description

In the gasification process, the feedstock is subjected to high temperatures and pressure with controlled quantities of oxygen and/or steam, sometimes in the presence of a catalyst. These conditions cause a conversion of the solid fuel to a mixture of gases, called synthesis gas or syngas, and char, a solid by-product.

Gasification is a multistage thermochemical reaction. When introduced into the reactor, the feedstock is rapidly heated and dried. Following further temperature increase, pyrolysis occurs. During this stage, volatile compounds are released from the solid feedstock in the form of gases, oils, and tars. The remaining solid portion—char—is carbon rich. Gasification occurs through multiple chemical reactions between the char, steam, liquid and gaseous volatile compounds, and the resulting gases from these reactions. The product gas contains a mixture of methane, hydrogen, carbon monoxide, and carbon dioxide. The relative portions of the produced gases, as well as the quantity and energy content of the char by-product, can be controlled by manipulating the reaction conditions. Operated at lower temperatures, the reaction can be limited to the pyrolysis process, resulting in an output stream mixture of volatile gases, oils and tars, and char.

Applications

The process of gasification is not a new concept. The use of coal-derived syngas became widespread during the latter half of the 19th century. "Town gas" was used extensively for municipal and domestic lighting, cooking, and heating. By-products were used as chemical feedstocks, such as in the production of dyes and fertilizer. This process continued to be used into the mid-20th century. During World War II, Germany employed syngas as a feedstock for the production of synthetic liquid fuels via the Fischer-Tropsch process because of shortages in petroleum supply. By the mid-20th century, however, technical advancements allowed for the exploitation of low-cost natural gas reserves, leading to the decline of the coal gasification industry.

Gasification processes are the subject of renewed interest because of their potential to reduce the environmental impacts of fossil fuel use. Synthesis gas can be used at higher efficiency than solid fuels, can be cleaned of pollutant-forming compounds to reduce air pollution when combusted, and is more easily integrated with carbon capture and sequestration systems to greatly reduce greenhouse gas emissions. Gasification also offers a more effective use of resources. Rather than simply combusting feedstocks to produce thermal energy and waste ash, higher-quality fuels and chemicals can be extracted from a variety of low-value feedstocks. This allows domestically available feedstocks to substitute for imported petroleum in a variety of both energy and material applications, thereby improving energy security. Waste products, such as agricultural residues and municipal solid waste, can be input to gasification processes, producing a consistent product stream with numerous higher-value applications. At present, a number of potential applications of gasification technology have been proposed to meet objectives of air pollution reduction, greenhouse gas mitigation, and improved energy security.

Gasification in Electricity Generation

The products of gasification can be employed as a fuel for electricity generation. Using coal in integrated gasification combined cycle (IGCC) power plants can reduce air pollutant and greenhouse gas emissions compared with traditional coal-fired generation. Pollutant-causing components, including mercury, are mostly removed from the synthesis gas stream, reducing emissions of nitrous oxides and sulfur oxides, for example, by one to two orders of magnitude when compared with conventional coal generation. Higher efficiency can be attained with IGCC when compared with direct coal combustion, somewhat reducing greenhouse gas emissions associated with coal use. Far greater greenhouse gas reductions can be attained by integrating IGCC power production with carbon capture and sequestration. Because gasification products can be separated to concentrate carbon dioxide, IGCC plants can be more easily adapted to capture carbon dioxide than traditional generating plants.

Using coal, IGCC technology has several advantages over alternatives. Coal is a plentiful and relatively cheap resource globally, allowing for the widespread deployment of this technology without overtaxing the available supply in the near or medium term. Unlike most renewable energy sources that only intermittently provide power, IGCC can deliver base load electricity. However, there are also some disadvantages to coal IGCC generation. Until technology to capture and store carbon dioxide is proven on a large scale, greenhouse gas emissions from IGCC plants will not be significantly less than those of conventional coal plants. One should also consider the environmental and social burden of coal mining, including the destruction of ecosystems at mine sites and the high risk of injury and death experienced by coal miners. Using more sustainable feedstocks for gasification can improve the overall environmental performance of electricity generation with IGCC technology but may limit the scale of potential adoption.

Liquid Fuels from Gasification

Synthetic diesel fuels can be produced from gasification-derived synthesis gas via the Fischer-Tropsch process. During this process, numerous competing chemical reactions occur simultaneously, coproducing a variety of outputs. The production of liquid fuels consumes hydrogen and carbon monoxide from the gasification process. The Fischer-Tropsch process is proven and has been employed on a large scale in South Africa for over 50 years.

Where domestically available feedstock is used as an input to gasification, producing fuels in this manner can reduce energy imports and increase energy security. However, any greenhouse gas benefit is dependent on the feedstock used in gasification. Where carbon-intensive fuels such as coal are used as the gasification input, the greenhouse gas emissions of Fischer-Tropsch fuels can be more than double those of conventional gasoline as a result of the higher carbon content of coal relative to traditional fuels. Using sustainably produced biomass or waste products as inputs to gasification can greatly reduce the greenhouse gas intensity of Fischer-Tropsch fuels.

Gasification of Biomass Sources

When biomass is used as a gasification feedstock, the output products are referred to as biogas, bio-oil, and biochar. Utilizing sustainably produced biomass feedstock in gasification processes can provide low greenhouse gas fuels. Properly managed, biomass feedstock can be considered a carbon neutral resource. Carbon dioxide emissions resulting from biomass oxidation during gasification and eventual use are not considered to contribute to global warming since an equivalent quantity of carbon dioxide is sequestered again in subsequent biomass growth. This, however, does not account for any fossil fuel emissions from biomass inputs like fertilizer, or transportation-related emissions.

Incorporating carbon capture and sequestration with biomass gasification can therefore result in negative greenhouse gas emissions. Biomass renewal will absorb additional carbon dioxide from the atmosphere, thereby actively reducing atmospheric greenhouse gases. Biochar offers an additional pathway to sequester carbon. This material, also useful as a soil enhancer when applied to agricultural and forest land, is able to store carbon for hundreds to thousands of years. However, increasing the yield of biochar from biomass pyrolysis correspondingly reduces the production of gaseous and liquid fuels, limiting the quantity of fossil fuel that can be substituted.

See Also: Alternative Fuels; Biomass Energy; Coal, Clean Technology.

Further Readings

Bridgwater, A.V., et al. "A Techno-Economic Comparison of Power Production by Biomass Fast Pyrolysis With Gasification and Combustion." *Renewable and Sustainable Energy Reviews*, 6 (2002).
Jaccard, Mark. *Sustainable Fossil Fuels: The Unusual Suspect in the Quest for Clean and Enduring Energy*. Cambridge: Cambridge University Press, 2005.

Rezaiyan, John and Nicolas P. Cheremisinoff. *Gasification Technologies: A Primer for Engineers and Scientists.* Boca Raton, FL: Taylor & Francis, 2005.

Jon McKechnie
Independent Scholar

GASOHOL

Alternatives to petroleum-derived transportation fuels are attractive because of the increasing demand and decreasing supply of conventional fossil fuels. Biofuels are attractive transportation fuels because they are renewable, can be locally produced, and are biodegradable, and their production and use may reduce net greenhouse gas (GHG) emissions compared with petroleum-derived fuels. Recently, ethanol produced from agricultural feedstocks such as corn and sugarcane has been used widely as an alternative fuel in Brazil, the United States, and other countries.

Ethanol can be mixed with gasoline in various ratios for use in light-duty vehicles. Low-level ethanol blends can be used in unmodified gasoline engines. High-ethanol blends can be used in flexible fuel vehicles. Gasohol, a blend of 10 percent ethanol mixed with 90 percent gasoline, can be used in gasoline internal combustion engines without need for modifications to the engine and the fuel system. All vehicle manufacturers in the United States and Canada approve the use of up to 10 percent (by volume) low-level ethanol blends in current vehicles.

Ethanol, also called ethyl alcohol, is a colorless, volatile, and flammable liquid. The molecular formula of ethanol is C_2H_5OH. On a volume basis, ethanol contains about two-thirds as much energy as gasoline. However, pure ethanol has a high octane value, which is used as a measure of the resistance of fuels to engine knocking in spark-ignition internal combustion engines. High-performance engines typically have higher compression ratios (the ratio of the volume of an engine's combustion chamber from its largest capacity to its smallest capacity) and are therefore more prone to engine knocking, so they require higher-octane fuel. Adding ethanol to gasoline increases its octane level and helps reduce the likelihood that engine knock problems will occur.

How Ethanol Differs From Gasoline

Unlike gasoline molecules, ethanol molecules contain oxygen, so ethanol fuel is often referred to as an "oxygenate." The oxygen in ethanol can improve the fuel combustion process, helping reduce air pollutants such as carbon monoxide and ozone-forming hydrocarbons. However, advances in emission control technology over the years have reduced the relative advantage of ethanol as a cleaner fuel. In addition, there are some concerns over potential human exposure to certain emissions related to the use of ethanol-blend fuel, including acetaldehyde, formaldehyde, and peroxyacetyl nitrate.

The production of ethanol from sugar and starch-based crops is a mature technology, with the former dominating in Brazil and the latter dominating in the United States.

The most common feedstock used for ethanol production in Brazil is sugarcane, which is grown in large areas in the northeast and center-south regions. After the sugarcane is harvested and transported to the ethanol plant, it is chopped and shredded before going through a series of milling equipment to extract the juice, which contains sucrose. The juice then flows to fermenters, where the sucrose is converted to wine. The resulting wine is distilled to separate ethanol from vinasse, which is left as a waste. The energy consumed by the process is supplied by combined heat and power systems that use the bagasse (i.e., the fibrous residue remaining after sugarcane stalks are crushed to extract their juice) as process fuel; the ethanol plant usually has excess energy such as electricity, which can be sold to the grid.

Most of the ethanol currently produced in the United States uses corn as a feedstock. Two processes, dry mill and wet mill, can be employed for converting corn to ethanol. In the dry mill process, the corn is ground and mixed with water to form a mash. The mash is then cooked with enzymes added to hydrolyze the starch to sugar. Then yeast is added to ferment the sugar, producing a mixture containing ethanol and solids. This mixture is then distilled and dehydrated to create fuel-grade ethanol. The solids remaining after distillation are often dried to produce distillers' dried grains with protein and are sold as animal feeds. Compared with dry milling, the wet milling process is more energy and capital intensive, as the grain must first be separated into its components, including starch, germ, fiber, and gluten. Corn is steeped to break the starch and protein bonds. Then the corn is coarsely ground to break the germ loose from the other components. The ground corn, consisting of fiber, starch, and protein in a water slurry, flows to the germ separators, where the low-density germ is separated from the slurry. The germs, containing about 85 percent corn oil, are pumped into screens and washed to remove any starch left in the mixture. The oil is then refined into finished corn oil. The germ residue is sold as animal feeds. The remaining corn is further segregated into gluten and starch. The gluten is readily for use in animal feeds, while the starch is fermented to ethanol. The fermentation process is similar to that in the dry mill process.

New Technologies

New technologies are being developed to convert lignocellulosic biomass such as agricultural and forest residues to ethanol using either biochemical or thermochemical processes. However, none of these conversion processes has been applied on a commercial scale.

The bioconversion of lignocellulose to ethanol is carried out by enzymatic hydrolysis of the cellulose fraction and fermentation of the cellulose and hemicellulose sugars of the lignocellulosic feedstock. The conversion process begins with the pretreatment of the lignocellulose substrate, which separates the lignin from the cellulose and hemicellulose and dissolves the hemicellulose sugars into sugar monomers that can be fermented to alcohol. Pretreatment precedes enzymatic hydrolysis and acts to prepare the otherwise recalcitrant cellulose for hydrolysis. During hydrolysis, a cocktail of advanced cellulose enzymes are used to convert the cellulose to glucose. During fermentation, yeast, bacteria, or fungal organisms convert the cellulose and hemicellulose sugars to ethanol. The lignin portion of the lignocellulosic biomass generated as a residue from this process can be used as a fuel to produce steam and electricity required for operating the facility.

The thermochemical conversion transforms all of the nonmineral constituents of the biomass into a common form—synthesis gas (or syngas)—through gasification. The syngas, which is rich in carbon monoxide and hydrogen, is conditioned and cleaned up and then passed over a catalyst, which converts it into mixed alcohols. The mixed alcohols are sent to a series of equipment such as a molecular sieve and distillation columns to separate the ethanol from other components.

During the last decade, considerable research has been conducted to investigate the environmental and energy impacts associated with the production and use of ethanol as an alternative fuel, focusing primarily on the GHG metric and the net energy of ethanol (i.e., whether manufacturing ethanol takes more nonrenewable energy than the resulting fuel provides). Overall, scientists agree that producing one megajoule (MJ) of ethanol from corn consumes less than one MJ of nonrenewable energy. In particular, producing one MJ of ethanol requires far less petroleum than is required to produce one MJ of gasoline. However, the GHG metric illustrates that the environmental performance of ethanol varies, depending greatly on production processes. For example, one study found that the GHG emission effects of corn ethanol could vary from a 3 percent increase (if coal is used as the process fuel) to a 52 percent reduction (if wood chips are used). Looking to the future, lignocellulosic ethanol could have the potential to improve the environmental performance of ethanol by significantly reducing the GHG emissions and having a better net energy balance, because the production of lignocellulosic biomass often requires lower agricultural inputs compared with that of corn and sugarcane. However, for lignocellulosic ethanol to be a viable substitute for fossil fuels, it should not only have superior environmental benefits over the fossil fuels it displaces but also be economically competitive and be producible in sufficient quantities to have a meaningful effect on energy demands.

See Also: Alternative Fuels; Biomass Energy; Flex Fuel Vehicles; Fossil Fuels; Gasification.

Further Readings

Cheremisinoff, P. Nicholas. *Gasohol for Energy Production*. Ann Arbor, MI: Ann Arbor Science Publishers, 1979.
Hunt, V. Daniel. *The Gasohol Handbook*. New York: Industrial, 1981.

Yimin Zhang
University of Toronto

GEOTHERMAL ENERGY

Iceland is able to exploit its significant geothermal energy resources through geothermal power plants like this one in the Krafla volcano region.

Source: iStockphoto.com

The Earth holds vast quantities of residual heat from the gravitational collapse that occurred during its formation, as well as radiogenic heat from the natural decay of radiogenic isotopes and possibly from natural fission reactions that occurred during its history. Although unevenly distributed, most of this heat resides very deep in the Earth. A small fraction, which is still a considerable amount, lies within easy access by man. Magmatic activity and flowing hydrothermal fluids bring localized bodies of hot rock and water to, or close to, the surface.

The earliest uses of geothermal energy involved hot water associated with natural hot springs and geysers. Since 1913, however, with the construction of the world's first commercial geothermal power generator in Larderello, Italy, experts have explored ways to obtain inexpensive electricity from the Earth's heat—with considerable success.

Although generally called a "renewable energy source," the scale of most power plants means that they "mine" the heat; that is, extract it from the localized area far faster than any natural recharge. Overall, however, the resource base is so immense that no one believes that man will exhaust it in the foreseeable future. Production from individual fields, however, may decline because the fluid is being extracted faster than the natural recharge (demonstrated by changes in gravity). In some cases (e.g., The Geysers area in California), water from a variety of sources is now being injected into the reservoir to maintain pressure. Water then picks up heat from the hot rock at depth and can be brought to the surface again, extending the practical life of each well.

Geothermal energy was one of the original beneficiaries of the U.S. Public Utilities Regulatory Policies Act of 1978, which required public utilities to buy power from small producers at their "avoided cost." This opened new opportunities for geothermal developers who had no grid or distribution network of their own. Many other countries have now added "feed-in" laws of their own, providing a major benefit for renewable energy to compete in the marketplace.

Geologic Settings

Heat from deep within the Earth can be brought to economical drilling depths by either magmatism or deep hydrologic circulation. Many commercial geothermal fields lie close to a recently active volcanic complex. Another common association is with deep faulting, such as in the U.S. Basin and Range Province, where the faults provide permeable pathways that allow hot water at depth to rise close to the surface.

Most geothermal prospects have been identified by hot springs at the surface or by nearby recent volcanism. A variety of geophysical techniques have been applied to find heat at depth. Measurements of heat flow in shallow wells provide one of the more consistent predictors of heat at depth. Areas with hot fluids can sometimes be identified because of resistivity or other geophysical anomalies, but only wells drilled directly into the subsurface formation can confirm a reservoir's temperature and ability to flow. Wells capable of producing commercial fluids vary in depth from a few hundred meters to over 4 kilometers. Hot dry rock system wells could even exceed 5 kilometers.

Categories of Geothermal Applications

Flowing Hot Water (Temperatures <100 Degrees C)

The earliest uses of geothermal heat revolved around natural hot springs, where the heat was used for bathing, greenhouses, or space heating. The use of direct heat from these resources for other commercial processes (e.g., food drying) has been added to this suite of activities more recently. Where hydrology permits, shallow wells have also been used. Iceland has taken major advantage of this resource for space heating.

Dry Steam Resources

In a few places, wells drilled into the subsurface will flow essentially dry steam (after priming by removing any liquid water adding pressure to the column). The Larderello system in Italy was the first such resource exploited, producing electricity from steam as hot as 200 degrees C. The Geysers geothermal area in northern California followed in 1921, producing dry steam at more than 180 degrees C. The Geysers now produces more electricity than any other geothermal field in the world. Such dry steam systems, however, are rare.

Flash Steam

More commonly, hot geothermal fluids (generally 170 degrees C or hotter) are liquid in the subsurface but may flash (boil) when the confining pressure is relieved. When the reservoir's heat energy fails to flash the entire flow of water coming to the surface, a combination of steam and liquid water arrives at the surface. The separated steam can drive a turbine generator directly. Use of a dual-flash cycle generally improves the economics of flash plants by separating the steam at two different pressures. A dual-flash cycle can often produce 20–30 percent more power than a single-flash system with the same fluid flow. The remaining liquid is usually returned to the reservoir.

Binary Cycle

The binary cycle uses heat exchangers to transfer heat from the extracted geothermal fluids to other working fluids for driving a power turbine. The secondary working fluid can be chosen to (1) boil during the exchange of heat and drive a generating turbine, (2) condense at surface conditions, and (3) be recyclable with no or minimal loss. This approach can be applied beneficially for geothermal systems from about 170 degrees C down to as low as 85 degrees C. The geothermal fluid can be kept under pressure and returned directly to the reservoir, if desired. Examples of binary cycle geothermal power plants in the United States can be found at Mammoth Lakes, California; Steamboat Springs, Nevada; and Hilo, Hawaii.

Hot Dry Rock (Also Known as Enhanced
Geothermal Systems or Hot Fractured Rock)

The systems described above all depend on natural fluid in the subsurface that can be brought to the surface by drilling wells. Many regions of the subsurface hold considerable heat but do not have hydrologic systems that can be tapped. In the early 1970s, Los Alamos National Laboratory ran a Hot Dry Rock program, which devised and conceptually validated the concept of drilling into impermeable hot rock at depth, creating fractures by hydrofracturing (injecting fluids at pressures), drilling a second well that intersects the fracture system, and creating a circulating system for injected water passing through the fractures (picking up heat as it passes) to be extracted from the second well. The extracted fluids can then be used just as in the previously discussed systems. Many analyses suggest that this resource far exceeds that from systems with existing hydrologic systems, but development costs exceed those of natural hydrothermal systems.

Ground Source Heat Pumps (GSHP)

Although GSHP systems can use heat from the Earth, the operation of GSHPs can use ground temperatures at equilibrium with average annual surface temperatures. Furthermore, they can also use the Earth for "dumping" heat when a heat pump operates in a cooling mode. These systems will not be considered further in this section.

Advantages and Disadvantages

All geothermal applications depend critically on the site-specific resource. Sites with economically producible resources are relatively limited. In the United States, they lie mostly in the western third of the country. Volcanically active countries such as Iceland, Japan, and the Philippines have many more sites and have been actively developing them.

Flowing hot water systems present the simplest geothermal resources to exploit, but applications are limited. Low-temperature (<90 degrees C) process heat opportunities may expand, but such use must take place close to the resource.

The higher-temperature systems mentioned above all can support generation of electrical power, which can then be transported easily via an existing power grid, often with minimal feed-in lines. Geothermal power systems, however, require considerable space for well drilling platforms, steam or hot water lines, the power facilities themselves, and a variety of access roads. In some cases, much of that land can serve multiple purposes, such as cattle grazing or even raising crops, though steam plants tend to be fairly noisy.

Unlike wind or solar, the geothermal resource for power plants is not intermittent, and geothermal plants function best when run continuously. Starting up and shutting down flow in piping that carries hot fluids causes expansion and shrinkage problems. This means that geothermal plants are best run at "baseload"; that is, running 24 hours/day for long periods of time between required maintenance. Because the load on any power system varies with time of day, with weekday versus weekend, and with seasonal needs, other energy sources must be used in conjunction with geothermal to follow the power demand. Hydropower with a reservoir provides an excellent complementary power source because its power can be varied quickly with no loss of potential future power production. Where hydropower is not available to meet this load-following requirement, natural gas often fills this need, as it can also easily be turned on and off; however, the fuel costs exceed many other systems.

Dry steam systems are rare. The Geysers system in California remains the largest commercial geothermal power generation in the world. Because of the simplicity of dry steam systems, The Geysers provides the cheapest electricity on the Pacific Gas and Electric's power grid. When considering environmental consequences, it must be noted that most dry steam geothermal reservoirs produce some carbon dioxide and sulfur dioxide along with "dry (water) steam." The sulfur dioxide is mostly being controlled by scrubbers. The carbon dioxide contributes to greenhouse warming but contributes far less than an equivalent fossil fuel–burning plant would.

Flash steam systems, which are more common, have similar problems to dry steam plants, in that other gases may be coproduced. Furthermore, wastewater, which can be highly mineralized, must be disposed of—most simply by reinjection into the reservoir. As a benefit, however, coproduced hot water from many geothermal power systems can be employed for combined heat and power, which increases overall energy efficiency.

Binary plants almost always keep the geothermal fluid separated from the surface environment and returned to the subsurface. The additional separate piping for the working fluid presents slightly increased costs. On the plus side, however, a working fluid can be chosen that optimizes the thermodynamic performance of the turbine without the corrosive potential of natural geothermal fluids.

Table I   2004 Worldwide Installed
Geothermal Generating Capacity
and Annual Electrical Energy
Produced

Country	Installed MW	Estimated Annual Energy Produced (GWh/a)
Australia	<1	3
Austria	<1	5
China	32	100
Costa Rica	162	1,170
El Salvador	105	550
Ethiopia	7	30
France (Guadalupe)	4	21
Germany	<1	2
Guatemala	29	180
Iceland	200	1,433
Indonesia	807	6,085
Italy	790	5,300
Japan	535	3,470
Kenya	127	1,100
Mexico	953	6,282
New Zealand	453	3,600
Nicaragua	78	308
Papua New Guinea	30	100
Philippines	1,931	8,630
Portugal (Azores)	8	42
Russia	100	275
Taiwan	3	15
Thailand	<1	2
Turkey	21	90
United States	2,395	16,000
TOTAL	8,771	54,793

Source: Adapted from Lund, 2004.

Although no commercial-scale power production has yet used the hot dry rock technology, interest is spurred by the large resource base—much larger than from natural hydrothermal systems. A 2008 resource assessment indicates that, if successfully developed, the "unconventional enhanced geothermal systems" in the United States could provide about half as much power as the country's current total installed electric power generating capacity. The real issue will be economics.

Hot dry rock systems require considerably more sophisticated technologies to locate an appropriate region of very hot rock (near 300 degrees C) without preexisting hydrothermal circulation (lacking either water or permeability), find the appropriate location for creating fractures and creating them, be able to intersect a sufficiently fractured reservoir, and establish circulation between wells such that not too much fluid is lost to the formation. These wells are typically deep—and expensive. Test operations so far have been capable of extracting around 90 percent of the injected water.

Worldwide Power Production

Table 1 lists the world's major geothermal power-producing countries as of 2004. By the first half of 2008, the world's total installed geothermal power capacity passed 10,000 megawatts and continues to increase.

Although no commercial-scale power production has yet used the hot dry rock technology, interest is spurred by the large resource base—much larger than from natural hydrothermal systems. A 2008 resource assessment indicates that, if successfully developed, the "unconventional enhanced geothermal systems" in the United States could provide about half as much power as the country's current total installed electric power generating capacity. The real issue will be economics.

Hot dry rock systems require considerably more sophisticated technologies to locate an appropriate region of very hot rock (near 300 degrees C) without preexisting hydrothermal circulation (lacking either water or permeability), find the appropriate location for creating fractures and creating them, be able to intersect a sufficiently fractured reservoir, and establish circulation between wells such that not too much fluid is lost to the formation. These wells are typically deep—and expensive. Test operations so far have been capable of extracting around 90 percent of the injected water.

Conclusions

Geothermal energy provides cost-effective direct heat energy and electric power. Between advances in technology, favorable terms for small power producers, and rising costs and environmental concerns for traditional energy sources, use of geothermal energy will most likely expand for some time. The resource base for deep heat from the Earth is immense, but economics will ultimately provide limits on the practicality of further expansion.

See Also: Feed-In Tariff; Hydroelectric Power; Solar Energy; Wind Power.

Further Readings

DiPippo, Ronald. *Geothermal Power Plants: Principles, Applications and Case Studies*, 2nd ed. Oxford: Elsevier Science, 2008.

Isherwood, William F. "Geothermal Reservoir Interpretation From Change in Gravity." *Proceedings: Third Workshop on Geothermal Reservoir Engineering*. Palo Alto, CA: Stanford University, 1977.

Lund, John W. "100 Years of Geothermal Power Production." *Geo-Heat Center Bulletin*, September 2004. http://pangea.stanford.edu/ERE/pdf/IGAstandard/SGW/2005/lund.pdf (Accessed January 2009).

William Isherwood
Chinese Academy of Sciences

GLOBAL WARMING

The phrase global warming refers to a phenomenon in which the Earth's surface temperature increases from its long-term averages generally because of an atmospheric blanket of greenhouse gases (GHGs; primarily carbon dioxide, methane, and chlorofluorocarbons) that serve to trap reradiated solar energy from escaping into space. This blanket of greenhouse gases is responsible for providing Earth a generally temperate, stable, and life-sustaining climate. In common parlance, global warming is often used interchangeably with climate change. In the present context, though, it is used in a more limited sense as a driver of global climate change.

The Science

In all of our solar system, Earth is the only planet known to support life. This uniqueness derives in great part from an atmosphere that regulates the Earth's surface temperature within a range conducive to the development of living organisms, including humankind. The explanation for this phenomenon was suspected as early as 1824, when French mathematician and

Core samples from the ocean floor can provide data on historical changes in the Earth's climate. This U.S. National Undersea Research Program diver is extracting a sample from a coral reef for climate studies.

Source: National Oceanic and Atmospheric Administration/Oceanic and Atmospheric Research/National Undersea Research Program

physicist Jean Baptiste Fourier postulated that gases in Earth's atmosphere might influence its surface temperature. In 1859, physicist John Tyndall suggested that changes in the concentrations of some atmospheric gases could result in changes to Earth's climate. The Swedish chemist Svante Arrhenius published an article in 1896 demonstrating that the amount of carbon dioxide in Earth's atmosphere would significantly affect its surface temperature. Arrhenius coined the phrase greenhouse effect and predicted that a geometric (nonlinear) increase in atmospheric carbon dioxide would result in an arithmetic (linear) increase in the Earth's surface temperature.

Based on this thermal blanket of greenhouse gases, scientists have long understood that the Earth has undergone a series of long-term cyclic cooling and warming phases. The former account, in part, for periods in which glaciers have covered vast areas of the planet; the latter for long periods of regional desertification during which man and beast have populated a greater portion of the Earth and during which their numbers have greatly multiplied. The most recent glacial period ended approximately 10,000 years ago; Greenland and Antarctica are vestiges of that period. The duration of our current interglacial period will be determined in no small part by the extent of warming caused by the greenhouse effect.

By testing polar and glacial ice cores at continuously increasing depths, scientists can determine the composition of Earth's atmosphere as a function of time. For example, an Arctic ice sample will contain minute pockets containing a small amount of air—and its constituent gases—that were trapped at the time the ice froze. Ice samples taken from deeper cores were formed earlier in time. Using these historic data, modern computer models of Earth's climate systems are able to calculate Earth's surface temperature over time. Using this and other proxy data, we have a longitudinal history of Earth's surface temperature stretching back hundreds of thousands of years.

An example of the type of data scientists are able to coax from ice core samples is shown in Figure 1. The top three curves show that present-day carbon dioxide and methane concentrations are comparable to, if not higher than, those seen over the past 420,000 years, and that Antarctic air temperatures correlate well with these concentrations. (Insolation is a measure of the total radiant energy received from the sun as a function of time and provides the baseline from which radiative forcing—and global mean surface temperature—is computed.)

Figure 2 presents data from the recent Intergovernmental Panel on Climate Change (IPCC) *Fourth Assessment Report* that shows the correlation between carbon dioxide gas concentrations and radiative forcing (the mechanism that drives global warming) for the 10,000 years before 2005. The spike in carbon dioxide concentrations since 1750—a now-standard year used in gauging recent anthropogenic climate impacts resulting from industrialization—is detailed in the inset. The explosion in use of petroleum-based fuels—and resulting carbon dioxide emissions—begins in approximately 1900 and accelerates dramatically through the present. The IPCC report documents similar growth patterns in concentrations of other naturally occurring GHGs.

Although global warming can be driven by natural processes (both continuous and sudden), surface temperature data measured and imputed over the past thousand years indicates that human (anthropogenic) forcing is playing a significant role in the present warming trend. These data indicate that the Earth's surface temperature has been increasing at an accelerating rate since the beginning of the Industrial Revolution in the mid-1800s.

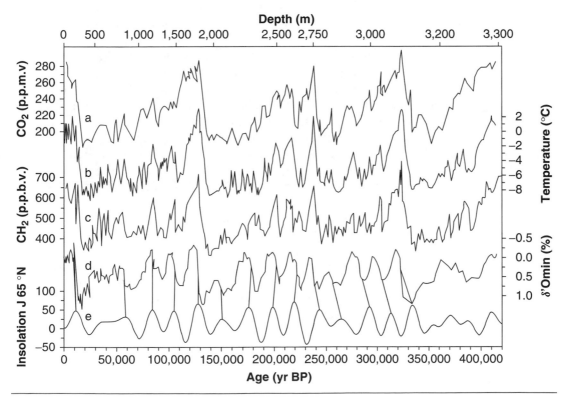

Figure 1   Carbon Dioxide and Methane Time Series From Vostok Station, East Antarctica, Ice Cores

Source: Petit et al., 1999.

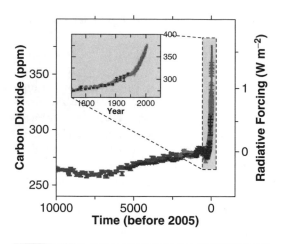

Figure 2   Atmospheric Concentrations of Carbon Dioxide, Methane, and Nitrous Oxide Over the Last 10,000 Years and Since 1750 (inset panel)

Source: Bernstein et al., 2007.

Analysis of atmospheric greenhouse gas concentrations and computer modeling of interacting land, ocean, and atmospheric systems has led scientists to conclude beyond any reasonable doubt that greenhouse gas emissions from human activity are the primary drivers in this dramatic increase in global warming.

Radiative forcing is a measure of the net solar and infrared radiation trapped in the Earth's atmosphere. Because of its direct relation to global mean surface temperature—positive forcing drives warming, negative forcing drives cooling—scientists use it to study the effects of natural and anthropogenic causes of global warming. Natural sources of radiative forcing include the earth, clouds, and naturally occurring greenhouse gases. Anthropogenic contributions to radiative forcing include greenhouse gas (positive forcing) and particulate (negative forcing) emissions caused by human activity, and land use changes resulting from human development activities.

A closely related concept is albedo, a measure of the extent to which a surface reflects light from the sun. Dark-colored objects (e.g., oceans and forests) have low albedo values because they absorb a greater percentage of solar energy than do lighter-colored bodies; light-colored objects (e.g., glaciers and ice caps) have high albedo values, as they reflect a greater proportion of solar radiation. A change in terrestrial albedo—resulting from the melting of a glacier to reveal the darker land mass underneath or from human urbanization or conversion of forests to agricultural uses—can have a significant effect on radiative forcing by increasing the amount of solar energy absorbed in the visible spectrum and reradiated as heat. A decrease in albedo generally correlates with an increase in radiative forcing and global mean surface temperature.

Data from the IPCC *Fourth Assessment Report* show the results of global climate model simulations using natural and anthropogenic radiative forcing mechanisms and their resulting temperature changes (Figure 3). The now-irrefutable conclusion from these data is that human activity has overwhelmed natural radiative forcing (warming) processes and is driving an unprecedented increase in global surface temperatures. IPCC scientists predict that unless GHG emissions are reduced by 80 percent no later than 2050, the global mean surface temperature will likely rise 2.4–6.4 degrees C.

Consequences of Global Warming

Although the geophysical and atmospheric drivers of global warming have been known for many years, and the consequences suspected, they were thought to be long term in nature—on the order of thousands of years. However, in testimony before the U.S. Congress in 1987 and 1988, climate scientist James Hansen characterized global warming as a real and present threat to the stability of Earth's climate system. He confidently stated

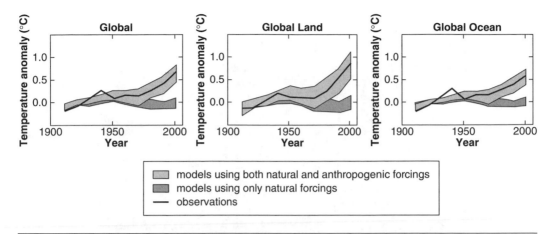

Figure 3   Comparison of Observed Global-Scale Changes in Surface Temperature With Results Simulated by Climate Models Using Natural and Anthropogenic Forcings

Source: Alley et al., 2007.

that the Earth 20 years hence would be warmer than it had been in the past 100,000 years. Over those 20 years, four assessment reports by the IPCC have left little doubt that we are seeing undeniable evidence of that warming.

In its *Fourth Assessment Report* the IPCC noted that the years 1995–2006 have ranked among the warmest since 1850 and that the linear temperature trend over the 100-year period 1906–2005 was greater than the corresponding trend over the 100-year period 1901–2000 (given in the *Third Assessment Report*). Overall, the number of hot days (measured by the heat index) has increased, while the number of cold days (on which frost was measured) has decreased.

At the same time we have been observing significantly warmer day- and nighttime temperatures, especially at higher Northern Hemisphere latitudes, a troubling rise in sea levels has been measured. Sea level rise is attributable to both the thermal expansion of water as it warms and to the melting of polar ice caps and glaciers. Over the past generation, scientists have measured increased rates of glacier melt in the Arctic, on Greenland, and on Antarctica. In 2008, the area of the Arctic ice cap was at an all-time recorded low, raising environmental fears and economic hope for the ultimate opening of the so-called Northwest Passage. In spite of any economic gains, the summer disappearance of this polar ice cap would be environmentally and ecologically devastating.

One consequence of the melting of so much frozen fresh water from the Arctic ice is the potential disruption of the oceanic thermohaline circulation (Figure 4) driven by water temperature and salinity. In the Atlantic Ocean, this circulation is responsible for the Gulf Stream, which delivers warmer, equatorial waters to northern Europe, moderating its mean surface temperatures and climate. Scientists are concerned that the melting of the Greenland ice cap caused by global warming will interfere with the Gulf Stream by dumping enormous volumes of fresh water into the North Atlantic, thereby decreasing the salinity (and density) of waters that normally sink to the ocean floor and return to the Southern Hemisphere.

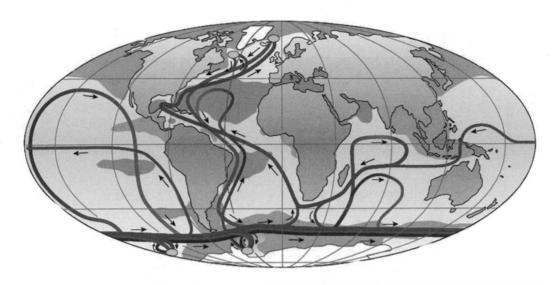

Figure 4 Thermohaline Circulation of the World's Ocean Systems

Source: Rahmsdorf, 2002.

At the rate it is presently occurring, global warming will have a significant effect on environmental and human systems over the next century and beyond. Increased surface temperatures will result in sea level rise that will threaten the death and displacement of many millions of the world population who live in low-lying, densely populated coastal areas including Bangladesh and Indonesia. Threatened coastal regions also contain many of the world's major economic and population centers including London, New York, and Shanghai. The resulting large-scale population relocation will have a dramatic effect on areas that do not have the political, economic, physical, or social infrastructure necessary to support the population influx. The rate of sea level rise has been shown to be accelerating in the 20th century.

In addition to having significant change on regional weather patterns, global warming will drive the shift of ecological habitats in time frames that will overwhelm the evolutionary capacities of many flora and fauna species. Scientists at the Audubon Society are already documenting the northward migration of many North American bird species. Their study finds a correlation in the distribution of bird species with the variation of latitudinal surface temperature and between rates of bird population change and rates of temperature change.

As global warming drives habitat migration, many disease vectors will also migrate. Public health experts expect an increase in the incidence of diseases such as malaria and dengue fever in previously unaffected areas. The increased incidence of the West Nile virus in North America is a stark reminder of the ability of diseases to travel with their hosts. Allergies will also become more pronounced as flora and their pollens migrate northward.

Global warming will result in significant changes in global precipitation patterns. IPCC reports predict that drought will become much more widespread in many areas, whereas rainfall amounts will dramatically increase in others. This regional redistribution of rainfall will contribute to the relocation of ecological habitats and disease vectors and will have a significant effect on agricultural distribution and productivity.

Scientists are concerned that poorly understood feedback mechanisms could result in the amplification of global warming effects. Computer simulations of atmospheric, oceanic, and terrestrial interactions have indicated that past a certain point—as of yet unknown—runaway global warming could be a distinct possibility. Increasing anthropogenic carbon dioxide emissions are exacerbating global mean temperature increases. In turn, these warming trends are melting glaciers and ice caps, increasing the Earth's albedo, enhancing the warming effect. The melting of ice caps and permafrost layers will release tons of methane—a greenhouse gas 20 times as potent as carbon dioxide—into the atmosphere, thus dramatically increasing the greenhouse effect and global mean temperatures. More melting will result in higher sea levels and greater social dislocation.

Of particular concern is the possibility that these feedback mechanisms will drive our climate processes to a tipping point at which they will shift rapidly to a new—and unpredictable—equilibrium that is far less conducive to supporting ecological systems as we have come to know them.

Global Warming Mitigation

In 2005, the Kyoto Protocol to the United Nations Framework Convention on Climate Change went into effect, specifying enforceable greenhouse gas emission reduction targets for signatory countries. It places the onus of emissions reductions on developed (Annex 1) industrialized countries. In addition, it outlines approaches that countries might employ to help them achieve those targets. These include the establishment of emissions trading markets—the European Union and several U.S. states have implemented such markets—and the Clean Development Mechanism. Recognizing the global nature of the greenhouse gas emission threat, the Clean Development Mechanism allows countries striving to attain their emissions reduction goals to undertake projects in developing countries, for which the signatory country can claim reduction credits.

Recognizing that naturally occurring phenomena can result in atmospheric cooling, some scientists have proposed counteracting the global warming effects of anthropogenic greenhouse gas emissions in the short term through the application of geo-engineering technologies. Initiatives to artificially increase the Earth's albedo include increasing Earth's albedo by floating large chunks of reflective polystyrene in the oceans, injecting sulfur aerosols into the upper troposphere to reflect more solar radiation, manufacturing clouds to increase the Earth's reflectance of solar radiation, and enhancing the oceans' carbon dioxide absorptive capacity by encouraging algal blooms with iron fertilization.

However, the feasibility of these suggested mitigations remains highly suspect by virtue of either their limited geographical scope—and thus their inability to have a truly global impact—or their obvious environmental side effects. Inserting into natural ecosystems any foreign elements, be they artificial ice floes or artificial clouds, could have unpredictable consequences in addition to the pollution effects.

The successor to the Kyoto Protocol—due to expire in 2012—is currently being negotiated and is expected to include many developing countries that were previously excused from emissions reduction responsibilities (e.g., China and India). In the meantime, significant attention has been drawn to the importance of reducing our reliance on carbon-based fuels, the main source of radiative forcing and global warming agents. Alternative energy sources—including wind, solar, and geothermal—are being explored but have yet to prove practical on a scale that will yield any significant effect on GHG emissions in the short term.

To the extent that trends in global warming are the result of the course taken by human development over the past 250 years, the increase in global mean temperatures has an inertia that may prove resistant to our best efforts to reverse them. The relatively long atmospheric lifetimes of GHGs—especially carbon dioxide and chlorofluorocarbons—imply that warming will continue even though we might drastically reduce our emissions levels.

See Also: Chlorofluorocarbons; Climate Change, Greenhouse Gases; Intergovernmental Panel on Climate Change; Kyoto Protocol; Radiative Forcing.

Further Readings

Alley, Richard B., et al. "Climate Change 2007: Working Group I—Summary for Policymakers." Intergovernmental Panel on Climate Change, 2007.

Bengtsson, L. "Geo-engineering to Confine Climate Change: Is It All That Feasible?" *Climatic Change*, 77/3-4 (2006).

Bernstein, Lenny, et al. "Climate Change 2007: Synthesis Report—Summary for Policymakers." Intergovernmental Panel on Climate Change, 2007.

Church, John A. and Neil J. White. "A 20th Century Acceleration in Global Sea-level Rise." *Geophysical Research Letters*, 33/L01602 (2006).

Forster, Piers, et al. "Changes in Atmospheric Constituents and in Radiative Forcing." In *IPCC Fourth Assessment Report: The Physical Science Basis*. Cambridge: Cambridge University Press, 2007.

Greer, Amy, et al. "Climate Change and Infectious Diseases in North America: The Road Ahead." *Canadian Medical Association Journal*, 178/6 (2008).

Niven, Daniel K., et al. "Birds and Climate Change: Ecological Disruption in Motion." http://www.audubon.org/news/pressroom/bacc (Accessed February 2009).

Petit, J. L., et al. "Climate and Atmospheric History of the Past 420,000 Years From the Vostok Ice Core, Antarctica." *Nature*, 399 (1999).

Rahmstorf, Stefan. "Ocean Circulation and Climate During the Past 120,000 Years." *Nature*, 419 (2002).

Watson, Robert T., et al. "Climate Change 2001: Synthesis Report—Summary for Policymakers." In *IPCC First Assessment Report*. Cambridge, MA: Cambridge University Press, 2001.

Kent Hurst
University of Texas at Arlington

GREEN BANKING

Green banking practices range from reducing paper and greenhouse gas emissions within banking institutions to lending practices that prioritize resource conservation or offer superior interest rates on loans for environmental purposes, such as buildings that reduce energy waste, renewable energy projects, or community development initiatives. Customer enthusiasm for green banking has risen sharply in recent years. As banks compete to attract customers, there is concern that some institutions may exaggerate what are largely superficial eco-friendly components of their operations. Yet, simultaneously, there are a number of credit unions and lending institutions that have made sustainability a top priority.

A simple environmental change is making checks from recycled paper. Citizens Bank of Rhode Island has a debit card made with recycled plastic. Yet the most common environmental change within banks is the move toward paperless transactions (e.g., Internet banking, direct deposit, and mobile banking, whereby mobile telecommunication devices are used for account transactions). Banks may even offer financial incentives to customers who switch to paperless banking because it saves them money on paper, printing, and postage. Nevertheless, to further promote the transition to paperless banking, banks must address customer concerns about security. Surprisingly, studies now suggest that going paperless may actually reduce a consumer's risk for fraud and identity theft because these crimes commonly occur when printed statements get in the wrong hands. As Internet and mobile banking expand, financial institutions will need to assure customers that security safeguards also have been intensified to prevent unintended interception of account information.

According to a 2008 study, over 600,000 tons of paper could be saved annually if every household in the United States stopped receiving paper bills and statements. According to the U.S. Postal Service, the average U.S. family receives approximately 19 paper bills and statements each month and makes seven monthly payments using paper. Processing these transactions electronically, the average household would save approximately seven pounds of paper, 63 gallons of water, and 24 square feet of forest annually. Electronic banking also saves customers from driving to the bank and in some instances waiting in traffic at a drive-through. Moreover, paperless transactions reduce the amount of mail transported by truck or plane between cities and could eventually lower the number of mail-processing facilities required, thereby notably reducing the Postal Service's carbon footprint.

Criticism has emerged that some banks are trying to paint themselves as being committed to the environment by merely promoting paperless banking. A concern of the green banking industry is that greenwashing, or exaggerating accomplishments and programs to display an unjustified environmentally friendly public image, may create backlash against the whole sector if it is not exposed and eliminated. Research shows that many consumers want banks to go beyond superficial modifications and public relations gimmicks to leverage real environmental change. Studies also demonstrate a significant increase in customer loyalty to financial institutions with substantive environmental protection programs.

Environmentally responsible banking, or ecobanking, is similar to socially responsible banking and community investment banking. Bank funds may be used to promote forest conservation, sustainable or organic agriculture, recycling, ecotourism, or clean energy production such as solar and wind power. In some instances, eco-loans are financed through customers voluntarily agreeing to maintain bank accounts with lower rates of return. The bank then lends the money that it normally would have returned to these customers had they enrolled in accounts with higher interest rates to community development programs or eco-friendly projects.

An increasing number of banks provide better lending rates on car loans for alternative fuel or electric vehicles, mortgages for energy-efficient buildings, and home improvement loans for energy-conserving windows or water-saving appliances. When banks invest in eco-projects, the bank management usually remains uninvolved in project planning and implementation. However, in some cases, such as with the eco-lending leader ShoreBank Pacific in Washington and Oregon, the bank may employ a scientist to advise project leaders on environmental impact, resource efficiency, landscape conservation, community diversity and stability, and other areas. Another frequently cited green investment leader is the Permaculture Credit Union (PCU). The PCU, a not-for-profit operation located in Santa Fe, New Mexico, states on its website that it is dedicated to care of the Earth as well as reinvestment of surplus to benefit the Earth and its inhabitants. According to the PCU, investments should be based on a code of ethics rather than the common practice of focusing on the bottom line. PCU therefore invests its funds in the local community and boycotts destructive technologies or industries. A popular PCU program uses favorable second mortgage loans to pay for solar heating, photovoltaic energy systems, building weatherization, and rainwater collection, among other options.

Customers increasingly look to select their bank on the basis of ecological criteria, such as the environmental improvements an institution has made to its buildings or the percentage of loans that are awarded to environmentally focused companies or initiatives supporting sustainable technologies. Increasingly, commitment to address global warming is an initiative bank customers desire. In Bank of America's green lending initiative, it changed its underwriting criteria to favor businesses that make less-carbon-intensive products. Bank of America's carbon credit trading program advises customers how to manage their carbon operations, take carbon positions, and trade credits. Likewise, many commercial banks have increased their portfolio of services to include carbon emission risk management, advisory services, and credit origination services.

Opportunities to apply for green credit and debit cards also are increasing. Wells Fargo now has a rewards card that allows cardholders to use the points that they earn with purchases to support renewable energies—in particular, wind power—to offset their personal carbon output. Rabobank, a Netherlands-based bank, has introduced a climate credit card in partnership with the World Wildlife Fund. Rabobank works through the fund to offset the carbon footprint of purchases with an equivalent amount of carbon credits for renewable energy projects in developing countries. There has been considerable customer interest in this credit card, which is widely lauded as an example of corporate responsibility. Nevertheless, some environmentalists criticize green reward cards for encouraging consumption instead of providing incentives for people to purchase less.

Recent green business initiatives that are rapidly gaining popularity in both theory and practice include wetland, biodiversity, or pollution mitigation banking. These "banks" should not be confused with traditional savings or loan institutions. In this instance the word "bank" is being used to describe either an ecological offset or a database of tradable credits. For example, in the case of mitigation banking, which allows a person or company to degrade a wetland or habitat at one site if a wetland or habitat area at another site is improved, the "bank" is the area that has been improved. Similarly, although several governmental and nongovernmental organizations maintain pollution emission trading "banks," in reality these are merely databases or websites that list available credits by location and pollutant type.

See Also: Carbon Footprint and Neutrality; Emissions Trading; Renewable Energies.

Further Readings

Cogen, Douglas G. *Corporate Governance and Climate Change: The Banking Sector*. Boston, MA: Ceres, 2008.

GreenBankingCentral.com. http://greenbankingcentral.typepad.com (Accessed January 2009).

Schwanhausser, Mark. *The Four E's of Green Banking: Educate, Enable, Make it Easy—and Be Earnest*. Pleasanton, CA: Javelin Strategy and Research, 2008.

Mary Finley-Brook
University of Richmond

GREEN ENERGY CERTIFICATION SCHEMES

Although the need to increase the use of renewable energy sources has been widely accepted, lively debate still exists about the best instruments to reach this goal. Today, the two most commonly used support systems are feed-in tariffs and green certificates. Feed-in tariffs involve an obligation on the part of electricity retailers to purchase green electricity that is supplied at a fixed price (above the market price). Until now, it has been the dominant policy instrument for the development of renewable energy, especially in Europe. However, with the tendency to develop new voluntary and market-based instruments, green certificates have now won the favor of many actors in the electricity sector.

Green certificates, also known as renewable energy certificates (RECs) or green tags, provide a mechanism for recognizing the "greenness" of energy; in other words, its environmental (and sometimes social) positive externalities. They have been particularly developed for the power sector. Because electrons are directly injected into the grid and mixed with other electrons when they are generated, one can hardly guarantee the consumer that the electricity provided comes from a renewable energy power plant.

Green certification aims to balance this uncertainty and offers proof that a specific quantity of renewable energy has been generated and injected into the grid. When renewable energy generators produce electricity, they create one certificate for each unit (usually 1 megawatt hour) placed on the grid. The certificate, in the form of a paper or electronic document, includes at least the following attributes:

- A unique identification number
- Generator
- Date of issuance and period of production covered by the certificate
- Unit (or amount)
- Geographic location of the generator (country, region)
- Renewable fuel source (solar, wind, hydro, etc.)
- Technology
- Capacity of the plant
- Expiration date of the certificate (if any)
- Financial support (direct or indirect) received for the production of renewable electricity

There are two major types of configuration for the use of green certificates: a mandatory market or a voluntary market. In the first case, the system is coupled with an obligation mechanism (e.g., renewable portfolio standards [RPS]) but offers flexibility to electricity producers and consumers to meet their obligations. First, they can decide to produce green power themselves. Second, they can buy green power from another producer to get the certificates. Third, they can buy certificates in a specific market at a price fixed by supply and demand. Last, producers and consumers can decide to pay a penalty if they do not reach the mandatory targets. Such a system exists in some U.S. states (e.g., Texas, Massachusetts) and European Union (EU) member states (the United Kingdom, Italy, Sweden).

When no compulsory renewable energy targets have been defined, green certificates remain a voluntary instrument that can be compared to green marketing. Consumers can decide to reduce the effect of their energy consumption on the environment. In this context, green certificates can be used by firms advancing social and environmental responsibility goals.

Green certification schemes have been set up in a few European countries (e.g., Sweden, Belgium, the United Kingdom, and Italy), in Australia, and in some U.S. states. Depending on geographical areas, and sometimes within these areas, different organizations are competing to define the standards for certificates.

In the United States, the first green offers appeared in the 1990s. The largest certification program, Green-e, was launched in 1997 in California and is managed by the Center for Resource Solution. It only recognizes RECs that are not used to meet a regulatory requirement and that are sourced from installations that were built after 1997. Since then, it has certified more than 50 percent of all voluntary renewable energy purchases; that is, 1,300 GWh. The second major certification scheme was introduced in 1998 by Environmental Resources Trust with its "Eco Power Program." This program may be used to certify electricity that has been used to fulfill a RPS if the state RPS legislation allows the same REC to be sold in both the compliance and voluntary markets.

In the EU, a Renewable Energy Certificate System was launched at the beginning of the 2000s as a private initiative from major electricity utilities, with the support of the European Commission. Considering that the persistence of national support systems will distort competition in the internal markets, these actors aimed at promoting the use of green certificates and initiating a harmonized EU market to trade these certificates. In 2001, a European directive laid the foundation for a common framework to support green electricity, and it also created a "guarantee of origins" to facilitate green electricity trading. As a consequence, since 2003, EU member states are obliged to deliver guarantees of origin to renewable energy producers that require it.

This was supposed to open up the possibility of creating an EU green certification trading system. However, because of strong disagreement between member states, the commission has decided to postpone the decision to do so. Without waiting for this harmonization, Renewable Energy Certificate Systems set up a green certification scheme in 19 member states and sponsored the establishment of the Association of Issuing Bodies, an umbrella organization for national bodies that issue and monitor green certificates. Green certification schemes are not exclusive. At the European level, a network of environmental nongovernmental organizations created a new green energy standard—EUGENE—that is presented as being more exacting with regard to environmental positive externalities. However, the superposition of certification schemes remains one of the major challenges for this type of support mechanism. First, the systems tend to become too complex to be understood. Second, it creates the danger of double counting, which occurs when the attributes of green power (e.g., emissions savings) are sold or accounted for more than once.

It remains unclear whether green certification schemes will overcome these limits and be of the same importance in EU renewable energy policy as it is in the United States. However, it can already be observed that the introduction of these schemes has shifted the debate on renewable energy support mechanisms toward market compatibility. Although green certificates are often presented as the most cost-efficient of instruments and praised for their coherence with liberalized market rules, they also carry some threats. In particular, they are criticized for creating price volatility and economic uncertainty that threaten the speed of development of renewable energy. From this perspective, the choice of policy instruments is definitely more than just a technical decision. The future of green certificates will be a great indicator of the orientation given to renewable energy policy in the world.

See Also: Emissions Trading; Feed-In Tariff; Green Power; Renewable Energies.

Further Readings

Söderholm, Patrik. "The Political Economy of International Green Certificate Markets." *Energy Policy*, 36/6 (2008).
Willstedt, Heiki and Veit Bürger. *Overview of Existing Green Power Labelling Schemes*, Report for the CLEAN-E Project, 2006.

Aurélien Evrard
Sciences Po Paris/Centre for Political Research (CEVIPOF)

GREENHOUSE GASES

Since the late 1980s, increasing global scientific and political attention has been focused on the related issues of global warming and global climate change. At the core of these investigations and policy debates lie a small number of atmospheric gases, most of which are naturally occurring, that are responsible for the maintenance of Earth's temperate, life-sustaining climate. However, now-irrefutable evidence exists that human contribution to the atmospheric concentrations of these few gases is likely causing unnatural and possibly irreversible warming or destabilization of our global climate. These anthropogenic greenhouse gases are inextricably linked to growth-oriented global economies, consumption patterns, and ways of life (at least in industrialized countries). Our inability to control their emission ultimately results in irreversible damage to our global ecosystems.

Background

Earth's atmosphere consists primarily of nitrogen (78 percent) and oxygen (21 percent), along with small amounts of a variety of other gases. Several of these gases—including water vapor, carbon dioxide, methane, nitrous oxide, and man-made chlorofluorocarbons (CFC-11 and CFC-12)—act to trap solar radiation in the lower regions of the Earth's atmosphere by absorbing infrared energy reflected from the planet's surface. When present in equilibrium concentrations, these gases (with the exception of CFCs) help the Earth retain sufficient solar energy to keep its atmospheric temperature in a reasonably narrow, stable, and habitable range. Any lower concentrations would result in the loss of heat and the cooling of the Earth's atmosphere; higher concentration of one or more of these gases would cause the atmosphere to retain more of the sun's radiant energy, thus warming not only the air but also the land and oceans. This would result in what has become known as global climate change.

Naturally occurring greenhouse gases exist in an equilibrium that maintains temperatures in a range that generally supports life on this planet. By analyzing gas samples found in ice cores and compounds bound in various sediments, scientists have shown that the concentrations of various greenhouse gases have varied over Earth's history. This variation has contributed to the oscillation of our climate between warm and cold periods. The advance and retreat of glaciers have been shown to be, in part, a result of atmospheric warming caused by increased concentrations of these gases, especially carbon dioxide.

However, with the coming of the Industrial Revolution and the attendant boom in human population beginning in the late 18th and early 19th centuries, man started to have a disproportional effect on the amounts of these gases collecting in Earth's atmosphere. The burning of coal to power steam engines began to liberate tons of additional carbon dioxide into the atmosphere. Later, the introduction of the internal combustion engine and petroleum-based fuels in the late 19th and early 20th centuries significantly contributed to this greenhouse gas pollution.

As manufacturing-based economies boomed, so did the quality of life for large segments of Earth's inhabitants. The Earth's population has grown dramatically and continues to consume the products of our industrial economies. As population- and wealth-driven demand for manufactured products that depend on the burning of fossil fuels has increased, so too has the concentration of carbon dioxide in our atmosphere and the resulting greenhouse-effect warming. As Earth's surface temperature has increased—as a result of natural atmospheric cycles or to human intervention—methane trapped in melting permafrost is being liberated, contributing even more significantly to the quantity of greenhouse gases in the atmosphere.

The Science

The degree to which an agent can cause atmospheric changes (e.g., warming) is known as "forcing." In this context, anthropogenic forcing occurs when an excess of a particular agent emitted by human activity precipitates the warming (or cooling) of the atmosphere. Research by James Hansen at the National Aeronautics and Space Administration Goddard Institute for Space Studies has demonstrated that different greenhouse gases have significantly different atmospheric warming effects. Even though it is approximately 20 times as potent as carbon dioxide as a greenhouse gas (as a function of emission quantity), methane has approximately a third of the climate forcing effect as does carbon dioxide. Nitrous oxide and CFCs have a still smaller forcing effect. Together with its central role in our carbon-based economy and its long lifetime in the atmosphere, carbon dioxide has become the most important—and controversial—target in greenhouse gas mitigation policy debates.

Atmospheric concentrations of principal greenhouse gases have increased substantially over preindustrial levels. Though significant amounts continue to be absorbed by vegetation and oceans, carbon dioxide concentrations have increased from approximately 280 parts per million (ppm) to 387 ppm in 2007 primarily because of the use of fossil fuels. Methane concentrations have increased to 1,774 parts per billion (ppb) in

2007, from only 700 ppb, attributable to rice cultivation, animal husbandry, biomass burning, and landfills. Concentrations of nitrous oxide have increased from 275 ppb to well over 300 ppb largely because of industry and agriculture. The only reasonably bright spot is the dramatic reduction in CFC emissions (as a result of the Montreal Protocol), the atmospheric concentrations of which are expected to decrease dramatically over the next century.

Unfortunately, even if we were able to attain the needed emissions reductions to stabilize or reverse global warming, the atmospheric persistence of the greenhouse gases already emitted would ensure a continuation of atmospheric warming for scores of years to come. Methane is estimated to be 10 years; nitrous oxide, 100 years; CFC-11 and CFC-12, 50 and 102 years, respectively; and a substantial fraction of carbon dioxide will remain in the atmosphere for decades to centuries, with a smaller portion persisting for thousands of years. However, a recent study by Susan Solomon and her collaborators at the National Oceanic and Atmospheric Administration suggests that the anthropogenic climate change resulting from emissions of these greenhouse gases (and related climate change processes) may very well be irreversible.

Responding to Greenhouse Gas Emissions

As the role of greenhouse gases in the warming of our atmosphere has become better understood and the ramifications clearer, significant attention has been devoted to the issue of anthropogenic greenhouse gas emissions. Although the science of the greenhouse effect and the global effect that increasing atmospheric densities of greenhouse gases would have on our climate has long been understood, more localized industrial pollution sources dominated national and global policy discourse before the 1990s. For example, acid rain and toxic waste associated with industrial manufacturing technologies and modern consumption habits were generally addressed on a local, case-by-case basis. The sources of these pollutants were generally isolated and could be eliminated by applying end-of-pipe process modifications (e.g., sulfur dioxide scrubbers and community recycling, respectively), and the large-scale technological political economy remained unexamined as the effects were generally jurisdictionally defined.

One of the first global environmental issues arose when scientists determined that CFCs—compounds used as coolants in refrigeration systems, dry cleaning chemicals, and spray can propellants—were responsible for the erosion of the ozone layer that protects the Earth from harmful ultraviolet radiation. When scientists detected a hole in this layer, situated largely over the Antarctic, global political interests combined to forge the Montreal Protocol phasing out the production and use of CFCs. Although this accord demonstrated that a global environmental threat could be successfully addressed in a global forum, the source of that threat was sufficiently limited that the scientific community was able to develop new, ozone-friendly compounds that substituted for CFCs. Again, the underlying political economy continued unchallenged.

Likewise, methane emissions in the petrochemical industry have been reduced by capturing and recycling by-product gases (instead of "flaring" them), and scientists are in the process of developing livestock feeds that reduce the production of methane resulting from animal digestion. In neither of these cases, though, has the underlying activity been fundamentally reevaluated to determine whether a greener alternative is possible.

The identification of carbon dioxide as the greenhouse gas most requiring emissions reductions derives from the global economy's reliance on carbon-based fuels. From the mid-19th century to the present, we have predicated our industrial advancement and wealth accumulation on the combustion of carboniferous fuels: coal, petroleum (in all its various refined forms), and natural gas. Pharmaceuticals, chemicals, plastics, and other consumer products require petroleum and natural gas as feedstocks in their manufacture.

A Global Policy Response

Unlike many preceding environmental challenges, anthropogenic greenhouse gas emissions have a global climate effect. As a consequence, concerted action is required to reduce these emissions by all parties to "the global atmospheric commons," especially the largest emitters. Reductions taken by one country or region could be counterbalanced by increases in another. In addition, global cooperation is required if the economic costs of greenhouse gas reduction are to be fairly apportioned.

The United Nations Conference on Environment and Development, held in Rio de Janeiro, produced the United Nations Framework Convention on Climate Change, an international agreement to significantly reduce greenhouse gas concentrations. Adopted in 1997, the Kyoto Protocol to the United Nations Framework Convention on Climate Change established legally binding targets for the reduction of major greenhouse gases by Annex I (industrialized) countries. In addition to encouraging shifts to more sustainable economies and development practices, the protocol established a greenhouse gas inventory that would account for greenhouse gas emissions and reductions at the national level. It also advocated the establishment of a market mechanism that would set a price for carbon emissions. (The European Union has implemented such a mechanism; the United States has yet to do so, though isolated, state-specific exchanges have been established.)

The Kyoto Protocol will expire in 2012, and negotiations are underway to create a successor agreement. It is likely that this treaty, if ratified, will specify targets for a larger number of major greenhouse gas emitters, including China and India. In the meantime, the *Fourth Assessment Report* from the Intergovernmental Panel on Climate Change has removed all reasonable doubt that human activity is predominantly responsible for the majority of the climate changes we are now seeing and that we are likely to experience. National, as well as global, action is urgently and immediately needed. Testifying before House Select Committee on Energy Independence, James Hansen urged immediate action to reduce greenhouse gas emissions before "it will become impractical to constrain atmospheric carbon dioxide, the greenhouse gas produced in burning fossil fuels, to a level that prevents the climate system from passing tipping points that lead to disastrous climate changes that spiral dynamically out of humanity's control."

This has implications for our immediate and long-term choices of energy sources, the globalization of our supply networks, development of our cities, and our personal consumption habits. In spite of coal's relative abundance, do we continue to build coal-fired power plants that emit a disproportionate fraction of global anthropogenic carbon dioxide, or do we invest in "green" energy sources (such as wind and solar) that are emission free? Do we continue to develop global supply chains that depend on petroleum-based transportation fuels, or do we develop regionally and locally sustainable economies in which the need for long-distance transportation of goods is reduced? Do we continue to build low-density, sprawling cities that depend on the automobile as the dominant form of personal transportation, or do we embrace "smart" development practices that emphasize compact development and transit?

Most fundamentally, do populations in industrialized countries continue to consume goods and services at unsustainable rates that drive greenhouse gas emissions ever higher, or do we rethink our consumption habits to emphasize quality of life over its quantity?

See Also: Climate Change; Global Warming; Intergovernmental Panel on Climate Change; Kyoto Protocol.

Further Readings

Bernstein, Lenny, et al. "Climate Change 2007: Synthesis Report—Summary for Policymakers." Intergovernmental Panel on Climate Change, 2007.

Forster, Piers, et al. "Chapter 2: Changes in Atmospheric Constituents and in Radiative Forcing." *Fourth Assessment Report: Working Group I Report—The Physical Science Basis.* Cambridge, MA: Cambridge University Press, 2007.

Hansen, James. "Global Warming Twenty Years Later: Tipping Points Near." http://www.columbia.edu/~jeh1/2008/TwentyYearsLater_20080623.pdf (Accessed February 2009).

Hansen, James E., et al. "Climate Forcings in the Industrial Era." *Proceedings of the National Academy of Sciences of the USA,* 95 (1998).

Ledley, Tamara S., et al. "Climate Change and Greenhouse Gases." *EOS,* 80/39 (1999).

Solomon, Susan, et al. "Irreversible Climate Change Due to Carbon Dioxide Emissions." *Proceedings of the National Academy of Sciences of the USA,* 106/6 (2009).

Tans, Pieter. "Recent Monthly Mean CO_2 at Mauna Loa." NOAA/ESRL. http://www.esrl.noaa.gov/gmd/ccgg/trends (Accessed February 2009).

Kent Hurst
University of Texas at Arlington

GREEN POWER

Energy generated from environmentally friendly and non-polluting sources is known as green energy. The main sources of green energies are geothermal, solar, wind, hydroelectric, and biomass. Nuclear energy is also considered by some to be green energy. These sources are considered green because they produce low carbon emission while generating power and create less pollution in the environment. It is a fact that global warming (GW) and subsequent climate change (CC) is happening due to the emission of greenhouse gases and particulate black carbon as a consequence or by product of energy generation. Greenhouse gases are mostly produced from the burning of fossil fuels like gasoline (petrol, diesel, kerosene, and other petroleum products), coal, and natural gases to produce energy.

Green energy is also known as renewable energy (RE) because most of the green sources are renewable. Green power is commonly used for electricity, heating, and cogeneration purposes. According to the U.S. Environmental Protection Agency, green power is a subset of RE and is the electricity produced from solar, wind, geothermal, biogas, biomass, and low-impact small hydroelectric sources. Electricity production from green power sources is accomplished with an environmental profile superior to conventional power-generation technologies. The production process does not produce any anthropogenic greenhouse gas

Switchgrass is a renewable agrofuel with the potential to produce twice as much ethanol as corn, according to geneticists with the U.S. Department of Agriculture (USDA) Agricultural Research Service.

Source: USDA Agricultural Research Service/Brett Hampton

emissions. To understand green power, it is essential to understand RE as a whole.

According to the International Energy Agency (IEA), RE is derived from natural sources that are replenished constantly; that is, the sources restore themselves over a short period of time and do not diminish. In its various forms, RE is derived directly from the sun, from heat generated deep within the earth, from wind movement, through the movement of the ocean current, with the use of moving water, from organic plant and waste materials (biomass), and from hydrogen derived from renewable resources. RE usage is a very old concept.

Wood has been a major energy source since the beginning of civilization, and until most recently (by the end of 19th century), wood supplied up to 90 percent of the world's energy needs. The use of wood or biomass (which is quite renewable, unlike fossil fuel) as an energy source has decreased in the last century as a result of the low price of fossil fuels and their conveyance advantage.

Concern about global warming and climate change, along with the rising cost of fossil fuel in the 21st century, has once again given rise to RE production and usage. Worldwide, government support is being provided to produce RE and generate green power. According to S. S. Panda, about 13 percent of the world's primary energy comes from renewable sources such as the burning of traditional biomass, 2–3 percent from hydropower, and less than 1 percent from modern technologies like geothermal, wind, solar, and marine energy.

Green Power/RE Energy Sources and Generation Mechanisms

There are many proven technologies available to produce RE, and they have been abundantly used all over the world since the 1970s. Some of the technologies are under development. Following are the some of the green power or RE sources, along with their working mechanisms to generate green energy.

Biomass

Electricity is generated with the direct burning of biomass to produce heat energy or steam to run turbines. In an indirect form, liquid or gaseous biofuel is produced from biomass of various forms such as wood, grasses, crops, and crop residues. Even hay, crop waste, and waste products from animal feedlots are used in heat generation to run turbines to produce electricity. These biomasses are also converted to energy through thermal or biological conversion processes to produce liquid or gaseous biofuels; these fuels are then used in motor vehicles. Thus, liquid biofuels are an alternative to the greenhouse gas–producing fossil fuels. Biofuel is not as clean as the other RE sources like solar, wind, geothermal, tidal, hydrogen, or nuclear but is better than the energy generated from fossil fuels. Biomass itself is carbon neutral, but when it is used for biofuel production, it blocks methane emission and thus becomes carbon negative. To counter its disposal problem, biomass like manufacturing waste, rice hulls, agricultural waste, and black liquor from paper production now is regularly being burned to produce electricity and biofuels.

Agrofuels, an alternative to fossil fuel, are also called biofuels. They are produced from agricultural products such as crops, seeds, crop residues, and so on. Crops with higher sugar, starch, and oil content are perfect for agrofuel production. Soybeans, sunflowers, palms, jatropha, *pongamia pinnata*, and algae are crops, plants, or other living biomass that are rich in oil and are renewable. These oils are heated to reduce their viscosity and are directly used in diesel engines as fuels. Sugarcane, sugar beet, and sweet sorghum are crops that have high sugar content; corn contains very high starch content. They are fermented with yeast to produce ethyl alcohol, or ethanol, a volatile, flammable, colorless liquid. The flammable property of the ethanol makes it a perfect alternative to fossil fuel for transportation vehicles.

Anaerobic Digestion

With the use of microorganisms, biodegradable materials are broken down in the absence of oxygen. Human, animal, and bird excreta; wastewater sludges; and organic wastes are biodegradable materials. Because of anaerobic digestion, these materials have their volume and mass reduced and in consequence produce methane- and carbon dioxide–rich biogas suitable for energy production. Essentially, this gas is a RE source. An anaerobic digestion–integrated waste management system reduces the emission of landfill gases into the atmosphere. The nutrient-rich solids produced as a by-product after the digestion is used as fertilizer. The by-product solid produced after cattle manure digestion is five times richer than the original manure itself.

The digestion process starts with bacterial hydrolysis, in which the input materials like insoluble organic polymers such as carbohydrates are broken down to make them available for other bacteria. Then the acidogenic bacteria convert the sugars and amino acids into carbon dioxide, hydrogen, ammonia, and organic acids. Next, the acetogenic bacteria convert these resulting organic acids into acetic acid and additional ammonia, hydrogen, and carbon dioxide. Finally, methanogens convert these products to methane and carbon dioxide.

In developing countries, the anaerobic digestion system is widely used. In countries like India, Nepal, Bangladesh, Laos, Cambodia, Vietnam, Bhutan, and China, many single-family households have their own biogas plants in their backyard. Government-backed programs promote the adaptation of biogas plants by families, as most rural families have their own cattle. Cattle manure is used in the incinerator on a daily basis to produce biogas. The gas is used for cooking and electricity generation for the household. It brings self-sufficiency to the community in those countries.

Wind

Wind is one of the most promising RE sources for electricity generation with a proven technology. With this technology, airflows are used to run wind turbines to produce electricity. It is as cost effective as fossil alternatives in good wind regimes like the vast Midwest plains in the United States and in any coastal plains in the world. Particularly when economic or environmental concerns are considered, wind energy is far better than fossil fuel sources. In general, modern wind turbines range from around 600 KW to 5 megawatts (MW) of rated power; the most common wind turbines for commercial use are of 1.5- to 3-MW capacity.

The only problem associated with wind energy is that its production capacity varies throughout a year. In typical conditions, in a year, windmills run at 25–35 percent of capacity. Therefore, other sources of green power such as hydroelectricity should be paired with it to match supply with demand. Hydroelectric power production can fluctuate (stop and start) as required. Large-scale wind farms are connected to local electric power grids, and small wind mills provide electricity to isolated locations like farmhouses. Utility companies generally buy back the excess electricity produced by windmills on a given day.

Wind energy is a green power widely accepted by environmentalists because it is plentiful and can be found everywhere, renewable, clean, and widely distributed. It releases minimal or almost no greenhouse gases to the atmosphere. Offshore wind farms are popular and produce high efficiency.

Solar

Solar power is the cleanest form of energy that has been harnessed by civilization since its inception. Solar power technologies produce electricity through heating/heat engines and photovoltaic (PV) systems. Depending on the way solar technologies capture, convert, and distribute sunlight as power, they are generally characterized as either passive solar or active solar. Active solar technologies use PV panels and solar thermal collectors with electrical or mechanical equipment that converts sunlight into direct electricity. Passive solar techniques include constructing a building oriented to the sun, selecting favorable thermal mass materials for homes, and designing spaces with natural air circulation.

A PV module is composed of multiple PV cells or arrays that convert sunlight directly into electricity. PV power is also widely considered to be cost competitive with wind power. With the dropping cost of PV cells, there should be a solar energy boom in the near future. Another point to be noted is that solar radiation is free, and most of the places on Earth with large human populations get a huge amount of solar radiation throughout the year. The Earth receives around 174 petawatts of solar radiation in the upper atmosphere.

Solar energy is widely used for off-grid applications, such as telecommunications, lighting road signals, and supplying power to villages in remote and rural areas. In general, solar energy is of two types: solar thermal and solar electric. At present, commercial solar electric technologies produce from a few kilowatts to hundreds of megawatts of electricity. Similar to wind power, solar power is nondispatchable, so a hybrid application of solar technology with a fossil fuel or hydroelectric power source is more economically competitive. Once solar energy is mass produced, it will immensely reduce the use of greenhouse gas–emitting technologies.

Geothermal

Geothermal power is the energy that is generated from the heat stored in the Earth or underground. Geothermal technology is a very old technology that was tested in 1904 at the Larderello dry steam field in Italy. Geothermal energy is obtained by tapping the heat of the Earth underground, as it is quite warmer than the surface itself. It is expensive to build a power station to convert geothermal energy to electricity or heat, but once the station is constructed, the operating costs are very low, resulting in low energy costs for suitable sites. If the hot water or steam resource from the geysers is at a high temperature near the surface of the earth, the unit cost of geothermal energy would be pretty low, as electricity generation from geothermal energy is a baseload technology. Some of the greenhouse gases are emitted through the geysers, but once converted to geothermal power, these gases cease to escape. Thus, geothermal power is carbon negative.

The most common geothermal technology is in the form of binary cycle power plants. In binary cycle power plants, moderately hot geothermal water is passed by a secondary fluid with a much lower boiling point than water, generating vapor (secondary fluid flashes to vapor) that drives the turbines. The Philippines and Iceland are two major countries that produce 15–20 percent of their energy from geothermal sources.

Hydropower

Hydropower, hydraulic power, or waterpower is another clean energy source and is the most mature form of RE because it is perhaps the oldest technology for mass-scale electricity production. It has a significant share of electricity generation worldwide, with a supply of 715,000 MW or 19 percent of the world's electricity. Because water is denser than air, water has more energy-producing ability than wind.

Hydroelectric power is very inexpensive compared with electricity generated from fossil fuels. Hydropower stations have many added advantages in addition to electricity generation—areas with hydropower units attract industries, and a city can easily be settled around them because hydropower units help in flood control, irrigation, pisciculture, navigation, and recreation. One of the major disadvantages associated with hydropower plants is that the reservoirs associated with them flood a huge amount of land containing precious natural resources.

Tidal

Energy from the ocean in the form of tidal forces, ocean currents, wave power, and thermal gradients has the ability to produce electricity or other useful forms of power. It is known as tidal power. Tidal power is not widely used but it has a high potential to meet the world's soaring energy demand and at the same time preserve the environment with no production of greenhouse gases. More predictable than wind and solar power, tidal power can be reliably used in the future. Tidal power is generated using technology similar to underwater windmills. Though a new concept, it is being deployed in many countries. Portugal now has the world's first wave farm, which generates 2.25 MW of electricity using three Pelamis P-750 machines.

Nuclear

Nuclear power is also considered an RE or green power, except that the nuclear waste must be managed properly. Nuclear power is generated from atomic nuclei via controlled nuclear reactions. Nuclear fission is the only method that is used today. Utility-scale nuclear reactors produce heat as a result of controlled nuclear fission that generates steam and that in turn produces electricity. When a relatively large fissile atomic nucleus, such as uranium-235 or plutonium-239, absorbs a neutron, fission of the atom often occurs, thus splitting the atom into two or more smaller nuclei. Kinetic energy or heat is produced as a reaction. The cooling system in the reactor removes heat from the reactor core and transports to another area, where the heat is harnessed to produce electricity or another form of power.

Until last year, nuclear energy supplied 14 percent of the world's electricity. By 2007, there were 439 nuclear power reactors operated in the world, with the United States, France, and Japan being the major producers, producing more than 56 percent.

Hydrogen Fuel Cell

Hydrogen can act as a crucial storage medium and carrier of RE, as hydrogen fuel cells are an electrochemical conversion device. A hydrogen cell produces electricity from fuel on the anode side and an oxidant on the cathode side that react in the presence of an electrolyte. Hydrogen cells use hydrogen as fuel and oxygen as oxidant; thus, unlike fossil fuel technologies, the by-products from this technology are not greenhouse gases. Therefore, hydrogen fuel cell technology or hydrogen power is a green power.

According to the IEA, hydrogen, along with RE technologies, is a major potential contributor to the sustainability of the energy sector. In the longer term, if costs can be dramatically reduced, hydrogen can act as the crucial storage medium and carrier of energy produced from renewables. At this time, hydrogen is being used in vehicles, thus limiting greenhouse gases. In the longer term, with proper research and development, hydrogen, battery-powered electricity, and solar energy will fully replace fossil fuels as the energy source for vehicles.

Green Power Potential

Many RE technologies have yet to be widely adopted so their potential to meet the world's ever-growing energy demand is untested, and the market is growing rapidly for many forms of RE. Electricity-generation capacity from RE sources reached an estimated 240 gigawatts worldwide in 2007—an increase of 50 percent over 2004. Green power or RE excluding hydroelectric and nuclear sources represents 5 percent of global power capacity and 3.4 percent of global power generation.

Total global installed wind power capacity by the end of 2008 was 120,791 MW. Wind energy is the fastest-growing RE in the world and almost doubled in the three years from 2005 to 2008. Since 1993, wind energy production has grown, with an annual average of 30 percent worldwide. Europe is in the forefront of wind power production. Several European countries, along with the United States, produce the largest percentage of wind energy. Denmark is the world leader in this technology, generating almost 20 percent of its total electricity need with wind power. U.S. wind power capacity grew by 45 percent to 16.8 gigawatts in 2007 and more than 50 percent in 2008. By the end of 2008, the United States had an installed wind-power capacity of 25,000 MW.

The present worldwide manufacturing output of the PV industry is more than 2,000 MW per year, with Japan, Germany, and the United States contributing 90 percent of all PV installations in the world. Solar energy use is also growing at a faster rate in developing countries. Worldwide photovoltaic production growth has averaged 40 percent per year since 2000, and installed capacity reached 10.6 gigawatts at the end of 2007. Developing countries like Kenya, China, and India make huge application of solar power. In Europe, the growth of power generated from solar source is anticipated to be around 20 percent per year. China has the highest possible growth of solar energy production and use in recent years. Solar heating and drying potential worldwide is estimated to be 600–900 picojoules.

The Philippines and Iceland are two major producers of geothermal power. At The Geysers in California, a 750-MW geothermal power plant is the largest in the world. Geothermal resources have been identified in more than 80 countries, with a potential 600,000 exajoules of energy.

Brazil has one of the largest RE programs in the world. It has the largest ethanol fuel production industry in the world, using mainly sugarcane. Gasoline sold in Brazil contains more than 25 percent ethanol, and more than 90 percent of new cars sold in Brazil are run with ethanol-mix fuel. By 2006, 16.3 billion liters—33.3 percent of the world's total ethanol or 42 percent of the world's ethanol used as fuel—was produced by Brazil from sugarcane. Ethanol fuel produced from corn is also widely available in the United States. Most of the countries in the world are mandating the use of more than 10 percent ethanol or other biofuel as a mixture with fossil fuel.

Although large hydropower plants are environmentally unfriendly, there is still tremendous potential for small hydropower plants in the world. Canada is the largest producer of hydroelectricity, followed by the United States and Brazil. Approximately two-thirds of the economically feasible hydropower generation potential remains undeveloped. Untapped hydropower resources are abundant in Latin America, Central Africa, India, and China.

RE production and its use are very popular in rural settings where grid-based electric supply is absent. Thus, RE is equally popular among the poor and rich countries of the world. According to IEA, the potential of bioenergy is 200–300 exajoules per annum, up from the current potential of 50 exajoules per annum. The potential is high because of the abundant presence of a resource base for its production in all countries. Bioenergy crops and municipal wastes could be used abundantly in coming years throughout the world to produce bioenergy.

Theoretically, wind and solar energy are capable of supplying a large percentage of global energy needs, as wind and solar radiation is ample everywhere. However, the practical potential is limited because of cost, variability and intermittency, and siting factors, but researchers do not agree with this argument. They believe that more than 1,000 percent of total electricity supply is possible without compromising grid reliability. It is expected that total long-term wind energy potential is five times the total current energy production.

Ocean energy systems offer the promise of low-cost reliable electricity for coastal regions, but low-cost technology is needed to make it economically viable. Hydrogen energy potential is immense because of serious research and development programs in Europe and the United States.

It is envisioned that green power production has a bright future in coming years with proper research and development, strong government initiation and help, heavy commercialization, and public education about the vagaries of energy generation through fossil fuel technologies. Most governments are promoting the expansion of green power to combat global warming and climate change.

Other Future Green Power Sources

According to physicist Bernard Cohen, uranium is effectively inexhaustible. Fast breeder reactors fueled by seawater-extracted uranium could produce and supply energy for billions of years. Nuclear fusion technology is in the development stage, and it is expected that by 2015 it will be used to produce electricity in abundance, thus limiting the use of fossil fuels for energy production. Nuclear fusion technology is one of the best prospective green power sources because it produces less nuclear waste and has fewer containment issues than nuclear fission technology.

See Also: Alternative Energy; Biomass Energy; Ethanol, Corn; Ethanol, Sugarcane; Fuel Cells; Fusion; Geothermal Energy; Global Warming; Greenhouse Gases; Hydroelectric Power; Hydrogen; Nuclear Power; Renewable Energies; Solar Energy; Three Gorges; Tidal Power; Wave Power; Wind Power; Wood Energy.

Further Readings

"Anaerobic Digestion." http://www.waste.nl/content/download/472/3779/file/ WB89-InfoSheet(Anaerobic%20Digestion).pdf (Accessed February 2009).

"Biofuels: The Promise and the Risks, in World Development Report 2008." The World Bank, 2008. http://siteresources.worldbank.org/INTWDR2008/Resources/2795087-1192112387976/WDR08_05_Focus_B.pdf (Accessed February 2009).

Global Wind Energy Council. "Global Installed Wind Power Capacity (MW)—Regional Distribution." http://www.gwec.net/fileadmin/documents/PressReleases/PR_stats_annex_table_2nd_feb_final_final.pdf (Accessed February 2009).

International Atomic Energy Agency. "Nuclear Power Plants Information: Number of Reactors in Operation Worldwide." http://www.iaea.org/cgi-bin/db.page.pl/pris.oprconst .htm (Accessed February 2009).

International Energy Agency Renewable Energy Working Party. "Renewable Energy into Mainstream." Report prepared by International Energy Agency, Netherlands. (Accessed February 2009).

Italian Geothermal Union. "The Celebration of the Centenary of the Geothermal-Electric Industry Was Concluded in Florence On December 10th, 2005." IGA News, 64 (April–June 2006).

James, L. and A. Dicks. Fuel Cell Systems Explained, 2nd ed. London: J. Wiley, 2003.

Lawbuary, J. "Biogas Technology in India: More Than Gandhi's Dream." http://www .ganesha.co.uk/Articles/Biogas%20Technology%20in%20India.htm (Accessed February 2009).

Panda, S. S. "Biomass." In Encyclopedia of Global Warming and Climate Change., S. G. Philander, ed. Thousand Oaks, CA: Sage, 2008.

Panda, S. S. "Energy, Renewable." In Encyclopedia of Global Warming and Climate Change. S. G. Philander, ed. Thousand Oaks, CA: Sage, 2008.

Phillibert, C. "The Present and Future Use of Solar Thermal Energy as a Primary Source of Energy." International Energy Agency. http://www.iea.org/textbase/papers/2005/ solarthermal.pdf (Accessed February 2009).

Reel, M. "Brazil's Road to Energy Independence." Washington Post (August 19, 2006).

U.S. Department of Energy. "Geothermal Basics." Office of Energy Efficiency and Renewable Energy. http://www1.eere.energy.gov/geothermal/geothermal_basics.html (Accessed February 2009).

U.S. Environmental Protection Agency. "Green Power Defined." http://www.epa.gov/grnpower/gpmarket/index.htm (Accessed February 2009).

U.S. Geological Survey. "Hydroelectric Power Water Use." http://ga.water.usgs.gov/edu/wuhy.html (Accessed February 2009).

Worldwatch Institute. "Renewables 2007: Global Status Report." http://www.ren21.net/pdf/RE2007_Global_Status_Report.pdf (Accessed February 2009).

World Wind Energy Association. "120 Gigawatt of Wind Turbines Globally Contribute to Secure Electricity Generation." http://www.wwindea.org/home/index.php?option=com_content&task=view&id=223&Itemid=40 (Accessed February 2009).

Sudhanshu Sekhar Panda
Gainesville State College

GREEN PRICING

Green pricing is commonly found in energy markets and finances environmentally friendly alternatives to conventional utilities. Interested customers pay an additional fee per kilowatt-hour to purchase clean energy from hydroelectric, wind, geothermal, solar, and biomass sources. Green power markets are still new, and to ensure quality and verify delivery, many utilities apply for certification from independent organizations. Renewable energy credits (RECs) are another method to assist utility companies in financing green energy investments. Although the REC purchaser does not directly buy electricity, REC sales may subsidize renewable energy production.

Research from the National Renewable Energy Laboratory of the U.S. Department of Energy documents that 750 utility companies—a quarter of the national total—offer green pricing programs to over 70 million customers. Yet fewer than one million customers purchase green energy, and sales make up less than 1 percent of total electricity sales. Nonetheless, an increasing number of U.S. customers—an estimated 20–30 percent more each year—would like access to such programs. Sales to nonresidential customers are increasing at twice the rate of sales to residential customers.

Premium pricing for renewable energy has fallen in recent years as a result of the lowering of production costs. Even so, fear of expensive premiums dissuades some customers from purchasing green energy. However, green premiums often only cost 1.5–2 cents more per kilowatt hour. Based on the national average, U.S. households purchase approximately 875 kilowatt-hours a month, so customers who purchase half of their electricity through green pricing should pay less than $10 per bill. Moreover, under certain circumstances, green energy may actually cost less than energy generated from fossil fuels. As one example, Austin Energy, the largest green pricing utility in the country, purchased wind energy for its Green Choice Program under a 10-year, fixed-price contract. As natural gas prices rose, the wind price remained fixed, and by comparison, rates for green energy customers were cheaper. Nonetheless, the goal of green pricing is not necessarily to reduce energy costs, and when the wind contract sold out, Austin Energy renegotiated its premiums. Still, a major motivation for some green pricing costumers is the fixed-price contracts that can insulate them from fossil fuel price increases. Yet, in some instances, green pricing programs may create abnormally high costs for consumers. Costs may not reflect energy prices but, rather, a lack of preparation for the transition to renewable energies. One example is Florida Power and Light, which canceled its green power program after spending upward of $100 on marketing and recruitment per acquired customer, in contrast to the national average of $38. Investment and marketing constraints to the initiation of green energy programs have encouraged concentration in the U.S. market: the 10 largest programs sell 70 percent of all green power and have 60 percent of customers. Although some utility companies have green energy customer participation rates ranging from 5–17 percent of their total customer base, to achieve high rates of involvement they must invest in marketing and outreach.

Because of mixed results in green energy pricing returns and delivery, independent third-party certification has grown. Green-e, the largest green energy certification organization in the United States, is administered by the Center for Resource Solutions. Green-e's role extends beyond project verification and quality control to stimulate demand for green energy. Green-e participates in market research, promotes green energy technologies and carbon offset markets, and encourages participation by utility companies. Green-e also certifies RECs or "green tags" (also known as renewable energy certificates) for quality purposes, much like their European counterpart RECS International.

RECs have become a popular method to promote green energy without direct connection to a renewable-energy grid. Companies that use green tags split green energy into two commodities, so that the electricity and the RECs are sold separately. Because they help to finance infrastructure, REC purchasers can claim to support green energy even if they are not purchasing the product. RECs are generally sold for between US$1.50 and $20 per megawatt-hour on the retail market, although they have been reported to cost as much as $90 in some instances. RECs now make up more than half of all green energy sales, surpassing traditional green pricing.

The popularity of RECs stems from the perception that they may be used to provide offsets for companies and individuals targeting carbon neutrality, but in fact there is often no accurate quantification of any subsequent emission reduction. In voluntary markets, RECs are often loosely regulated and vary significantly. Some utilities profit from the misuse of RECs by selling credits for preexisting projects without demonstrating "additionality," or that the credits make possible green energy that would not have been otherwise feasible. To address this problem, Green-e oversees a certification system that limits the number and life span of RECs given to a qualifying company. Greene's accreditation may also increase the price of quality RECs and promote their purchase among consumers. Currently REC critics argue that the credits do not play a significant role in the creation of new alternative power projects, as a federal renewable energy tax credit granted during the first decade of production greatly outweighs REC bonuses. Nevertheless, REC sales in the United States increased 50 percent from 2006 to 2007.

Green energy supply is led by a handful of companies because of investment constraints and an institutional learning curve. Green pricing provides important financial support to budding renewable energies, but customer enthusiasm and participation may wax and wane in response to turbulent fossil fuel markets. RECs have outpaced traditional green pricing within electricity contracts, even though REC investment in alternative energy projects has the potential to be watered down, and it often remains difficult to verify emission reductions in projects lacking independent certification.

See Also: Electricity; Environmentally Preferable Purchasing; Green Energy Certification Schemes; Green Power; Power and Power Plants; Public Utilities; Renewable Energies.

Further Readings

Barcott, Bruce. "Green Tags." *World Watch Magazine,* 20/4 (2007).
Bird, Lori, et al. "Green Power Marketing in the United States." National Renewable Energy Laboratory Technical Report TP-6A2-44094, 2008.
Englander, David. "Green Pricing Programs." http://www.marketwatch.com (Accessed January 2009).

Mary Finley-Brook
Charles Kline
University of Richmond

GRID-CONNECTED SYSTEM

Many renewable sources of energy are designed to be separate and stand-alone, providing their owners with electricity that supplements what they draw from the utility electric grids. Considerable financial investments required for these stand-alone systems are often responsible for their slow diffusion into society. Grid-connected systems, in contrast, as the name implies, are integrated into the public electricity distribution system and are designed to eliminate some of the financial disincentives that stand-alone energy systems create.

Key aspects of the grid-connected system are:

- Small, distributed centers of electric energy, often from renewable energy sources, are connected to the public electricity distribution system.
- The need is eliminated for costly household investments in energy storage in the form of batteries by using the electric grid as an energy sink.
- Energy generated in surplus of household demand is bought by the public utility, which supplies electricity when local generation is below demand.
- They require a specific policy environment that creates financial incentives for both utility companies and households.
- Some new technical and economic issues are introduced as a result of this interconnection.

The architecture of the electric power system that is predominant in most countries is marked by three clearly differentiated technical and spatial domains: generation, transmission, and distribution. The generation domain is dominated by massive power plants that generate electricity far away from customer loads. The transmission domain conveys the large volumes of energy generated to customer centers in populated urban and industrial areas. Finally, in the distribution domain, the electric energy conveyed over long distances is distributed to industrial, commercial, and domestic customers in load centers. As historians of large technical systems have concluded, this specific spatially differentiated architecture of the power system has evolved as a result of particular social choices made by pioneering inventors and developers. Over the years, these initial choices have acquired considerable momentum that prevents easy transitions in the path of energy development.

Grid-connected systems, also referred to as distributed or dispersed generation, are notable because they represent a movement away from the historically constituted model of the explicit spatial separation between generation of electric energy and its distribution. Rather than generation being located at large distances away from customers, through distributed generation, customers who own sources of energy can not only draw energy from the distribution grid but also supply energy into it. That this shift is significant is evident from the seriousness with which grid-connected distribution generation has been received within the electricity utility industry. It is the concurrence of several factors that have enhanced the viability such a schema.

Environmental concerns are a primary reason for the popularity of grid-connected systems. These systems dovetail well with efforts that tap into cheap fuel opportunities from recycling, from combined heat and power plants, and with efforts to limit greenhouse gas emissions. Each of these opportunities find grid-connected systems to be a cost-effective means to reduce the power system's overwhelming dependence on dirty, fossil fuel generation plants. Another reason for the interest in distributed generation arises from changes in national regulatory frameworks that govern the organization of utilities and the economic logic that underlies their electricity services. The deregulation of the electricity industry in North America and Europe from its natural monopoly status has been accompanied by the increasing competition in electricity energy provision and the unbundling of the vertically integrated organizational structure of generation, transmission, and distribution domains. The increased organizational flexibility and the competitive pricing of electricity have made cheap, small sources of energy such as grid-connected systems particularly attractive to utilities because they provide a means of deferring costly capital investments in large electric power plants.

The basic technical components in a grid-connected system are similar to a stand-alone generation system. Both systems possess small energy sources (often a generator or photovoltaic array) that create electric energy. Grid-connected systems range from traditional microscale turbines that are powered by natural gas, wind, or cogeneration to nontraditional sources such as photovoltaic arrays and fuel cells. Although stand-alone systems store the energy they generate locally in expensive batteries, in grid-connected systems, energy generated by the customer is injected into the electric power grid. In the case of a grid-connected system, the surplus energy from the source is routed to the public distribution grid via the owner's electricity connection through some sophisticated hardware called the "balance of system" that interfaces between the customer's generating system and the public electricity distribution system. The flow of electric energy on the customer's connection in both directions brings to the fore several technical and economic issues.

Technical issues that arise from the interface between the customer side and the utility side of the connection do so from a desire to isolate each side so as to minimize the damage that the connection could inflict on either side. Although the equipment on the customer side is vulnerable to high-voltage surges that could damage it, the utility side is equally vulnerable to distortion in the quality of the power from the energy supplied by the consumer. In addition, grid-connected systems have engendered an important economic issue—net metering. Net metering allows a system owner to receive some retail credit for the excess energy they feed back into the grid. Thus, as a consumer, the owner pays the retail rate for the energy they consume from the grid, but in addition, the utility purchases excess energy generated by the owner at the wholesale rate. In the United States, major energy policy legislation such as the Public Utility Regulatory Policies Act of 1978 and the Energy Policy Act of 2005 have made it mandatory for public utilities to provide net metering to all customers who request it. The power produced by the owner may be valued at a static or dynamic rate. Static forms of net metering assign a flat rate based on the net difference between energy inflow and outflow. Dynamic forms of net metering rely on smart electric meters that calculate the price of net energy flow on dynamic factors such as the time of use or market fluctuations.

In conclusion, grid-connected systems represent an interesting new paradigm in energy production and distribution that holds great promise for the spread of renewable energy.

See Also: Batteries; Energy Storage; Green Power; Green Pricing; Power and Power Plants; Public Utilities; Renewable Energy Portfolio; Smart Grid.

Further Readings

Ackermann, Thomas, et al. "Distributed Generation: A Definition." *Electric Power Systems Research,* 57 (2001).

Budhraja, Vikram S. "The Future Electricity Business." *The Electricity Journal* (November 1999).

El-Khattam, W. and M. M. A. Salama. "Distributed Generation Technologies, Definitions and Benefits." *Electric Power Systems Research,* 71 (2004).

Hughes, Thomas P. *Networks of Power—Electrification in Western Society 1880–1930.* Baltimore, MD: Johns Hopkins University Press, 1983.

Pepermans, G., et al. "Distributed Generation: Definitions, Benefits and Issues." *Energy Policy,* 33/6 (2005).

U.S. Department of Energy. "Connecting Your System to the Electricity Grid—A Consumer's Guide to Energy Efficiency and Renewable Energy." http://apps1.eere.energy.gov/consumer/your_home/electricity/index.cfm?mytopic=10520 (Accessed January 2009).

Govind Gopakumar
Rensselaer Polytechnic Institute

Heating Degree Day

Several heating- and cooling-based indices have been developed to reflect the strong relationships between energy use and atmospheric conditions. One such index is the heating degree day, which can describe the demand for energy used to heat interior spaces. The heating degree day has been employed for energy management or forecasting purposes by utility companies and other institutions or organizations, as well as by heating engineers and financial professionals.

The heating degree day is predicated on the premise that when average daily temperatures fall below a certain threshold or base level, the need to heat homes and other buildings begins. The outdoor air temperature, then, has a direct and significant influence on the amount of fuel used to warm building interiors. Very cold days require the consumption of more energy to maintain comfortable indoor temperatures than merely cool days. The base temperature employed varies, depending on geographic location. For example, the United States, which follows a Fahrenheit temperature scale, uses a base temperature of 65 degrees F (approximately 18.3 degrees C). The United Kingdom uses 10 degrees C, 15.5 degrees C, or 18.5 degrees C. A value of 18 degrees C has also been used. Although these are typical base temperatures, the computation of a heating degree day can, theoretically, be made with any number of base values.

Calculating the heating degree day requires knowing the daily mean temperature, which in its simplest form is the sum of the daily maximum and minimum, divided by 2; daily means are computed from a time period starting at midnight and ending at the following midnight. Although other methods to compute average daily temperatures exist, this is the most common, as daily maximum and minimum temperature data are relatively accessible. To compute the heating degree days, this mean value must be lower than the base temperature. The calculated average daily temperature then is subtracted from the base temperature, and the result is the number of heating degree days. For example, a day with a high of 70 degrees F and a low of 50 degrees F would have a daily mean of 60 degrees F. Subtraction from a base temperature of 65 degrees F results in a 5 degree difference, or 5 heating degree days.

The heating degree day concept does not reflect perfectly the relationships between ambient air temperature and energy demand for interior spaces. Variables independent of the outdoor environment affect energy usage for heating; these variable are founded on human circumstances and building characteristics and vary by individual location. Unaddressed factors include sensitivity and acclimation to cooler interior temperatures by humans inhabiting a said space, differing building materials and design, building orientation and incident solar radiation, and variations in the ability to afford fuel. Using the heating degree day as a theoretical foundation to analyze energy consumption over a more spatially coarse area, such as a city instead of an individual home, might provide a clearer overall picture of energy use. This index, however, can still be used by individuals to establish relationships between their own energy consumption and the local thermal environment. Utility company statements often include information about the number of heating degree days for the most recent billing period, which allows individuals to calculate the amount of energy used per heating degree day. With longer-term information, these results can be employed to assess the performance of home improvements (e.g., added insulation or a more energy-efficient furnace).

Energy consumption to maintain interior temperatures is not limited to the cold-weather heating season. A complementary index—the cooling degree day—describes energy usage based on the use of air conditioning. Computation of the cooling degree day is similar to that of the heating degree day, except the base temperature serves as the bottom threshold for cooling buildings. After average daily temperatures exceed the base temperature, implementation of air conditioning is assumed. The cooling degree day, then, is indicative of the likelihood of energy consumption to bring indoor temperatures down to a level of thermal comfort.

Heating degree days can be calculated for various time scales, such as weekly, monthly, or seasonally. These aggregated totals of heating degree days provide valuable information regarding energy consumption, as shown in previous studies. The heating degree day and the cooling degree day have been positively correlated with energy usage, regardless of the source of the energy (i.e., electric, oil, natural gas), as air temperature, on which these indices are based, is a principal determinant of residential energy consumption.

Recognizing the economic effect of weather and climate, the heating degree day also makes an appearance on financial markets in the form of weather derivatives, a type of financial instrument. Sellers of weather derivatives assume the risk of negative climatic conditions in exchange for a payment, allowing companies to hedge against weather and climate condition. The seller profits when the atmospheric state is positive or remains the same. In the United States, weather contracts are based on heating degree days (for winter) and cooling degree days (for summer), often using aggregated monthly totals. Weather contracts for European cities also use heating degree days for winter; summer contracts use the Cumulative Average Temperature, a summation of average temperatures for a given city.

Heating degree days, summed for the course of a complete cold-weather season, show relationships with both latitude and altitude. In the continental United States, for example, the highest cumulative totals of heating degree days are found along the northern tier states of North Dakota and Minnesota (along the Canadian border), as well as the higher elevations of the Rocky Mountains in Colorado and Wyoming. This reflects the effect latitude has on temperatures; the decrease in temperature with altitude also contributes to larger values. However, additional factors act to determine the temperature (and thus, the degree day). The relationships between latitude, altitude, and heating degree days are not perfect. For example, temperatures at coastal locations are moderated by the proximity of large water bodies. The city of Vancouver (British Columbia, Canada), at latitude 49 degrees N, has approximately 2,900 heating degree days in a year (based on 18 degrees C). Winnipeg

(Manitoba, Canada) is located at a comparable latitude (50 degrees N), but is located in the interior of a large land mass; its annual heating degree days average over 5,700. These variations reflect the differences in specific heat (the ability of a substance, such as land or water, to store heat.) This effect is even more pronounced at higher latitudes. Similar results are found with cooling degree days.

Average values of heating (and cooling) degree days can be used to predict potential energy demand. In the United States, the National Center for Climatic Data releases 30-year averages, called "normals," for both degree day indices. The National Center for Climatic Data publishes monthly and annual data for individual station locations as well as the state as a whole. The latter are weighted by population, ensuring that a state's average is inclined toward areas of higher population concentrations; energy demands would naturally be higher in such locations. These normals can be compared to current heating (cooling) degree day trends.

See Also: Forecasting; Home Energy Rating Systems; Public Utilities.

Further Readings

Ahrens, C. Donald. *Meteorology Today*, 9th ed. Belmont, CA: Brooks/Cole, 2009.
Canadian Council of Ministers of the Environment. "Heating and Cooling." http://www .ccme.ca/assets/pdf/cc_ind_people_htng_clng_e.pdf (Accessed December 2008).
Fusaro, Peter. "Weather Derivatives Starting to Rocket." *Energy Hedge,* http://www .chicagomercantileexchange.org/files/energyhedge_rocket.pdf (Accessed December 2008).
Sailor, D. J. and J. R. Muñoz. "Sensitivity of Electricity and Natural Gas Consumption to Climate in the U.S.—Methodology and Results for Eight States." *Energy*, 22/10 (1997).

Petra Zimmermann
Ball State University

HEAT ISLAND EFFECT

A classic example of unintentionally modifying a local climate, the heat island effect was first written about in 1820. This effect refers to an atmospheric phenomenon in which the air temperatures within an urban environment are measurably higher than those of a proximate rural setting. Isotherms (lines of equal temperature) show regions of warmer air above urban locations—hence the term *island*. Urban heat islands have been identified over settlements with populations of as few as 1,000 people. Some have used population as a proxy for an estimate of the heat island effect. Some have used population as a proxy for an estimate of the heat island effect. A rough relationship between population and temperature can be expressed by *T(urban – rural) = P log p*, where *T* is the temperature, *P* is the constant of proportionality (based on urban form), and *p* is the population of the city. Physically, the heat island effect is a product of the intricate relationships between urban geometries, reductions in green space, urban surface and building materials, urban energy consumption, and local weather and physiographic conditions. Built environments store copious amounts of heat energy, much of it from absorbed solar radiation, which is reradiated as longwave thermal infrared energy. Profiles of temperature plotted against distance from an urban core show higher temperatures above areas with denser populations—even suburban locales have warmer temperatures than rural ones, though the temperatures are less warm than those above the dense urban center.

Regional weather patterns and the local physical environment also affect the development of an urban heat island. Local airflow and circulation patterns increase/decrease the severity of the heat island effect. Cities built in ways that enhance airflow benefit from the ability to move warmed air away from the urban center.

Complex interactions between the built environment and the atmosphere lead to the generation of sensible heat. Sources for this heat energy include the following:

- Materials that absorb energy, such as blacktop or concrete
  - Low-albedo surfaces both absorb and re-reradiate longwave energy into the ambient atmosphere
  - Many urban surface materials have high heat capacities (ability to store heat)
- Heat generated by anthropogenic sources, such as air-conditioning systems and cars
  - The operation of heating/cooling systems or subways, for example, generates heat energy that is released into the surrounding atmosphere
- A reduction in the amount of vegetation in an urban environment
  - Decreased vegetation reduces the amount of evaporation and transpiration (evapotranspiration), processes that cool the surrounding environment
- Physical barriers, such as buildings or landforms, which can dampen the magnitudes of the winds
  - Winds serve to transport sensible heat; lowered wind speeds (from surface friction) result in less heat moved from the urban environment
- Geometric frameworks of urban areas
  - Buildings in proximity to one another absorb some of the longwave radiation emitted by other buildings, so that energy remains in the ambient environment
- Preponderance of paved surfaces
  - Paved surfaces increase precipitation runoff, effectively "waterproofing" the surface, resulting in less infiltration and subsequent evaporation
- Air pollution
  - Undesirable materials released into the atmosphere include greenhouse gases and aerosols that contribute to the increase in air temperatures

Surface and atmospheric heat islands behave differently. The surface heat island, based on temperatures of surface materials, is strongly dependent on solar radiation, registering its greatest magnitude during the day; dry urban materials become much warmer than the corresponding moister surfaces of rural regions. The intensity of the atmospheric heat island (founded on temperatures of the ambient air) also differs diurnally. Although the atmospheric heat island effect can be observed throughout the entire day, the effect is strongest at night. The difference between the daily maximum temperature of an urban core and that of its rural hinterland is smaller than the difference between the daily minimum temperatures (observed during nighttime hours). This difference reflects the convection that typically develops during days; convection, which moves warm air, is diminished or absent at night. Detection of surface and atmospheric heat islands is accomplished by measuring both air and surface temperatures. In situ air temperatures are measured at a height of approximately 1.5 meters. Remote sensing tools, such as satellite data, are used to sense the temperatures of surface materials. Sensors mounted on aircraft or satellites detect thermal energy.

The magnitude of the heat island effect varies with the seasons and meteorological conditions. Summer tends to lead to the development of stronger heat islands than winter, though this is not true of all places. Geographic location may also affect the seasonality of the heat island effect. Some low-latitude locations show greater differences between wet and dry seasons, rather than winter and summer. Heat islands are also stronger under high-pressure conditions; similarly, cloud cover, often associated with low pressure, dampens the intensity of heat islands.

The heat island can affect energy consumption considerably, particularly during summers and in low-latitude urbanized areas. A warmer ambient environment raises demand for air conditioning. Acting as a positive feedback, this additional energy usage further heats the surrounding air, magnifying the heat island effect. Conversely, warmer air temperatures reduce the need for wintertime heating, which reduces the consumption of energy related to heating. Additional negative effects of the heat island include increases in heat-related illnesses and a reduction in water quality.

Mitigating heat island effects can take many forms. Planting trees removes carbon dioxide from the urban atmosphere, reducing the levels of this greenhouse gas; these same trees also contribute to cooler temperatures and shading, lessening the need to heat the urban space. Reducing the amount of electricity used decreases the amount of anthropogenic heat created. Architectural innovations, such as green (vegetated) roofs, can decrease energy consumption as well as cool the air; green roofs increase evapotranspiration, which has a cooling effect on both the building surface and the surrounding air. Traffic policies that discourage unnecessary use of automobiles reduce both pollution and human-generated heat. Incorporating green space and environmentally friendly landscaping lessens the magnitude of the heat island. Heat island mitigation efforts at the community level yield significant benefits, and some cities have now adopted strategies to reduce the unwanted effects of the heat island. Overcoming the negative consequences of the heat island effect requires a multipronged approach.

See Also: Climate Change; Combustion Engine; Electricity; Fossil Fuels; Global Warming; Greenhouse Gases.

Further Readings

Arnfield, A. John. "Two Decades of Urban Climate Research: A Review of Turbulence, Exchanges of Energy and Water, and the Urban Heat Island." *International Journal of Climatology,* 23/1 (2003).

Robinson, Peter J. and Ann Henderson-Sellers. *Contemporary Climatology,* 2nd ed. Essex: Longman, 1999.

U.S. Environmental Protection Agency. "Heat Island Effect." http://www.epa.gov/heatisland/index.htm (Accessed November 2008).

Petra Zimmermann
Ball State University

HOME ENERGY RATING SYSTEMS

A Home Energy Rating is a measurement of a home's energy efficiency, used primarily in the United States. Home Energy Rating Systems (HERS) ratings make use of a relative energy-use index called the HERS Index—a HERS Index of 100 represents the energy use of the "American Standard Building," and an index of 0 indicates that the proposed building uses no net purchased energy (a zero energy building). Other countries also have similar schemes—in Australia it is known as the House Energy Rating and is based on a 5-star rating, and in the United Kingdom, the Energy Performance Certificates have a rating from A to G.

HERS provide a standardized evaluation of a home's energy efficiency and expected energy costs. The evaluation is conducted in accordance with uniform standards and includes a detailed home energy-use assessment, conducted by a state-certified assessor, using a suite of nationally accredited procedures and software tools. The rating can be used to judge the current energy efficiency of a home or to estimate the efficiency of a home that is being built or refurbished. A rating done before construction or improvement is called a "projected rating"; a rating that is used to determine a home's current efficiency is referred to as a "confirmed rating." The varying climatic conditions in different parts of the country are taken into account and are benchmarked according to average household energy consumption particular to a specific climatic region.

A home energy rating can also qualify a homeowner or buyer for an energy efficient mortgage or an energy improvement mortgage. It can also be used for both existing and new homes. A home energy rating of an existing home allows a homeowner to receive a report listing potential options for upgrading the home's energy efficiency. The owner can then use the report to determine the most efficient ways in which to improve the home's energy efficiency. The home energy rating of a new home allows buyers to compare the energy efficiencies of different homes that they are considering buying.

The Rating and Assessment

A confirmed rating requires an inspection of the home by an energy rater or assessor. The rater assesses the home to identify its energy characteristics, such as insulation levels, window efficiency, wall-to-window ratios, the heating and cooling system efficiency, the solar orientation of the home, and the water heating system. Performance testing, such as a blower door test for air leakage and duct leakage, is often also part of the rating. The data collected is entered into a RESNET software program and translated into a rating score between 1 and 100, depending on its relative efficiency. The home's energy rating is then equated to a star rating—one star for a very inefficient home, five stars for a highly efficient home. The U.S. Department of Energy–recommended Home Energy Rating report will typically contain the following:

- Overall rating score of the house
- Recommended cost-effective energy modifications
- Estimates of the cost, annual savings, and useful projected life of the modifications
- The potential improved rating score after the installation of modifications
- The estimated annual energy costs for the home, before and after modifications

History

Home Energy Ratings date back to 1981, when a group of mortgage industry leaders established the National Shelter Industry Energy Advisory Council with a goal of establishing a measurement system factoring the energy-efficient features of a home into the mortgage loan. This resulted in the formation of the Energy Rated Homes of America, a national nonprofit organization that has grown steadily and now has member rating programs operating in the majority of U.S. states. In response to the 1992 Energy Policy Act, the U.S. Department of Energy collaborated with the newly established HERS Council to develop a common Home Energy Rating System.

Energy Mortgages

The Home Energy Rating is a recognized tool within the U.S. mortgage industry. An energy mortgage is a mortgage that credits a home's energy efficiency in the home loan. For an energy-efficient home, for example, it could mean giving the home buyer the ability to buy a higher-quality home because of the lower monthly cost of heating and cooling the home. For homes in which the energy efficiency can be improved, this concept allows the money saved in monthly utility bills to finance energy improvements. Two main types of energy mortgages exist:

- The energy improvement mortgage, which finances the energy upgrades of an existing home in the mortgage loan, using monthly energy savings
- The energy efficient mortgage, which uses the energy savings from a new energy-efficient home to increase the home buying power of consumers and capitalizes the energy savings in the appraisal

The U.K. Energy Performance Certificates Scheme

Energy Performance Certificates are part of Home Information Packs and have been in effect since August 1, 2007, in England and Wales for domestic properties with four or more bedrooms. The scheme was extended to include three-bedroom homes from September 10, 2007. Since October 2008, Display Energy Certificates are also required for larger public buildings, showing how energy efficient the countries' public buildings are. The Display Energy Certificate must be displayed at all times in a prominent place clearly visible to the public, and it is accompanied by an Advisory Report that lists cost-effective measures to improve the energy rating of the building.

The introduction of both Home Information Packs and Energy Performance Certificates has been controversial, however, and has been opposed by many in the U.K. housing industry, such as the Royal Institute of Chartered Surveyors. In practice, there are also many problems with the credibility of the system, as well as with the accuracy of the energy audit itself. The energy survey is performed by a Domestic Energy Assessor, who visits the property and then inputs the observed information into a software program that then calculates the energy efficiency of the building. The program provides a single number for the rating of energy efficiency and a recommended value of the potential for improvement. There are similar figures for environmental impact. A table of estimated energy bills per annum is also presented, but without any reference to the owner's actual bills. The inspection is often superficial and based on personal opinion, but the certificates themselves do not allow any margin of error. Energy Performance Certificates present the energy efficiency of dwellings on a scale of A–G. The most efficient homes—which should have the lowest fuel bills—are in band A. The certificate uses the same scale to define the impact a home has on the environment. The average property in the United Kingdom is in bands D–E for both ratings. The certificate also includes recommendations on ways to improve the home's energy efficiency.

See Also: Energy Audit; Energy Payback; Green Energy Certification Schemes; Insulation.

Further Readings

"Energy Performance of Buildings: Planning, Building and the Environment." http://www
.communities.gov.uk/epbd (Accessed February 2009).
Fairey, P., et al., *The HERS Rating Method and the Derivation of the Normalized Modified
Loads Method*, Research Report No. FSEC-RR-54-00. Cocoa, FL: Florida Solar Energy
Center, 2000.
National Home Energy Rating. http://www.nher.co.uk (Accessed February 2009).
Nevin, Rick, et al. "Construction and the Appraiser: More Evidence of Rational Market
Values for Home Energy Efficiency." *The Appraisal Journal* (October 1999).
Residential Energy Services Networks. http://www.resnet.us (Accessed February 2009).

Chris Goodier
Loughborough University

HUBBERT'S PEAK

Hubbert's peak is the point at which an oil field or geographic region reaches its maximum production capacity. Hubbert's peak is named after Marion King Hubbert, a highly
respected geophysicist and petroleum geologist. According to Hubbert, oil production
from a given field plotted over time resembles a bell curve. Production from a newly
exploited field can rise quickly because of the natural pressurization of the field; as oil is
extracted over time, the field's pressure drops. In addition, the light, sweet oil is drawn
off early, leaving behind the heavy oil that is both harder to extract and harder to refine.
At some point in time, depending on how aggressively the field is exploited and on the
specific geological characteristics of the field, oil production peaks. Once oil extraction
reaches Hubbert's peak, production from a given oil field cannot be increased, no matter
how much effort is applied toward doing so. Hubbert's peak can be estimated for a given
oil field, for a geographic region, or for the entire globe. Oil helped fuel the Industrial
Revolution, and because it remains a primary energy source fueling the modern global
growth economy, peak oil in nations, regions, and the world has important economic and
geopolitical implications. According to Colin J. Campbell and other respected petroleum
geologists, soon global demand for oil will outstrip global extraction and supply. The
long-term implications of declining oil production are immense.

The concept of Hubbert's peak was developed in the 1940s and 1950s. In 1949, using
statistical and physical methods, Hubbert estimated total world oil and natural gas supplies and documented their rapid consumption. In 1956, he predicted that oil production
in the United States would peak in the late 1960s or early 1970s. U.S. domestic production
did in peak in 1970. After predicting the U.S. oil production peak, Hubbert forecast that
global oil production would peak sometime between 1990 and 2000. His prediction might
have proven correct in the absence of the oil price shocks of the 1970s, which raised prices
and reduced demand, so that global oil production decreased for the first time in history.
Many petroleum geologists predict that global oil production will peak before 2020.

Hubbert's peak also applies to natural gas production. Oil and gas are produced under
the same geological conditions and can occur together in geological formations. Fields that
contain higher proportions of natural gas tend to lie deeper in the Earth's crust than do

fields that predominantly contain oil. Natural gas can also be produced from coal beds—a form of gas known as coal-bed methane. Because natural gas can be converted to liquids, the timing of peak gas is likely to be influenced by demand for liquid hydrocarbon fuels.

Social and Economic Implications of Hubbert's Peak in Oil and Gas Production

Peak oil and gas have the potential to create devastating effects in terms of food production, particularly if the world population continues to grow. Natural gas is an essential feedstock for synthetic agricultural fertilizers. Without fertilizers, agricultural productivity would likely drop sharply as a result of soil damage that has resulted from postwar, Green Revolution agricultural methods. Current industrial agricultural methods also require the application of petroleum-based pesticides. Although environmental and health improvements could arguably be achieved by eliminating the use of petrochemical inputs to agriculture, unplanned shortages of these inputs or rising prices for them would likely result in food shortages.

Declines in energy supplies and available chemical feedstocks derived from oil and natural gas also have the potential to devastate the global economy that requires growth. Natural gas is heavily used in the electricity generation necessary to power homes and businesses, and it serves as a critical feedstock for the manufacture of plastics. Oil is an essential fuel for the global transportation system—a precondition for globalization. Modern people and societies rely on movement of goods over long distances, and in automobile-based societies like the United States, suburbs have developed with the car as the assumed mode of transportation. Decreasing car dependence in these societies will prove challenging.

Ameliorating social and economic effects of peak oil and gas would mean reducing dependency and developing alternative energy sources. Discovering enough oil and gas to support modern lifestyles and the global economy indefinitely is not an option, and worldwide oil production is unlikely to increase enough to avoid near-term supply shortages—barring a long-term, global economic recession reducing demand. Oil production lags behind discoveries, and worldwide oil discoveries peaked in the mid-1960s. Discoveries have declined extensively in recent decades, making it difficult for even large finds to reverse this trend. Recent discoveries typically do not compare in size to earlier ones made in significant oil regions such as the Middle East, and the gap between discoveries and consumption has grown since the 1980s. If the oil discovery trend could be reversed, there is little doubt it would have been during the past 45 years. Even if new discoveries could delay global peak oil production for many years, these discoveries would only delay responding to inevitable declines in oil production. Such a delay would further socioeconomic dependence on oil-based production, transportation, and development patterns and increase the probability of catastrophic climate change.

Net energy is important to consider in relation to both oil production and alternative energy sources. Extracting oil from post-peak fields requires increasing energy investments in the form of new extraction methods and technologies. The energy profit from producing this harder-to-get oil declines, thereby exacerbating the supply problem. If extraction continues, lifting oil from declining fields eventually becomes an energy-losing proposition. Hydrogen, often cited as a potential contributor to a new energy economy, is a net energy loser. Hydrogen must be refined from natural gas or electrolyzed from water (a process that requires electricity). The second law of thermodynamics states that energy can be converted from one form to another, but in the process, some of it is dissipated as heat. Therefore, hydrogen captured through these processes actually has less energy available to be applied to work than was available from the electricity or the natural gas used to capture the hydrogen. Although hydrogen may be useful as an energy storage medium, possibly for excess energy generated from renewable sources, it is not an energy source. Similarly, some biofuels such as ethanol offer a negative to minimally positive net energy return. It would be difficult for any biofuel to match the net energy ratio of conventional oil, and growing the crops necessary to fuel global transportation would likely mean competing with food production.

Oil shale and oil sands are inefficient in terms of net energy profit compared with liquid petroleum, and their production is environmentally damaging, especially in terms of water. Continued exploitation of these deposits is likely as conventional oil supplies decline, but they will not make up in net energy terms for conventional petroleum. Furthermore, the processing of oil sands releases a great deal of carbon dioxide and thereby exacerbates climate change.

Forms of renewable energy cited as replacements for oil (including wind and solar power generation) produce electricity. The global transportation infrastructure requires liquid fuels, so transportation remains the most difficult problem in terms of finding replacements for oil. The global fleet numbers hundreds of millions of cars and trucks. Although electric vehicles can be produced, a high energy investment would be needed to produce the vehicles and the infrastructure necessary to transition all automobiles, trucks, and trains to electricity, and it is doubtful electricity could fuel global shipping and air travel. Coal can be liquefied to produce synthetic petroleum, but as with oil shale and oil sands, there is currently little to no infrastructure globally for refining this liquid fuel. Furthermore, dependence on coal for transportation would require massive mining and would contribute heavily to climate change, and societies would be relying on yet another depletable source of energy for transportation.

Technology and energy optimists often cite energy-efficiency gains over time as a potential solution to peak oil and gas, noting that, as technology improves, more work can be done with lower quantities of energy, but it is important to note that efficiency means little in the depletion picture without reduced total usage. Given worldwide population growth and newly emerging industrial nations, the global appetite for energy was rising quickly before the global economic recession. Rising efficiency has also been correlated with increased energy density of primary energy sources. Increased efficiency levels will be harder to achieve using renewable energy sources and coal, which are less energy dense than petroleum.

Given the implications of fossil fuel depletion, Hubbert's peak is likely to signal significant social, technological, and environmental change.

See Also: Alternative Fuels; Fossil Fuels; Natural Gas; Oil.

Further Readings

Deffeyes, Kenneth S. *Hubbert's Peak: The Impending World Oil Shortage*. Princeton, NJ: Princeton University Press, 2001.
The End of Suburbia. Dir. Gregory Greene. Electric Wallpaper, 2004.

Heinberg, Richard. *The Party's Over: Oil War and the Fate of Industrial Societies*. Gabriola Island, Canada: New Society, 2005.

Tina Evans
Fort Lewis College

HYDROELECTRIC POWER

Hydroelectric power is the electrical power generated from the gravitational potential energy of falling water. This well-established technology catapults water to the status of leading renewable energy resource for electrical generation in the United States and worldwide. According to the Energy Information Administration, it accounted for 71 percent of renewable generation in the United States and 63 percent worldwide in 2007. These amounts correspond to 6 percent of the total electrical energy generated in the United States and 24 percent of the world total. Hydroelectric power's significant contribution to the electrical energy supply results from the simple concepts for its plant design, its economic and environmental advantages over other resources, and its ability to meet consumer energy needs immediately on demand. With all the advantages of hydropower, there are also social and environmental concerns resulting from the construction of dams and the spinning of underwater turbines. These impacts are well understood and can be prevented with proper technologies and planning.

Hydroelectric power facilities like the massive Hoover Dam, pictured here, provide up to 24 percent of the world's total electrical energy supply.

Source: U.S. Department of Agriculture, National Resources Conservation Service/Lynn Betts

Generation

Generating hydroelectric power from falling water is a mechanically simple process. Water released from a dammed reservoir or on its natural path downstream falls from a certain height to flow through turbines. As it falls, the water's gravitational potential energy is converted into the kinetic energy of its motion. The water flows through a series of turbines that are connected by a common shaft to the rotor of an electrical generator. The mechanical energy of the rotating turbines is converted into electrical energy by the generator through the principle of electromagnetic induction. The generator consists of electromagnets mounted on its rotor and coils of copper wire on its stationary stator. As the magnetic field rotates, it induces a current in the stator coils that supplies electrical power to the national grid to be transmitted and distributed to consumers. The amount of electrical power generated from the falling water depends on its flow rate, the amount flowing through a given volume over time, and the head—the height difference from the reservoir to the turbine. The total power output of the world's hydroelectric plants is 675,000 MWe, enough to supply the annual energy needs of 1 billion people.

Hydropower plants fall under six major technologies: impoundment, run-of-river, kinetic hydropower, microscale, diversion, and pumped storage. All systems except pumped storage are considered conventional hydropower. Impoundment hydropower schemes use a dam to store water in a high reservoir and release it on demand to flow down through a penstock, through turbines, and finally into the river beyond the dam. In contrast, run-of-river systems work with the natural flow of water from higher to lower regions. Because the power of falling water is proportional to both its flow rate and the head, run-of-river schemes can easily be integrated into the existing layout of the river. The same amount of power is produced with either combination of small flow rate and high head or large flow rate and low head. A similar scheme is kinetic hydropower, which puts turbines on riverbeds. Both technologies are attractive alternatives to impoundment because they have minimal to zero visibility in the natural landscape. Microscale projects produce up to 100 kW of power with minimal environmental consequences as opposed to larger designs. Diversion hydropower plants divert a portion of a river through a canal or penstock and may or may not require a dam. Finally, pumped storage designs are net energy consumers but offer the unique ability to store large amounts of excess grid energy in the form of hydropower.

Hydropower is an attractive renewable resource because of its capability to be stored and to respond quickly to unexpected surges in demand. The ability to store hydropower is achieved through pumped storage schemes. Hydroelectric plants incorporating this method have a high-level reservoir and a lower reservoir with the same volume capacity. Excess electrical energy from the grid generated during periods of low demand is used to pump water into the higher reservoir. This excess energy generally comes from coal-fired or nuclear power plants. Certain designs enable generators to work in reverse as pumps, converting excess electrical energy from the grid to mechanical energy of the turbines pumping water back into the higher reservoir. Although energy losses are inevitable, efficient turbine generators can recover up to 80 percent of the original input. During peak demand, the water is released to flow through the turbines into the lower reservoir. The response time is extremely fast: Turbines kept spinning in air can produce at full capacity in a matter of seconds. Even nonmoving turbines can be brought to full speed in one minute. At this time, pumped storage plants provide the most practical and cost-effective means of large-scale grid power storage. Other means of grid power storage are industrial-sized battery systems at power plants using energy input from intermittent renewable resources such as solar and wind.

Economic Advantages

Hydropower plants are an economically practical investment because of their long lifetime and minimal operational costs. The capital cost for an electrical power plant includes the costs of up-front engineering and construction, purchase of land, permit and legal expenses, fuel cost, operation, and maintenance. The lifetime and generating capacity of the equipment and the available subsidies for capital loans and tax credits are also important factors. Considering all these, the capital cost per kilowatt of power capacity over the lifetime of the plant is used to gauge the economic benefits of investing in a particular type of power production. For hydro, as with most renewable resources, the fuel costs are zero, so that immediately helps it compete with fossil fuels. The lifetime of dam structures is typically 50–100 years, and the equipment for hydropower schemes have lifetimes ranging from 25–50 years. Compare this with the lifetime of a nuclear power plant, which is only 30–40 years. Operation and maintenance costs are almost negligible for hydropower plants, averaging about 0.7 cents/kWh of electrical energy produced. Many hydropower plants are automated or require only a few employees. The major factor for the capital cost of hydropower is the engineering and construction of the site. Greenfield development, or flooding a valley for the reservoir rather than building a dam on an existing river, is the most expensive option. It is also the only option for countries that have exhausted their available hydropower potential and for pumped storage projects.

Incentives provide renewable energy technologies, including hydropower, a way to be more competitive in the market with fossil fuel and nuclear generation schemes. The Energy Improvement and Extension Act of 2008 allocated $800 million toward clean renewable energy bonds. Business entities such as electrical cooperatives, local governments, and certain lenders can use clean renewable energy bonds as no-interest loans to finance the construction of renewable energy projects. The U.S. Department of Energy also issues loan guarantees with over $10 billion in authority to any projects that reduce greenhouse gas emissions and air pollutants. The Energy Policy of 2005 granted $10 million in appropriations over a decade for hydroelectric plants to receive payments of 1.8 cents/kWh of electrical energy generated and sold up to $750,000 per year. Green certification programs assist hydropower production by attracting business from consumers who opt to use only electrical energy generated from renewable sources. With 70 planned projects under way, these incentives are expected to increase hydroelectric generation by 11 percent over the next decade.

Hydroelectric energy continues to flourish because there is still untapped potential capacity, and its wholesale price is the most affordable in the market. According to a study by the U.S. Department of Energy, the total undeveloped hydropower potential capacity in the United States is 30,000 MW. A 2007 report by the Energy Information Administration compared the capital cost of constructing power plants based on the resource and technology, listing conventional hydropower as $1,500/kW. This estimate is higher than the capital costs of a coal-fired plant with scrubber ($1,290/kW) and a wind power plant ($1,208/kW). However, hydropower fares much better against other alternative technologies such as solar thermal ($3,149/kW), geothermal ($1,880/kW), and advanced nuclear ($2,081/kW). The lowest capital cost of all the existing technologies is for an advanced open cycle gas turbine using natural gas at only $398/kW. This type of plant is also one of the fastest to construct, taking only two to three years to go online from groundbreaking. Thanks to incentives and long lifetimes, hydroelectric energy actually costs less for utilities to purchase—1 cent/kWh compared with 2–3 cents/kWh from fossil fuel plants. Although the economic advantages over other plant technologies are debatable, the environmental implications of hydroelectric power compared to fossil fuels and nuclear are much cleaner.

Environmental and Social Advantages

Using hydroelectric power helps to offset the emissions resulting from combustion of fossil fuels and reduces reliance on finite and imported fuel sources. Hydropower is based on the natural hydrological cycle driven by the sun, so it is a renewable energy source, and there are no related fuel costs. Water is a clean fuel source, so it immediately trumps fossil fuels and nuclear power in terms of air pollution concerns. Hydroelectric power is virtually emissions free beyond initial construction of the plant. Its generation releases no carbon dioxide, and other greenhouse gas emissions such as oxides of sulfur and nitrogen are negligible. In addition, hydroelectric generation produces no particulates or chemical compounds directly harmful to human health, and there is no danger of radioactive waste, as with nuclear power.

Hydropower plants increase the diversity, reliability, and affordability of a nation's electrical energy supply. The grid storage capability of hydroelectric pumped storage plants enables generation from intermittent renewable resources such as solar and wind to be more economically feasible. The ability to control the flow of water and the rapid response time to demand mean hydroelectric power increases the reliability of the electrical energy market. Because hydropower does not depend on the fluctuating market prices of a fuel, the price of the electrical energy it generates is less expensive and more stable. Water for fuel also means countries may generate electrical energy domestically using local resources without having to depend on or negotiate with volatile governments. Domestic hydropower generation boosts rural economies that need it the most, creating jobs and expanding access to health and education.

Flood control, irrigation, and recreation are indirect social benefits of hydropower dams and ultimately depend on the location and planning of the site. Multiuse dams provide a constant supply of irrigation water for agricultural use or for drinking water. Reservoirs can provide a back-up water supply in times of drought. Large hydropower dams can be used to control floods, which protects human populations and ecosystems living downstream of the plant. Many impoundment hydropower reservoirs are enjoyed by the public for recreational activities such as fishing, swimming, and boating. In some cases, a visual improvement to the landscape is achieved with the addition of these lake reservoirs. Some hydroelectric installations such as the Hoover Dam even become popular tourist attractions.

Social and Environmental Concerns

Fish mobility and mortality are affected by the dams and turbines of hydropower plants. The upstream migration of fish such as salmon to spawn sites is hindered by the presence of dams. However, this problem can be minimized through the installation of fish ladders or elevators or by trapping the fish and manually transporting them farther upstream. Fish migrating downstream toward the ocean also can be injured or killed during their passage through turbines. To prevent these mortalities, hydropower plants divert fish with screens, lighting, and sound devices in front of the penstock or turbine intake.

Fish are not the only creatures whose lives and habitats are endangered by the installation of hydroelectric plants. The displacement of entire human populations because of the building of dams is a major social issue. People living in valleys used for the reservoir site must be completely relocated, losing their livelihoods and communities. The World Commission on Dams estimates that anywhere from 40 to 80 million people have been forced to relocate because of the construction of dams. Others who live on the banks of rivers that overflow because of hydropower dams face potentially devastating floods. Catastrophes resulting from dam failures are responsible for hundreds of thousands of human deaths. The 1975 failure of the Banqiao Dam in China as a result of a monumental surge of rainfall led to the deaths of 26,000 people from the flooding, and another 145,000 died from the ensuing epidemics and famines. Human civilizations and ecosystems are equally at risk.

Maintaining the river's water quality and its ecosystems are fundamental to minimizing the environmental impact of hydropower. Impoundment hydropower plants must ensure that a minimum water flow is met to maintain the ecosystems along riverbanks. The amount of dissolved oxygen in water is reduced by underwater turbines, creating dead zones in riverbank habitats and further endangering marine life. Projects are underway to develop advanced turbine technology that will prevent fish injury and mortality and improve compliance with water quality standards. Aeration techniques can also be applied to oxygenate the water. In some locations, diversion of the water supply can cause wells to dry, vegetation to die, and wildlife to suffer. Dams can disrupt the natural flow of soil and nutrients into an ecosystem. The buildup of this silt can also lower the effective volume of the reservoir and subsequently reduce the capacity of the hydropower plant.

When land is artificially flooded for the installation of a hydropower reservoir, the carbon that was trapped inside the plants is released. When the plant matter decays anaerobically under water, dissolved methane is produced and later released into the atmosphere as it passes through the turbines. This process does not just occur during the initial flooding. There may be a continuous supply of decaying plant material along the reservoir banks, depending on the weather and seasonal changes in the water depth. Hydropower plants in tropical climates are especially at risk because the weather is more conducive to decay-related greenhouse gas emissions. These findings call into question the renewable nature of hydropower, which critics claim may result in more greenhouse gases than fossil fuel plants of the same size. However, proper sustainable planning and construction such as clearing the forests before flooding can prevent these decay-related emissions. They are not an inherent factor of hydroelectric generation.

See Also: Green Energy Certification Schemes; Microhydro Power; On-Site Renewable Energy Generation; Renewable Energies; Three Gorges Dam.

Further Readings

Energy Information Administration. "Electricity Data, Electric Power Capacity and Fuel Use, Electric Surveys and Analysis." http://www.eia.doe.gov/fuelelectric.html (Accessed January 2009).

Energy Information Administration. "Estimated Capital Cost of Power Generating Plant Technologies." http://www.jcmiras.net/surge/p130.htm (Accessed January 2009).

U.S. Government. "2009 Ultimate Guide to Hydropower, Hydroelectric Power, Dams, Turbine, Safety, Environmental Impact, Fish Passage, Impoundment, Pumped Storage, Diversion, Run-of-River." *Progressive Management* (2008).

Von Schon, H.A.E.C. *Hydro-Electric Practice: A Practical Manual of the Development of Water Power, Its Conversion to Electric Energy, and Its Distant Transmission.* France Press, 2008.

Kristen Casalenuovo
Virginia Commonwealth University

Hydrogen

Gaseous hydrogen could have a major role in the future as a clean energy carrier. Unlike current fossil fuels, however, hydrogen gas cannot be harvested effectively from our environment without using energy from another source. Hydrogen can be used through either

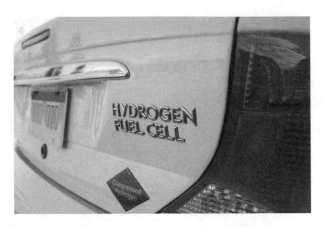

Hydrogen fuel cell vehicles like this prototype have the potential to function with greater efficiency than gasoline-powered vehicles and are becoming more viable with the introduction of hydrogen refueling stations.

Source: iStockphoto.com

direct combustion or fuel cells quite efficiently, without emitting pollutants detrimental to human health or carbon-based greenhouse gases. Hydrogen fuel cells have been used effectively since early in the U.S. space program and have now been developed for many special uses, including vehicular transportation, stationary heat and power, and portable uses.

The first major uses of hydrogen in fuel cells came with the U.S. space program. Cost and volume were less important factors than weight, reliability, and the availability of hydrogen. That space vehicles already carried hydrogen as fuel, the reliability of fuel cells with no moving parts, and the usefulness of heat, power, and water helped promote the use of hydrogen fuel cells. Since then, fuel cells with multiple purposes have quickened the use of hydrogen as an accepted form for storing and carrying energy.

Elemental hydrogen and its heavier isotopes (deuterium, with an added neutron; and tritium, with two neutrons) may also play a critical role in future fusion energy power plants. The article on fusion in this encyclopedia provides a discussion of future prospects of fusion power; this article discusses molecular hydrogen as a near-term energy carrier.

Properties

Hydrogen refers to both the element and the gaseous molecule (H_2). Hydrogen is the lightest element in the periodic table, consisting of one proton and one electron. Because of its high reactivity, hydrogen atoms in Earth's environment commonly combine into pairs as H_2 or bond into a variety of other common compounds such as water and hydrocarbons. The small size of the hydrogen molecule makes it comparatively difficult to contain, as it can pass through small leaks that larger molecules cannot. Furthermore, constant exposure to hydrogen causes a phenomenon known as hydrogen embrittlement in many materials. Because hydrogen embrittlement can lead to leakage or catastrophic failures in both metallic and nonmetallic components, additional care must be taken in storing and transporting hydrogen.

Hydrogen is odorless, colorless, and tasteless, and it is nontoxic. Hydrogen gas liquefies at a "boiling point" of minus 253 degrees C (20 degrees K) and cannot be kept liquid at greater than minus 240 degrees C (37 degrees K), even at high pressure (maximum at 12.4 barg or 195 psig). Ironically, water contains more than 50 percent more hydrogen atoms per volume than does liquid hydrogen (H_2).

Hydrogen gas's oxidation potential makes it extremely flammable over a wide range of concentrations. Hydrogen burns more intensely than gasoline, but hydrogen's explosive potential is reduced by its buoyancy, which results in rapid dispersal. (Even in the great Hindenburg disaster, the hydrogen burned but did not explode.)

As the lightest of all molecules, hydrogen has the highest energy density, meaning that it contains more chemical energy per weight than any other substance. This potential advantage is reduced, however, by its low volumetric energy density as either a gas or liquid. Even as a liquid, hydrogen has only about 7 percent of the density of gasoline. This low density means that large volumes are necessary for storing substantial amounts of energy (hydrogen has a very low "energy density"; see Table 1). Furthermore, hydrogen's volatility and propensity to leak through small openings necessitates additional weight for its containment as either a highly compressed gas or cryogenic liquid.

Hydrogen liberates about 2.5 times the amount of energy during its reaction than the heat of combustion of common hydrocarbon fuels (gasoline, diesel, methane, propane, etc.) on a mass (weight) basis. Therefore, a given energy requirement requires a mass of hydrogen only about a third of the mass of hydrocarbon fuel needed. But volumetrically, hydrogen's energy density is poor (since it has such a low density), even though its energy-to-weight ratio is the best of all fuels (because it is so light). Table 1 shows the energy density of common energy carriers (fuels).

Alternative containment and transport methods under investigation include sorption onto "hydrates," storage in other chemicals (such as sodium borohydride), or storage in "nanotubes." These have the prospect of increasing the amount of hydrogen per volume, but so far they have greatly added to the total weight of the total fuel material. Commercially suitable alternatives have not yet been fully demonstrated. For stationary uses (e.g., building combined heat and power), however, the volume of the hydrogen storage creates less of a problem.

Sources

Hydrogen is only as "green" as the energy sources used to produce it. Traditional sources of hydrogen include electrolysis of water, "reforming" of hydrocarbons, and its separation

Table 1   Volumetric Energy Density of Various Energy Carriers

Energy Carrier	Energy Density
Hydrogen	10.1 MJ/cubic meter; gas at 1 atm (1 bar) and 15°C
	1,825 MJ/cubic meter; gas at 200 barg and 15°C
	4,500 MJ/cubic meter; gas at 690 barg and 15°C
	8,491 MJ/cubic meter; liquid
Methanol	15,800 MJ/cubic meter; liquid
Methane	32.6 MJ/cubic meter; gas at 1 atm and 15°C
	6,860 MJ/cubic meter; gas at 200 barg and 15°C
	20,920 MJ/cubic meter; liquid
Propane	86.7 MJ/cubic meter; gas at 1 atm and 15°C
	23,489 MJ/cubic meter; liquid
Gasoline	31,150 MJ/cubic meter; liquid
Diesel	31,436 MJ/cubic meter minimum; liquid

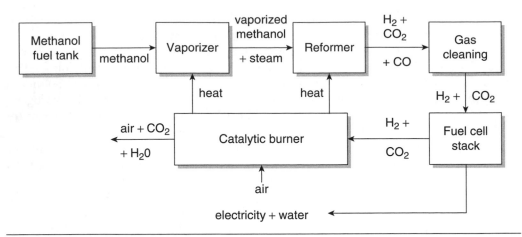

Figure I   Reformer (Methanol Example)

from intermediary steps of various chemical reactions (e.g., in coal or biomass gasification). Of these, the reforming of hydrocarbons is probably the most energy efficient at this time (see Figure 1). In addition to hydrogen, however, reforming of hydrocarbons also yields nitrous oxides, carbon dioxide, carbon monoxide, and trace gases. As a consequence, reforming of fossil fuels does not avoid the depletion of our limited supply of fossil hydrocarbon fuels nor prevent formation of carbon dioxide and carbon monoxide, though reforming potentially emits less of these than does traditional combustion. Most reforming processes convert the produced carbon monoxide to carbon dioxide using heat from the reforming process. If hydrogen comes from renewably produced biofuels instead of fossil fuels, the reforming process can be argued to also be "green," as the carbon involved comes from the active cycle and returns to it.

Hydrolysis uses electrical energy to break the hydrogen bond with oxygen in water, but electrical energy today is typically produced from fossil fuels. With the inefficiencies associated with traditional technologies, hydrolysis remains a relatively inefficient way to store energy. However, when the electricity for electrolysis comes from otherwise underused renewable sources, the efficiency arguments change significantly.

New sources of hydrogen are being developed and refined. For example, coal gasification (potentially with biomass) in a "clean coal" facility can produce a supply of hydrogen. Whether this source is environmentally green depends on carbon capture and storage, but the value of by-products of various polygeneration schemes may provide the economic incentive to pay the costs of carbon sequestration.

New technologies show promise for very efficient extraction of hydrogen from biofuels. For example, a method for extracting hydrogen from ethanol with a 90 percent yield factor has been demonstrated using inexpensive catalysts and moderate temperatures. Transporting liquid biofuels, rather than hydrogen itself, can simplify its distribution. The biofuel could be transported to fueling stations that make hydrogen on site for delivery to fuel cell vehicles. Alternatively, hydrogen can be carried on-board as sodium borohydride ($NaBH_4$), available from sodium borate (borax), and potentially be recycled by hydrolysis (processes being developed by Millennium Cell Inc. as their "Hydrogen on Demand™"

system). Hydrogen is also a component of syngas and can be produced from biomass using pyrolysis technologies.

Considerable promise has also been shown by new technologies involving combinations of solar/biologic/catalytic approaches to producing hydrogen more efficiently and without reliance on fossil fuels. For example, researchers have found microbes that produce hydrogen naturally, although their isolation and practical application have not yet been perfected. One such approach uses the naturally occurring microorganisms in sewage to degrade the waste while producing hydrogen, with waste treatment as a bonus.

Uses

Hydrogen represents a clean energy carrier, whatever its use. Although direct combustion shares many of the advantages, most attention has been given to hydrogen's use in fuel cells. Direct combustion could in some cases avoid the step of converting energy into electricity and back into heat or mechanical energy. But hydrogen combustion systems still have not achieved the overall efficiency of the fuel cell cycle. Other industrial and chemical uses of hydrogen abound but will not be discussed here.

One of hydrogen's biggest advantages comes from its lack of smog-producing compounds when combusted or reacted. Hydrogen contains no carbon or sulfur, so no carbon monoxide, carbon dioxide, sulfur oxide, or particulates are produced during combustion or reaction. (Lubricating oils used in accompanying mechanical components may produce some in trace amounts.) Hydrogen also allows for leaner combustion, resulting in lower combustion temperatures and very low nitrous oxide emissions (none from fuel cells because of their low operating temperatures). Hydrogen used in a fuel cell produces no harmful emissions. The only greenhouse gas emitted is water vapor; however, most experts consider it relatively benign as such, because its natural cycle includes cloud formation and rain, which govern atmospheric concentrations.

The Future of Hydrogen

Some scholars believe that the hydrogen economy will be real in as little as one decade, but others believe it at least 30–50 years in the future. Both Jeremy Rifkin and Joseph Romm agree that the ultimate goal should be for a "green hydrogen economy." A fossil fuel–based hydrogen economy can be achieved within the next few years, especially if it is based on natural gas. Technologies and equipment exist today for reforming natural gas (similar to the process in Figure 1) or deriving hydrogen from coal. But to reduce carbon emissions, such systems must include carbon dioxide capture and storage (sequestration). Technologies for obtaining hydrogen from renewable sources have been demonstrated, but the economics must be improved before likely widescale implementation.

California and British Columbia now tout their "hydrogen highways"—series of hydrogen refueling stations between major transportation destinations. Efficient fuel cell–compatible electric drivetrains continue to be refined by the current hybrid and electrical vehicle industries. These developments help promote the future use of hydrogen fuel cells for transportation, with prototypes coming available each year. This trend could help prevent automotive applications from being a direct source of pollution and greenhouse gases. As discussed above, however, production of hydrogen must also avoid the related pollution for any real achievement on this front.

Theoretically, hydrogen fuel cell propulsion systems should be capable of greater efficiency than the combustion of fossil fuels. One can still debate whether we have reached that goal yet. Hydrogen compression and/or liquefaction carry their own losses. Typical gasoline-powered cars lose over 80 percent of their power to engine losses and to consumption during standby and idle (a 20 percent efficiency). In comparison, various calculations for hydrogen fuel cell vehicles vary from approximately equal losses (from liquefaction, storage, air compression, manufacturing of the fuel cell, inverter, and the electric motor) to more optimistic reports of "tank to wheel" efficiencies of over 35 percent (or "power plant to wheel" efficiency of 22 percent).

Because hydrogen fuel cells give off no harmful pollutants, they can be located in homes and businesses close to the uses of the electricity. This allows waste heat to be used efficiently in the same structures (which is more difficult in a combustion plant, which creates noxious exhausts and noise). Overall efficiencies from hydrogen to the combination of useful heat and power conceptually reach 80 percent, but this does not include the production of the hydrogen.

See Also: Carbon Sequestration; Coal, Clean Technology; Combustion Engine; Fuel Cells.

Further Readings

Rifkin, Jeremy. *The Hydrogen Economy: The Creation of the Worldwide Energy Web and the Redistribution of Power on Earth*. New York: Penguin Putnam, 2000.

Romm, Joseph. *The Hype About Hydrogen: Fact and Fiction in the Race to Save the Climate*. Washington, D.C.: Island Press, 2004.

Williams, Robert H. *Toward Polygeneration of Fluid Fuels and Electricity via Gasification of Coal and Biomass*. Princeton, NJ: Princeton Environmental Institute, Princeton University, 2004. http://www.bipartisanpolicy.org/files/news/finalReport/IV.4.a%20-%20Toward%20Polygeneration.pdf (Accessed February 2009).

William Isherwood
Chinese Academy of Sciences

INNOVATION

Innovation is the successful application of new ideas, techniques, methods and technologies in order to produce some measure of positive benefit over incumbent concepts. Innovation involves using creativity to produce solutions that differ from the status quo; however, innovation is greater than the act of simply developing new ideas—it is the successful application and the diffusion of these ideas throughout society. Innovation will be key in sustainable energy systems if we are to move into a more sustainable mode of energy production and use.

The Concept of Innovation

An inventor is someone who develops a new technology or idea, whereas an innovator is a person who manages to make the technology or idea work in practice. This is explained well by Jan Fagerman, who states that "an important distinction is normally made between invention and innovation. Invention is the first occurrence of an idea for a new product or process, while innovation is the first attempt to carry it out into practice."

While innovation can broadly be described as the successful introduction of something new, Joseph Schumpeter defined innovation as being in one of five classes:

- New goods; introducing new products and services to the market.
- New methods; developing a new way of doing something.
- New markets; the development of a new marketplace for products or services.
- New sources; exploiting new resources of raw materials or components—whether that resource is one that exists and hasn't previously been utilized, or whether it has to be created.
- New organization; reorganizing or configuring a market, industry, or organization to work in a different way.

Valuing Energy Innovation

Innovation, in economic terms, is seen as a change to a product or method of delivering service that results in the creation of additional value. In the sphere of sustainable energy, value is something that is not construed as purely financial; it is defined more broadly in terms of an improvement that is appreciated by society. Value can also be defined in terms of better environmental performance or delivering energy services in a way that is more socially cohesive or in some way benefits the community. Many believe that the path of energy innovation will advance in such a way that the general structure changes from one of "centralization," where electricity and fossil fuels are centrally produced; to a world where energy generation is "distributed," with users of energy becoming producer-consumers with embedded generation in their premises contributing to power generation. Innovative distributed generation technologies have social benefits; as users generate energy they become more interested in its production. Furthermore, the rewards for energy generation are also distributed through communities, rather than being centrally funneled to a utility. This is an example of the social value that can be created through energy innovation.

Fields of Energy Innovation

There are a number of new innovative technologies that are currently in the process of, or very close to, commercialization. Conventional energy technologies benefit from a history of research and development investment and a long path of incremental innovation, however, it is clear that with growing awareness of the "Peak Resources" theory, and the effects of greenhouse gas emissions on the climate that there is a need to accelerate the pace of innovation.

Innovation in Energy Generation

The human race is very reliant on fossil fuels to provide our primary energy needs. Fossil fuels are the dominant mode of energy supply for most transportation types, and conventional thermal power plants rely on a combination of fossil fuel and nuclear power, both of which result in deleterious effects to the environment. New energy technologies will harness ambient energy flows from the environment to deliver "cleaner" forms of energy without depleting finite resources; i.e., solar energy, wind energy, hydro power (especially micro-hydro power), biomass generation, biofuel production (next generation biofuels), and geothermal energy.

Innovation in Energy Distribution

The infrastructure used to deliver energy from its point of generation to the end user is likely to change in the decades to come, and the methods through which "secondary energy" is distributed, and "energy vectors" are used to move and store secondary energy are likely to change as well.

In terms of electricity generation, the networks currently used to transmit electricity are considered by some to be "dumb" and "passive." In order to deal with an increase in the quantity of innovative renewable generation and create new "embedded generation" in the network, power grids will evolve to become "smart"—routing electrons more intelligently and reconciling varying supply with demand. Innovations likely to aid this need are: smart grids, hydrogen infrastructure, and fuel cell energy conversion.

Innovation in Energy Use

In order to conserve scare resources and make the best use of clean energy that is available, using improved technologies, such as more efficient lighting products, motors, and electronic products.

Models of Innovation

Linear Model of Innovation

The linear model of innovation suggests that technical change follows a process of:

Invention > Innovation > Diffusion

In the linear model of innovation; the reason for innovation is often presented as being of one of two classes: a "technology push," where new technological capabilities that arise from the process of research and development are translated into ideas that have practical applications and then brought to the market, or a market pull, where a demand from the marketplace stimulates the development of a new product or service.

Technology Adoption Lifecycle Model

Innovators

The first group adopts any innovation. They tend to be wiling to take risks with products and services, and occupy a younger demographic. They have disposable income, a high level of education, and comprise around 2.5 percent of the marketplace.

Early Adopters

Early adopters lead opinion, and tend to have a higher status than the majority. Early adopters comprise around 13.5 percent of the marketplace.

Early Majority

The early majority tend to be thoughtful and inquiring, and react to change more quickly than the average person, but require an established body of users to adopt an innovation before they are ready to commit themselves. The early majority comprise around 34 percent of the marketplace.

Late Majority

The late majority tend to be conservative users that are skeptical about innovation and change. They will only adopt an innovation when the majority of people have already proven that innovation to be useful. The late majority comprise around 34 percent of the marketplace.

Laggards

This group of individuals is resistant to change. They have an aversion to innovation and tend to be older than the other groups. They prize tradition over innovation, and tend to be limited in terms of financial resources and social status. Laggards comprise around 16 percent of the marketplace.

End-User Innovation

End-user innovation exists when an entity (a person, company or organization) identifies a problem with an existing product, or identifies a need that is not met, and works to meet this need by developing a solution. It is a form of innovation that is driven by the users of products, rather than by suppliers. Key to the diffusion of this innovation is that users are fundamentally social creatures and share their product with other users—spreading the concept further.

Open Innovation

Open innovation is a model that differs from the traditional paradigm of "closed innovation" where a company keeps control of its intellectual property by controlling how their ideas are generated and developed, ensuring that the whole new product development (NPD) cycle is kept in-house. With open innovation, intellectual property is shared, with companies donating their patents to an outside body that holds them in a "common pool." By allowing free unlimited use of this information, companies can collaborate within networks of open innovation.

Incremental Versus Disruptive Innovation

Innovation covers a broad spectrum of technological developments, from small improvements that subtly affect a product's performance to radical reenvisioning of the technologies used to deliver a particular feature or function.

Incremental innovations are those that gradually improve on an existing product through reworking or redesigning parts or components of the whole product to effect gradual improvement to the product. An example of an incremental innovation in energy can be found in motor vehicle internal combustion engines. Incremental innovation has allowed the fuel economy and emissions of internal combustion engines to gradually improve over time, however, the fundamental product remains recognizable.

Disruptive innovations are those that will completely reshape the marketplace of an existing sector by virtue of delivering amenity through entirely different modes, methods, or techniques. They often require substantial investments in infrastructure and production methods. Disruptive innovations often come from outside an industry sector, and as such are not always welcomed by existing producers that supply that sector.

See Also: Alternative Energy; Flex Fuel Vehicles; Renewable Energies.

Further Readings:

Goldstein, David B. *Saving Energy, Growing Jobs: How Environmental Protection Promotes Economic Growth, Competition, Profitability and Innovation.* Point Richmond, CA: Bay Tree Publishing, 2007.

Pogutz, Stefano, et al. *Innovation, Markets and Sustainable Energy: The Challenge of Hydrogen and Fuel Cells*. Cheltenham Glos, U.K.: Edward Elgar Publishing, 2009.

Praetorius, Barbara, et al. *Innovation for Sustainable Electricity Systems: Exploring the Dynamics of Energy Transitions. Series: Sustainability and Innovation*. New York: Springer, 2009.

Gavin Harper
Cardiff University

INSULATION

A first step toward reducing energy consumption is to install insulation and vapor barriers in buildings, to reduce as much as possible the pass-through of heat (thermal bypass) from inside a heated building to the outside, and the pass-through of hot ambient air into a cooled building. Insulation is not fully effective unless it is installed properly; that is, aligned with a contiguous air barrier across the entire building envelope, in full contact with the sealed interior and exterior air barrier(s), with all holes and cracks fully sealed and no gaps, voids, compression, or wind intrusion. This is needed to stop the flow of air (and thus heat) between the temperature-controlled ("conditioned") area (heated or cooled) and the external air. Such barriers also reduce moisture movement, resulting in a more comfortable, durable, and energy-efficient building. Insulation can also reduce sound movement.

The proper insulation of buildings is a crucial component of energy conservation. This builder is applying spray foam insulation to the interior walls of a remodeled home.

Source: iStockphoto.com

In climate zones 4 through 6, the most critical air flow is from inside the home to the outside during cold weather; therefore, an internal air barrier is needed. In climate zones 1 through 3, the internal air barrier closest to the conditioned space is not required because the air flows predominantly from the outside to the inside of the house in hot climates. In climate zones 5 and higher (colder climes), and any home with open web truss-joist floors, a vapor barrier needs to be installed on the warm side to prevent moisture from passing through the insulation.

Types of Insulation

There are several types of insulation available: batt, rigid, blown-in insulation (fiberglass or cellulose), and spray foam (open- or closed-cell), and factory-built insulated wall assemblies.

Batt insulation is manufactured out of fiberglass or rock wool into "blankets" sized for typical framing bays and manually fitted into place. It requires extra diligence during installation to ensure that no gaps, voids, compression, or misalignment occurs, particularly when placing the material into atypically shaped spaces and cut-throughs and around wiring and piping. It can be easily installed by homeowners, without any special equipment, with canned foam insulation used around wiring and plumbing. Batt insulation has been the standard insulation in woodframe residential buildings for many years; it is the pink, fluffy, itchy material attached to a paper or foil backing that you see today in many older homes.

Blown in insulation is typically made out of fiberglass or cellulose and blown into construction assemblies dry or wet to fill the entire framed assembly without any gaps, voids, compression, or misalignment. The advantage of blown-in insulation is that it inherently fills the entire wall cavity without any gaps, voids, or compression. It also can be distributed into areas that are difficult to reach and open web areas that would require substantial time to fill with batt insulation.

Wet-spray cellulose insulation is blown into wall assemblies with a mixture of water and glue that allows it to stay in place without falling out or settling. Because it goes in wet, it needs time to dry before the cavity is closed. Cellulose can be made from recycled paper, so it has the advantage of being environmentally friendly.

Spray foam insulation (both open- and closed-cell) is typically made from polyurethane. It is sprayed into construction assemblies as a liquid that expands to fill the surrounding cavity. Once dry, it functions as both an air barrier and a thermal barrier and fills holes and cracks for both a well-insulated and air-tight wall assembly. Therefore, it is not critical that the foam be aligned with the interior finish. It requires a trained installer wearing a protective mask and clothing, because of potential damage from its vapors to human lungs. Once dry, it is no longer harmful. Closed-cell spray foam is more dense than open-cell and functions as a vapor barrier to restrict the flow of moisture, thus eliminating the need to install a separate vapor barrier.

Factory-built wall systems such as structural insulated panels and insulated concrete forms integrate structural assemblies with insulating properties. Structural insulated panels are whole wall panels composed of insulated foam board glued to an internal and external layer of wood sheathing, such as plywood. It is typically manufactured with precut window openings and chases. Insulated concrete forms are blocks made from extruded polystyrene insulation with interlocking top and bottom edges. They are placed into position, steel reinforcing rods are added, and concrete is poured into the voids, creating an air-tight, well-insulated, and structurally sound wall. The insulation is inherently aligned with the exterior and interior vapor barriers, with no gaps, voids, or compression.

Insulating paint additives are designed to be mixed with ordinary paint, industrial coating, roof coating, epoxy, urethane, or high-temperature paint to achieve further energy savings. They work by refracting, reflecting, and dissipating radiant heat to reduce heat buildup and heat transfer through a building wall, ceiling, roof, or other coated surfaces. They can reduce exterior solar radiant heat gain by over 38 percent and interior heat loss through paint by over 10 percent. They are made from ceramic microspheres that have the appearance of a fine powder and add a light, textured feel to the paint finish, not noticeable to the eye. Insulating paint is a simple and inexpensive way for homeowners to increase their energy savings, particularly when it is difficult to add insulation behind existing walls.

The insulation material density and thickness affect is its R-value. R-value is a measure of the thermal resistance of a material. Higher R-values indicate better resistance to heat flow through material. Insulation should be installed to achieve a higher R-value in areas where heat would typically escape, such as through a roof. Insulation should be carefully fitted around piping, electrical wiring, and other holes cut between conditioned and unconditioned space, rather than being compressed in these areas, because compression reduces the R-value of the insulation.

Special Attention Needed

Some areas of a building require special attention in terms of insulation installation. It is imperative that a garage or other area containing noxious gases be completely sealed from the conditioned areas of a house or work space, not only for energy savings but also from a health perspective. Car exhaust, hazardous fumes, and dangerous materials must be kept from moving into living and work spaces. Also, cold and hot air in a garage can lead to thermal bypass if insulation is not properly installed between the garage ceiling and the subfloor below. Recessed lighting fixtures should be located in a dropped ceiling to prevent the heat from the lights from creating a natural draft, pulling air from a conditioned space into an attic. Caution should be used when installing insulation against potentially hot surfaces because both combustible and noncombustible insulation may present a fire hazard if caused to overheat.

Wind intrusion can occur at roof eaves through soffit vents. A baffle of solid material should be installed at all roof framing bays with soffit vents (ideally between all roof rafters and trusses) to direct air flow over and above attic insulation to prevent wind washing at attic insulation. In cold climates, exposed concrete foundation slab edges should be insulated with rigid insulation.

It is common practice to install HVAC (heating, venting, and air conditioning) ductwork and air handlers in attic space. One way to improve the performance of these systems and reduce heat loss and gain is to create an unvented, conditioned air attic with the HVAC system located inside the conditioned attic space. In this case, the sloped attic roof and all vertical walls should be insulated. Continuous top and bottom plates and an air barrier should be installed on the attic side of insulated walls, including exposed edges of insulation at joists and rafters.

A complete air barrier should be installed at the intersection of a porch roof and conditioned space, staircases adjoining exterior walls, garages or attics, attic hatches and drop-down stairs, and skylights. These areas are essentially thermal holes if not fully insulated and sealed. Be sure to allow safe access to the steps in drop-down stairs. Insulated box covers can be made or purchased to cover drop-down stairs, and gaskets, caulk, or foam can be added to fill the voids and provide a complete air barrier.

Installation of insulation should be performed by trained professionals, adhering to the standards set by their industry. Health and safety precautions must be followed, because of the danger of exposure to toxic chemicals emitted from wet and foam-sprayed insulation materials. Improper installation can also cause dangerous air quality conditions for the building's occupants (from back-drafting furnaces, for example), rotting of wood structures, and freezing water lines. Once hidden behind sheetrock and other finish materials, insulation is difficult to access, so care should be taken to do a thorough and careful job of sealing and insulating before any other materials are installed.

See Also: Home Energy Rating Systems.

Further Readings

Central Vermont Public Service Corporation. *Home Energy Savings Guidebook*. Rutland, VT: Conservation Energy Group, 1981.

Energy Star Qualified Homes. *Thermal Bypass Checklist Guide* (June 2007).

Elizabeth L. Golden
University of Cincinnati

INTERGOVERNMENTAL PANEL ON CLIMATE CHANGE

The Intergovernmental Panel on Climate Change (IPCC) was established by the World Meteorological Organization and the United Nations Environmental Programme. The IPCC was created to provide decision makers and other parties interested in climate change with an open, transparent source of objective information about climate change. Its reports are considered by many to be the "gold standard" on climate changes. Others dispute this and the IPCC's findings.

The IPCC is a multigovernmental scientific body in which many nations participate at various levels. When the IPCC meets to discuss findings, to hear reports, and to consider adopting reports, governments participate in the decision making and in subsequent reviews.

As an agency of the United Nations, the IPCC seeks to use the information it provides to promote its human development goals.

The IPCC's constituency is composed of governments of the member states that belong to the World Meteorological Organization and United Nations Environmental Programme. Those who contribute to IPCC reports include hundreds of scientists around the world. They contribute by being authors, by reporting research findings, and as reviewers.

The work of the IPCC is a division of labor performed by three panels that meet in plenary sessions. The panels are called working groups. The first working group reviews the scientific basis of climate change. Working Group 2 reviews impact, vulnerability, and adaptation arising from climate change. Working Group 3 deals with migration that arises from climate change. In addition, there is a Task Force on National Greenhouse Gas Inventories. Altogether, the membership and staff of these panels constitutes a sizeable bureaucracy.

Energy matters are the responsibility of Working Group 3 of the IPCC. It uses the staff research of the International Energy Studies Group (IESG). The IESG staff researches topics involving energy and environment for the IPCC to provide it with the data in compiles to make its energy reports. Most of the attention is on technologies, costs, and potential. Other factors include near-term and long-term issues.

IESG support for IPCC is analytical. It has contributed to the *Special Report on Emissions Scenarios* and the *Special Report on Methodological and Technical Issues in Technology Transfers*. To perform its work, the IESG staff attends meetings and gives presentations at IPCC plenary meetings. The staff also researches and prepares reports on topics dealing with end-energy use and greenhouse gas emissions.

The IPCC has called upon the IESG during negotiation sessions of the Climate Change Convention. It worked on the Good Practices Guidance for Land-Use Change and Forestry project. Staff reports also contributed to the *Third Scientific Assessment Report: Climate*

Change, 2001; *Special Report of Land-Use Change and Forestry* (2000); *Special Report of Methodology and Technological Issues in Technology Transfer 2000*; *Special Report on Emissions Scenarios, 2000*; *Special Report on Climate Change Response Strategies, 1991*; and the 1995 *Second Scientific Assessment Report*. Other reports include those in the Industrial Energies Studies category and in the category of Modeling Energy and Climate Change Futures. The papers developed have been extensive and have been used in IPCC reports.

Industrial Energy Studies of the IESG have included those that focus on improving energy efficiency. It is expected, through scenarios studies, that by 2030 two-thirds of carbon dioxide emissions generated by industrial emissions will be for industrial growth. Therefore it is viewed as critical that energy efficiency be radically improved to maintain current levels or to lower them. To promote energy efficiency, studies have been conducted toward developing an energy management standard for use in industrial facilities. Companies that have instituted energy savings of 20 percent or more on a voluntary basis are reaping benefits for their stockholders and for the climate. An International Organization for Standardization standard for energy management has been developed that has broad implications for usage.

Among the studies used by the IPCC have been those that contribute to an energy budget for planet Earth. An energy budget compares energy gains and losses for the planet as a whole. The energy gains are those from the sun. The losses include radiant energy from the Earth and its use of energy sources developed by humans. At the present time, the energy system for the Earth is in equilibrium.

The energy sector of the global economy includes a number of different energy sources. They are essentially renewable and nonrenewable. The renewable sources include solar, wind, hydroelectric, biogas, biofuels, and others. The nonrenewable sources are the fossil fuels—gas, coal, and oil.

The energy supply is organized into three economic subsystems of production, distribution, and consumption. In the area of fossil fuels, especially petroleum, the supply has increased in recent decades at a rate that seems to be less than the increases in the rate of consumption. The issue of supply of petroleum is often stated in terms of Hubbert's peak. The peak is the time in history when the maximum amount of oil will have been discovered. In essence, this means that half of all the oil in the world will have been discovered. Oil is used in the modern world for a vast number of products, the most consumptive of which is gasoline, or other fuels. The danger is that burning the world's oil supply as automobile gasoline will eventually rob people of all the other beneficial products derived from petroleum.

Another danger from the use of petroleum as gasoline or fuel derivatives such as aviation fuel is greenhouse gas emissions. The IESG measures these (as do other agencies) for the IPCC. The goal is the reduction of greenhouse gas emissions in the interests of preventing climate change (global warming in particular). The goal is the decarbonization of the energy supply by developing alternatives. However, the supply of fossil fuels is not likely to affect this goal in the near future.

Other fossil fuels—natural gas and coal—are usually used for heat and to generate electricity. These sources of energy also have a carbon "footprint." They are in abundant supply; however, they are also used in a wide range of products from fertilizer to cosmetics.

IPCC work with IESG has included study of the use of energy in manufacturing in California. It has also examined the use and disposal of products in California. The goal of reducing energy usage is important because it reduces costs for manufacturers, for consumers, and for the public and its agencies such as state and local governments. To develop an energy use profile, a state energy balance was developed. Attention was given to assessing the industrial sector's energy efficiency potential. Accompanying these studies were examinations of greenhouse gas emissions, especially carbon dioxide. Other aspects of California's energy consumption were part of the overall concern over climate change. These included its large wine industry, food processing industry, and petroleum industry. In the case of wineries, benchmarking water and energy usage standards are important developments. In the case of food production, reducing processing costs while maintaining sanitation and food safety standards was vital. In the case of petroleum production, reduction of energy use is a reduction in costs.

Other energy studies were case studies of innovative technologies and how they are adopted. Barriers to energy efficiency adoption and energy efficiency management are cultural practices that are similar to the adoption of any technology throughout history.

Examination of the industries that supply energy and the technologies they use has shown that these generators and their technologies are costly and have long life cycle usages. Therefore, to radically change the current system will require decades extending to as long as 50 to 100 years.

Of special concern to the IPCC are renewable energy supplies as a way to mitigate climate change. On January 21–25, 2008, the IPCC conducted a workshop on Renewable Energy Sources and Climate Change Mitigation in Luebeck, Germany. The workshop presenters were experts from 63 countries. In addition, the IPCC also asked an additional 200 experts to attend. About 120 of the experts were from developing countries. Very few of those who attended were nominated by organizations as attendees.

The workshop was organized into several broad areas of renewable energy use. The areas were defined as questions, arising from (1) the question of complex system integration of renewable energy source, (2) questions of environmental and social impacts, (3) the kind of policies needed, (4) how to further advance technology diffusion, and (5) the question of the rather different regional application possibilities. About 120 experts were invited.

A special report was issued documenting the accelerating rate of climate change, the effect climate change is having, and measurable growing confidence in the scientific community that climate change is anthropogenic. Using information provided by the participants, data were included in the report about the economic potential for renewable energy sources to provide supplies to meet growing demand at a manageable cost principally because of dynamic market forces.

Sources of renewable energy examined included bio-energy, which is expected to provide 15 percent of total energy used in the future. Included in the discussions and presentations were considerations of the markets, industry developments, and resource potentials in different regions.

Direct solar energy is expected to provide 10 percent of energy needs. The technology uses thermal—photovoltaics—concentrated solar power that is applied to supplying heating, cooling, lighting, cooking, desalination, fuel, and electrical uses. The regional nature of the resource was examined. This type of energy can be used even in the polar regions in summer time.

Geothermal energy is expected to provide 5 percent of energy. It is a resource that is present in volcanic regions. Prospects and cost trends were also included for discussion.

Hydropower is expected to provide 5 percent of electrical usage. It is a well-exploited resource in many regions of the world. However, building dams has had negative effects on fish and other wildlife. There is also a cultural and environmental loss when wild rivers are reduced to silting lakes.

Ocean energy is also expected to provide 5 percent of the energy. The types of technologies being tested include wave, tidal, ocean thermal, and osmotic. The results from these applications are still in trial stages for many aspects of their technology. The results are expected to be greater in some areas than in others. For example, the tides in the Bay of Fundy in Nova Scotia are the highest in the world. The tidal surges can produce enormous forces for energy usage, but in many areas of the world the tides are much smaller and not as easy to use.

Wind energy is also expected to provide energy in the future. The costs are significant because windmill sites have to be tested extensively before being adopted and connected with expensive transmission lines to the electrical grid. Wind is not abundant in many areas on a steady-enough basis to be cost-effective. Also, wind generators are opposed by many people who find them aesthetically unattractive and will organize to oppose local wind power generators.

Renewable energy sources were described in the report as part of a system of sustainable development. The environmental impacts and the socioeconomic impacts were also included. Knowledge gaps and future research questions were included as well, as were mitigation costs for implementing renewable energy technologies.

Part of the IPCC's *Fourth Assessment Report Summary* was issued in November 2007; the full report was issued not long afterward. It declared that global warming, based on temperature changes recently observed, is now "unequivocal." The projected climate change is expected to have significant implications for energy supplies and usage in the future.

In June 2009, IPCC's energy priorities summary was used by participants of "Connectivity Week" at Santa Clara, California. The multiconference had begun several years earlier as "BuilConn." Its focus was on the union of information technology and energy—whether as a cost or as a resource. The vision of creating a "smart grid" by using the vast growing number of personal computers as a source of energy that feeds the national and local power grids is seen as providing exciting opportunities for business and the possibility of reducing climate change.

Connectivity Week's displays, speeches, and presentations provided attendees with a vision that sees software development as key to energy management in the future. The use of computer chips to save, store, and then reuse energy in a cyclical fashion was especially favored at the conference. Future buildings will be climate controlled with this technology.

The IPCC's *Fifth Assessment Report* is due for finalization in 2014. Information from the 30th Session of the IPCC (April 21–23, 2009, in Antalya, Turkey) will be used, as will data from the scoping meeting of experts to define the outline of the fifth report, scheduled for July 13–17, 2009. Information from conferences such as those at Connectivity Week will be included.

See Also: Climate Change; Global Warming; Greenhouse Gases; Hubbert's Peak.

Further Readings

Bolin, Bert. *History of the Science and Politics of Climate Change: The Role of the Intergovernmental Panel on Climate Change*. Cambridge: Cambridge University Press, 2007.

Conde, Cecelia, et al., eds. *Climate Change*. Herndon, VA: Earthscan/James & James, 2009.

Gray, Vincent. *Greenhouse Delusion: A Critique of "Climate Change 2001."* Essex: Multi-Science Publishing, 2004.

Maslin, Mark. *Global Warming: A Very Short Introduction.* New York: Oxford University Press, 2004.

McCarthy, James J., et al., eds. *Climate Change 2001: Contribution of Working Group II to the Third Assessment Report of the Intergovernmental Panel on Climate Change.* Cambridge: Cambridge University Press, 2001.

Metz, Bert, et al., eds. *Carbon Dioxide Capture and Storage: Intergovernmental Panel on Climate Change.* Cambridge: Cambridge University Press, 2005.

Metz, Bert, et al., eds. *Climate Change 2001: Contribution of Working Group III to the Third Assessment Report of the Intergovernmental Panel on Climate Change.* Intergovernmental Panel on Climate Change, Working Group III Staff. Cambridge: Cambridge University Press, August 2001.

Metz, Bert, et al. *Methodological and Technological Issues in Technology Transfer: A Special Report of the Intergovernmental Panel on Climate Change.* Cambridge: Cambridge University Press, 2000.

Michaels, Patrick J. *Shattered Consensus: The True State of Global Warming.* Lanham, MD: Rowman & Littlefield, 2006.

Parry, Martin and Timothy Carter, eds. *Climate Impact and Adaptation Assessment: A Guide to the IPCC Approach.* Herndon, VA: Earthscan/James & James, 1998.

Reay, Dave. *Climate Change Begins at Home: Life on the Two-Way Street of Global Warming.* New York: Palgrave Macmillan, 2005.

Watson, Robert T., ed. *Climate Change 2001: Third Assessment Report of the Intergovernmental Panel on Climate Change.* Cambridge: Cambridge University Press, 2002.

Zedillo, Ernesto, ed. *Global Warming: Looking Beyond Kyoto.* Washington, D.C.: Brookings Institute, 2007.

Andrew J. Waskey
Dalton State College

Internal Energy Market

Energy products and energy-related services once under the exclusive jurisdiction of governments have been privatized and are traded in various markets. Such markets may operate at national, regional, and international levels. Internally, large countries such as the United States and Canada have domestic markets for energy. Traditionally, energy transactions between the United States and Canada have also been in force, operating to redistribute energy internationally. In such circumstances, it has been found to be more cost-effective to distribute energy internationally than domestically. For instance, the Canadian Province of Quebec typically exports electricity to the northeastern United States, while the western Canadian Province of Ontario imports electricity from the United States more cost effectively for all parties than transmitting energy internally within each country. Following the development of the European Union (EU) as a single market, a privatized energy system and market internal to EU members has been established and extended by agreements with EU nonmembers. Effective July 1, 2006, the Energy Community Treaty has further expanded the European internal energy market. This internal market now includes the member states of the EU and Croatia, Bosnia and Herzegovina, Serbia, Montenegro, the former Yugoslav Republic of Macedonia, Albania, Bulgaria, Romania, and the United Nations Interim Mission in Kosovo.

The EU established an integrated market; energy products and services being essential to life in today's world received particular treatment. Even before the Energy Community Treaty, EU membership was not a prerequisite to participate in this energy market. By including the countries that themselves border the substantial reserves of the Caspian Sea and the Middle East in the Energy Community Treaty, the internal energy market is of economically strategic significance.

With privatization of the energy supplies in Europe, the formation of an internal energy market promoted market competition and gave consumers a choice of sources for energy. Market forces ought to operate to keep prices reasonable and at the same time provide access to the energy grid to all parties regardless of organizational size. Measures to avoid distorting the market include restrictions on subsidization and discrimination.

Originally a U.S. requirement for the Kyoto Protocol, a carbon dioxide emissions trading market is an integral part of the internal energy market. By design, this market includes carbon sources and sinks and encourages alternative energy sources and diversification of sources to advance reliability and supply security in the network or grid.

Under the European internal energy market scheme, electricity and gas are differentiated. The gas market parallels the electricity market and is defined in relation to the gas transportation system. "Gas" then includes natural gas, liquefied natural gas, biogas, gas from biomass, and other types of gas.

The European Parliament and of the Council of 26 June 2003 Directives 2003/54/EC (which replaced Directive 96/92/EC) and 2003/55/EC (which replaced Directive 98/30/EC) provide common rules for the internal market in electricity and natural gas, respectively. Regulations (EC) No. 1228/2003 and No. 1775/2005 address access to the electricity networks and gas networks, respectively. These are amended pursuant to a Codecision procedure (COD/2007/0196) with the addition of an energy regulators cooperation agency. The regime does provide some derogations or exceptions to allow for specified circumstances including transitional provisions, a market crisis, or threats to the physical safety of persons, installations, or system integrity. Annual reports at the European Commission level provide benchmarking for the market. Decision No. 1364/2006/EC of the European Parliament and of the Council of 6 September 2006 in laying down guidelines for trans-European energy networks provides financial assistance for projects of common interest and for priority projects and projects of European interest.

The process of privatizing the market involved the organizational and legal unbundling of various elements of the energy system into distinct energy-related goods and services, production, transmission, and distribution systems, open to different and multiple suppliers. Where vertically integrated energy companies are permitted to own the transmission network, they must have it managed by a completely independent company or body. The open-market approach is intended to provide universal service with quality, consumer protection, and secure supplies. The supply of electricity and gas must meet a minimum quality standard at a reasonable price with secure supplies. Market rules for tendering for new capacity are specified, and transparency is required. Directive 2004/17/EC coordinated the procurement procedures in the water, energy, transport, and postal services sectors. Directive 94/22/EC complements it with the procurement procedures for prospection, exploration, and production of hydrocarbons.

Transmission or distribution system operators must be independent. They are therefore not permitted to participate in the integrated market. The operator is entrusted with the obligation of ensuring a number of essential features of the market. For instance, the operator is responsible for ensuring nondiscrimination between system users. Energy efficiency and sustainable practices linked to climate protection measures are required. Energy flows need to be managed on the demand side, as well as the supply side. Energy reserves and the interconnections are necessary for the integrity, efficiency, and reliability of the system as a whole. With gas broadly defined, the operators' task of ensuring integration and nondiscrimination of the system at all points is more complex, and quality, technical specifications, and safety standards also apply.

The regulatory protections for consumer protection and vulnerable sections of society that include small businesses nuance the market. Each consumer has the legal right to choose a supplier simply and effectively armed with the facts. An energy consumers' charter includes essential rights: the right to relevant, easily accessible, and understandable information on the different suppliers and supply possibilities; the right to a straightforward procedure for changing suppliers; protection against energy poverty for the most vulnerable consumers; protection against unfair commercial practices; and assistance in avoiding disconnection of energy supplies, among others. These rights have a corresponding service level obligation for the suppliers.

The incorporation of a greenhouse gas emission allowance trading scheme aligns the energy market with the Kyoto Protocol. This formal inclusion entrenches the accounting of waste from energy production, transmission, and usage. If the market does operate as intended, the objectives of improved efficiency as well as sustainability of energy production and usage could be advanced. It may be noted that lower energy prices may not reflect the full cost of energy and may be counterproductive to reducing emissions and consumption.

Taxation as a part of the economics of the internal energy market allows for more complex linkages with societal interests, as they may be addressed by governments. Energy products and electricity are only taxed when they are used as fuel for motor vehicles, for heating, or as electricity by the end user. The minimum tax rates previously applied to mineral oils apply to coal, natural gas, and electricity to reduce distortions of competition between them. Domestic taxes and tax advantages may encourage energy efficiency and modify consumption of energy and greenhouse gas emissions.

References to security of supply are conceived of primarily as stability within the internal market. Where sources of energy supplies are external to the same market, security involves international stability. In a globalized economy, primary sources of fuel—being hydrocarbon reserves—are not always internal to consumer market jurisdictions. National sovereignty and international agreements, as well as private multinational or transnational business enterprises, become relevant issues.

The EU internal energy market as described is concerned with electricity and gas energy as is usable by the end users—the retail user and the home user, rather than commercial users. This exclusion from the internal energy market does not preclude industries from other market and regulatory regimes. The evolution of the internal energy market continues with the EU Reform Treaty of October 2007. In this reform, free and undistorted competition is moved from the list of the EU's objectives in the main body of the treaty to a schedule.

See Also: Carbon Tax; Carbon Trading and Offsetting; Emissions Trading; Green Pricing; Public Utilities; Renewable Energies.

Further Readings

Eikeland, Per Ove. "EU Internal Energy Market Policy: New Dynamics in the Brussels Policy Game?" *FNI Report 14/2008*. Lysaker, FNI, 2008.

EU Emissions Trading. http://europa.eu/scadplus/leg/en/lvb/l28012.htm (Accessed March 2008).

EU Energy. http://europa.eu/scadplus/leg/en/s14002.htm (Accessed March 2008).

Filipovic, Sanja and Gordan Tanic. "The Policy of Consumer Protection in the Electricity Market." *Ekonomski Anali/Economic Annals*, 53/178–179:157–82 (2008).

Lea, Ruth. "An Economically Liberal European Union Will Not Be Delivered by the EU Reform Treaty." *Economic Affairs*, 28/1:70–73 (March 2008).

Warnig, Matthias. "A Link for Security of Energy Supply." *OECD Observer*, 267:76–78 (May/Jun 2008).

Lester de Souza
Independent Scholar

INTERNATIONAL RENEWABLE ENERGY AGENCY

The first steps toward the creation of an international organization dedicated to renewable energy were made more than 25 years ago. At the end of this long process, and under the leadership of a few European countries, an International Renewable Energy Agency (IRENA) was finally set up January 26, 2009, in Bonn, Germany, by about 50 countries. This new organization aims to promote the use of renewable energy sources, support national policies, and reinforce international cooperation on these issues.

A Long Path to IRENA

The idea of an international body in charge of the development of renewable energy was pushed forward for the first time in 1981 at the United Nations conference on new and renewable energy in Nairobi. However, it was continually postponed on the grounds that other organizations in charge of energy or environmental matters could take responsibility for this new issue. In fact, some international institutions currently focus on renewable energy. They include the International Energy Agency, World Bank, Renewable Energy Policy Network for the 21st Century, Renewable Energy and Energy Efficient Partnership, and several United Nations agencies.

This situation was criticized specifically by the proponents of a new organization. First, according to them, the international renewable energy regime lacked coherence and efficiency because it was highly fragmented. Second, it has been argued that it was necessary to balance existing institutions, such as the International Energy Agency or International Atomic Energy Agency, which were very much focused on conventional energy sources. Furthermore, supporters of IRENA wanted this new agency to deal not only with climate change issues but also with other dimensions of energy policy, such as security of supply, health issues, or peacekeeping. Finally, they imagined an organization outside the United Nations system, so that it would not be blocked by international treaties. It would develop according to a step-by-step approach, where some pioneers would create this agency and would progressively enlarge it to include other interested countries.

This whole process has been led by German political actors, with strong support from other EU member states such as Spain and Denmark, which together with Germany are the world's leading producers of renewable energy technologies. In 2001, the World Council for Renewable Energy and EUROSOLAR initiated a conference to reinforce lobbying for the creation of IRENA. Both institutions are very much influenced by personalities like Hermann Scheer or Hans-Josef Fell, members of the German Parliament. In 2004, the International Parliamentary Forum on Renewable Energy, hosted by the German Parliament, raised the issue again.

Finally, in 2008 the creation process accelerated. In April, a new conference was organized in Berlin, with 170 delegates from 60 countries. Its task was to discuss the objectives, activities, structure, and finance of the future organization. These discussions were furthered in workshops and materialized in a Final Preparatory Conference held in Madrid in October 2008. At this conference, 51 states reached agreement on the statutes of the international agency. They include a very broad majority of European countries, as well as Australia, Argentina, Brazil, India, Indonesia, and the United Arab Emirates.

These states finally signed the founding treaty at the Conference on the Establishment of IRENA on January 26, 2009, in Bonn. However, discussions regarding the future headquarters of the agency are still in progress.

Goals and Organization

The general goal of IRENA is promote the rapid adoption and the use of renewable energy worldwide. More precisely, it aims to contribute to the removal of barriers to the development of renewable energy and to create a favorable political environment not only in industrialized nations but also in developing countries.

To reach these goals, the agency will engage in different activities, according to the needs of member states:

- Analyze, monitor, and systematize current renewable energy practices: policy instruments, incentives, investment mechanisms, best practices, available technologies, integrated systems and equipment, and success-failure factors
- Initiate discussion and ensure interaction with other governmental and nongovernmental organizations and networks
- Provide policy advice and assistance to its members and stimulate international discussions on renewable energy policy and its framework conditions
- Improve knowledge and technology transfer between member states
- Offer capacity building
- Provide advice on the financing for renewable energy and support the application of related mechanisms
- Stimulate and encourage research, including on socioeconomic issues, and foster research networks, joint research, and development and deployment of technologies

- Provide information about the development and deployment of national and international technical standards in relation to renewable energy
- Disseminate information and increase public awareness on the benefits and potential offered by renewable energy

Any state that is a member of the United Nations can join IRENA. The organization aims to guarantee equality between all members from both industrialized and developing countries. The agency remains an intergovernmental organization. It encompasses three major organs:

- The assembly is the supreme organ of IRENA. It adopts the budget and work program of the agency, submitted by the council. It makes decisions relating to all financial matters of the agency.
- The council is elected by the assembly. It is responsible and accountable to it. Its members are elected for two years. It considers and submits the draft work program, budget, and annual report to the assembly.
- The secretariat assists the assembly and the council in the performance of their functions. It includes a director-general, appointed by the assembly on the recommendation of the council for a term of four years.

IRENA will have a budget of $50 million, funded mainly by mandatory contributions from its member states—based on the United Nations assessment scale—and by voluntary contributions.

The extensive fragmentation of international policy regimes on renewable energy made it crucial and urgent to create a new and permanent institutional framework. This appeared as the best solution to reinforce cohesion, expertise, and information sharing. During the discussion process on these new institutional arrangements, some actors proposed establishing an international sustainable energy agency, which would have included the management of energy efficiency issues rather than focusing only on renewable energy. Proponents of this alternative have already criticized IRENA, considering that it continues to look at the energy supply side only.

See Also: Energy Policy; Green Power; Intergovernmental Panel on Climate Change; Kyoto Protocol.

Further Readings

Federal Ministry for the Environment, Nature Conservation and Nuclear Safety. *Founding an International Renewable Energy Agency (IRENA): Promoting Renewable Energy Worldwide*. Berlin, 2008.
Steiner, Achim, et al. "International Institutional Arrangements in Support of Renewable Energy." In *Renewable Energy: A Global Review of Technologies, Policies and Markets*, Dirk Assmann, et al., eds. Sterling, VA: Earthscan, 2006.

Aurélien Evrard
Sciences Po Paris/Centre for Political Research (CEVIPOF)

KYOTO PROTOCOL

The Kyoto Protocol is an environmental pact that addresses the issue of global warming and climate change, specifically by setting emission targets for greenhouse gases. Signatories to the Kyoto Protocol include most of the world's countries, except the world's largest emitter to U.S. The reduction of greenhouse gases is accomplished using a market-based approach that aims to lower emissions in the cheapest way possible. This market operates on a cap and trade basis, which gives signatories limits on greenhouse gas emissions. The Kyoto Protocol establishes target reductions for participants and institutes and establishes various mechanisms through which these targets can be met.

The Intergovernmental Panel on Climate Change issued its first assessment report in 1990, increasing international concern about climate change and global warming. In 1992, participants at the United Nations Conference on Environment and Development (also known as the Earth Summit or Rio Summit) crafted the United Nations Framework Convention on Climate Change (UNFCCC). Its aims included the reduction of greenhouse gas emissions into the atmosphere, thereby lessening the chances of anthropogenic disruption of the Earth's climate system. The greenhouse gases in question are carbon dioxide (CO_2), methane, nitrous oxide, sulfur hexafluoride, hydrofluorocarbons, and perfluorocarbons. The UNFCCC includes terms for periodic updates regarding emissions limits. Although the UNFCCC is a legally nonbinding treaty, its most well-known update, known as the Kyoto Protocol, is a legal instrument. The Kyoto Protocol was adopted on December 11, 1997, and the agreement was named after the city in which the formal adoption took place—Kyoto, Japan. This protocol came into force on February 16, 2005. To further the goals of reducing greenhouse gas emissions, the Kyoto Protocol put forth various measures, called mechanisms—market-based initiatives that enable involved parties to meet their target reductions.

For the Kyoto Protocol to come into force, two conditions needed to be met. First, at least 55 countries needed to ratify the pact. Second, those ratifications needed to include countries making up at least 55 percent of total CO_2 emissions; the countries in the "55 percent" clause had to be from industrialized nations with established emissions targets. The first condition was met in May 23, 2002, when Iceland ratified the treaty. After Russia ratified the Kyoto Protocol on November 18, 2004, the second condition was met. The protocol took force 90 days later, making the emissions targets legal and requiring signatories to meet them.

Legal Provisions

The UNFCCC recognized the varied economic histories and situations of the Kyoto signatories by dividing the respective countries into Annex I (industrialized countries) and non-Annex I (developing countries). This was an effort to introduce equity among parties. Historically, industrialized nations were responsible for most of the anthropogenic greenhouse gas emissions, whereas developing countries emitted comparatively little, so the former was given greater reductions obligations. According to the Kyoto Protocol, Annex I countries, which include the United States and various members of the European Union, are to engage in both the reduction of greenhouse gas emissions and the development or expansion of sinks to remove these gases from the environment. These activities should have little or no negative effect on other countries. Additionally, Annex I countries are to aid financially the greenhouse gas reduction efforts of developing countries. All parties to the treaty (Annex I and non-Annex I) are to unite efforts to develop and disseminate appropriate technologies, as well as to educate citizens on climate change issues. The Kyoto Protocol sets greenhouse gas emissions limits for the Annex I countries; these limits are legally binding and serve as the "cap" of the cap-and-trade emissions exchanges. Non-Annex I countries have no specific targets. Rather, they abide by a general commitment to decrease the amounts of greenhouse gas emissions. The group of non-Annex I countries includes developing countries and some rapidly industrializing economies, such as China and India.

Signatories who have ratified this agreement must legally abide by its rules, limits, and caps. Those signatories make up nearly all of the world's industrialized and developed countries (Annex I); the United States is the sole industrialized Annex I country not to have ratified the treaty (as of December 2008).

Annex B of the Kyoto Protocol (amended in 2006) lists the reductions (as percentages) for the parties required to decrease emissions. These parties are referred to as Annex B countries. Those target percentages are relative to a base year or period and are set for a commitment period between 2008 and 2012. The base year for most parties is 1990, although those with transitional economies do have some flexibility with the establishment of the base. The largest of these reduction targets is an 8 percent decrease (for many European Union member states). The United States has a 7 percent target reduction. Although most parties listed in Annex B have targets that commit them to decreases in greenhouse gas emissions, several are permitted increases. These countries typically emit few gases, which warrants these exceptions. For example, Iceland's emissions target permits an increase of up to 10 percent from the base; Australia is permitted an 8 percent increase. In addition, there is flexibility with respect to the targets themselves. The Kyoto Protocol permits parties to change targets based on land use, land-use change, and forestry activities if those activities curb emissions or act to sequester carbon. An example of a land use, land-use change, and forestry action is the planting of forests.

The targets set for each country can be met through various means. Countries may develop their own methods for the reduction of greenhouse gas emissions. In addition, the Kyoto Protocol establishes three flexibility mechanisms to meet targets. The term *flexibility* refers to the transfers of financial investments and technologies that permit parties to satisfy their targets. These mechanisms are emissions trading, clean development mechanism, and joint implementation, and they are founded on market concepts. That is, carbon units are established as tradable commodities, and the trading of these units among parties constitutes an economic market based on greenhouse gas reduction measures. The trading units may be the result of direct emissions reductions or sequestration and removal activities undertaken by participants. They may also originate from project-based actions between countries. Both types contribute to the efforts to reduce greenhouse gas emissions that are the foundation and aim of the Kyoto Protocol. To participate in the mechanisms, countries must meet certain eligibility requirements. Each mechanism is associated with its own requirements.

Emissions Trading

The first of the Kyoto Protocol mechanisms, emissions trading, is addressed in Article 17, and effectively treats carbon as a commodity. Typically, a limit or cap is placed on the emission of a certain pollutant, and those countries emitting less than their limit can sell the remaining emissions (in the form of carbon credits) to those emitting more. The cap is the allowable level of emissions or assigned amount and is expressed as Assigned Amount Units. An Assigned Amount Unit is equivalent to one metric ton of CO_2 equivalent and is the basis for the emissions credits. Emissions trading is sometimes referred to as the "carbon market" because of CO_2's position as the major atmospheric greenhouse gas, despite there being five additional gases or groups addressed in the Kyoto Protocol.

Emissions trading, as established by the Kyoto Protocol, could have important energy implications, which would help accomplish the aims of the protocol. Power generated by environmentally cleaner sources (such as wind or solar radiation) is favored over that provided by fossil fuels, such as coal, which send much CO_2 into the atmosphere. This renders the "dirty" fuel less competitive. Indeed, emissions trading may lead to new opportunities for businesses that develop clean energy, produce or enhance carbon sequestration technologies, or further carbon abatement in current technologies.

Although the concept and practice of emissions trading predates the coming into force of the Kyoto Protocol, the pact's mechanisms can be joined to existing trading schemes. The European Union Emissions Trading System, already in place, has nonetheless adopted the flexibility mechanisms within the Kyoto Protocol. The so-called linking directive permits credits from the Kyoto Protocol's flexibility mechanisms to be used within the European Union Emissions Trading System during its second phase (2008–12).

Clean Development Mechanism

Article 12 of the Kyoto Protocol defines the second mechanism—Clean Development Mechanism (CDM)—as one permitting Annex B countries to undertake activities that enable developing countries to reduce their greenhouse gas emissions or remove them from the atmosphere. This would have the benefit of lowering emissions from countries that are rapidly industrializing by giving them an economic incentive to engage in using "clean" technologies. The Annex B parties can then credit these external reductions or removals to themselves, effectively meeting their own targets by reducing emissions in other places where such reductions or removals may be cheaper economically. CDM projects must meet established requirements, and they must demonstrate that emissions decreases are in addition to any occurring in the absence of said project. That clause requires the development of a baseline or "business as usual" scenario. Setting the baseline follows approved approaches and may be founded on actual, preproject emissions or average emissions from similar projects. In addition, participation in a CDM project by an Annex B country does not absolve it from the obligation to reduce its domestic emissions.

Emissions-reduction projects in developing economies earn them a type of carbon credit called Certified Emission Reduction (CER) credits. A CER amounts to a reduction in emissions equivalent to one metric ton of CO_2, and it reflects the difference between baseline and actual emissions. It is these CERs that are traded on the carbon market. The cash flow, then, travels from the Annex I party to the non-Annex I party. CERs must undergo a validation process before they can be traded. This process is carried out by an accredited and independent third party, known as a Designated Operational Entity (DOE). The CDM Executive Board certifies the DOEs, and their names are publicly available. Functions of the DOEs are the validation of proposed CDM projects and the substantiation of emissions reductions of said projects. The sharing of information between buyers and sellers operating within the CDM arena is aided by the CDM Bazaar, a Web portal designed to facilitate dialogue between parties interested in partaking of CDM project opportunities.

Registration of a CDM project signifies approval. The approval is granted by the CDM Executive Board. Registered CDM projects, which surpassed 1,000 in April 2008, include among them wind farm/wind power schemes, methane recovery projects, and renewable energy generation from hydropower stations. Information about the projects, including descriptions, involved parties, and registration and validation procedures, is publicly available. Estimates of greenhouse gas emissions reductions are a vital part of a CDM's project design and description; these are also made public.

There has not been complete acceptance of CDM projects, and some of them have been met with criticism. Objections have been raised regarding the social costs of the projects and whether they truly contribute to sustainable development. One such concern addresses whether local, unempowered populations are harmed by the projects, such as being subject to involuntary displacements from their homes. Additional criticisms question whether these projects actually subsidize certain industries (e.g., fertilizer) in a select group of developing countries, rather than reducing greenhouse gas emissions. Others question the inclusion of large hydropower generating plants or the building of harmful dams as CDM projects, as they may be neither sustainable nor environmentally beneficial.

Joint Implementation

The third mechanism is Joint Implementation (JI), and its requirements are addressed in Article 6 of the Kyoto Protocol. JI allows for the transfer or acquisition of emission reduction units (ERUs) between Annex 1 countries; emissions caps of these countries are listed in Annex B. One ERU is defined as one metric ton of CO_2 equivalent. Similar to CDM, JI activities aimed at reducing or removing greenhouse gases must be in addition to any that would arise in the absence of these activities. Certain eligibility requirements must be met for any project to qualify under JI. These requirements differ, depending on which of the tracks a JI project follows.

Verifying emissions reductions under JI may take one of two paths—track 1 or track 2. Under track 1 (also referred to as "fast track"), a participating country may authenticate its own reductions or removals from JI projects; these must be in addition to any reductions or removals that would occur in the absence of said projects. Track 1 host parties must meet six conditions to be able to participate. These conditions are (1) being a party to the Kyoto Protocol; (2) creating a methodology to estimate reductions or removals in accordance with the Kyoto Protocol; (3) developing a national registry of the credits (both creation and trading of them), based on certain guidelines; (4) properly calculating its assigned amount of CO_2-equivalent emissions; (5) submitting its inventory results yearly;

and (6) properly submitting any additional or supplemental information. Track 1 projects are also referred to as "party-verified," denoting their validation by the host party. Countries engaging in track 1 projects include Germany, Romania, Hungary, and New Zealand. Examples of track 1 projects include a Romanian and Danish effort to introduce sawdust as a fuel source, which employs a CO_2-neutral source of energy. New Zealand's 2008-approved Tararua Wind Farm projects, also track 1, are jointly implemented with nonhost parties France and Japan.

Track 2 projects follow a less restrictive path. The host parties need not meet all of the criteria set forth for track 1. Economies in transition, such as those of Russia or central Europe, benefit from this track, as they can pursue projects without first fulfilling all of the track 1 requirements. Instead of verification by the host party, verification is done independently, but with the approval of the JI Supervisory Committee; track 2, then, has international oversight, whereas track 1 is supervised by the host party. Under track 2, ERUs are transferred to the party that has invested in the project by the host party.

CDM and JI have some parallels. The concept of additionality, in which projects should generate emissions reductions on top of those from business-as-usual scenarios, is common to both. Despite sharing some similarities, differences do exist. The major difference between JI and CDM projects involves the shifting of credits toward the cap. JI project credits are deducted from the host party and added to the other party. There is no net effect on the caps of Annex B participants, because the credits move solely between Annex B countries. With CDM, the credits are generated externally (from non-Annex B parties).

See Also: Carbon Trading and Offsetting; Emissions Trading; Global Warming; Greenhouse Gases; Intergovernmental Panel on Climate Change.

Further Readings

United Nations. "Kyoto Protocol to the United Nations Framework Convention on Climate Change." http://unfccc.int/resource/docs/convkp/kpeng.pdf (Accessed December 2008).
United Nations Framework Convention on Climate Change. "Kyoto Protocol." http://unfccc.int/kyoto_protocol/items/2830.php (Accessed December 2008).
Wara, Michael. "Measuring the Clean Development Mechanism's Performance and Potential." Program on Energy and Sustainable Development Working Paper #56, July 2006. http://iis-db.stanford.edu/pubs/21211/Wara_CDM.pdf (Accessed January 2009).

Petra Zimmermann
Ball State University

LANDFILL METHANE

Methane is an organic compound with a chemical formula of one carbon atom and four hydrogen atoms (CH_4). It is a naturally colorless, odorless gas that will explode violently in dense pockets or enclosed spaces if a spark is present or introduced. After the March 18, 1937, natural gas explosion in New London, Texas, that killed nearly 300 high school students and their teachers, an obnoxious odorant was added to commercially used methane.

Methane is often found with petroleum deposits that are pumped from oil fields. For much of the history of the petroleum industry, methane was an unwanted by-product that was usually flared off (burned). Today it is captured, if possible, for commercial use. Natural gas fields that contain methane are usually found at much great depths than oil. In nature, methane is often accompanied by other gases. It is also found in coal deposits and is a great threat to coal miners in deep-shaft mining. Explosions of

This capped vacuum well is sited above a former landfill and provides a means to extract methane and other gases from decomposing matter for fuel and other applications.

Source: iStockphoto.com

firedamp in coal mines (methane and other gases) have killed many miners over the centuries.

Methane is the main gas in natural gas and in biogas that is produced by biochemical processes from organic materials. Methane in biogas is generated from organic matter in the absence of oxygen. It is produced by the anaerobic digestion of bacteria in a fermentation process. Landfill gas (LFG) is a product of the decomposition of solid waste materials such as food waste, paper, yard cuttings, and other organic materials. LFG is a mixed gas. About half is carbon dioxide, and the other half is methane. In addition, there are trace amounts of other gases, most of which are also organic compounds.

Studies of the decomposition of waste materials in municipal solid waste landfills have demonstrated that they are the largest source of anthropogenic methane emissions in the United States. In 2007, LFG amounted to nearly a quarter of the methane emissions. As greenhouse gases, both methane and carbon dioxide emissions are damaging to the environment because they are contributors to global warming. However, the failure to capture these emissions has been a significant waste of a resource for energy and other purposes.

Harvesting LFG accomplishes a number of environmental and energy goals. Environmentally, LFG does not escape into the atmosphere to add to the problems of smog and global warming. In addition, capture of LFG reduces odors and smog in the communities where landfills are located. Finally, capture allows the methane portion to be used for fuel or other purposes. The carbon dioxide portion has numerous industrial uses, from dry ice to the enhancement of oil well production.

The technology for extracting LFG has continued to develop with technology borrowed from the natural gas industry. LFG is extracted from a landfill with wells that are drilled into the landfill. The wells are capped with a wellhead. Extraction equipment is installed, and the wells are linked to either a collection point or to some electrical generation system.

There are two major parts to the energy grid in the United States: the transportation system, which is heavily dependent on fossil fuels, and the electrical grid. The electrical grid is much more diversified because fuels used include all of the fossil fuels as well as nuclear, geothermal, solar, and water power, with LFG now available as an energy contributor.

In the United States in 2008, according to U.S. Environmental Protection Agency statistics, nearly 500 LFG energy projects were in operation. In addition, there were nearly 600 more potential sites. LFG can be used directly or indirectly as an energy source for both electrical and transportation systems. LFG is currently being used either on-site or for sale to the electrical grid, which consumes about two-thirds of current production. On-site uses can be fuel for internal combustion engines, turbines, and other types of engines, some of which are still in the experimental stage of development. In the next several decades, the wastes—formerly a source of pollution—will be converted into energy, reducing costs and pollution.

In some projects LFG is being used as a fuel substitute for commercial natural gas or other fossil fuels. In the case of boilers, dryers, kiln, greenhouses, fish farms, and heating operations, the use of LFG is an efficient use of a cheap resource. In remote locations it supplies energy that would be expensive to obtain from other sources.

Because the methane in LFG is a material for the manufacture of a number of products including fertilizer, it is now being used in a wide range of industries including automobile manufacturing, chemical production, food processing, pharmaceuticals, cement and brick manufacturing, wastewater treatment, consumer electronics and products, paper and steel production, and prisons and hospitals, among others.

Methane is often used as an energy source in cogeneration projects. Using methane as an LFG in cogeneration has a number of advantages. Burning the gas drives electrical generating turbines to produce electricity. The burning gas also produces heat, which is used as thermal energy to create steam or hot water that can be used to heat buildings or other facilities. Because these cogeneration processes are efficient producers and users of energy from a cheap source (LFG), they have been instituted in a number of industrial centers. Publicly announced plans to consider or to build LFG-supplied cogeneration centers have been increasing steadily in recent decades.

Other energy uses of LFG are delivering it to the natural gas pipeline system for use at locations remote from the landfill generating the gas. Condensing the LFG into a type of compressed natural gas or compressing it further into liquefied natural gas has allowed its use as a vehicle fuel. Plans are being developed to create centers for converting LFG into methanol, which can be used as a fuel or as a fuel enhancement.

The Environmental Protection Agency has developed a Landfill Methane Outreach Program. Participation by public or private entities is voluntary. The goal is to promote the development of landfill methane projects that recover and use LFG (http://www.epa.gov/lmop/overview.htm). Not only is energy generated, but methane is used, rather than being allowed to leak into the atmosphere, where it raises the levels of greenhouse gases that cause global warming.

The Evergreen Renewable Energy Program, operated by the Dairyland Power Cooperative in La Crosse, Wisconsin, is a notable energy project using landfill methane. Its methane source is cow and pig manure. The East Kentucky Power Cooperative operates the EnviroWatts program, which uses five landfill gas-to-electricity-generating facilities that produce enough electricity to power 12,000 homes. The Short Mountain Methane Power Plant in Eugene, Oregon, operated by the Emerald People's Utility District, produces 2.5 megawatts annually. Exelon Power operates a landfill-generating station in Pennsylvania that generates 6 megawatts of power.

See Also: Alternative Energy; Fossil Fuels; Global Warming; Greenhouse Gases.

Further Readings

Emcon Associates. *Methane Generation and Recovery from Landfills*. London: Taylor & Francis, 1980.

Environmental Protection Agency, Office of Air and Radiation. *EPA Landfill Methane Outreach Program Opportunities for Utility Allies*. Washington, D.C.: U.S. Environmental Protection Agency, Office of Air and Radiation, 1994.

Isaacson, Ron, ed. *Methane From Community Wastes*. New York: Elsevier Applied Science, 1991.

Pawlowska, Lucjan, ed. *Management of Pollutant Emission From Landfills and Sludge*. Boca Raton, FL: CRC Press, 2007.

Peer, Rebecca Lynn. *Development of an Empirical Model of Methane Emissions From Landfills Project Summary*. Research Triangle Park, NC: U.S. Environmental Protection Agency, Air and Energy Engineering Research Laboratory, 1992.

Andrew J. Waskey
Dalton State College

LEED STANDARDS

The Leadership in Energy and Environmental Design (LEED) standards are criteria for environmentally sustainable or "green" construction. LEED provides both a metric for evaluating green buildings through the certification of construction projects and an accreditation for construction and maintenance professionals. LEED standards are developed through a transparent process by which technical criteria are proposed by LEED committees and approved by professional building organizations that make up the United States Green Building Council (USGBC). By providing a point rating system in six separate areas of the design, construction, and operation processes, LEED standards afford construction professionals measurement tools and procedural advice for sustainable building and design.

History of the LEED Certification

Originally initiated by Robert K. Watson, a senior scientist and director of the International Energy and Green Building programs at the National Resource Defense Council, LEED is now overseen by the USGBC, a Washington, D.C.–based nonprofit coalition comprising professional members of the building industry. The initial impetus for the development of LEED was to establish one single standard for new green construction projects while promoting a more integrated design process, recognizing professionals with environmental leadership in the construction field, and stimulating public and private awareness of the growing need to consider the natural environment during and after construction. Over the years, and by including a wide range of building professionals (from engineers to facility managers), the USGBC has allowed LEED to grow considerably since its inception in 1994. Criteria for certification now include six comprehensive standards of green building that require attention during all phases of planning and development. The articulation of concepts for green building, as well as the establishment of a corresponding rating system, has given professionals interested in reducing a building's environmental impact a planning and certification process by which this goal might be systematically obtained.

Point Rating System for Project Certification

Standards for LEED certification are based on a point rating system. In addition, LEED standards organize the building process into readily understandable strategies and steps and help to establish which professionals within that process should take on leadership roles and at what times (Table 1). By developing these standards, buildings can gain LEED status at four levels: Certified (26–32 points), Silver (33–38 points), Gold (39–51 points), and Platinum (52–69 points), based on total points awarded in the six areas of consideration (see below). A number of U.S. and international buildings have been awarded platinum status; they range in use from education (the California Academy of Sciences in San Francisco and the Donald Bren School of Environmental Science and Management) to government (California Department of Education) to nonprofit organizations (New York Academy of Sciences, National Resources Defense Council Robert Redford Building). LEED certification begins with an application that documents compliance with the point rating system as well as payment of associated fees. This process, and subsequent certification, is governed solely by the USGBC.

The point system comprises six areas: planning and development aspects for the site, water, energy, construction materials, indoor environmental quality, and innovation and design.

The first area, sustainable sites, refers to the consideration of the location of a structure and the manner in which it is initially planned for development to reduce the environmental impact to the surrounding area. The goals of site planning involve developing only those areas that are appropriate by promoting the reuse of existing buildings, protecting ecologically and agriculturally important areas, and reducing the need for automobile use

Table 1   LEED Standard Construction Phases

Construction Phases	Construction Purpose and Team Members
Phase I: Pre-design	Develop building programming needs;
	LEED Accredited Professional (LAP), Civil Engineers, Building Architect, Contractor
Phase II: Schematic Design	Design, process planning;
	LAP, Buildings and Landscape Architects, Mechanical Engineer, Civil Engineer,
	Lighting Engineer, Commissioning Agent,
	Contractor
Phase III: Design Development	Design refinement;
	LAP, Mechanical Engineer, Buildings Architect, Contractor, Building Management
Phase IV: Construction Documentation	Refinement of Construction Operation;
	Contractor, Buildings Architect
Phase V: Construction Administration	Overseeing Construction Process;
	Commissioning Agent, Mechanical Engineer, Contractor
Phase VI: Occupation/ Operation	Monitoring System Performance; LAP, Building Management

Note: LEED Standards offer recommendations on what type of professionals should be involved during each phase of the construction process.

and transportation. This first category of the LEED standards requires (with no points awarded) that pollution from construction activity be minimized by controlling soil erosion, waterway sedimentation, and airborne dust generation. Additional points are awarded for reducing light pollution, reducing heat island effect, and building on sites where development is complicated by real or perceived environmental contamination (or brownfields). Site consideration constitutes potentially 14 points toward certification.

Credit Point	
Construction Activity and Pollution Prevention	Required
Site Selection	1 point
Development Density and Community	1 point
Brownfield Redevelopment	1 point
Alternate Transportation	1–4 points
Site Development	1–2 points
Stormwater Design	1–2 points
Heat Island Effect	1–2 points
Light Pollution Reduction	1 point

The second area is water efficiency. The goal of water efficiency is to ensure that construction is mindful of the ways in which water is used in buildings. This includes reducing the quantity of water needed for building operation as well as reducing its effect on municipal water supply and treatment burden. This section includes a total of 5 points in the areas of water-efficient landscaping, innovative wastewater technologies, and water use reduction. LEED standards promote these goals by eliminating the use of potable water for irrigation, recycling water, and the use of water-conserving fixtures.

Water-Efficient Landscaping	1–2 points
Innovative Wastewater Technologies	1 point
Water Use Reduction	1–2 points

The third area is energy and the atmosphere: A major concern of LEED criteria is the abatement of the built environment contributing to global warming; this is reflected in the inclusion of the energy and atmosphere section into the standards. Energy and atmosphere consideration and planning potentially encompass the most points that a building can earn to become certified (17 points). The goals of this section are to establish energy efficiency and system performance, optimize energy, support ozone protection protocols, and encourage renewable and alternate energy sources. Similar to site selection, Energy and Atmosphere also include some criteria for which no points are awarded because these activities are considered as prerequisites. The Energy and Atmosphere standards promote establishing a minimum level of energy efficiency for the structure, encouraging on-site renewable energy self-supply and rigorous and ongoing energy monitoring programs.

Fundamental Commissioning of the Building Energy Systems	Required
Minimum Energy Performance	Required
Fundamental Refrigerant Management	Required
Optimizing Energy Performance	1–10 points
On-site Renewable Energy	1–3 points
Enhanced Commissioning	1 point
Enhanced Refrigerant Management	1 point
Measurement and Verification	1 point
Green Power	1 point

The fourth area is materials and resources. This section outlines the importance of the building materials used and takes into account the waste stream from the building during construction and in its operation and inhabitance. The goals of materials and resources are to reduce the overall amount of materials needed, to promote the use of materials with less environmental impact, and to reduce and manage waste. This section comprises potentially 13 points and highlights awareness of the manufacturing process and life cycle of materials used in buildings.

Storage and Collection of Recyclables	Required
Building Reuse	1–3 points
Construction Waste Management	1–2 points
Materials Reuse	1–2 points
Recycled Content	1–2 points
Regional Materials	1–2 points
Rapidly Renewable Materials	1 point
Certified Wood	1 point

The fifth area is indoor environmental quality. Indoor environmental quality (15 points) refers to the active methods of promoting safe internal environments, such as increased ventilation, as well as passive design methods meant to improve inhabitants' connection to the outdoor environment. The goals of indoor environmental quality are to establish good indoor air quality; eliminate, reduce, and manage sources of indoor pollutants; ensure thermal comfort and system controllability; and provide occupants' connection to the world outside the building.

Minimum Indoor Environmental Quality Performance	Required
Environmental Tobacco Smoke Control	Required
Outdoor Air Delivery Monitoring	1 point
Increased Ventilation	1 point
Construction of Indoor Environmental Quality Management Plan	1–2 points
Low-emitting Materials	1–4 points
Indoor Chemical and Pollutant Source Control	1 point
Controllability of Systems	1–2 points
Thermal Comfort	1–2 points
Daylight and Views	1–2 points

The final area is innovation and design process. LEED provides additional credit points for innovative design and operation that are awarded to projects that exhibit an exemplary level of creativity with regard to the ways in which LEED criteria are applied. These points can be awarded in areas like design for deconstruction (or potential disassembly) and recyclability of the structure.

LEED Accreditation

Where built structures are LEED certified, construction and maintenance professionals can become LEED accredited. The accreditation process requires only a passing grade on a USGBC exam, with no other educational or training requirements. After passing, accredited professionals are then validated to participate in LEED standard activities, including access to a professional USGBC community and development of potentially LEED-certified projects. Further, one point is awarded to buildings that include an accredited professional during the planning and construction process, ostensibly serving the dual purpose of promoting LEED accreditation as an occupational endeavor and ensuring the inclusion of accredited professionals within future green building projects. Some have criticized the lack of additional requirements to become LEED accredited beyond a passing grade on the exam. Critics and other construction professionals have suggested adding professional work standards similar to those imposed on engineers and architects, requiring a specified minimum number of work hours that must be accrued in the field before examinations may be taken.

Drawbacks of LEED

There are some criticisms of the LEED standards. Although the environmental advantages of LEED certification have been touted through studies conducted by independent researchers and the USGBC, there are potential drawbacks to the process. The LEED process requires considerable coordination and knowledge that may slow down the construction process and lead to misunderstandings between team members (e.g., LAP and contractors). Beyond planning and coordination, these standards also rely on the availability of sometimes expensive (when compared to traditional methods) and not readily available construction knowledge and building supplies. These added expenses usually initially raise construction costs; however, the USGBC has outlined the delayed benefits of long-term operational costs such as healthier working environments, decreased energy costs, and increased inhabitant productivity. Further, because LEED buildings are often more expensive to construct and demand a specific knowledge set, many LEED buildings are found in more metropolitan areas and not distributed uniformly throughout the United States and the world.

Others have criticized the LEED standards' focus on efficient use of fossil fuels rather than solely the promotion of alternative energy sources. More than half of the points for certification are assumed by some critics to pertain specifically to the areas of energy efficiency by way of reduced fossil fuel use, whereas fewer points are awarded in the areas of the promotion of reusable energy. Further, some building materials like leather are encouraged based on the criterion that they do not emit volatile organic compounds; however, little consideration is given to the unsustainable practices such as chemical treatment processes, as well as the poor treatment of cattle in industrialized farm settings, that go into the generation of these products.

Future of LEED

LEED continues to evolve and expand into several areas of environmental planning with the inclusion of new ideas and a widening range of professionals seeking a more sustainable approach to their practice. Although new areas of LEED certification are in their infancy relative to LEED guidelines for new construction and existing buildings, other areas of LEED certification now include commercial interiors, core and shell projects, homes, neighborhood development, schools, and retail space. The next version of LEED (3.0) is expected to include requirements for carbon footprints and significant reductions in greenhouse gas emissions. Over time, LEED standards are expected to change with new discoveries in sustainability science. Each new LEED area of interest and its adaption to LEED standards is intended to provide refined measures for sustainable development.

See Also: Appliance Standards; Daylighting; Geothermal Energy; Green Power; Heat Island Effect.

Further Readings

Greenbuild. "LEED Reweighted; Certification 'Tiered'." *ASHRAE Journal*, 51/1:6 (2008).
Klara, R. "The Green Giant (LEED Certification)." *Architecture*, 94/8:27–28 (2005).
Miller, S. L. "USGBC Takes LEED to the Next Level: Refinements Will Make Certification System More User-friendly." *Architect*, 96/14:14–15 (2007).

Steven A. Gray
Rutgers University

Lindsay L. Gray
Alexander Gorlin Architects

LIFE CYCLE ANALYSIS

Life cycle analysis, also known as life cycle assessment, is the examination of the environmental impact of a service or product. It involves measuring all resources, including both energy and raw materials used and waste generated, involved in the manufacture, use, and disposal of a product. For this reason it is sometimes known as "cradle-to-grave analysis," as it aims to account for all resources used at all stages of a product's life.

Life cycle analysis is an outgrowth of energy audits and global modeling studies that became popular in the late 1960s and early 1970s as people came to realize that the Earth's resources were not infinite and thus became interested in studying the environmental implications of different aspects of human behavior, including consumer purchases. An early term for this process in the United States was *resource and environmental profile analysis*, whereas in Europe the term *ecobalance* was used. The field is still in development, as there are many issues to be settled regarding what must be included in evaluating the effect of a single product, and methods are not completely standardized.

Frequently, the interest is in comparing two alternative methods of achieving a goal: for instance, should a university cafeteria provide disposable paper cups for the students, or only glasses? Although initially the paper cups would seem to be the more resource-intensive choice, since they must be manufactured, shipped, and disposed of after a single use, a study might reveal that the resources involved in washing the glasses and replacing those that are broken actually have a greater environmental impact.

Stages in a Life Cycle Analysis

The first step in a life cycle analysis is to collect the necessary data. Although this may be a lengthy and detailed process, if boundaries and methodology have been clearly defined and followed and reliable data are available, this step (sometimes called the "life cycle inventory") should not cause contention among researchers.

The second step, sometimes called "life cycle assessment," requires making value judgments that may be arguable: different researchers looking at the same data could draw quite different conclusions. The reason for this is that it requires comparing not just different quantities of the same thing but different quantities of different things. To take the cafeteria example above, the effect of heating and using a given quantity of water must be evaluated against the effect of disposing of a given quantity of solid waste. And all factors must be considered, including initial transportation costs (perhaps the paper cups were ordered from the cheapest supplier, who is located halfway across the country, but the glassware supplier is local), the chemicals used to wash the glasses, and so on. Factors such as the return rate must also be considered (e.g., if some glassware is delivered broken, or if the wrong size of cup is shipped). Many researchers try to aggregate all costs for a product or service into a single numerical score, which allows comparison but leaves the conclusion open to argument because no universally accepted and standardized method of arriving at these numbers has yet been developed.

Recycling introduces a number of new considerations. For instance, some goods such as steel can be recycled almost indefinitely, although of course this requires energy (and may not be economically feasible in some circumstances). Paper, in contrast, can be reprocessed only four or five times. In any case, the energy required to transport materials for recycling is an important factor: for a rural facility located far from a recycling plant this may be considerable, whereas for another, recycling may exist within the same city, making the calculations for two otherwise comparable cafeterias quite different. Also, the price paid for recyclable materials such as paper and aluminum fluctuates widely, and it may not be reasonable to assume that because something can be recycled, it will be, unless the force of law is behind the recycling process; otherwise, the same goods may be recycled when it is economically favorable to do so (i.e., when the value of the materials outweighs the costs of recycling them) and disposed of as garbage at other times.

Reasons to Conduct a Life Cycle Analysis

Life cycle analyses are conducted for a variety of reasons. One reason is to allow a municipality to plan for future waste disposal needs and to evaluate the feasibility of a recycling program. This requires predicting what products residents will be purchasing and disposing of in the future, the costs and potential revenues to be gained by recycling particular materials, the costs or savings in terms of transporting materials as either garbage or recycling, and the monetary and environmental costs of disposing of goods, whether by incineration, landfill, or some other method.

Another purpose may be for a particular manufacturer who seeks evidence to label his product as more environmentally friendly than that of a rival manufacturer. A related reason may be that some consumers prefer to buy products that they believe do the least possible damage to the environment. However, such claims must be evaluated carefully: Given the complexity of the process and the lack of standard procedures to make the final evaluations, they are often virtually impossible to prove and have frequently been attacked by environmental groups. An egregious example is an early study by a plastics industry trade group known as the Council for Solid Waste Solutions that found that plastic shopping bags were better for the environment than paper bags because they produced less solid waste. Upon closer examination, it was found that the group had not taken into account the time it would take the two products to degrade—only their initial bulk.

A third reason may be to identify stages in the life cycle of a product in which the environmental burden could be reduced and to evaluate which changes would produce the greatest benefits at the least cost. For instance, a city might want to evaluate how much cost is associated with disposal of the packaging of particular goods, and what benefits could be gained by adding a sales tax on goods considered to be overpackaged.

Finally, businesses may conduct a life cycle analysis to reduce waste and increase profitability or to reduce environmental impact in their manufacturing or other processes. For instance, if a business set the goal of reducing overall carbon dioxide emissions, a life cycle analysis could evaluate the relative effect expected by reducing electrical usage in a plant, choosing more efficient delivery vehicles, reducing thermostat settings in an office building, or purchasing electricity generated by solar or wind power.

Tools for Conducting a Life Cycle Analysis

Because of the complexity of conducting a life cycle analysis, a number of tools have been developed to make the process easier. The U.S. Environmental Protection Agency provides a Web page with introductory material and a detailed method for conducting an analysis. It breaks the process down into four steps—goal definition and scoping, inventory analysis, impact assessment, and interpretation, with multiple detailed questions to be considered within each step.

Carnegie Mellon University provides an Internet-based tool to conduct a life cycle analysis. It defines a product life cycle as a progression through five or six stages: raw materials are extracted from the Earth, processed into finished materials, manufactured into a product, used for a designated purpose by a consumer, and disposed of by the consumer (either recycled or discarded), and possibly a sixth stage that accounts for transportation of materials and products between the stages. Ten standard models are available (seven from the United States plus models from Canada, Germany, and Spain) and it is also possible to build custom models: in any case, the process of conducting an analysis is facilitated by making choices on a series of point-and-click computer screens.

The Web page of the European Commission Joint Research Center provides links to a number of tools to conduct a life cycle analysis. Over 50 such tools were included on the Web page in June 2009, representing a variety of approaches and languages including English, German, Japanese, Italian, Spanish, Dutch, and Chinese.

See Also: Carbon Footprint and Neutrality; Environmental Stewardship; Recycling; Sustainability; Waste Incineration.

Further Readings

Carnegie Mellon University. "EIO-LCA: Free, Fast, Easy Life Cycle Assessment." http://www .eiolca.net (Accessed June 2009).

Environmental Protection Agency. "Life Cycles Assessment: Principles and Practice." Contract No. 68-C02-067, May 2006. http://www.epa.gov/NRMRL/lcaccess/lca101.html (Accessed June 2009).

European Commission Joint Research Center, "LCA Tools, Services and Data." http://lca.jrc .ec.europa.eu/lcainfohub/directory.vm (Accessed June 2009).

Global Development Center. "Life Cycle Analysis and Assessment." http://www.gdrc.org/uem/ lca/life-cycle.html (Accessed June 2009).

Henrickson, Chris T., et al. *Environmental Life Cycle Assessment of Goods and Services: An Input-Output Approach*. Washington, D.C.: Resources for the Future, 2006.

Meier, Barry. "Consumer's World: Life-Cycle Studies: Imperfect Science." *New York Times* (22 September 1990). http://www.nytimes.com/1990/09/22/style/consumer-s-world-life-cycle-studies-imperfect-science.html?pagewanted=1 (Accessed June 2009).

Mithrarathne, Nalanie, et al. *Sustainable Living: The Role of the Whole Life Costs and Values*. Amsterdam: Butterworth-Heinemann, 2007.

Sarah Boslaugh
Washington University in St. Louis

M

METERING

The typical electric meter is a piece of 19th-century technology that is poised for a 21st-century transformation. Today, most consumers are not able to locate their meters, and their interest in the information the meter provides is limited to the amount due on their utility bills. But new "smart meters" are ready to play a crucial role in making utility service more efficient and reliable, while helping consumers reduce costs and conserve resources.

The electromechanical induction meter was invented in 1888 by Elihu Thomson and is still being used in more than 94 percent of the homes in the United States. The Thomson meter has a series of dials and a spinning metal disk encased within a glass cover. The disk spins faster or slower according to the amount of electricity being used. The numbers on the dials provide the "reading" that is recorded by a utility meter reader. More sophisticated meters are also in use in commercial and industrial settings. These meters measure electricity use at the time of day that power is being used (usually in 15- or 30-minute intervals) and by the demand, or level of power that is being drawn at any given moment. Business customers then are charged more complex rates wherein their demand for electricity is calculated, as well as their overall usage.

Until recently, residential electrical demand was considered negligible compared with that of commercial/industrial users, so most residential electricity customers were charged only for their total usage over time, or kilowatt-hours, as measured by the Thomson analog meter. The primary evolution in metering technology was driven by the utilities' interest in reducing the personnel costs associated with manual meter reading. Some newer meters can be read remotely.

Today, electrical demand and the costs of providing it, both economically and environmentally, are rising rapidly. Smart meters have been identified as a primary mechanism to help use the electricity that is produced more efficiently. Smart meters are solid-state electronic meters, equipped with digital displays and small computers that measure electricity use precisely, in units as small as a few minutes at a time. Smart meters can be read remotely and provide precise information on power outages. Smart meters can provide two-way communication to both the utility and the customer. This ability, together with upgrades to the electrical grid system that will add information, automation, and sensing capacities, make up what has come to be known as the smart grid.

The smart meter's communication capacities offer a wide range of new opportunities to consumers and utilities. For the utility, the ability to reduce electricity use at times of high system demand is a valuable tool for ensuring the grid system's reliability. An example of this is automatic cycling of air conditioners, which is already in operation in many utilities. On hot summer days, when electrical demand is high, air conditioners that have been identified as available for power reductions can have their cooling units shut off for short periods of time. Customers who permit cycling are compensated financially, and the utility can avoid power outages and buying or generating additional high-cost power to meet demand. At this time, this is done with fairly low-tech radio-controlled switches. Smart meters will be able to improve the delivery of air conditioning cycling programs as well as communicate with other home appliances.

Smart meters combined with home energy display devices or sophisticated Web-based tools can also inform customers in real time about their electricity use. This means customers are able to identify and change their usage—a process that was impossible when electrical use data were only available a month after it had occurred, when the bill was delivered. Smart meters also enable customers to take advantage of price differences in electricity, which vary seasonally and by time of day, with lower prices in off-peak times such as nights and weekends and higher prices in the summer and times of high demand. With rates that reflect the actual costs of electricity, customers can shift their loads and pay less for power. Variable electricity rates are still not common in the United States, but where they exist, the combination of information and pricing incentives has been shown to reduce electricity consumption.

Smart meters also provide new opportunities for distributed generation of electricity. For example, the amount of electricity produced by a photovoltaic array connected to the grid was previously accounted for by allowing the electric meter to "spin backwards." Smart meters that record the time of day electricity is generated can more accurately assign a value to the power that is produced. A small-scale system operator, such as a homeowner with a photovoltaic system, can sell to the grid when power is expensive and buy from the grid when power is cheap.

The widespread deployment of plug-in hybrid electric vehicles will also rely on smart meters. The infrastructure to support this vehicle of the future does not exist yet, but when it does, not only will cars obtain power from the grid, but their batteries will also be able to supply power back to the grid when it is needed. Smart meters will measure the flow of electricity in these transactions.

Although smart meters hold promise, some significant challenges remain. First, the United States lags far behind other industrialized countries in the wide-scale deployments of smart meters. The Mid-Atlantic and portions of the Midwest and South regions (the states of Pennsylvania, Wisconsin, Connecticut, Kansas, and Idaho) have the highest penetration, and installations are in early stages or still under consideration in other states. Pilot programs to test how consumers will react to new pricing options for electricity that can cut demand have been undertaken in many states, notably, Illinois and California. However, the entire package of how these pieces fit together, and how they can save on energy and lower costs, is still more of a vision than reality.

Second, consensus on the right technology to deploy has not been reached. There are competing versions of smart meters on the market today, and which process should be used to provide communication with the grid is still being debated. Some smart grid proponents advocate broadband over power line. In this model, the electric grid becomes another wire into homes and businesses to provide high-speed Internet access. Broadband over power line then is also used for the smart grid's operations and communications with the smart meters. Others propose using more conventional radio, wireless, and other communications standards to talk to the meters, without any additional Internet access.

Third, the transition to smart meters will be expensive. Like many new technologies, the costs for smart meters have continued to drop, but billions of dollars will still need to be spent. If smart meters are well deployed, with the full range of programs and options that can reduce energy use and consumers' bills, smart meters and the smart grid could pay for themselves over time. However, in the short term, the huge capital expenditures that will be required remain a significant hurdle. Although the 2009 federal stimulus bill includes significant funding for smart grid activities, ultimately most utility regulation of the smart grid will be hashed out state by state and utility by utility. As a consequence, the timetable for replacing the 19th-century predecessors with their 21st-century counterparts remains uncertain. Establishing this advanced metering infrastructure is crucial to the efficient and cost-effective delivery and consumption of energy.

See Also: Electricity; Energy Audit; Photovoltaics (PV); Plug-In Hybrid; Public Utilities; Smart Grid.

Further Readings

Federal Energy Regulatory Commission. "Assessment of Demand Response and Advanced Metering." http://www.ferc.gov/legal/staff-reports/demand-response.pdf (Accessed February 2009).

National Energy Technology Laboratory. "The Modern Grid Strategy." http://www.netl.doe .gov/moderngrid (Accessed February 2009).

Marjorie Isaacson
Anthony Star
CNT Energy, Center for Neighborhood Technology

METRIC TONS OF CARBON EQUIVALENT (MTCE)

Carbon dioxide (CO_2) is considered to be the main greenhouse gas, but other gases—methane, nitrous oxide, perfluorocarbons, hydrofluorocarbons, and sulfur hexafluoride—present in traces, significantly affect the radiative budget of the atmosphere. Different greenhouse gases have different atmospheric residence times and specific radiative forcing and thus cannot be comparable on a ton-for-ton basis. On a molecular basis, many non-CO_2 gases have more radiative forcing potential than CO_2. The Kyoto Protocol sanctions reduction in emissions of any greenhouse gases to be credited in terms of carbon equivalency. Signatories of this protocol have to control the emission of greenhouse gases on the basis of CO_2 equivalent. Carbon equivalency of different greenhouse gases allows scientists and policy makers to compare the contribution of each greenhouse gas. Consideration of non-CO_2 greenhouse gas emissions could be effective in controlling the emissions impact and reducing the costs rather than considering only CO_2 strategy.

Table 1   Lifetimes and Global Warming Potentials (GWP) Relative to CO_2 for Selected Greenhouse Gases (GHG)

Industrial or Common Name	Chemical Formula	Lifetime (Years)	GWP for Given Time Period	
			20-Years	100-Years
Carbon Dioxide	CO_2	N/A	1	1
Methane	CH_4	12	72	25
Nitrous Oxide	NO_2	114	289	298
CFC-11	CCl_3F	45	6,730	4,750
Halon-1301	$CBrF_3$	65	8,480	7,140
HFC-23	CHF_3	270	12,000	14,800
Sulfur Hexafluoride	SF_6	3,200	16,300	22,800
Nitrogen Trifluoride	NF_3	740	12,300	17,200
HFE-125	CHF_2OCF_3	136	13,800	14,900

Source: Adapted from Table 2.14, IPCC, 2007.

Metric tons of carbon equivalent (MTCE) is a common approach to expressing the emission or pool of greenhouse gases. Greenhouse gases with long atmospheric residence times (e.g., CO_2, methane, nitrous oxide, perfluorocarbons, hydrofluorocarbons, and sulfur hexafluoride) are evenly distributed throughout the atmosphere, and their global average concentrations can be computed easily. However, concentration of gases with short residence times such as water vapor, carbon monoxide, tropospheric ozone, ozone precursors (e.g., NO_x), and tropospheric aerosols vary regionally, and it is hard to quantify their atmospheric effect. These gases cannot be expressed in terms of carbon equivalent.

Inventory of each greenhouse gas is converted to carbon equivalency using the individual global warming potential (GWP) value of greenhouse gases (Table 1). GWP value of individual greenhouse gases reflects the globally averaged relative radiative forcing. According to the Intergovernmental Panel on Climate Change, GWP is defined as the ratio of the time-integrated radiative forcing (both direct and indirect) from the instantaneous release of 1 kg of a greenhouse gas relative to that of 1 kg of CO_2 over a period of 100 years. Direct radiative forcing takes place when the particular greenhouse gas absorbs solar radiation. Indirect radiative forcing happens when chemical alterations involving the original gas produce one or more greenhouse gases or when a gas influences radiative associated processes such as residence time of other gases. Radiative forcing of a particular gas depends on its ability to capture infrared radiation emitted by Earth's surface and atmosphere. The concept of GWP reduces the complexity in calculations and variations among different greenhouse gases. This approach also undermines chemical interactions among the greenhouse gases and other gases. Values of GWP for different greenhouse gases also depend on the selection of time horizon for integrating the effects of radiative forcing. For example, methane has a GWP value of 21 under a 100-year time scale and 56 under a 20-year scale.

It is common in international standards to express emissions in CO_2 equivalent rather than carbon equivalency, whereas U.S. sources have expressed emission in terms of carbon equivalent. The potential of emission reduction programs has often been mentioned in million metric tons of carbon equivalent. This approach considers the carbon in the CO_2 molecule, contrary to the entire CO_2 molecule.

Calculation for the Carbon Equivalent of a Greenhouse Gas

Conversion of greenhouse gas to CO_2 equivalent = Million metric ton (MMT) of GHG (10^{12} g) $\times$ GWP

Conversion from CO_2 equivalent to carbon equivalent = CO_2 equivalent $\times$ 0.27

The conversion between CO_2 equivalent and carbon equivalent is based on the ratio of the atomic mass of a carbon dioxide molecule (C [12] + 2 O [2 $\times$ 16] = 44) to the atomic mass of a carbon atom (12). For conversion of CO_2 equivalent to carbon equivalent, CO_2 equivalent value of greenhouse gas is multiplied by 0.27 (12/44).

1 metric ton of CO_2 equivalent = 0.27 metric tons of carbon equivalent

Or, 1 metric ton of carbon equivalent = (44/12) = 3.667 metric tons of CO_2 equivalent.

For example:

2 MMT N_2O $\times$ 120 (GWP of N_2O) = 240 MMT CO_2

240 MMT CO_2 $\times$ 0.27 = 64.8 million metric ton carbon equivalent.

See Also: Greenhouse Gases; Intergovernmental Panel on Climate Change; Kyoto Protocol; Radiative Forcing.

Further Readings

Intergovernmental Panel on Climate Change. *The Science of Climate Change*. Cambridge: Cambridge University Press, 1996.

Reilly, J., et al. "Multi-gas Assessment of the Kyoto Protocol." *Nature*, 401:549–55 (October 1999).

Shine, K. P. and W. T. Sturges. "CO_2 Is Not the Only Gas." *Science*, 315:1804–5 (2007).

Amitava Chatterjee
University of California, Riverside

Microhydro Power

Microhydro is a subcategory of hydroelectric power generation commonly understood to have a generating capacity of 100 kW or less. As with all forms of hydroelectric power, microhydro emits no air pollutants and is carbon neutral, yet because of its small size, it has additional environmental, social, and economic advantages. Three such advantages that are particularly important include its negligible effects on river and surface-water ecologies, its ability to be used in extremely remote communities, and its cost-effectiveness relative to other sources of energy. There are no firm estimates regarding the extent of microhydro around the world, but thousands of these systems have been installed in Southeast Asia, especially in China, Nepal, Vietnam, and the Philippines.

Hydroelectric power generation is classified broadly into two divisions—large and small scale. There is no universally accepted distinction between large and small, but small hydropower is typically considered to have a power generating capacity of 10 megawatts (MW) or less. (Note that 1 MW is equal to 1,000 kilowatts.) Small hydropower is further classified as minihydro (less than 1 MW), microhydro (less than 100 kW), and picohydro (less than 5 kW). Installed picohydro systems are usually between 200 watts and 1 kilowatt, however.

To put these small power plants in context, consider that the Hoover Dam has a generating capacity more than 20,000 times greater than the largest microhydro system, or what a typical household hairdryer requires 1 kW of power to operate. Microhydro therefore is not used as part of large-scale centralized power production or to meet the needs of energy-intensive households typical in high-income countries. Its primary benefit is for applications in remote areas with no access to electricity and for small electric power loads. Microhydro is most commonly adopted in rural areas of low-income countries for limited residential and community use (lights, radios, televisions, and so forth), for small agricultural equipment, and for small industries located on good river resources. In spite of its limitations, research indicates that the potential for microhydro is considerable, especially for picohydro. A World Bank report described the global potential for picohydro as "massive," with more than a million technically feasible household sites in the South American Andes alone.

Technical Considerations

Microhydro systems cannot be installed everywhere there is flowing water in a river or stream—physics and engineering govern whether microhydro systems are feasible. The potential for microhydro is determined by site-specific characteristics such as water flow and topography (the physics) and the ability to match appropriately designed equipment to site conditions (the engineering).

All scales of hydropower rely on water pressure to spin a turbine, a device that converts mechanical energy into electricity. Water pressure for microhydro is determined by two primary factors: (1) the amount of water that is flowing, and (2) the vertical distance over which the water flows. Vertical distance is referred to as "head height," and the amount of flowing water as "flow rate." Microhydro systems can be installed in a range of stream conditions from high head/low flow to low head/high flow conditions. Low head height is generally understood as less than 33 feet (10 meters).

Microhydro systems divert water from a stream or river into a pipe called a "penstock." The flow of water through the penstock is then directed to a turbine. Depending on the site, a penstock can be just a few feet in length or can run several hundred yards to achieve the necessary head height for water pressure. Water diversion is accomplished through "run-of-river" systems that do not require impoundment dams to create large upstream reservoirs of water.

Microhydro turbines have been designed to maximize efficiency through a combination of head height and flow rate. The most common turbines for microhydro are "impulse turbines" that spin as nozzles of water are directed at them. These are best used with low flow rates over varying head heights and include pelton, turgo, and cross-flow turbines. "Reaction turbines" are used when head heights are low, but flow rates are comparatively strong. Unlike impulse turbines, reaction turbines are fully immersed in water. Common types of reaction turbines include Francis, propeller, and Kaplan turbines. Electricity generated by microhydro systems can be used to directly power devices, or it can be stored in batteries. Microhydro systems are capable of generating electricity in alternating current, as well as at high voltages (e.g., 220 volts). This is more cost-effective than wind and solar technologies, which often produce low-voltage direct current that must be converted for direct use.

Advantages and Disadvantages

The environmental advantages of microhydro are significant. Because these systems use water and not fossil fuels to power turbines, they emit none of the air pollutants associated with electric power production and are a carbon-neutral energy source. Because microhydro is a run-of-river system, it does not create the extensive environmental impacts associated with large hydroelectric power dams and reservoirs.

The social and economic advantages of microhydro systems are equally notable, especially in terms of their affordability and ability to alleviate energy poverty. Micro- and picohydro can provide much-needed electricity to isolated communities, creating new income-earning opportunities and displacing fossil fuels, particularly kerosene for lighting and diesel for small combustion generators. A World Bank study indicates that picohydro is readily affordable by all but the poorest communities and can cost as little as $25 for a 200-watt system or $1,000 for a 3-kW community pelton system. Overall, microhydro has proven to be a particularly appropriate technology in low-income nations, where experiences in China, Nepal, Kenya, and Vietnam demonstrate that these systems can be locally manufactured, operated, and maintained.

Microhydro is not without limitations. Appropriate sites are not common, and because it is so site-specific, there are no standardized, "turnkey" systems that can be installed, although much progress has been made in this regard with respect to picohydro. There are seasonality issues with water flow (dry seasons, snowmelt, monsoons), which constrain the peak output of microhydro systems, and their small power output limits electric power loads. Although they are somewhat simple systems, they still require training to effectively apply, operate, and maintain them.

See Also: Carbon Footprint and Neutrality; Hydroelectric Power; Renewable Energies.

Further Readings

Harvey, Adam. *Microhydro Design Manual.* London: Intermediate Technology Publications, 2000.

Microhydropower Internet Portal. "Microhydro Definitions," "Turbines," "Introduction to Hydropower," and "Civil Work Components." http://www.microhydropower.net (Accessed January 2009).

World Bank. "Stimulating the Picohydropower Market for Low-Income Households in Ecuador," ESMAP Technical Paper 35993 (December 2005). http://www.worldbank.org (Accessed January 2009).

Maria Papadakis
James Madison University

MOUNTAINTOP REMOVAL

Mountaintop removal is the process of mining coal where the top of a mountain, called "overburden," is blasted away to reveal the underlying coal seams. It began in the 1970s as a response to the oil crises, in an effort to achieve energy independence. The site most heavily mined by mountaintop removal has been Appalachia, specifically in West Virginia and Kentucky. Two other states with extensive mountaintop removal are Virginia and Tennessee.

This 1995 photograph of Samples Mine in West Virginia illustrates the conspicuous effects of mountaintop removal processes that include deforestation, blasting, dumping rubble into "valley fills," and dredging of coal from blasted areas.

Source: Tending the Commons: Folklife and Landscape in Southern West Virginia, American Folklife Center, Library of Congress/Lyntha Scott Eiler

The first step in mountaintop removal is clearing the land of woody growth. The deforestation products are sold as lumber or burned. Topsoil is supposed to be saved for later reclamation of the land, once the mining has been completed, but legal loopholes allow for topsoil substitute to be used instead. The original batch of topsoil, then, is dumped into "valley fills" or nearby streams. Next, miners use explosives to remove the top 1,000 vertical feet off the mountain, allowing access to coal seams. Multiple blasts on a mountain can be performed to mine even deeper for further coal seams. Excess land and rock that have been blasted off are temporarily stored in already-mined sites or dumped into the valley fills, which are also called "hollow fills." Once a mountaintop site has been completely mined, it is to be designated as postmining land usable for several options as outlined by federal law (Surface Mining Control and Reclamation Act). Topsoil or substitute must be laid down, followed by fertilized grass seed. Further reclamation procedures are up to the agreement between the landowner and the mining company. One option is to restore the land as pastureland, as it has now been cleared of uneven terrain and forestry.

The Surface Mining Control and Reclamation Act is overseen by the Office of Surface Mining Reclamation and Enforcement, an agency with a mission to "carry out the requirements of the Surface Mining Control and Reclamation Act in cooperation with States and Indian Tribes." The Office of Surface Mining Reclamation and Enforcement requested funding of $246,014,000 for the Abandoned Mine Reclamation Fund in 2006.

Proponents of mountaintop removal claim it has economic benefits and only minor, if any, environmental and social downsides. One environmental upshot is that the coal found here burns cleaner than deep mountain coal, because of its lower sulfur content. In this sense, the Clean Air Act indirectly supports mountaintop removal because of its caps on sulfur emissions from coal burning. In addition, mountaintop removal is more affordable because it does not rely on vast teams of miners and the underground development of mine shafts. The surface-type mining, as opposed to underground work, is safer for the workers too, aside from the explosives. In terms of providing energy to U.S. households, coal power provides the majority of electricity for U.S. homes. Furthermore, coal can be converted into fuel for cars and trucks, adding further support to the coal industry and its fostering of energy independence.

Despite the economic benefits of mountaintop removal that include cheaper energy sources, a cheaper mining technique, and improved safety for workers, there are downsides. The easier technique of mountaintop removal compared with underground mining results in fewer mining jobs. In addition, the people most directly affected by the mining do not see the economic benefits. Although the mining takes place in Appalachia, local residents are among the poorest in the nation. According to a 2006 census, 17.3 percent of individuals living in West Virginia were living below the poverty level, accounting for 12.7 percent of the state's families. Eastern Kentucky is also heavily mined via mountaintop removal; 17 percent of this state's individuals were living below the poverty level in 2006, accounting for 13.1 percent of families. In these communities, often a house is the only asset a family has, and home values can drop strikingly after mountaintop removal has taken place nearby.

A recent study by the Health Policy Institute at West Virginia University, led by Michael Hendryx, Ph.D., claims that in fact, the economic turn of coal mining is actually a negative one. Hendryx calculates that because of premature deaths and illnesses, the net cost of coal mining is five times as great as what is earned from mined coal. The illnesses for which residents in coal-mining areas are at elevated risk include chronic obstructive pulmonary disease and hypertension, or high blood pressure. Importantly, this study includes both underground and mountaintop mining and looks at mining only, excluding the act of burning coal.

According to Appalachian Voices, an organization that works to foster Appalachian culture, mountaintop removal blasts can take place quite close to homes (within 300 hundred feet), even throughout the night, and it causes floods and landslides that can damage property as well. Critics of mountaintop removal argue that permits for removal do not require strict enough review of endangered and native species' habitats in the region to be destroyed. According to the Environmental Protection Agency, the lands reclaimed after mountaintop removal are not as healthy as they were before, and streams that have been dumped into have increased mineral levels—they may even fill up with soils, and the damage may go so far as to have negative effects both socially and culturally.

In addition, drinking water can be contaminated in the process of mountaintop removal. Although excess soil and rock are supposed to be dumped into the valley fills, sometimes mining companies receive permits to dump waste into streams. On several occasions there have been lawsuits raised against the companies. At this time, the Massey Energy Company is in a legal battle but is allowed to continue mining until the ruling is made. The contamination of water is significant because Appalachian waterways connect with such major rivers as the Mississippi, affecting populations and water life as far away as New Orleans, Louisiana.

Many people have spoken out against mountaintop removal, but the mining companies maintain that it is beneficial to the people, not harmful to the environment, and leaves more useful land afterward. Critics argue that the mining companies have enough wealth to find loopholes in laws and continue unchecked. They may have found support from President Barack Obama, who is quoted to have said "We're tearing up the Appalachian Mountains because of fossil fuels" in his August 2007 campaign speech in Lexington, Kentucky, clearly supporting alternate means of energy to mountaintop removal.

See Also: Coal, Clean Technology; Electricity; Environmental Measures; Environmental Stewardship.

Further Readings

Appalachian Voices. "Mountaintop Removal Homepage." http://www.appvoices.org (Accessed June 2009).

Hendryx, M. and M. M. Ahern. "Mortality in Appalachian Coal Mining Regions: The Value of Statistical Life Lost." *Public Health Reports* (July/August 2009).

Loeb, P. *Moving Mountains: How One Woman and Her Community Won Justice From Big Coal.* Lexington: University Press of Kentucky, 2007.

"OSMRE Bureau Highlights, Fiscal Year 2006." http://www.osmre.gov/topic/budget/docs/FY06a.pdf (Accessed June 2009).

Pancake, A. *Strange as This Weather Has Been.* Berkeley, CA: Counterpoint, 2007.

Shnayerson, M. *Coal River.* New York: Farrar, Straus & Giroux), 2008.

"Surface Mining Control and Reclamation Act: A Response to Concerns about Placement of Coal Combustion By-Products at Coal Mines." http://www.mcrcc.osmre.gov/PDF/Forums/CCB3/5-8.pdf (Accessed June 2009).

U.S. Census Bureau. "2006 American Community Survey, B17001. Poverty Status in the Past 12 Months by Sex and Age" and "B17010, Poverty Status in the Past 12 Months of Families by Family Type by Presence of Related Children Under 18 Years by Age of Related Children." http://factfinder.census.gov (Accessed June 2009).

U.S. Environmental Protection Agency. "Downstream Effects of Mountaintop Coal Mining: Comparing Biologic Conditions Using Family- and Genus-level Macroinvertebrate Bioassessment Tools." http://www.epa.gov/region03/mtntop/index/htm (Accessed June 2009).

Claudia Winograd
University of Illinois at Urbana-Champaign

NATURAL GAS

Natural gas is mostly formed from plankton—tiny water-dwelling organisms, including algae and protozoans—that accumulated on the ocean floor as they died. These organisms were slowly buried and compressed under layers of sediment. Over long periods of time, the pressure and heat generated by overlying sediments converted this organic material into natural gas. Natural gas frequently migrates through porous and fractured reservoir rock with petroleum and subsequently accumulates in underground reservoirs. Because natural gas and petroleum are formed by similar natural processes, these two hydrocarbons are often found together in underground reservoirs. After gradually forming in the Earth's crust, natural gas and petroleum slowly flow into the tiny holes of nearby porous rocks that serve as reservoirs. Because this porous reservoir rock is often filled with water, the natural gas and petroleum, both of which are lighter than water and less dense than the surrounding rock, migrate upward through the crust, sometimes for long distances. Eventually, some of these upward-moving hydrocarbons become

Liquid natural gas tankers dock at specialized port facilities with liquefaction plants nearby. Natural gas is transported in liquid form because its volume is reduced by 600 times compared with its gaseous state.

Source: iStockphoto.com

trapped by an impermeable (nonporous) layer of rock, known as the cap rock. Natural gas is lighter than petroleum, so it forms a layer over the petroleum. This layer is called a gas cap.

Natural gas may also come from other sources. Coal beds contain appreciable quantities of methane, the principal component of natural gas. Thus, natural gas may also form

297

in coal deposits, where it is often found dispersed throughout the pores and fractures of the coal bed. This type of natural gas is often referred to as coal-bed methane. Coal-bed methane is natural gas found in seams of coal. Another source is the methane produced in landfills. Landfill gas is considered a renewable source of methane because it comes from decaying garbage. The gas from coal beds and landfills accounts for 10 percent of the total gas supply today. This percentage is likely to increase. As its name implies, natural gas emerges from the ground in gas form, which is difficult to transport and store. Natural gas is presently transported through extensive networks of pipelines or is liquefied and transported by ship.

The ancient people of Greece, Persia, and India discovered natural gas many centuries ago. They were mystified by the burning springs created when natural gas seeping from cracks in the ground was ignited by lightning. About 2,500 years ago, the Chinese recognized its usefulness. They piped the gas from shallow wells and burned it under large pans to evaporate sea water to get the salt. Natural gas was first used in the United States in 1816 to illuminate the streets of Baltimore with gas lamps. In 1821, William Hart dug the first successful American natural gas well, about 27 feet deep, in Fredonia, New York.

What Is Natural Gas?

Natural gas is a mixture of hydrocarbons with a small proportion of nonhydrocarbons. Methane is the principal hydrocarbon component of natural gas. Other hydrocarbon components are ethane, propane, butanes, and a very small proportion of pentanes and heavier hydrocarbons. The nonhydrocarbon components of natural gas include carbon dioxide, hydrogen sulfide, nitrogen, inert gases (particularly helium), and water vapor. In rare cases, hydrogen occurs in natural gas. The nonhydrocarbon constituents of natural gas are classified as diluents and contaminants. The diluents are noncombustible gases that reduce the heating value of the gas; they include nitrogen, carbon dioxide, and water vapor. Contaminants include hydrogen sulfide and other sulfur compounds. The contaminants can damage production and transportation equipment. The actual composition of natural gas varies from place to place. Even natural gases from different underground reservoirs in the same area vary in their compositions.

Methane is the product of diagenesis of organic matter and catagenesis of kerogen. Other alkanes in natural gas are produced by thermal cracking reactions of organic matter during catagenesis of kerogen. Nitrogen arises from the thermal decomposition of proteins. Main sources of hydrogen sulfide include anaerobic reduction of the sulfate anion supplied by the saline sea water that occasionally floods the accumulating organic matter in the stagnant water of the swamp, and thermal decomposition of organosulfur compounds during catagenesis. Carbon dioxide evolves from a variety of reactions in the various stages of the fossil fuel formation process, including anaerobic decay reactions, decarboxylation reactions of esters and organic acids, and thermal decomposition of carbonate rocks. Hydrogen arises, though rarely, from dehydrogenation reactions. The occurrence of hydrogen in natural gas is rare because hydrogen is very mobile and easily escapes from the confining rock strata. Helium is believed to be a product of the natural radioactive decay of uranium, thorium, and radium in granite rocks deep inside the Earth.

There are three systems of classification of natural gas. It may be classified as associated or nonassociated, wet or dry, or sweet or sour natural gas. Associated natural gas is found together with liquid petroleum in a reservoir. It is always wet. Nonassociated gas is not found together with liquid petroleum. It is isolated in natural gas fields. There are two types of associated natural gas; namely, gas cap gas and dissolved gas. A certain amount of natural gas almost always occurs in connection with oil deposits and is brought to the surface together with the oil when a well is drilled. Such gas is called casing-head gas. Certain wells, however, yield only natural gas.

Dry gas has a very high methane content, whereas wet gas contains a considerable amount of hydrocarbons of higher molecular weight such as ethane, propane, and butane. The terms *dry* and *wet*, used in this context, are not related to the moisture content of the gas but to the absence or presence of condensable alkanes. In general, a dry natural gas has more than 90 percent methane. Dry natural gas contains less than 0.013 L/M^3; that is 0.013 liters of condensable hydrocarbon liquid per cubic meter of natural gas. Wet natural gas is formed by thermal cracking during catagenesis, which gives a variety of hydrocarbon products including the light alkanes. Residue gas is the gas remaining (mostly methane) after the alkanes have been extracted from wet gas. Sour natural gas contains hydrogen sulfide, whereas sweet natural gas contains no hydrogen sulfide. Sour natural gas is undesirable because hydrogen sulfide has an unpleasant smell and dissolves in water to give a mildly acidic solution that is corrosive to the pipes, valves, and metallic components of gas handling and storage systems.

Production of Natural Gas

Natural gas is commercially produced from oil and natural gas fields. The natural gas industry is producing gas from increasingly more challenging resource types such as sour gas, tight gas, shale gas, and coal-bed methane. The world's largest proven gas reserves are located in Russia, with $47.57 \times 10 \times 10^{12}$ cubic meters. Russia is the world's largest producer of natural gas. In 2006, the major proven natural gas resources in billion cubic meters were located as follows: world, 175,400; Russia, 47,570; Iran, 26,370; Saudi Arabia, 6,568; and United Arab Emirates, 5,823. Qatar recorded 25,790 in 2007. The world's largest gas field is located in Qatar's offshore North field. It is estimated to have 25 trillion cubic meters of gas—enough to last more than 200 years at optimum production levels. The second-largest natural gas field is also connected to Qatar's North field. It is the South Pars gas field in Iranian waters in the Persian Gulf, having estimated reserves of 8 to 14 trillion cubic meters of gas.

It may be difficult to find natural gas because it is usually trapped in porous rocks deep underground. Scientists employ many techniques to discover gas deposits. The surface rocks could serve as clues about underground formations. Scientists may set off small explosions or drop heavy weights on the earth's surface and record the sound waves as they bounce back from the sedimentary rock layers underground. Gravitational pull of rock masses within the Earth could also be measured. Drilling usually follows promising test trials. Natural gas wells average 6,100 feet deep and are expensive to drill. Hence, it is important to choose sites carefully. About 48 percent of exploratory wells produce gas. In contrast, 85 percent of the developmental wells—that is, wells drilled on known gas fields—yield gas.

Natural gas contains over 90 percent methane. The aim of natural gas processing is to remove hydrogen sulfide (sweetening), carbon dioxide, moisture, and the condensable hydrocarbons (ethane, propane, butane, and pentane), leaving only methane. The product of natural gas processing, methane, is sometimes called stripped gas. When liquefied, it is called liquefied natural gas (LNG). When natural gas is cooled to a temperature of approximately minus 260 degrees F at atmospheric pressure, it condenses to LNG, which weighs less than one-half as much as water. The volume of the natural gas is reduced 600 times when it is liquefied. LNG is odorless, colorless, noncorrosive, and nontoxic. It is easier to store than the gaseous form because it takes up much less space. It is also easier to transport. Shipping LNG by tanker is ideal, especially in situations where pipeline export is not advisable because of technical, economic, or geopolitical reasons. The process is expensive, however, and the LNG chain requires heavy capital investment in liquefaction and regasification plants, refrigerated storage tanks, and port facilities for loading and unloading. There are also specially designed LNG tankers.

Once natural gas has been extracted from the ground, it is usually transported by pipeline to a refinery, where it is processed. Natural gas is processed in an extraction unit to remove the nonhydrocarbon compounds, especially hydrogen sulfide and carbon dioxide. Two processes used for this purpose are absorption and adsorption. Absorption uses a liquid that absorbs the natural gas and impurities and disperses them throughout its volume. In a process known as chemisorption, the impurities react with the absorbing liquid. The natural gas can then be stripped from the absorbent, while the impurities remain in the liquid. Common absorbing liquids are water, aqueous amine solutions, and sodium carbonate.

Adsorption is a process that concentrates the natural gas on the surface of a solid or a liquid to remove impurities. A substance commonly used for this purpose is carbon, which has a large surface area per unit mass. For example, sulfur compounds in natural gas collect on a carbon adsorbing surface, and the sulfur compounds are then combined with hydrogen and oxygen to form sulfuric acid, which can be removed. After the impurities have been removed in the extraction unit, the natural gas is transported to a processing plant, where the other alkanes and substances are separated and removed for different uses. Natural gas can be converted into liquid fuels, including gasoline, by using gas-to-liquids technology, which links methane into larger hydrocarbon molecules. Methane that is joined to form carbon chains or rings can be processed into gasoline, diesel fuel, and jet fuel. Adding steam and oxygen to methane links methane carbon atoms and produces synthesis gas. This synthesis gas is then brought together with hydrogen at high temperatures in the presence of a catalyst, and the resulting liquid synthetic fuels are typically clean-burning, high-quality fuels.

How Natural Gas Is Used

Applications of natural gas can be considered under two general headings: the fuel uses and the nonfuel uses. A greater proportion of natural gas is used as fuel. Natural gas may be used as fuel to produce steam, which turns turbines in the generation of electricity. Particularly high efficiencies can be achieved through combining gas turbines with a steam turbine in combined cycle mode. Combined cycle power generation using natural gas is the cleanest source of power available using fossil fuels; this technology is widely used wherever gas can be obtained at a reasonable cost. Fuel cell technology may in the future provide cleaner options for converting natural gas into electricity. A fuel cell is similar to a battery but uses a chemical process rather than combustion to convert the energy content of a fuel into electricity. The chemical process of producing electricity from natural gas is much more energy efficient than combustion, and it emits no air pollutants. In its fuel uses, natural gas has some comparative advantages over petroleum-derived fuels and coal and is considered a premium fuel. Natural gas burns cleaner than other fossil fuels and produces less carbon dioxide per unit energy released. For an equivalent amount of heat, burning natural gas produces about 30 percent less carbon dioxide than burning petroleum and about 45 percent less than burning coal. Also, it may be used as fuel in the cement industry to heat kilns. It could be used in the glass-making industry and in various other fuel-consuming industries. Natural gas is also used as a domestic or commercial cooking gas. It is used in ovens, clothes dryers, boilers, furnaces, and water heaters. It is used primarily in commercial and residential buildings for heating water and air, for air conditioning, and as fuel for stoves and other heating appliances.

There are presently natural gas vehicles in many countries such as the United States, Argentina, Brazil, India, Pakistan, Italy, and Iran. Compressed natural gas is a cleaner alternative to other automobile fuels such as gasoline and diesel. Natural gas is the leading automotive fuel alternative to gasoline. The energy efficiency is generally equal to that of gasoline engines. Natural gas can be used in any vehicle with a regular internal combustion engine, but the vehicle must be outfitted with a special carburetor and fuel tank.

Natural gas has attracted global attention as the transition fuel to a more sustainable/renewable energy mix worldwide. Natural gas is a major feedstock in the Haber process for the production of ammonia for use in fertilizer production. Liquid methane is used as a jet engine fuel. It has more specific energy than the standard kerosene mixes and its low temperature can help cool the air, which the engine compresses for greater volumetric efficiency. Natural gas can be used to produce hydrogen. Hydrogen is a primary feedstock for the chemical industry. It is a hydrogenating agent, an important commodity for oil refineries, and a fuel source in hydrogen vehicles. Natural gas is also used in the manufacture of petrochemicals, fabrics, glass, steel, plastics, paint, and other products. It is a cleaner fuel than gasoline and costs less.

Natural gas liberates more heat per unit weight than coal and petroleum products. Being a gas, natural gas leaves no residue of ash when burned, unlike coal, which leaves ashes when burned. Natural gas is the most environmentally friendly fossil fuel. It produces fewer contaminants than the burning of other fossil fuels. Being a gas, natural gas is easy to meter and regulate in gas combustion systems. This contrasts with the labor-intensive drudgery of shoveling coal. The principal nonfuel use of natural gas is the manufacture of synthesis gas. Synthesis gas is used for the production of ammonia, methanol, hydrogen, aldehydes, alcohols, and acids. Natural gas is also used for manufacturing carbon black, which is used to reinforce polymers, manufacture printing ink, black paint, lacquers, carbon papers, typewriter ribbons, and acetylene black. Synthetic rubber and fibers like nylon are also made from chemicals derived from it.

Challenges in Using Natural Gas

The main challenges in the use of natural gas as fuel are encountered in its storage, long-distance transportation, and incomplete combustion of methane. Because of its very low density at room temperature and atmospheric pressure, storage of natural gas requires its liquefaction at very low temperatures. Storage of natural gas is, therefore, expensive in comparison with that of liquid petroleum products and solid coal. Also, because gas leakages may not immediately be noticed, a potentially hazardous accumulation of gas may build up from the storage. Hence, storage of natural gas is potentially hazardous. Storage acts as a buffer between the pipeline and the distribution system, however, allowing distribution companies to serve their customers more reliably by withdrawing more gas from storage to meet customer demands during peak use periods. It also allows the sale of fixed quantities of natural gas on the spot market during off-peak periods. Having local storage of gas also reduces the time necessary for a delivery system to respond to increased gas demand. Storage also allows continuous service, even when production or pipeline transportation services are interrupted.

Intercontinental transportation of natural gas is also very expensive and potentially hazardous, as it involves liquefaction of large volumes of LNG by cooling to minus 162 degrees C. The manufacture of LNG is expensive in terms of both the process plants and the sophisticated ships that carry the cargo. The use of natural gas pipelines is impractical across oceans. The fuel uses of natural gas involve combustion of methane. However, in a limited supply of oxygen, incomplete combustion of methane takes place, giving off carbon monoxide and water. Incomplete combustion is undesirable because it produces less heat and more carbon monoxide, an extremely poisonous gas. It also produces soot, which is undesirable in a gas combustion system.

Even though natural gas is comparatively the cleanest fossil fuel, it does contribute substantially to global emissions. It is more potent than carbon dioxide when released into the atmosphere but is not of serious concern because of the small amounts in which this occurs. The price of natural gas varies greatly depending on location and type of consumer. The typical caloric value of natural gas is about 1,000 British thermal units (Btus), depending on gas composition. Natural gas is sold in cubic feet and measured in Btus. One Btu is the amount of heat needed to raise the temperature of one pound of water 1 degree Fahrenheit. The current deregulation and restructuring of the pricing and transmission of natural gas is allowing the development of new markets in all sectors for natural gas. New technologies and policies have caused natural gas to be a fuel of choice, leading to the development of natural gas reserves in unconventional reservoirs such as buried coal beds.

See Also: Electricity; Flaring; Fossil Fuels; Pipelines.

Further Readings

American Gas Association. http://www.aga.org (Accessed February 2009).
LNG information. http://www.ch-iv.com/lng/lngfact.htm (Accessed February 2009).
Natural Gas. http://www.naturalgas.com (Accessed February 2009).
Obuasi, P. The Origin of Petroleum, Natural Gas and Coal. Nsukka: University of Nigeria, 2000.

Akan Bassey Williams
Covenant University

Nitrogen Oxides

Anthropogenic activities increase the concentrations of nitrogen oxides in atmosphere, resulting in water, air, and soil pollution and human health concerns. Nitrogen is the most abundant element found in the environment, but it exists mostly in an inert state, as N_2. Combustion of fossil fuels and extensive use of inorganic nitrogenous fertilizer have profound influence on the nitrogen cycle and transform inactive atmospheric N_2 into reactive nitrogen forms (Nr). From 1860 to 2005, formation of Nr increased from ~15 Tg (= 10^{15} g) nitrogen in 1860 to 187 Tg nitrogen in 2005, with a steady rise in population and energy demand. Main sources of nitrogen oxides include combustion of fossil fuels, biomass burning, lightning, oxidation of ammonia, microbial processes in soils, and input from the stratosphere.

Nitrogen oxides include nitric oxide (NO), nitrogen dioxide (NO_2), and nitrous oxide (N_2O). Combustion of fossil fuel can oxidize atmospheric nitrogen (N_2) to produce NO in the environment. The dominant source of nitrogen oxides in the air is combustion processes; 90–95 percent of these nitrogen oxides are usually emitted as NO, and 5–10 percent as NO_2. NO may be oxidized to NO_2 by atmospheric oxygen according to the reaction:

$$NO + NO + O_2 \rightarrow 2NO_2$$

NO can also react with atmospheric ozone:

$$NO + O_3 \rightarrow NO_2 + O_2$$

In addition, ozone is formed when NO_2 is photolyzed, forming NO and an O atom, and then reacts rapidly with O_2 to form ozone:

$$NO_2 + h\nu \rightarrow O + NO$$

$$O + O_2 \rightarrow O_3$$

NO_2 also can be generated as a result of soil microbial mineralization of organic matter. Soil microbe–mediated processes, nitrification, and denitrification release NO and N_2O into the atmosphere. These processes accelerate when nitrogenous fertilizers are added to promote crop yield. Among these nitrogen oxides, the least reactive is N_2O, but it is an important greenhouse gas because of its long atmospheric residence time—about 120 years—and its high global warming potential (310). Moreover, in the atmosphere, N_2O slowly diffuses into the stratosphere and generates NO as a result of photodissociation by ultraviolet radiation. Formation of NO decreases the concentration of stratospheric ozone, so an increase in atmospheric N_2O would pose a threat to the ozone layer.

The global increase in atmospheric nitrogen oxides has three main causes: (1) application of nitrogenous fertilizer to boost agricultural production for the increasing population; (2) widespread cultivation, particularly of legumes and rice, which increases the atmospheric N_2 fixation by symbiotic and nonsymbiotic bacteria; and (3) combustion of fossil fuels, which forms nitrogen oxides. According to J. N. Galloway et al., an increase in nitrogen oxides concentration can lead to the following serious environmental problems:

- Formation of tropospheric ozone and aerosols that can cause respiratory trouble, cancer, and cardiac disease
- Deposition of nitrogen aberrantly increases or decreases the forest and grassland productivity and alternate biodiversity
- Nitrogen deposition in water bodies can cause acidification and loss of biodiversity
- Nitrogen deposition can pollute coastal water by eutrophication, hypoxia, and habitat degradation
- Nitrogen oxides contribute to an increase in greenhouse gases and depletion of the ozone layer, which consequently has potential for global warming

Agricultural and industrial activities have increased biologically available nitrogen oxides formation and influence the nitrogen cycle. Increased nitrogen deposition can cause injury to plant tissues and alter biomass production and vulnerability to secondary stress factors. At the ecosystem level, deposited nitrogen modifies interspecies relationships and results in loss of biodiversity. Critical loads for the most sensitive ecosystems are estimated at 5–10 kg N ha^{-1} $year^{-1}$; more average for ecosystems is 15–20 kg N ha^{-1} $year^{-1}$. Deposition of nitrogen oxides on water bodies can increase the level of NO_3^- and NO_2^- in water, which may cause algal blooms and deplete the oxygen from water bodies.

Elevated concentration of nitrogen oxides in ambient and indoor air may have a negative effect on human health. Short-term NO_2 exposure causes decreases in lung function and increased airway responsiveness. Other effects include decreases in host defenses and alterations in lung cells and their activity. Long-term exposure to NO_2 is associated with respiratory illness. Individuals with asthma and chronic obstructive pulmonary disease are more susceptible than healthy individuals. Children aged 5–12 years constitute a subpopulation potentially susceptible to an increase in respiratory morbidity associated with NO_2 exposure.

Increased concentration of nitrogen oxides in the environment can be controlled by decreasing the rate of nitrogen oxides production during energy and food production and converting nitrogen oxides to inert N_2. Nitrogen oxides are formed inadvertently during fossil fuel combustion through oxidation of fossil-organic nitrogen or through the oxidation of atmospheric N_2. Generation of nitrogen oxides from fossil fuel combustion can be mitigated by using an alternative method to provide energy (e.g., hydrocarbon-based fuel cells) or by eliminating nitrogen oxides from the combustion products. Formation of nitrogen oxides during food production can be minimized by increasing the nitrogen-use efficiency in crop and animal agriculture, recycling crop and animal residues within the system, encouraging nitrogen addition through biological fixation, and providing incentives to farmers for less use of reactive nitrogen. Another promising way to minimize the global nitrogen pollution problem is to convert reactive nitrogen forms to inert N_2, or denitrification. However, current knowledge of the denitrification process is not sufficient for protection and promotion of denitrification hot spots and needs further development in measuring denitrification. Reducing nitrogen oxides from the environment is challenging, but it is possible and of prime importance. Strategies to check extreme dependency on nitrogenous fertilizer and fossil fuels need to be considered to protect the environment from the harmful effects of nitrogen oxides.

See Also: Fossil Fuels; Global Warming; Greenhouse Gases.

Further Readings

Food and Agriculture Organization of the United Nations (FAO). "FAO Statistical Databases. (2006) http://faostat.fao.org/default.aspx (Accessed January 2009).
Galloway, J. N., et al. "The Nitrogen Cascade." *Bioscience*, 53:341–56 (2003).
Galloway, J. N., et al. "Transformation of the Nitrogen Cycle: Recent Trends, Questions, and Potential Solutions." *Science*, 320:889–92 (2008).
Kulkarni, M.V., et al. "Solving the Global Nitrogen Problem." *Frontiers in Ecology and Environment*, 6:199–206 (2008).

Logan, J. A. "Nitrogen Oxides in the Troposphere: Global and Regional Budgets." *Journal of Geophysical Research*, 88:10785–807 (1983).
World Health Organization. *Environmental Health Criteria 188, Nitrogen Oxides*, 1997.

Amitava Chatterjee
University of California, Riverside

NONPOINT SOURCE

Water pollution occurs mostly through two major sources: point sources and nonpoint sources (NPS). Point source pollution occurs at identifiable locations where pollution is discharged into water bodies, such as sites with industrial and municipal wastewater discharge, solid waste disposal runoff (or leachate), concentrated animal feedlots, urban storm drains, and commercial ships. Pollution from diffuse sources (e.g., large-scale agricultural runoff, wind-borne pathogens) that affect water bodies and is not readily identifiable is known as NPS pollution. According to the U.S. Environmental Protection Agency (EPA), nonpoint source pollutants refer to any water pollution sources that are not explicitly identified in Section 502(14) of the Clean Water Act of 1987.

NPS pollution generally occurs from precipitation, runoff from land use or land cover, infiltration, drainage, seepage, or atmospheric deposition. The runoff from rainfall and snowmelt transports natural pollutants and human-induced pollutants such as fertilizers, insecticides and pesticides, and other chemicals in agricultural lands into streams, lakes, wetlands, groundwater, oceans, and other water bodies. Windborne debris blown to and deposited in water bodies is another NPS pollution process. NPS pollution also takes place as a result of the animal, bird, and human excreta flow with runoff to water bodies. Even nutrient-rich decomposed forest litter serves as NPS pollution material that is transported to streams with runoff.

Point source pollution can be regulated, but NPS pollution is difficult to regulate, as it is derived from various sources with no easy accounting for it. According to the EPA, NPS pollution, occurring mostly from agricultural land runoff, is the primary source of water pollution in United States. According to state reports, it contributed up to 48 percent of the impaired stream miles and 41 percent of the impaired lake acres in the 48 contiguous states of United States. Urban runoff is another important source of NPS pollution. Oil, fertilizers, and chemicals applied to lawns, and sewer and septic tank leaks to streams, are examples of urban runoff pollution. Table 1 shows the leading sources of water quality impairment caused by various NPS.

Nitrogen and phosphorous are two major agricultural land nutrients that wash away with runoff to streams and degrade water quality. Potassium, secondary nutrients, and other micronutrients from agricultural land wash away as well. These nutrients enhance aquatic plant growth. This is known as eutrophication of water bodies. The excessive aquatic plant growth, in turn, produces more nutrients in the water as a result of its dying and decaying. It depletes the oxygen availability in water and thus destroys fish habitat. Highly nutrient enriched water stimulates algae growth. Thus the amount of turbidity is enhanced, along with discoloring of water. A foul smell comes from the water, making it unusable by humans and animals. Harmful secondary nutrients from agricultural land like zinc, arsenic, iron, and so on cause serious diseases in humans and animals. Eroded soil

Table I   Leading Sources of Water Quality Impairment From NPS for Rivers, Lakes, and Estuaries

Rivers and Streams	Lakes, Ponds, and Reservoirs	Estuaries
Agriculture (48%)	Agriculture (41%)	Municipal point sources (37%)
Hydrologic modifications (20%)	Hydrologic modifications (18%)	Urban runoff/storm sewers (32%)
Habitat modifications (14%)	Urban runoff/storm sewers (18%)	Industrial discharges (26%)
Urban runoff/storm sewers (13%)	Misc. nonpoint source pollution (14%)	Atmospheric deposition (24%)

particles or sediments are the main NPS pollutants, which also bring those nutrients along with it. Sediment itself affects the water use in many ways, such as reducing the availability of sunlight for aquatic plants, covering fish spawning areas and food supplies, smothering coral reefs, clogging the filtering capacity of filter feeders, and clogging and harming the fish gills.

According to the EPA, the nonlegal communities, who do not necessarily agree with water program managers about the regulation of NPS pollution, illustrate NPS pollution as (1) NPS discharge is intermittent and mostly related to meteorological events, (2) NPS pollutants are generated over a large extensive land area or watersheds and move overland to reach the surface water bodies or infiltrate groundwater, (3) NPS pollution is related to uncontrollable climatic conditions and the topography and land cover of the watershed, and (4) the extent of NPS pollution is difficult to monitor and is expensive.

With the implementation of the National Pollutant Discharge Elimination System, since 1972, point source pollution has been decreased a great deal with the use of pollution control strategies in the federal Clean Water Act. However, this point source pollution reduction gain did not make the water bodies of the United States totally pollutant free because the NPS pollution was not controlled. Therefore, Congress passed an amended version of the Clean Water Act, referred to as the 1987 Water Quality Act. The fundamental principle of this act was: "It is the national policy that programs for the control of nonpoint sources of pollution be developed and implemented in an expeditious manner so as to enable the goals of this Act to be met through the control of both point and nonpoint sources of pollution."

Section 319 of the 1987 Water Quality Act established a national program to address NPS of water pollution. For environmentally impaired watersheds, the NPS program also includes regulations on the maximum amount of a pollutant a water body can receive before compromising water quality standards, a calculation known as the total maximum daily load (TMDL). The TMDL for NPS pollutants addresses maximum inputs for nutrient and sediment load, pathogens, and metals (e.g., mercury, copper, and lead) among others. Although there is no federally regulated enforcement of the NPS program as with point sources, there has been some significant progress in limiting NPS pollution in water bodies of United States since the implementation of Section 319. Federal programs designed to protect natural resources in the United States and reduce NPS pollution include the national estuary, pesticides, coastal nonpoint pollution control, source water protection, rural clean water, environmental quality incentive, conservation reserve, conservation security, conservation reserve enhancement, wetlands reserve, wildlife habitat incentive, forest land enhancement, and grazing reserve programs.

Best practices management, management practices, accepted agricultural practices, management measures, best practices systems, management practice systems, resource management systems, and total resource management systems are the processes currently pursued in the United States to reduce NPS pollution. Watershed management decision support systems are a comprehensive procedure to lessen point and nonpoint source pollution. Decision support systems such as the Beaver Lake Watershed Management system are designed using geospatial technology, watershed models, computer programs, and Web design techniques to analyze and model the watershed parameters for water quality control and provide the public with easy access to information for point and nonpoint source pollution prevention.

Several high-end software programs and models have been developed since the 1970s to quantify the nonpoint source pollution from watersheds. On an easy to complex and comprehensive scale, they are USLE (Universal Soil Loss Equation), RUSLE, AGNPS (Agricultural Nonpoint Source), HSPF (Hydrologic Simulation Program-Fortran), BASINS (Better Assessment Science Integrating Point and Nonpoint Sources), and SWAT (Soil and Water Assessment Tool).

See Also: Best Management Practices; Environmental Measures.

Further Readings

Panda, S. S. and H. J. Byrd. "Geo-spatial Model Development for 12-digit HUC Based NPS Pollution Prioritization Mapping for Fannin County, GA." *Proceedings of Georgia Water Resources Conference, April 27–29, 2009.* Athens, Georgia.

Panda, S. S., et al. "Development of a GIS-based Decision Support System for Beaver Lake Watershed Management." *Proceedings of American Water Resources Association (AWRA) Spring Specialty Conference, May 17-19, 2004.* Nashville, Tennessee.

Thornton, Jeffrey A., Walter Rast, Marjorie M. Holland, Geza Jolankai, and Sven-Olof Ryding. eds. *Assessment and Control of Nonpoint Source Pollution of Aquatic Ecosystems: A Practical Approach.* Pearl River, NY: Parthenon Publishing Group, 1999.

U.S. Environmental Protection Agency. "National Management Measures to Control Nonpoint Source Pollution from Agriculture." Document No. EPA 841-B-03-004 (July 2003).

U.S. Environmental Protection Agency. "2000 National Water Quality Inventory." Washington, D.C.: U.S. Environmental Protection Agency, Office of Water, 2002. http://www.epa.gov/305b/2000report (Accessed February 2009).

Sudhanshu Sekhar Panda
Gainesville State College

NONRENEWABLE ENERGY RESOURCES

Energy is commonly referred to as the ability to do work; it enhances man's ability to convert raw materials to useful products. Energy is of two forms and sources; namely, renewable and nonrenewable sources of energy. The renewable energy resources are those that are naturally generated and can be obtained from the sun, water, winds, waves, tidal, and other replenishing natural sources. These resources are nondepleting, abundantly available, and naturally occurring. The nonrenewable energy resources are those sources that cannot be replenished by natural phenomenon over a short period of time. They account for over 75 percent of the energy sources used worldwide. Nonrenewable energy sources are those sources that have been formed over many millions of years; they naturally exist and are organic in nature—mostly plants and animals that died millions of years ago and thus have fully decomposed, releasing all their carbon contents back to the Earth and becoming a usable source of useful energy. They emanate from the ground either as liquids (e.g., crude oil), solids (e.g., coal), or gases (e.g., associated and nonassociated gases) and exist in fixed amounts per location.

The nonrenewable energy resources are basically of two types: fossil fuels and nuclear fuels. Each of these resources can be used to generate electricity, and they power the majority of today's industrial and domestic processes.

Fossil Fuels

The fossil fuels derived their name from the archeological term *fossil*, which means "dug up." They are formed as a result of the geological processes that have taken place on dead and decomposed plants and animals. They have a rich energy resource base and naturally exist deep down under the earth. They can only be brought to the surface of the earth by exploration. Chemically, fossil fuels consist largely of hydrocarbons, which are compounds composed of hydrogen and carbon. However, some fossil fuels contain additional compounds of substances such as nitrogen and sulfur. The most commonly used fossil fuels are those of petroleum products, coal, and natural gas.

Petroleum products are refined extracts of crude oil. They include gasoline, diesel fuels, kerosene, fuel oils, petrochemicals, liquefied petroleum gas, asphalt, tar, and so on. Petroleum products are the most consumed of the different examples and components fossil fuels. The majority of transportation systems run on petroleum products, and most industrial and domestic processes are powered by energy derived from it; in addition, the world economy has so been built around petroleum products that changes in their availability or price determine fluctuation in economic variables.

Coal is a black combustible solid formed by the decomposition of traditional biomass of plant origin in the absence of oxygen. It has the same chemical constituent of carbon and hydrogen as other fossil fuels but also contains oxygen and small amounts of sulfur and nitrogen. It can be used directly as fuel, giving an enormous amount of energy and heat when burned; it is a chemical reactant and a source of organic chemicals, and it can be converted to gaseous and liquid fuels. Its use as a means of producing electricity has come a long way. Today some countries depend on coal-fired turbines for electricity generation. Apart from using coal to generate electrical energy, it is also employed in the making of fertilizers, pesticides, solvents, and so on. It can be heated in the presence of steam and oxygen to produce synthesis gas that can be used directly as fuel or refined into burning gas.

Natural gases are gaseous mixtures of saturated hydrocarbon. They contain at least about 75 percent methane gas, along with other hydrocarbons of the alkane group, which include ethane, propane, butane, and pentane, in very small proportions. They can be used domestically for heating and cooking and industrially for different processes ranging from fuel in the baking of bricks and ceramic tiles, in the production of cement, in boilers to generate steam, and in the glass making and food processing industries. Natural gases also play a major role in petrochemical industries, where they are used in the making of fertilizers, detergents, and plastics. Importantly, natural gases can be used to power gas turbines for the generation of electricity. Many countries like Nigeria depend a great deal on power generated from gas-fired turbines. The combustion of natural gases produces huge amounts of energy and can be useful to power automobiles, airplanes, trains, ships, industries, and homes.

The use of fossil fuels has various advantages and disadvantages. Some of the advantages include being a very dependable source of commercially generated electricity. More than 70 percent of the world's energy is generated from fossil fuels yearly. This may not be reduced if major concentration is not put into developing alternative sources of power generation. Also, fossil fuels are one of the cheapest sources of energy, being easy to harness and able to be stored for a long time. The majority of today's industrial, commercial, and domestic systems and processes, and even our transportation system, have been designed to run on fossil fuels or fossil fuel–derived energy. However, the use of fossil fuels poses some major challenges to the global environment, to individual humans, and even to national economies. On the environmental front, burning fossil fuels for energy releases a lot of carbon monoxide, carbon dioxide, water vapor, and other deleterious compounds of nitrogen and sulfur that could include methane, nitrous oxides (NO_x), and sulfurous oxides (SO_x). These products of fossil fuel combustion are active gases that terribly affect the world's environment, causing global warming and leading to climate change. These gases are part of a group of gases collectively called greenhouse gases. In addition, the combustion of coal can lead to the production of fly ash and smog, which is harmful to health. Acid rain is another harmful effect of fossil fuel combustion. Sulfur dioxide released to the atmosphere can undergo serial combination to form sulfuric acid, which is an acid rain and has a corrosive effect on buildings, roofing materials, and paints. It is also reported that acid rains have damaging effects on crops, forests, streams, lakes, and rivers. When humans are exposed to large volumes of some of these gaseous products of methane, NO_x, and SO_x, respiratory infections can develop. Nationally, depending on fossil fuels for energy can lead to foreign exchange depletion, as part of it will be used to fund the purchase of petroleum products for national consumption. In addition, the phenomenon of oil spills on national water bodies is something that will continue. This invariably will affect aquatic life and fish farming and will affect revenue generation either directly or indirectly.

Nuclear Fuels

Nuclear fuels are the other type of nonrenewable energy resource and one of the more controversial energy resources, being constantly subjected to world politics and debate. Unlike fossil fuels, they are produced as a result of the radioactive and nuclear reaction of some radioactive elements. The nucleus of the radioactive material spontaneously disintegrates over a period of time to release energy and produce fast-moving particles that can strike other atoms along the way. The striking together and breaking down of these radioactive particles when allowed to take place in a controlled and monitored environment produces nuclear reactions that lead to the generation of enormous heat that can be used for positive work. The controlled environments are commonly referred to as nuclear reactors.

Nuclear fuels—mainly uranium—are found in rocks. Uranium is a chemically reactive and radioactive element that occurs naturally in a free state and in combination as oxides or complex salt. Uranium is being developed around the world as a rich source of electricity production. The technology exists around the world in such places as the United States,

Ukraine, and Iran. Uranium is used in power plants through the process of fission reaction in a reactor to provide heat, which is used to generate electricity. The heat is produced in the nuclear reactor when neutrons strike uranium atoms, causing them to split in a continuous chain reaction, thereby continuously producing heat that is used to boil water to produce steam. The steam is then used to drive turbine generators to produce electricity. Control rods, made of boron, are placed intermittently along the core of the fuel assembly to regulate the rate of fission reaction. The rods are made of boron because boron absorbs neutrons and thus can reduce the amount of the neutrons available for the process when the rods are present. When the rods are removed, more neutrons become available for fission reaction, and so the reactions take place at a very fast pace, producing more heat. This reaction is continuous as long as the reactive substances are present.

A major advantage of uranium and nuclear fuels is that they do not produce greenhouse gases during use and thus do not contribute to global warming and climate change. However, their waste is a major challenge. Nuclear waste is still highly radioactive for a long time (over a hundred years) and can be dangerous to human, animal, and plant life if the wastes are not properly stored in such a way as to discontinue reactivity. Nuclear accidents can be very damaging.

Conclusively, the use of nonrenewable energy resources has been the mainstay of the world's energy supply to date. Its availability is undoubted, but its depleting nature and contribution to environmental harm are major concerns. Gaseous products from its burning and exposure to certain volumes of its direct radiation are not human friendly. Its effects as pollutants are not encouraging. The global political landscape has gradually been shaped to deviate from the present worldwide dependence on its use to concentrating on developing better, richer, much more available and nondepleting, naturally available clean energy resources.

See Also: Climate Change; Fossil Fuels; Global Warming; Greenhouse Gases; Natural Gas; Nuclear Power; Uranium.

Further Readings

Abrahamson, D. E., ed. *The Challenge of Global Warming*. Washington, D.C.: Island Press, 1989.

Bailey, R. A., et al. *Chemistry of the Environment*. London: Academic Press, 2002.

Schwaller, A. E. *Transportation, Energy and Power Technology*. New York: McGraw-Hill, 1989.

Starr, C. and G. T. Miller, Jr. *Environmental Biology*. New York: McGraw-Hill, 2004.

Ajayi Oluseyi Olanrewaju
Covenant University, Nigeria

NUCLEAR POWER

Nuclear power is a relatively recent energy source, beginning approximately half a century ago with both civilian and military applications. Currently, nuclear power is generated through atomic fission (a nuclear reaction that splits the nucleus of the atom into smaller, energized components), but new technologies are being developed to make nuclear power development safer and more controllable. In comparison to fossil fuel–based energy sources, nuclear power has relatively low carbon emissions and, as such, is regarded as a green energy source. However, accidents such as the 1986 Chernobyl disaster and the growing problems of radioactive waste management contribute to maintaining a public fear around the widespread development of

Second-generation reactors like the 1988 Torness Nuclear Power Station near Dunbar, Scotland, shown above, increased nuclear energy capacity to 300 GW in the late 1980s.

Source: iStockphoto.com

nuclear power plants. These civic concerns coupled with a spatially and temporally limited uranium supply do not make nuclear power a renewable nor sustainable energy source.

History

Both the energy and military nuclear power stories follow one single path: since the discovery of radium by Pierre and Marie Curie, scientists, experts, politicians, and industries have always contributed to the development of nuclear power. By nuclear power, we mean here civil nuclear power, excluding military nuclear power and naval and space population.

According to the International Atomic Energy Agency (IAEA), the 436 operational nuclear reactors in the world provide 18,368 TWh (18,368 billion kWh). This represents 14.2 percent of the world's electricity production and somewhere between two to six percent of the total global energy production. The greater part of electricity is provided by thermal power stations through the use of coal, petrol, or gas.

The first atomic cell, as a micro nuclear power generator, was built by Enrico Fermi and Lea Szilard in 1942. It was called "Chicago Pile 1" and provided power of only 0.5 W but was mainly used to contribute to the Manhattan Project and the development of the Hiroshima and Nagasaki nuclear bombs. French scientists Lew Kowarski and Frédéric Joliot-Curie also developed an atomic cell in 1948, called Zoé, with the same goal—to anticipate military nuclear application.

After the end of World War II, the "Atoms for Peace" discourse, announced in December 1953 by President Dwight Eisenhower, furthered the use of nuclear power for energy, not only for war. The United States had already built a nuclear reactor capable of generating electricity in December 1951, but the Soviet Union's Obninsk nuclear power plant was the first to provide an electricity grid in 1954, with a capacity of 5 MWh. Then other nuclear plants opened: in 1956, Marcoule (France) and Sellafield (Great Britain) were the first commercial plants, followed by Shippingport (United States) one year later; 1957 also saw the creation of the IAEA, which compiles data on nuclear development. The IAEA's Power Reactor Information System, for example, presents data on quantity and evolution of the whole world reactor set. The IAEA's Incident Reporting System compiles information on both technical and human factors related to events of safety significance that occur at nuclear plants.

The Framatome organization, created in 1958, is also an important organization, showing the early links between the two countries highest in number of nuclear power plants—the United States and France, now with, respectively, 104 and 59 operational reactors. The "Franco-Américaine de construction atomique" gets together engineers of the four mother companies. Nowadays, it is part of the French AREVA multinational industrial conglomerate.

Power Capacity and Operating Cycle

IAEA's data illustrate the growth of the development of nuclear capacity from almost nothing in 1960 (1 GW or less) to 100 GW by 1980, to 300 GW by 1990 and to approximately 400 GW presently. This increase was made possible by the development of second-generation reactors. After the first attempts during the 1950s and 1960s (atomic cell and reactor of the first generation), a new wave of nuclear technologies was developed (second-generation reactor), mostly represented by pressurized water reactors (about 85 percent of reactors) and, at a lower level, by boiled water reactors and Russian high-power channel-type reactors.

As thermal power stations use fossil fuels to heat up water and activate turbines, nuclear power plants use controlled nuclear chain reactions to generate steam to make turbines and generators run to produce electricity. Nuclear chain reactions result from fission of the atomic nucleus, generally uranium-235, by the absorption of a neutron. Fission divides an atom into two or more smaller nuclei and releases other free neutrons. These neutrons can be absorbed by other fissile atoms and create more fission. Nuclear technology is no more than the use of this chain reaction: the controlled chain reaction is used for nuclear power plants while the uncontrolled version is used for nuclear bombs. Such fission produces a very high quantity of energy: about 20 Mega eV (electronvolt, an energy measure) resulting from the fissile products' kinetic energy (the kinetic energy of new neutrons and new nuclei resulting from fission). This kinetic energy produces the very heat used in the reactor to pressurize or boil water in another closed circuit, depending on the selected technology. Turbines are then activated by pressurized or boiled water and active generators that produce electricity.

As a consequence, nuclear power plants need a source of cool water that will be heated again by the reactor. This is the purpose of the enormous concrete aerorefrigerated towers that are the most recognized aspect of nuclear power plants. The towers use ambient air to cool primary circuit water, inducing a lot of evaporation. Other sources of cooling can be used, such as a river, and still remain radioactive fall-out free. Moreover, water coolant systems induce cooling of the reactor, which reduces the speed of neutrons' interactions with uranium-235 and maintains a relatively controllable chain reaction.

Third and fourth generation reactors are also being developed. Reactors of the third generation will still use fission-produced electricity, such as the European pressurized water reactor (EPR) under construction in France and a similar project in Finland under STUK (the Finnish Radiation and Nuclear Authority). These newer reactors will have a capacity of 1,600 MW—one of the highest production capacities to date. In contrast to earlier generation reactors, fourth generation nuclear reactors propose to use controlled fusion energy. Whereas fission splits the atom, fusion assembles two atomic nuclei to form a heavier nucleus, with extraordinary energy liberation. Using "tokamak" technology, the fourth-generation ITER corporation prototype project may show a controlled fusion use in order to fuel a new nuclear reactor. According to the American National Science Foundation, "proposed advanced (generation IV) nuclear power plants aim to incorporate a suite of new technologies that will produce nuclear power in a manner that is sustainable, economical, safe, reliable, and proliferation resistant, [but] additional research, development, and analysis of advanced nuclear power is needed." The Generation IV International Forum (GIF) website provides detailed information on these projects, explaining six others technologies planned for development.

Controversies and Nuclear Fear

All these new technologies also have known failures. The examples of the Three Mile Island and Chernobyl accidents are well known. The Three Mile Island (Pennsylvania, United States) reactor number two, a PWR type, was subject to a partial fusion of its central core on March 28, 1979. This provoked an important contamination within the confinement enclosure. However, radioactive rejects were limited and seemed not to affect the population or the environment. The Chernobyl nuclear reactor incident would not be though, and would not be without severe consequences.

On April 26, 1986, the Soviet-era RMBK Chernobyl (modern Ukraine) reactor number four, as a consequence of technical and human failure, had an exceptional jump rise in its power output (about a 100-fold increase in four minutes) that resulted in multiple explosions and long-lasting fires. The heart of the reactor burned for six days, engendering a radioactive cloud containing mainly iodine-131 and cesium-137. According to Marie-Hélène Labbé, some 600,000 to 800,000 "liquidators" were recruited to stop the fire and shut down the reactor, but they were not equipped to not be irradiated—or even not contaminated. Then the radioactive cloud drifted over Europe, provoking a still-ongoing controversy over its effects. For example, French authorities announced that fallout of cesium-137 was no more than 5,400 becquerels (radiation unit of measure) in countries to the east and southeast, while the independent French Commission for the Independent Research and Information on Radioactivity (CRIIRAD) expertise revealed 30,000 to 35,000 becquerels per square meter.

After the first period of production of electricity by nuclear power, the Chernobyl accident has applied a relative brake on nuclear plants' extension. Indeed, a relative decline of production is characteristic of the period from the late 1980s until now. Since 2006 and 2007, however, Chinese demand for nuclear power has reactivated construction of reactors and new grid connections. Consequently, the attention has recently focused on present and potential radioactive resource reserves, primarily uranium. In 2007, this necessitated some 70,000 metric tons of uranium extracted from mines, but also from nuclear stock and military resources. World production of uranium was about 41,000 metric tons in 2006, which represents an increase of 14 percent in comparison with 2000 or 2001. Uranium is mostly extracted from Canada (10,000 metric tons), Australia (7,500 metric tons), and Kazakhstan (5,500 metric tons). Uranium reserves area estimated at 5.5 million metric tons with a supplementary potential of 10 million metric tons were already detected. However, consumption will rapidly grow to 94,000 to 122,000 metric tons, depending on the scenario, and the IAEA only plans for one century of reserve. Some nongovernmental organizations, though, think that this resource will be more rapidly used by 2030.

As a result, uranium is now a strategic resource for nuclear countries. For example, France is very dependent on uranium and tries to ensure continued supplies to produce its electricity and export it. Recently, France has invested in mines in the Congo. China also tries to predict its future growth of uranium consumption, also looking in the direction of its African and Asiatic neighbors' reserves.

Uranium consumption is also problematic regarding its waste production. In fact, nuclear wastes have always been paradoxical: Since nuclear production in the 1970s, production of waste has only grown, but no stocking solution was chosen, revealing the complexity of the problem. In fact, only a specific kind of waste is really problematic: high-level waste, or waste of category C. Although category C wastes are only 1 percent of the total volume of radioactive waste, they represent 95 percent of total radioactive waste activity of all categories. Depending on experts, the decrease of its radioactivity will be effective in 1,000 to 10,000 years. This kind of waste must be isolated from humans and their environment. If waste reprocessing permits limiting their quantity, this cannot be considered to be an optimal solution because of the incapability to reprocess all wastes. Moreover, waste reprocessing plants seem to have effects on their environment, as some studies show concerning, for example, the La Hague complex in France.

Technical characteristics and decision-making difficulties concerning wastes permit the introduction of opacity about decisions on nuclear issues. Since the first elaborations of atomic cells and bombs, "secret" seems to be a necessary step for each nuclear development. The effect of Chernobyl has been convincing people that public stakeholders are sometimes hiding facts. For example, French authorities have tried to persuade their citizens that the westward spread of the Chernobyl radioactive cloud stopped at the Franco-German border.

On a broader scale, Marie-Hélène Labbé indicates that, associated with its military use, the nuclear "big fear" is linked to four elements relative to nuclear essence: atoms' smallness, their invisibility, their almost unlimited lifetime, and the easiness of their propagation. Such characteristics call to mind other big fears such as biological terrorism. Experts consider that high-level accidents such as Chernobyl have a probability of 1/100,000 by year and by reactor (it was 1/10,000 20 years ago), and technical or human-induced incidents are counted each year. If they represent mostly limited dangers for the population and for workers, they constantly remind individuals of the possibility of a Chernobyl-like accident. Paradoxically, studies have shown that the nearer people live to a nuclear power plant, the less they are afraid of a probable accident. Françoise Zonabend says residents near nuclear plants live in "the land of the denied death," acting in everyday life as if the nuclear plant were not there and trying to never name it. In a similar way, nuclear station workers claim their risk-taking as an occupational hazard that reinforces their identity and their symbolic vision of their work.

All of this refers to the modernity dialectic: on one hand, postindustrial modernity pursues its complex scientific and technical development, trying to provide a better quality of life, while on the other hand, uncontrolled, unlimited, thoughtless progress and uncertainty engender risk, undesirable effects, and unexpected feedback. Nuclear power is also nuclear weakness. Preoccupation with prolongation of the life of nuclear reactors from 20 years to 40 years in spite of some recurring problems, or with the continued existence of Chernobyl-like reactors or other unsafe reactors (such as the Russian VVER 230, a Pressure-Water Reactor), which have to be secured, calls to mind these weaknesses and maintains the population's faith in its fear of nuclear energy.

Moreover, this fear is reactivated during nuclear waste transportation, with the help of antinuclear movements that draw media attention to this specific point. Nuclear waste transports between Germany and France, for example, regularly see protestors halt movement by chaining themselves to the railway tracks. One of nuclear wastes' main problems is its long lifetime. Although reprocessing enables the minimization of long-term radioactive waste (category C waste), the potential for radioactive pollution resulting from the remaining waste cannot be excluded. Two main solutions are advocated: (very) long-term stocking, planning on the difficulty of reprocessing this waste in the future and on the decrease of their radioactivity in about 10,000 years, depending on prediction. The U.S. debate on underground storage at Yucca Mountain comes under this kind of solution. Located approximately 90 miles northwest of Las Vegas in Nye County, Nevada, Yucca Mountain is a long-term repository for spent nuclear fuel and high-level radioactive waste, particularly from used fuel rods from nuclear reactors. President Barack Obama has proposed to eliminate federal funding for Yucca Mountain and investigate alternative solutions for nuclear waste management; however, the nearby Waste Isolation Pilot Plant in Carlsbad, New Mexico would remain a functioning long-term repository. Yucca Mountain and other places that stock nuclear waste are in remote or less desirable locations, chosen from the NIMBY (Not In My Backyard) mentality, and as such, the nuclear waste will likely remain rather than be transported elsewhere. That is why, in this perspective, wastes are only stocked for a few years and remain easily accessible when science and technology will be able to use them. However, such a vision is a promethean one: science will not necessarily be able to find a solution, and nuclear waste will be a preoccupation—if not an important problem—for the next generations.

Works of Hans Jonas may be taken into account: the German philosopher explains that we have to choose a responsibility principle coupled with a fear ethic, avoiding displacing our responsibility on the next generation. That is the main point that antinuclear movements criticize, because of the strong responsibility that falls to future generations.

Is Nuclear a Green, Sustainable, or Renewable Energy?

Green energies are generally characterized by their non-emission of carbon dioxide (CO_2), the primary greenhouse gas responsible for recent climate change, and their long-term resource capability. The nuclear lobby has profited from the new, big environmental problem since the 1990s—greenhouse gases—calling to mind its ecological aspects, its power, and its security, avoiding debates on wastes or plant incidents. Moreover, some NGOs would like to take into account in the carbon balance the transport of resources and waste, which is important. However, regarding the future of energy, most experts think only nuclear plants will be able to provide enough power for the third millennium.

Looking now in terms of sustainable energies, we introduce a social dimension that cannot be associated with nuclear energy. The anthropological fear evoked here and the responsibility that falls to future generations avoid any attempt to make nuclear energy sustainable. Nuclear capacity necessitates a high level of technology that is controlled only by specialist engineers and technicians and is impossible to develop at the local level that is preferred by sustainable principles. On a global scale, controversies about the potential use of nuclear power by countries such as Iran or Pakistan underline the considerable gap between already nuclearized countries and the rest of the world. There is also a significant amount of cement used in nuclear power plant construction and cement production is a significant source of carbon dioxide emissions.

Finally, the question most asked remains this one: Is nuclear energy a renewable energy or not? It is often put forward that there is huge resource of uranium, as well as of tritium, for future-generation reactors. However, using a strict definition, is it a renewable resource or an energy that is inexhaustible? Nuclear energy uses a limited resource, and there is no complete waste reprocessing. As uranium or other nuclear fuel remains not considered as a renewable resource, experts and NGOs generally consider nuclear power as nonrenewable.

The fascination that surrounds the nuclear question cannot erase difficulties in controlling its extremely complex mechanism nor erase problems too often neglected, such as the probability of accidents and waste management. But the actual energy consumption of mostly northern and rich countries does not permit us to go without nuclear power. However, as renewable and sustainable energies are more and more developed, one can see a departure. Even if energy policies are quite independent of such questions, the 2008 financial crisis calls to mind the better short-term profitability of these new energies, while nuclear power necessitates a long-term investment. If nuclear human experience is a relative progress, contemporary reflexivity and caution principles will probably encourage forsaking such fantastic power, even if it necessitates a significant reduction of energy consumption.

See Also: Chernobyl; Nonrenewable Energy Resources; Nuclear Proliferation; Power and Power Plants; Three Mile Island; Uranium; Yucca Mountain.

Further Readings

Bodansky, David. *Nuclear Energy: Principles, Practices and Prospects.* New York, NY: Springer-Verlag, 2008.

CRIIRAD, Commission de Recherche et d'Information Indépendantes sur la Radioactivité [Independent Research and Information Commission on Radioactivity]. http://www .Criirad.org (Accessed May 2009).

Generation IV International Forum (GIF). http://www.gen-4.org (Accessed May 2009).

Harper, Gavin D. J. *A Nuclear Waste: Nuclear Power, Climate Change and the Energy Crisis.* London: Zed Books, forthcoming.

International Atomic Energy Agency. *Energy, Electricity and Nuclear Power Estimates for the Period up to 2030.* Vienna, 2008. http://www.iaea.org (Accessed May 2009).

National Science Foundation. *Building a Sustainable Energy Future.* NSB-09-35 http://www .nsf.gov/nsb/publications/2009/comments_se_report.pdf (Accessed April 2009).

Philippe Boudes
University Paris West Nanterre la Défense

Nuclear Proliferation

The term *nuclear proliferation* refers to the acquisition of nuclear weapons or technologies and materials enabling the production of nuclear weapons by nations that have not possessed them previously. Horizontal proliferation refers to the spread of nuclear capabilities to new nations, whereas vertical proliferation refers to the expansion of existing nuclear arsenals. Proliferation can be quantitative (the total number of weapons possessed by a nation) or qualitative (acquisition of new types of weapons with new capabilities).

Although proliferation has historically been viewed as an activity of nations, new concerns are emerging about the potential acquisition of nuclear weapons by nonstate actors such as terrorist groups. From a green energy perspective, nuclear proliferation is a concern to the extent that nuclear power is considered a green energy source. The environmental effect and sustainability of nuclear power are matters of debate, but irrespective of those debates, there remain questions concerning the links between commercial nuclear power and nuclear proliferation. The potential for facilitating proliferation can be an impediment to the adoption, expansion, and international exchange of technologies, materials, and knowledge related to commercial nuclear power.

The development of nuclear weapons by the United States during World War II and their use in 1945 produced immediate concerns about the acquisition of similar weapons by other nations. The Soviet Union tested its first nuclear weapon in 1949, followed by England (1952), France (1960), and China (1964). Those nations, which became the five permanent members of the United Nations Security Council in 1946, were the first members of the so-called nuclear club. Their status as the five authorized nuclear weapons states was formally legitimated by the Treaty on the Nonproliferation of Nuclear Weapons, also known as the Nonproliferation Treaty or NPT, which opened for signature in 1968 and became effective in 1970. Most of the world's nations have signed the NPT, with the exceptions of India, Israel, and Pakistan. North Korea asserts that it withdrew from the treaty in 2003. India, Pakistan, and North Korea have tested nuclear weapons, and Israel is widely believed to possess them, although it has not acknowledged doing so. South Africa is believed to have possessed nuclear weapons in the past and to have later eliminated them. Libya is believed to have made progress toward developing nuclear weapons before agreeing to give up its effort. Questions surround the suspected efforts of a number of other nations including Syria.

Under the NPT, the five authorized nuclear weapons states are entitled to keep their weapons and associated infrastructures but have pledged to work toward their eventual elimination. Despite official claims and the practical difficulties surrounding disarmament efforts, many non-nuclear-weapons states and nongovernmental organizations have been critical of the nuclear states' commitments to disarmament. In exchange for their license to possess nuclear weapons, the authorized states are expected to share nuclear technologies, materials, and knowledge with other nations when doing so does not enable proliferation. In practice, however, it is often difficult to determine the proliferation potential of specific technologies and to obtain consensus regarding those determinations.

Article IV of the NPT asserts the "inalienable right of all the Parties to the Treaty to develop research, production and use of nuclear energy for peaceful purposes." Iran is an example of a nation that has repeatedly appealed to that statement as a warrant for its nuclear program, although its peaceful intentions are disputed. Article III of the NPT authorizes the International Atomic Energy Agency (IAEA), an independent international agency affiliated with the United Nations, to verify the non-nuclear-weapons states' compliance with their treaty obligations. At the same time, the IAEA serves as a vehicle for the promotion and international dissemination of nonweapons nuclear technologies. As both a promoter and regulator of nuclear technologies, the IAEA plays a complex and often conflicted role in the global nuclear system.

Many analysts maintain that the most difficult step in developing nuclear weapons is the acquisition of fissile materials, also called weapons-grade materials or special nuclear materials. Those materials are the fuel for atomic weapons such as the ones dropped on Hiroshima and Nagasaki and for the fission-driven primary stages of thermonuclear, or hydrogen, weapons. Weapons-grade materials include suitably enriched uranium, containing sufficient amounts of the fissile isotope uranium-235, and plutonium-239. Technologies for uranium enrichment include gas centrifuge systems, electromagnetic separation, and gaseous diffusion. All of these technologies require large industrial plants, advanced technology components, and specialized materials. Plutonium is produced in nuclear reactors, which can be configured to maximize plutonium production or power production or to balance those two missions. Reactors are also used to produce tritium—a radioactive isotope of hydrogen that boosts the power of fission weapons and provides part of the fuel for thermonuclear weapons. Uranium enrichment facilities and reactors are examples of dual-use technologies that can support peaceful nuclear energy programs as well as nuclear weapons programs; the United States cancelled its plutonium breeder reactor program in part because of its proliferation potential as a dual-use technology. Proliferation concerns also surround the U.S. debate regarding mixed oxide reactor fuels. Advocates claim that the use of mixed oxide fuels offers opportunities to dispose of excess plutonium, whereas opponents claim that it opens a path to broader circulation of plutonium and associated problems of control.

Questions surrounding proliferation involve the full nuclear fuel cycle—a series of processes including uranium mining and milling, conversion of uranium to forms suitable for processing, uranium enrichment, reactor fuel fabrication, reactor operation to produce plutonium, reprocessing of irradiated reactor fuel (also known as spent nuclear fuel) to extract plutonium, and storage and disposal of the products of all those operations. Self-sufficient nuclear weapons programs must master many of those steps to produce uranium weapons, and all of them to produce plutonium weapons. Alternatively, nations or groups may bypass some or all of those steps by acquiring materials or completed weapons from external sources. Accordingly, the IAEA maintains a system of standards, safeguards, regulations, and inspection regimes to monitor the activities of the nations that have signed the NPT.

Related questions surround the governmental and commercial activities making up the increasingly globalized nuclear economy. Beginning in the 1950s with the Atoms for Peace program and continuing through the Global Nuclear Energy Partnership initiated in 2006 and a cooperation agreement reached with India in 2008, the United States has shared nuclear energy technologies with other nations. The Department of Energy is the principal U.S. agency engaged in international nuclear programs. The European Atomic Energy Community and the Russian Federation Ministry for Atomic Energy facilitate nuclear cooperation among their respective member states. Transnational corporate vendors including AREVA, British Nuclear Fuels Limited, United States Enrichment Corporation, and URENCO promote and sell nuclear technologies in a global market subject to international controls. Episodes including the unauthorized international sharing of nuclear weapons technologies by the Pakistani scientist A. Q. Kahn and his associates, the acquisition of nuclear weapons capabilities by North Korea, and the controversies surrounding Iran's nuclear program illustrate the complex relationships between nuclear power and nuclear proliferation.

See Also: Department of Energy, U.S.; Energy Policy; Nuclear Power; Uranium.

Further Readings

Corera, Gordon. *Shopping for Bombs: Nuclear Proliferation, Global Insecurity, and the Rise and Fall of the A.Q. Khan Network*. New York: Oxford University Press, USA, 2009.

Diehl, Sarah J. and James Clay Moltz. *Nuclear Weapons and Nonproliferation: A Reference Book*. Santa Barbara, CA: ABC-CLIO, 2002.

International Atomic Energy Agency. "Information Circular #140: Text of the Treaty on the Non-proliferation of Nuclear Weapons." http://www.iaea.org/Publications/Documents/Infcircs/Others/infcirc140.pdf (Accessed February 2009).

Leventhal, Paul, et al., eds. *Nuclear Power and the Spread of Nuclear Weapons: Can We Have One without the Other?* Washington, D.C.: Brassey's, 2002.

Lodgaard, Sverre and Bremer Maerli. *Nuclear Proliferation and International Security (Routledge Global Security Studies)*. New York: Routledge, 2009.

William J. Kinsella
North Carolina State University

OFFSHORE DRILLING

The term *offshore drilling* commonly refers to both exploratory boreholes and final producing wells accessing such underwater resource reservoirs of oil and gas deposits. Oil and gas analysts concur that natural limits place a predictable ceiling on the Earth's aggregate quantity of oil and gas. Roughly one-half of the remaining undiscovered crude oil and natural gas lies under the oceans and seas. Compelled by harsh conditions and demands for environmental protection, safety, and cost control, offshore drilling relies on continuous innovations in technology, management systems, and legal regulations.

Offshore-related matters often are framed in terms of water depth: shallow water starts at the water's surface, deepwater begins at a depth of 1,500 feet (456 meters), and ultradeepwater offshore drilling exceeds 1 mile (1,600 meters) in water depth. At this time, the deepest ultradeepwater production wells operate in water depths of just over 7,500 ft. (2,250 meters); exploratory boreholes have been drilled beneath 10,000 feet (3,000 meters) of water. The foremost infrastructural requirement in offshore drilling is a platform or rig for basing equipment and labor. In typical shallow water depths to about 500 feet (150 meters), self-erecting jack-up platforms stand on the seabed. The tallest fixed standing platforms reach about 2,000 feet (610 meters) in height, most of which remains underwater. In some shallow

The Devil's Tower offshore drilling platform in the Gulf of Mexico, completed in 2003, is a floating platform anchored to the sea floor with suction pile anchors in 5,610 feet of water.

Source: iStockphoto.com

water, most deepwater, and all ultradeepwater, either drillships or semisubmersible structures serve as offshore drilling platforms. Both of these types of floating platforms can maintain precise locations via computer-aided dynamic positioning, but in production phases, semisubmersible platforms or floating production, storage, and offloading vessels are typically tethered to the seabed. In most locales, numerous production wells are drilled in extensive fields of offshore oil and/or gas. Such fields are populated by multiple wellheads and pumps connected by highly engineered cable and pipe that can reach more than a dozen miles (20 kilometers) to individual platforms; floating production, storage, and offloading vessels; or even land-based facilities.

Severe yet fragile conditions encountered in offshore drilling heighten the challenge of environmental protection throughout project lifetimes, from exploratory drilling to plugged and abandoned wellheads. Natural phenomena such as hurricanes and earthquakes threaten exploratory and production facilities, with spills and leaks being the most visible but not sole concerns. In arctic offshore settings, the sheer mechanical force of ice floes represents a major test for environmental protection, as do complicating factors of remote distances, lack of infrastructure, and frigid weather. Pushing the limits of human systems and technology increases probable contingencies and failures, which can result in ecological harm ranging from catastrophic blowouts to more subtle forms of degradation. For example, sonic seismology potentially stresses rare sea mammals, and most years witness the unplanned sinking of at least one platform or drillship, including those in transit. Even small spills can be locally catastrophic.

Between 2004 and 2008 in U.S. Gulf of Mexico waters, an aggregate 94 spills and leaks of oil or production chemicals occurred (50 barrels or more). Of these, 59 were caused by hurricanes. Drilling generates significant quantities of waste in the forms of discharged cuttings and over a dozen categories of industrial fluids including drilling mud, lubricants, and emulsifiers, which can contaminate the environment or resource reservoir. Among recent developments, environmental uncertainty surrounds the potential use of offshore reservoirs for carbon dioxide sequestration.

In light of these risks, in most jurisdictions, environmental regulations stipulate that offshore drilling projects must implement extensive controls and abatements. As a result, exploration activities have sought to mitigate the disturbance of sea mammals through measures such as underwater hydrophonic sensing and visual spotting from ship and sky. Emerging "smart" drilling systems combine computer-controlled drill-tip sensors and real-time pressure regulation of the borehole, which should dramatically reduce blowouts and damage to resource reservoirs. Separate cuttings barges have been employed to collect drilling waste, thereby protecting fragile coastal littorals. Newer-generation industrial fluids meet performance requirements while reducing likelihoods of contaminating resource reservoirs or surrounding environments. Gulf of Mexico hurricanes Katrina, Rita, and Ivan hastened efforts to plug and abandon unused offshore wells, thereby reducing risks of leaks and spills. Directional and extended-reach technologies enable drilling horizontally nearly 20,000 feet (6 kilometers), thus allowing onshore facilities to tap offshore fields and to overcome selected, although not all, environmental concerns in arctic settings.

Some of the most wide-reaching regulations regarding offshore drilling are by the United Nations Convention on the Law of the Sea (UNCLOS), whose original provisions began collecting member-state signatories in 1982 and became effective in 1994. UNCLOS has been ratified by 157 nations, with the United States and Venezuela being high-profile exceptions. Oil and gas, as well as undersea mining and fishing, were three primary concerns of the UNCLOS initiative. (Procedural codifications regarding the latter two industries became effective amendments in 1996 and 2001, respectively.) UNCLOS seeks to resolve an international hodgepodge of offshore territorial claims ranging from three miles to 200 miles and more. UNCLOS sets a new global standard by establishing an exclusive economic zone (EEZ) encompassing all waters, seabed, and subseabed up to 200 nautical miles (nm) beyond the breadth of each signatory nation's 12-nm territorial sea. This expansive realm for enterprise was offset by largely doing away with economic or territorial claims based on historical notions of the continental shelf, including those in the first (1958) United Nations Convention on the Law of the Sea. UNCLOS does allow coastal states with exceptional continental shelves to apply for relief, which if approved, extend their EEZs beyond the 200-nm baseline to either 350 nm or 100 nm from the 2,500-m (8,188-foot) depth, depending on demonstrable geographical and geological criteria. UNCLOS collects and redistributes a maximum 7 percent royalty on sales of any oil and gas recovered within these exceptions to the 200-nm EEZ baseline. U.S. offshore oil and gas claims remain based on historical notions of its outer continental shelf. Notably, these U.S. claims incorporate key UNCLOS definitions pertaining both to a 200-nm baseline as well as an extended zone of exclusive economic control.

See Also: Arctic National Wildlife Refuge; Caspian Sea; Energy Policy; Hubbert's Peak; Natural Gas; Oil; Oil Majors.

Further Readings

International Energy Agency. "2008 World Energy Outlook." http://www.worldenergy outlook.org/docs/weo2008/WEO2008_es_english.pdf (Accessed January 2009).

Offshore, 68/1–12 (January–December 2008). http://www.offshore-mag.com (Accessed January 2009).

United Nations Convention on the Law of the Sea. "Overview and Full Text." http://www .un.org/Depts/los/convention_agreements/convention_overview_convention.htm (Accessed January 2009).

U.S. Department of Energy, Energy Information Administration. http://www.eia.doe.gov (Accessed January 2009).

U.S. Department of Interior, Minerals Management Service. "Spills—Statistics and Summary 1996–2007." http://www.mms.gov/incidents/spills1996-2008.htm#2004-2007 (Accessed January 2009).

U.S. Geological Survey. "World Petroleum Assessment 2000 Description and Results." http://pubs.usgs.gov/dds/dds-060 (Accessed January 2009).

Hugh Deaner
University of Kentucky

Oil

The term *oil* encompasses substances that have similar physical properties rather than their chemical constitution. Besides viscosity, oils are typically immiscible with water and hydrophobic as well as miscible with other oils or lipophilic. Carbon- and silicon-based oils have analogous properties that vary principally according to the properties of their constitutive elements. Predominantly traceable to the actions or presence of various life forms, carbon-based oils may be found *in vivo* or in living tissues, in subterranean deposits or may be synthesized *in vitro*. Through photosynthesis diatoms produce oils indirectly or directly. At the end of their life, the same carbon-based life forms including diatoms, plants, and animals sometimes accumulate and fossilize in subterranean sedimentary rock formations. Over long periods of time under certain conditions of high pressure and temperature the fossilized carbon undergoes slow chemical reactions, which break down large, complex organic molecules into simpler, smaller hydrocarbon molecules. These hydrocarbons form fossil sources including the fossil fuels of natural gas, oil, and coal. In today's economy, oil in the form of petroleum is critical and ubiquitous. Synthetic oils including silicon oils have been developed for specialized purposes and may also be hydrocarbon free.

In earlier times, petroleum (Latin: *petra oleum*) or 'rock oil' may have seeped to the surface from subterranean locations. Crude petroleum is a complex substance including multiple hydrocarbon compounds in suspensions and in fluid states of various viscosities. When subjected to processes such as cracking and fractionation crude oil yields economically valuable substances. Under particular conditions of pressure and temperature fossil carbon is transformed into gaseous, liquid, and solid hydrocarbons. Organic materials including petroleum, coal, tar, methane, biomass and some polymers are carbon substances. Other solid carbon products are graphite, diamonds, carbines, fullerene, fibers of carbon and graphite, glassy carbons, carbon black, nanoparticles carbon, carbon nanotubes, and multiwall carbon spheres. Volatile hydrocarbons such as methane and alcohols are either liquid or gaseous.

Petroleum from fossil carbon is formed over long periods of time. In the early stages, the fossil deposit consisting mainly of larger (heavy) hydrocarbons has the thick, nearly solid consistency of asphalt. As large molecules breakdown, successively "lighter" hydrocarbons are produced. Lubricating oils, heating oils, gasoline and kerosene (jet fuel) are derived from thinned liquids. In the final stages, most or all of the petroleum is broken down further into very simple, light, gaseous molecules such as methane or natural gas.

Fossil deposits converted to today's fuels have been traced to the Carboniferous period of the Paleozoic Era (360 to 286 million years ago). Some coal deposits have been dated to the late Cretaceous Period (about 65 million years ago). Virtually no petroleum is found in rocks younger than one to two million years old. The rate at which geologic processes produce significant quantities of oil and gas is too slow to replace current consumption rates. Effectively, fossil oil and gas are nonrenewable energy sources.

Through industrial fractionation processes, crude oil is separated into its components including kerosene, gasoline, and tar. The hydrocarbon cracking process produces ethene, propylene, and other by-products through decomposition of heavy hydrocarbon. Diesel oil, a by-product of cracking, has 15–20 percent naphthalene, 20–25 percent methylnaphthalene, 17–20 percent dimethylnaphthalene, and 14–18 percent trimethylnaphthalene. These compounds are harmful to the quality of diesel oil and may be separated out by distillation. Bitumen or heavy oil, a form of petroleum with the consistency of tar may be found in mixtures of sand, clay, silt, and water popularly termed oil sands or tar sands and in oil shales.

Dynamic organizational, political, and socio-legal contexts make prospecting for oil a complex business. Laser induced fluorescence generating spectral 'signatures' for oils can be used to monitor their distribution in the field. Quantum electronics and applied laser spectroscopy used in remote sensing techniques can also provide reliable data for large volumes of the environment compared to typical point source analyses.

To extract oil from depleting underground reservoirs, inert gases are injected at high pressure to maintain a sufficient gradient. Injected gases mix with the oil and are forced out. Depending on pressure, temperature, and composition, several liquid and vapor phases can coexist. Phases rich in hydrocarbons are more economically attractive than others. The inert gases can be separated and reused in the extraction process. Carbon dioxide separation processes including engineered or specially formulated amine techniques are also used to enhance oil recovery applications in flue gases at industrial and domestic facilities.

Economic Implications

Hydrocarbons from fossil sources used for a wide variety of critical purposes are a feature of industrialized economies. Petroleum is the major feedstock for the chemical industry and essential to transportation and as fuel. Consequently, the availability of oil has been significant in defining social structures, events, and economic strength. Additionally, as existing fossil sources are depleted, the costs of extraction and recovery increase. Improved technology, economic recession, and reduced projected demand can extend the life of the fossil oil business but is insufficient to balance current consumption rates and geologic conversion rates from fossil deposits. Dependency on foreign supplies, vulnerability to price shocks and supply interruptions are continuing concerns.

The societal addiction to fossil fuels has consequences for redistributing carbon in the environment. Inertia in the current systems and adaptability limitations impede the implementation of non-fossil fuel energy systems.

Energy

In the last century, petroleum replaced plant biomass as the primary fuel. Oil alone is 39 percent of worldwide energy supply. Together, fossil fuels, oil, coal and gas fulfill 86 percent of total energy requirements. In 2006, imported and domestically produced oil in the EU alone was valued at about €250 billion ($367 billion) a year. By 2030, oil is projected to still be the single largest energy source and 33.8 percent of total energy consumption in the EU. Natural gas will account for 27.3 percent and coal 12.2 percent. At the same time, by 2030 U.S. domestic crude oil production is estimated to decline to 61 percent and in the North Sea oil production is projected to decline significantly.

The emerging global energy market features increased demand from China and India; instability on the supply side due to disputes in Iraq, Iran, Russia, Ukraine and parties who have oil interests in those regions; diminishing refinery capacity; shrinking spare capacity; reduced investment capabilities; recessionary pressures; limited credit facilities; and related economic realities. The attempt to control oil resources has been identified as a major reason for World War II.

The United States and the EU import the largest quantity of energy globally and have both advocated replacement of foreign suppliers with domestic ones but differed in their proposed solutions. The U.S. "National Energy Policy" plan (2001) emphasized accelerated development of domestic sources of oil. In its "Green Paper" (2000) the EU Commission advocated developing alternative energy sources and domestic alternatives to oil. Notwithstanding the U.S. and EU policies, it is unlikely that the goal of national self-sufficiency is achievable in a globalized oil market and improbable that any country or region can achieve energy security in isolation.

The policy direction of the EU and other jurisdictions along with environmental concern has raised interest in non-fossil derived fuels and chemicals. Research indicates ligno-cellulosic material from biomass (including oil palm stones, shells, and fibers, and coconut, almond and walnut shells) has a high potential as an alternative to the petroleum-based precursor for the production of carbon products.

Environment

Private enterprise, consumer preference and governmental regulatory regimes contribute to oil industry and environmental sustainability. Operation of the fossil oil economy has triggered social justice issues in some communities.

Producing petroleum generates a large amount of waste. Produced water or naturally occurring water and petroleum produced from wellbores are over 98 percent of the waste. Rigwash is spilled drilling fluid and cleaning fluids from the rig floor. Associated waste includes by-products of gas dehydration and sweetening treatment, oil-contaminated debris, filter media, and waste hydrocarbons. Petroleum refinery industries produce various gaseous, liquid, and solid wastes. For example, flue gases from catalyst regenerators and boilers/heaters contain SPM (coke and catalyst fines), SO_x, CO, and hydrocarbons. Wastewater from oil refineries is mainly sour water stripper condensate, contaminated process water, cooling tower blowdown, caustic wash water, desalter water, and oily cleaning water. Solid wastes from separation processes include spent filter clays used for lube oil and kerosene refining. Conversion process wastes originate from operation of the fluid catalytic cracker, hydrotreating operations, hydrocracking operations, and coking operations, and mainly spent catalysts from them.

Emerging initiatives can ameliorate the effects of oil usage. For example, automated systems can improve efficiency of oil use in buildings. The same oil industry, which has shaped the landscape, can remain an active participant in the restoration and development of spaces. For instance, the Kenneth Hahn State Recreation Area in Los Angeles is located in the Baldwin Hills area above the Inglewood Oilfield. The existence, availability, and development of the park have been closely linked to the production and operation of oil resources.

Remediation

Studies have investigated the treatment and remediation of environmental pollutants from oil and oil products. Algae and phytoplankton appear to access carbon by metabolizing organic pollutants from oil production. Green, red, diatomaceous algae and cyanobacteria have biodegraded some naphthenic acids from oil production processes. Other polycyclic aromatic hydrocarbons (PAHs) can also be degraded by light induced algal transformations. Leakages from fuel storage tanks and disposition of used motor oil have contaminated soil and groundwater. Soil humus adsorbs the components of the used oil according to the hydrocarbon properties of the soil. Plant species can phytoremediate soils according to particular microbial species' hydrocarbon degrading activity and carbon turnover in root formations.

The 2007 U.S. Air Force Center for Engineering and the Environment's *Final Protocol for In Situ Bioremediation of Chlorinated Solvents Using Edible Oil* recognizes an edible oil process, which can be used for the treatment of contaminated aquifers. When properly prepared and injected, the edible oil biodegrades slowly providing sufficient carbon for several years of reductive dechlorination. This single injection process has lower operation and maintenance costs than multiple injection processes, reduces longevity of the discharge source and contaminant mass, enhances natural attenuation, and controls dissolved plume migration.

Efficient bioconversion of palm oil by-products to green materials reduces wastage. Major palm oil by-products such as palm acid oil and palm kernel acid oil are used for animal feed. *Cupriavidus necator* H16 (ATCC 17699) (formerly *Ralstonia eutropha*) when fed these by-products as the sole carbon source have biologically converted or synthesized them to poly(3-hydroxybutyrate).

In Vivo Oils

Living tissue produces oils such as animal fats (used to make tallow, lard and butter), fish oils, and plant oils. Of total plant oils production, oil palm, soybeans, rapeseed and sunflower, together account for approximately 79 percent, about 14 percent is used chemically and 6 percent as feed material. According to the FAO, world production of plant oils increased between 1996 and 2006 by about 50 million metric tons to 127 million metric tons.

Fossil oil precursors such as ancient spores and planktonic algae are similar to plant oils' lipid-rich organic material. With few exceptions, such as the waxes of jojoba oil, plant oils consist almost entirely of triacylglycerol esters containing three fatty acids with Carbon chain lengths of 8 to 24 carbon atoms. Since plant triacylglycerols are chemically most similar to fossil oils, they have the greatest potential to replace existing feedstock. As renewable sources of high-value fatty acids, plant oils can provide for both the chemical and health-related industries. Cottonseed oil and sunflower oil are considered alternatives to fossil oils, diesel fuel and fuel oil.

In Vitro Oils

Engineered carbon and silicon oils may be analogous. The presence of silicon instead of carbon in chemical chains results in significantly different properties between analogous oils including flammability and heat transfer characteristics. Silicone oils may be used as diluents such as in silicone elastomers to obtain silicone rubber casts of respiratory systems and in processes to measure the intracellular amount of dissolved inorganic carbon in unicellular green algae *Chlamydomonas reinhardtii*.

Mixtures of silicon and carbon oils are used in specific consumer and industrial products. Silicon oils have replaced carbon oils in hydraulic fluids including brake fluid, diffusion pumps and in products such as spark plug boots in cars. As a hydrocarbon free substance, silicon oil is used for electrical insulation, and medical grade silicon oils have been used to treat vision problems. Halocarbon-based oils such as synthetic Polychlorotrifluoroethylene-based substances have been used as inert lubricants in chemically reactive environments. A typical friction modifier in an automotive engine lubricant is molybdenum dithiocarbamate, which may also contain zinc dialkyldithiophosphate in synthetic polyalphaolefin base oil.

Since oil in all its forms is crucial to all life globally, continuing attention will be essential for the foreseeable future.

See Also: Alternative Fuels; Automobiles; Combustion Engine; Energy Policy; Food Miles; Fossil Fuels; Greenhouse Gases; Petroviolence; Wood Energy.

Further Readings

Bahgat, Gawdat. "Europe's Energy Security: Challenges and Opportunities." *International Affairs*, v.82/ 5 (October 2006).
Byrne, Jason, et al. "The Park Made of Oil: Towards a Historical Political Ecology of the Kenneth Hahn State Recreation Area." *Local Environment*, v.12/2 (April 2007).
Dyer, John M., Sten Stymne, Allan G. Green, and Anders S. Carlsson. "High-Value Oils From Plants." *Plant Journal*, v.54/ 4 (May 2008).

Lester de Souza
Independent Scholar

OIL MAJORS

Oil majors is a term used to describe the leading investor-owned petroleum companies. The companies are joint-stock companies listed on the world's stock exchanges. As vertically integrated oil (and gas) companies, they do business in every stage of oil production from exploration to distribution at the service station pump. The stages include exploration, production, refining, trading, transportation, and marketing.

Their vertical integration is a management strategy that allows for ownership and control of as many aspects of the business as possible. The reality of business life for business managers is that, sooner or later, there are economic shocks to the economy that affect business. The shocks usually come as a surprise because of their totally unexpected nature, their location, historic timing, or some other factor. By creating vertical integration of their market share of the oil industry, the oil majors have sought to minimize losses resulting from economic shocks, whether caused by the weather, financial crises, wars, adverse political events driven by ideology, or some other factor.

Shell Oil Company is a major oil company that is an example of a company that has survived numerous shocks. It began in 1880 in the Dutch colony of Indonesia with the discovery of natural oil seepages. By 1885, the first well was producing, and it was operating as the Royal Dutch Company under a charter granted by the Dutch king William III. In 1907, Standard Oil and other oil companies were seeking to take over Royal Dutch. It was able to resist these efforts by merging with M. Samuel & Company of London, which owned a tanker fleet with open access to the British-controlled Suez Canal. Until the advent of the automobile, the oil business was the kerosene business, shipping kerosene in cans to consumers globally. Using tankers, Samuel could ship kerosene in bulk to be dispensed locally in cans. This type of vertical integration allowed the Royal Dutch Shell Company, formed by a merger with the Shell Transport and Trading Company, to master transportation costs. The merger of operations also gave it greater strength for competing with John D. Rockefeller's Standard Oil.

Shell celebrated its 100th birthday in 2007, having survived wars, depressions, nationalizations, competitors, and production and marketing challenges to be the second-largest oil company in the world. Today it has organized its vertical integration into five main businesses: exploration and production, gas and power, refining and marketing, chemicals, and trading/shipping. Its Shell Oil gasoline stations are seen throughout the United States and the world. Operating in over one hundred countries, it is vertically integrated to gain the maximum in economies of scale and efficiency in production and marketing.

Shell and other oil majors organize their subsidiary companies in a hierarchy. Each is producing a different product or service. The contributions of each subsidiary combine to satisfy the common goal of making profits for the company. Vertical integration is different from horizontal integration, in which a company may be the supplier of a product for many different companies across different market segments.

Vertical integration can produce a vertical monopoly, as occurred with Standard Oil Company. In the United States, it would then be vulnerable to antitrust action; however, in other countries it would be accepted as a cartel.

One problem that vertical integration can avoid is the "hold up" problem. The problem arises for smaller companies that must cooperate but are slow to do so because it may give the other an advantage that so empowers one company that it can exercise power over the other company with which it was seeking to cooperate for mutual benefit.

At their peak in the middle of the 20th century, the oil majors numbered seven and were called the "Seven Sisters." They were British Petroleum, Royal Dutch Shell (Shell), Texaco, Exxon, Mobil Corporation, Chevron, and Conoco-Phillips. In 2009 the Seven Sisters have been reduced to six super-majors through mergers in the 1990s: Exxon Mobil, Total S. A., British Petroleum, Shell, Chevron, and Conoco-Phillips; with the addition of Italian-owned Eni S.p.A., however, there are still seven.

As private, investor-owned companies, the oil majors are in stark contrast to the state-owned companies (national companies) in Russia, the Middle East, and Latin America. The biggest state-owned oil companies today are sometimes called the "Seven New Sisters." They are Saudi Aramco (Saudi Arabia), Gazprom (Russia), China National Petroleum Company (China), National Iranian Oil Company (Iran), Petroleos de Venezuela S. A. (Venezuela), Petrobras (Brazil), and Petronas (Malaysia). Only Brazil, among these state-owned oil companies, comes close to having a democracy. The others are run by authoritarian regimes.

The national oil companies are today the top 10 oil reserve holders. In contrast, the new super oil majors are well below the national oil companies in terms of their oil holdings. Many of the oil fields operated by the state-owned oil monopolies were originally discovered by the investor oil majors. Politics in the postcolonial era, and especially after the rise of Oil Producing and Exporting Countries following the 1973 war between Arabs and Israel, pushed the oil majors out in the name of nationalism or for other political reasons. Sometimes their assets were expropriated without sufficient or even any compensation for their investments.

Exxon Mobil is the world's largest publicly traded company in terms of revenue. It was formed November 30, 1999, by the merger of Exxon and Mobil. Both companies had been successors to parts of the Standard Oil Trust first created in 1870 by John D. Rockefeller. Standard Oil had been broken up as a monopoly by order of the U.S. Supreme Court in 1911. As with other major oil companies, production of crude oil is its lifeblood. At the beginning of 2008, it had proven reserves of 72 billion barrels of oil. Given current rates of production, these reserves are expected to be exhausted in 2022. However, successful exploration is expected to extend the pool of Exxon Mobil's reserves for additional decades.

In addition to oil exploration and production, Exxon Mobil also refines oil. The company currently operates 38 refineries in 21 countries with a daily capacity of 6.3 million barrels.

British Petroleum began just after 1900, with the discovery of oil in Iran. Because oil was seen by the British government as a strategic asset for its navy, it took a controlling interest in the company. Chevron began in California and, through a complex history with Standard Oil, Texaco, Gulf, and other companies, became one of the super majors. Conoco-Phillips was formed in 2002 with the merger of Conoco and Phillips Petroleum. Both had been among the larger oil companies before their merger.

Exxon Mobil and the other majors are often the targets of consumer complaints about gasoline prices and political accusations of exploitation by politicians. The reality is that today, its production is only 3 percent of global daily production. Its production is also exceeded by several of the state-owned petroleum monopolies. Its oil and gas reserves place it 14th in the world.

Some scientific organizations have attacked Exxon Mobil for not accepting their view of global warming. The major oil companies are much more visible targets in countries where market economies dominate with democratic governments than are the growing state monopoly companies in autocratic or authoritarian countries. In the latter, criticism of a state company wrong doing is unpatriotic at best and could be considered subversive or worse. This means that issues about economics, the environment, or energy supplies are not just economic or scientific—they are first and foremost political.

Other oil majors are threatened by efforts to deal with climate change through the imposition of taxes, which are hailed as controls on greenhouse gas emissions. The economic effect of cap-and-trade policies is as yet unknown, but these and other environmental worries—usually framed in ideological terms—may push all the oil majors into the alternative energy business still in the development stage.

Total S. A., an oil company based in Paris and listed on the New York Stock Exchange as TOT (Euronext: FP), is now one of the super-major oil companies. It engages in the businesses of a vertically integrated oil company from exploration to sales. It also has business interests in power generation, product trading, and chemical trading.

After World War I, French prime minister Raymond Poincare vetoed the proposal to form a partnership with Royal Dutch Shell. He elected instead to support the organization of a French oil company. Colonel Ernest Mercier organized the company (Compagnie française des pétroles) with the financial aid of 90 banks and other companies on March 28, 1924. Compagnie française des pétroles was seen as an economic defense strategy in the event of another war with Germany but was still an investor-owned company. Its original oil stake included the German share of the Iraqi oil fields ceded by Turkey and Germany as war reparations by the San Remo Conference.

In the years since World War II, the company has explored widely for oil in the Sahara and in other parts of formerly French colonial Africa. These are now areas of high political risk, but French investments have been in line with the long-term knowledge of these areas that it gained through its colonial experience. Attempts at exploration in other regions of the world such as South America have not been so successful.

The oil majors, especially the super-majors, are sometimes called "Big Oil," which is a pejorative term. It is used by people who are opposed to the majors' individual and collective economic power, believed by some to be a controlling factor in politics in the United States. Big Oil versus the consumer is also used (as in David versus Goliath) as an image for depicting the oil companies gouging consumers. Many people believe that the oil majors have the power to manipulate oil prices, profiteering at the expense of consumers in North America and Europe. Although the term *Big Oil* is used in the media, it is not used to describe the Oil Producing and Exporting Countries' manipulation of oil prices, nor the action of the state-owned oil companies for what are probably partisan or ideological reasons. The manipulation of state monopolies is ignored when they may have more to do with price gouging than the action of the investor-owned oil majors.

Because the oil majors are really in the energy business, all of them have considered alternative energy—some more than others. Although the fossil fuels of oil and gas have been their major moneymakers, there is the problem of an oil peak, at which time one-half of the world's oil supplies will have been consumed. The development of renewable energy sources such as ethanol is currently underway, with the introduction of ethanol into many brands of gasoline. However, because diverting corn from food to gasoline has had a negative effect on global food prices, it is expected that investigation of other alternative fuels such as methane, hydrogen, or other potential fuels will be part of the oil majors' research and development budgets for years to come.

See Also: Ethanol, Corn; *Exxon Valdez;* Greenhouse Gases; Oil; Organization of Petroleum Exporting Countries.

Further Readings

Bamberg, James. *British Petroleum and Global Oil 1950–1975.* Cambridge: Cambridge University Press, 2000.

Bamberg, James. *The History of the British Petroleum Company,* Vol. 2: *The Anglo-Iranian Years 1928–1954.* Cambridge: Cambridge University Press, 1994.

Cordesman, Anthony H. *Energy Developments in the Middle East.* Westport, CT: Praeger, 2004.

Ferrier, R. W. *The History of the British Petroleum Company,* Vol. 1: *The Developing Years 1901–1932.* Cambridge: Cambridge University Press, 1982.

Howard, Roger. *The Oil Hunters: Exploration and Espionage in the Middle East.* New York: Hambledon Continuum, 2008.

Petroleum Industry Research Foundation. *The Role of the Majors in Oil and Gas Exploration & Production.* New York: Petroleum Industry Research Foundation, 1989.

Roberts, Gwilym and David Fowler. *Built by Oil.* Reading, England: Ithaca Press, 1995.

Share, Jeffrey and Joseph Pratt, eds. *Oil Makers.* Houston, TX: Rice University Press, 1995.

Yergin, Daniel. *The Prize: The Epic Quest for Oil, Money, and Power.* New York: Touchstone, 1993.

Andrew J. Waskey
Dalton State College

OIL SANDS

In contrast to conventional forms of petroleum that include crude oil, unconventional sources of liquid petroleum include heavy oil, oil sands, and oil shale. Oil sands, also

This sample of oil sand, which is a combination of clay, sand, water, and bitumen, is a viable alternate source of fossil fuel and comes from the active mine site at Fort McMurray in Alberta, Canada.

Source: iStockphoto.com

known as tar sands, are technically called bitumen. Bitumen has the consistency of tar and can be processed into petroleum. Its occurrence in a mixture of sand, clay, silt, and water gives rise to the popular terms oil sands or tar sands.

A common method for extracting bitumen when deposits lie near the surface is strip mining or open pit mining. Current operating open-pit mines can be as large as 150 square kilometers and as deep as 90 meters. As is usual with strip mining, the area requires preparation and uncovering before the deposit can be acquired. Preparation involves clearing the forests, diverting watercourses, draining wetlands, and lowering the water table. Uncovering the overburden of soil, rocks, and silt then follows. To extract a barrel of bitumen, about four tons of overburden must be moved.

Since over 80 percent of oil sands are located too deep to be strip-mined, other techniques must be employed. An alternative to strip-mining is softening the bitumen with high-pressure steam and then pumping it to the surface. Under conventional recovery techniques, the mixture coming out of the mine is washed with hot water to wash the bitumen out of the sand mixture. After separation from the sand, the bitumen is upgraded to approximately synthetic crude oil quality by either adding hydrogen or removing carbon. For a refinery to process this synthetic crude from bitumen, it cannot be treated in the same manner as conventional liquid petroleum but must be specially configured.

Extracting the bitumen as currently practiced requires large amounts of water and energy in addition to the other environmental impacts from the mining process. The cost of accessing the oil from oil sands bitumen through conventional processes has been an impediment to commercial production. The first plant in Alberta, Canada, near bitumen deposit concentrations, started operations in 1967; the second unit opened in 1978, and the third in 1985. With increases in the price of conventional oil and global demand expected to continue for some time, there has been renewed interest and investment in the production from oil sands deposits.

In 2003, when the U.S. Department of Energy formally acknowledged reserves of crude bitumen as a source of oil, Canada's oil reserves placed it second only to Saudi Arabia globally. Elevated concerns for secure supplies of oil in the United States and Canada provided another incentive to bring the Canadian oil sands into production. U.S. domestic crude oil production has been declining, and estimates place foreign dependency at 61 percent by 2030. Canada is the largest oil supplier to the United States, and the oil sands are the obvious source of secure and stable future supplies.

With the capital-intensive processes required to commercialize production of oil from the bitumen, investors have considerable interest in the projects moving forward. This interest is met with similarly large concerns in opposition for reasons other than the financial return on the investment. One of these concerns the amount of energy required to extract the bitumen. In terms of energy, it takes about 1.3 million British thermal units (Btu) to produce a barrel of oil that yields 5.8 million Btu, resulting in an energy output to input ratio of 4.5. For some operations, the equivalent of one barrel of oil is required to produce three barrels of oil from the oil sands. By comparison, for a conventional oil well, the energy output to input ratio is between 11 and 17. This energy is used for various stages in the extraction and upgrading to synthetic crude quality. The bulk of the energy is derived from natural gas used in the heating phases of the production.

Other considerations that generate opposition include the damage perceived in open-pit mining at this scale, as well as the damage caused by the huge volumes of water and the discharges into the atmosphere. Displacement of the boreal forest necessary to access the bitumen, habitat destruction, loss of biodiversity, and human community displacement all serve to amplify the climate change effects of the oil industry. The United Nations Environment Programme has identified the oil sands mines as one of the top 100 spaces suffering from environmental degradation globally.

For the bitumen extraction operations in Alberta, Canada, the Alberta Environment agency estimates the area disturbed by oil sands mining is about 205 square miles. Although oil companies here are required to remediate or reclaim the disturbed land, it was only on March 19, 2008 that Syncrude Canada issued a press release announcing that it had received the industry's first land reclamation certificate. However, the impacts of production remain an ongoing concern and environmental disasters related to the production facilities have still occurred. For instance, in 2008 a flock of ducks was decimated when they landed in a tailings pond. Damage to nearby waterways, such as the Athabasca River, from industrial water use and waste has yet to be determined. Furthermore, industrial water exploitation remains high year-round, even during seasonal reduced waterflows and, hence, puts a strain on domestic and other commercial availability.

Clear-cutting of the boreal forest, which represents one-quarter of the world's remaining intact forests, negatively affects environmental oxygen and carbon levels. Wildlife affected includes bears, wolves, lynx, and the last large woodland caribou in the world. Waterfowl and songbirds rely on its wetlands and lakes. Roads, pipelines, and supply access to in situ drilling facilities result in fragmentation of the habitat and increase the edge effect zones.

Announced research on downstream remediation includes extracting minerals from oil sands tailings. It is proposed that such minerals can then be used in the manufacturing of products ranging from ceramic tiles and kitchenware to medical appliances and paints. The concept is to channel mine froth tailings into a separation plant to recover bitumen, titanium minerals, zircon, and naphtha. If such projects materialize, they may have environmental benefits including reduced carbon dioxide emissions and reduced disposal areas. Balancing out any potential gains are the real risks of environmental damage that would accompany the expanded transportation of production from the oil sands to the conterminous United States.

The recent economic crisis and reduced demand has resulted in a reversion to earlier market conditions and rendered oil sands less commercially viable. In the longer term, it is expected that consumers will prefer to pay more for continued access to the unconventional source of oil than to do without.

See Also: Oil; Oil Shale; Pipelines.

Further Readings

Allen, Erik W. "Process Water Treatment in Canada's Oil Sands Industry: I. Target Pollutants and Treatment Objectives." *Journal of Environmental Engineering & Science*, 7/2 (2008).

McGee, Bruce C. W. "Electro-thermal Pilot in the Athabasca Oil Sands: Theory Versus Performance." *World Oil*, v.229/11 (2008).

McGinnis, J. and E. Confrotte. "Oil Sands: Vision Resource for the Energy Industry." *Hydrocarbon Processing*, v.86/9 (2007).

Scales, Marilyn. "Growing for Black Gold." *Canadian Mining Journal*, 28/3 (2007).

Scales, Marilyn and Jane Weniuk. "Oil Sands Changing the Face of Canada." *Canadian Mining Journal*, 129/3 (2008).

Testa, Bridget Mintz. "Tar on Tap." *Mechanical Engineering*, 130/12 (2008).

Woynillowicz, Dan. "Tar Sands Fever!" *World Watch*, 20/5 (2007).

Lester de Souza
Independent Scholar

OIL SHALE

Oil provides about 40 percent of the world's energy needs and is vital for economic growth. It has also become one of the key global challenges in the 21st century. Economic development and global population rise have increased the demand for energy resources, and there are fears that available resources are becoming finite. This, coupled with the price volatility of conventional crude oil, has led to renewed interests in the exploration and extraction of unconventional petroleum resources like shale oil and oil sands. Oil shale is one such unconventional petroleum resource with a potential to supplement declining conventional oil production. Although it can yield substantial quantities of petroleum, high operating costs and adverse environmental effects have prevented its significant commercial exploitation. Efforts are ongoing within the industry to develop technologies that would address and reduce most of these concerns while remaining commercially and economically viable at the same time.

Oil shale refers to any sedimentary rock that contains solid bituminous materials (kerogen) that are released as petroleum-like liquids when the rock is heated. Kerogen is the first phase of geologic transformation from organic material to petroleum. The largest deposits of oil shale can be found in Wyoming, Utah, and Colorado in the United States. Estimates of the oil shale resource in these locations range from 1.5 to 1.8 trillion barrels. However, not all of this resource is recoverable. Other countries with equally large oil shale reserves are Brazil, Russia, Zaire, Australia, Canada, Italy, and China.

Kerogen is an organic matter that requires elevated temperature and pressure processing necessary to generate conventional light oil. It has a high hydrogen-to-carbon ratio, giving it the potential to be superior to heavy oil or coal as a source of liquid fuel. Although similar to heavy oils because they can be found near the surface up to depths of about 1,000 meters, they differ in their formation methods. Although heavy oils such as tar sands (or oil sands) originate from the bio-degradation of oil, heat, and pressure have not yet transformed the kerogen in oil shale into petroleum.

Two methods used by industry for producing shale oil are mining, followed by surface retorting, and in situ retorting. Retorting occurs in a vessel called a retort and is the process of heating oil shale to obtain oil from it. The two methods are as follows:

- Mining: This could be either underground mining using the room-and-pillar method or surface mining. At this time, both methods appear able to meet the requirements for the commercial development of oil shale. However, the commercial viability of the surface retorting technology is still questionable. These techniques, although simple, require expensive surface facilities and the disposal of vast quantities of spent rock, resulting in significant environmental issues.
- In situ retorting: This technique entails heating oil shale underground, extracting the oil, and transporting it to an upgrading or refining facility. It can potentially extract more oil from a given area of land than the mining and retort method, as it can access materials at greater depths than surface mines. As a result of its efficiency and minimal environmental impact, in situ retorting is now capturing industries' attention. At this time, Shell and Chevron are investing in the development and testing of in situ technologies. However, the technological, economic, and environmental viability of large-scale in situ retorting projects is still largely unknown.

Historically, shale oil has been exploited commercially for oil-based products and used in industry as fuel for the production of heat and power. Oil shale–fired power plants are already operating in Estonia, Israel, China, and Germany. Although not economically feasible, it has also been used as a substitute for natural gas. Also, the production of transportation fuels from unconventional (e.g., oil sands and oil shale) and renewable (e.g., biomass) energy resources has been generating worldwide interest over the last decade as a result of the price volatility of conventional crude oil from which transportation fuels are produced. By-products from shale oil production such as oil shale ash are being used in China as a raw material for the production of cement.

Although having the potential to yield substantial quantities of petroleum and supplement declining conventional oil production, the high operating costs and environmental impacts of using oil shale have prevented its significant commercial exploitation. The environmental impacts of its development include the following:

- Ecosystem displacement: A considerable amount of material mined and processed during surface retorting operations is usually piled above the ground, thereby creating an unnaturally elevated landscape and likely causing the displacement of preexisting flora and fauna.
- Groundwater contamination: Spent shale contains trace amounts of numerous heavy metals as well as salts. These could be leached into the soil and ground water aquifers by rainfall or snow unless proper waste management technology is implemented.
- Air pollution: Local air quality could be damaged by the particulate and gas emissions resulting from mining and retorting operations—creating serious health problems for residents in the locality. Also, because shale oil contains high levels of nitrogen and sulfur, its retorting and refining produces large quantities of nitrogen and sulfur oxides that could wash out of the atmosphere as acid rain.

Oil shale has the potential to contribute to the global energy fuel mix and also reduce the global reliance on conventional crude oil and its products. However, no matter which technological approach is used, oil shale development will definitely be accompanied by adverse land use and ecological impacts. The prospects for oil shale remain uncertain largely because of the major technical, management, and financial investments it requires; thus, in the immediate decades to come, conventional petroleum resources (e.g., oil) will continue to fulfill a vital role in bridging toward a global energy future.

See Also: Environmental Stewardship; Oil; Oil Sands.

Further Readings

Altun, N. E., et al. "Oil Shales in the World and Turkey; Reserves, Current Situation and Future Prospects: A Review." *Oil Shale,* 23/3 (2006).
Grunewald, Elliot. "Oil Shale and the Environmental Cost of Production." GP200A, http://srb.stanford.edu/nur/GP200A%20Papers/elliot_grunewald_paper.pdf (Accessed June 2006).
National Petroleum Council. "Oil Shales." http://www.npc.org/Study_Topic_Papers/27-TTG-Oil-Shales.pdf (Accessed July 2007).
RAND Corporation. *Oil Shale Development in the United States: Prospects and Policy Issues.* Washington, D.C.: National Energy Technology Laboratory of the U.S. Department of Energy, 2005.

Oyeshola Femi Kofoworola
University of Toronto

ON-SITE RENEWABLE ENERGY GENERATION

On-site renewable energy generation, is the use of renewable energy, such as solar and wind power, by a family, business, or organization onsite to provide for their own electrical energy requirements. Using renewable energy sources to generate electricity locally reduces reliance on the centralized utility grid for electricity, and thus reduces utility costs and pollution. Being that most utilities rely predominantly on the combustion of fossil fuels to produce electricity, decreasing reliance on the utility grid also reduces emissions of carbon dioxide and other greenhouse gases. Fossil fuel combustion for the generation of electricity, transportation, and industrial processes is the primary cause of air pollution and global warming. By generating electricity from on-site renewable energy sources, the demand for fossil fuels is reduced, and thus the concomitant smog, acid rain, climate destabilization, and pollution-related illness and death are also reduced. Throughout the world, many individuals, families, small and large companies, educational institutions, and government facilities now use on-site sources of renewable energy.

In addition to reductions of the environmental impact of energy use, on-site renewable generation possesses several other benefits. It can provide users with stable electricity costs, as these costs are dominated by the capital costs. When the source of on-site power generates more energy than is consumed on site, many regions allow the excess power to be returned to the electric grid for credit from the local electric utility (e.g., net metering). In addition, the use of on-site renewable energy can increase power reliability. In many parts of the world the grid is not reliable, and having a source of renewable energy greatly improves electrical operating time. Even in the industrialized world, an on-site power source is very useful for the rare occurrences of power outages resulting from our heavy reliance on electrical power to complete normal functions. On-site renewable energy generation also provides another technical advantage—decreased power transmission distances. At this time, industrial countries generate most of their electricity in large centralized facilities, such as coal-, nuclear-, hydro-, or gas-powered plants. These plants have excellent economies of scale but usually transmit electricity over long distances. The losses over these long distances can be substantial. In the United States, these losses average 6 percent, but in certain areas of the world it can be as high as 50 percent. Thus, the use of on-site renewable energy generation reduces the amount of energy lost in transmitting electricity because the electricity is generated very near where it is used. This geographically proximate production also has the tangential benefit to the network of reducing the size and number of power lines that must be constructed. Finally, many organizations choose to install on-site renewable energy systems to visually demonstrate their commitment to green power and overall sustainability. These organizations thus generate positive publicity and visibly demonstrate a civic commitment. This is becoming an increasingly important factor for businesses and universities, who are trying to attract youths, who tend to be more environmentally aware.

On-site renewable energy generation, because of its distributed nature, tends to have relatively low power outputs (a few kilowatts to a few megawatts) compared with the thousands of megawatts seen in commercial electricity production. These small-scale technologies, which organizations can use for generating green power on-site, include solar (photovoltaic panels), which converts sunlight into electricity. These systems are normally either built in or retrofitted on top of existing rooftops. The average home has more than enough roof surface area, if it is not shaded, to provide for the household's needs with photovoltaics. Because most of the world has access to reasonable sunlight, photovoltaic panels have the most potential to be widely adopted.

Also included are wind turbines, which convert wind into electricity. In general, these systems are not able to be as ubiquitous because not all regions of the world have high-enough sustained wind velocities to justify installation. These systems are more applicable to rural areas, although there are several companies working on small wind generators that can be integrated into buildings. Wind towers and generators have nontrivial insurable liabilities caused by high winds, but good operating safety. Wind also has the additional benefit of complementing solar; on days when there is no sun, there tend to be high winds, and vice versa. This encourages the design of hybrid systems.

If the source of hydrogen is green (e.g., wind or solar), fuel cells, which produce electricity and only by-products of water and heat, are another means of on-site green power. These systems can also be adapted to use the "waste" heat to benefit the building they are placed in—such as for space heating.

Microhydro technology also can be used to produce energy from moving water on small streams or rivers. On a small scale, it does not have the negatives associated with large-scale hydropower, but unfortunately it suffers from limited deployment because of the maintenance and availability of suitable waterways.

Finally, biomass combustion can also be used to produce on-site green electricity. Some facilities have also been demonstrated to work by using recovered methane gas from a landfill or from sewage treatment plants.

These technologies are often lumped together and referred to as distributed generation, on-site generation, dispersed generation, embedded generation, decentralized generation, decentralized energy, or distributed energy technologies. All of these technologies will promote an organization's or a facility's energy independence and improve fuel diversity, and thus security.

For the majority of homeowners, businesses, and organizations, the investment in an on-site renewable energy system requires a substantial capital investment. This is normally the critical factor that limits deployment of small-scale renewable energy systems. Recognizing the benefits of distributed renewable energy has enabled many governments to provide significant financial incentives to invest in such systems. In the United States, many states, as well as the federal government, provide both tax and funding incentives for on-site systems. The most comprehensive source on such incentives is the Database of State Incentives for Renewable Energy, provided free of charge on the Internet. It is a comprehensive source of information on state, local, utility, and federal incentives that promote renewable energy and energy efficiency. On-site renewable energy generation technologies, such as wind and solar power, are generally cost-stable and can provide and economic benefit in the long term. Particularly in regions where net metering is in place, renewable energy generator owners have the ability to spin their electric meters backward by offering excess electricity generated on-site back into the electric grid.

See Also: Combined Heat and Power; Feed-In Tariff; Geothermal Energy; Grid-Connected System; Metering; Photovoltaics (PV); Renewable Energies; Wind Turbine.

Further Readings

Clark, Woodrow W., II and Larry Eisenberg. "Agile Sustainable Communities: On-site Renewable Energy Generation." *Utilities Policy*, 16/4 (2008).
Database of State Incentives for Renewables and Efficiency. http://www.dsireusa.org/ (Accessed January 2009).
Pearce, Joshua. "Photovoltaics—A Path to Sustainable Futures." *Futures*, 34/7 (2002).
Pearce, Joshua M. and Paul J. Harris. "Reducing Greenhouse Gas Emissions by Inducing Energy Conservation and Distributed Generation From Elimination of Electric Utility Customer Charges." *Energy Policy*, 35 (2007).
U.S. Environmental Protection Agency. "Onsite Renewable Technologies." http://www.epa .gov/oaintrnt/energy/renewtech.htm (Accessed January 2009).
World Resources Institute. "Chapter 7: Planning an On-Site Renewable Generation Project." http://pdf.wri.org/guide_purchase_green_chap7.pdf (Accessed November 2008).

Joshua M. Pearce
Queen's University

ORGANIZATION OF PETROLEUM EXPORTING COUNTRIES

Formed in 1960, the Organization of Petroleum Exporting Countries (OPEC) is an international business association of oil-exporting countries formed to stabilize the petroleum market through unified petroleum policies of its member countries. At this time, OPEC consists of 12 major oil-producing countries—Angola, Ecuador, Gabon, Iran, Iraq, Kuwait, Libya, Nigeria, Qatar, Saudi Arabia, United Arab Emirates, and Venezuela;

Indonesia withdrew its membership in 2008 after becoming a net petroleum importer and following disagreement over recent production strategies. This assemblage controls nearly half of the crude oil and about a fifth of the natural gas traded internationally. OPEC membership is open to any nation that substantially exports oil and abides by the principles of the organization. Diplomats of OPEC member countries (or heads of delegation) meet periodically to manage their petroleum-marketing strategies, including pricing and output. During the first five years of operation, OPEC meetings took place at its headquarters in Geneva, Switzerland, but they were moved to its current location in Vienna, Austria, on September 1, 1965. The high demand for petroleum-based products by modern industrial society has established OPEC as a considerable economic and political power. However, the latest movement by more developed nations toward green energy development may weaken the future long-term position of OPEC in the international arena as postindustrial nations rely less on fossil-based fuels.

OPEC is one of the most lucrative commodity cartels in the world. A cartel is a formal agreement among producers to manage the prices and production of a particular good or service. This agreement may include stipulations on price setting, product distribution and output, market shares, profit sharing, customer share, and territory allocation, among others. In particular, the producers often limit product supply to raise prices above the competitive-market value. Cartels often arise in oligopolistic industries in which there are a limited number of sellers, with each member being aware and influenced by the market policies and actions of the other cartel members. Typically, cartels center on a narrow product line, such as petroleum and its derivatives (e.g., gasoline, natural gas, kerosene, jet fuel). The OPEC cartel manages 45 percent of the world's crude oil production, and forecasts show that this will increase to 50 percent or more within the next quarter of a century. Yet, in terms of known oil reserves, OPEC member countries possess approximately 78 percent of the world total of 1.2 billion barrels, with over half of this total owned by three Persian Gulf members—Saudi Arabia, Iran, and Iraq.

How It Began

Before the formation of OPEC, Middle Eastern and North African oil was controlled by foreign companies. In the wake of World War II, seven international corporations known as the "Seven Sisters" governed oil production, refinery, and distribution and were immensely profitable in an era of increasing oil consumption. These companies included British Petroleum, Royal Dutch Shell (Shell), Texaco, Exxon, Mobil Corporation, Chevron, and Conoco-Phillips, and collectively they diminished the market influence of developing-world competitors. OPEC was formed as a counter to the supremacy of the Seven Sisters.

During the Baghdad Conference of September 10–14, 1960, four Persian Gulf countries (Iran, Iraq, Kuwait, and Saudi Arabia) and Venezuela founded OPEC with the overriding goal of acquiring higher prices for Middle Eastern oil and that of other developing nations. The five founding members were later joined by an additional eight members over the years, with the last being Angola in 2007. The Baghdad Conference was prompted by the 1959 U.S. Mandatory Oil Import Quota (MOIP) program that restricted the foreign importation of petroleum products into the United States and favored non–Persian Gulf nations such as Mexico. As a result of MOIP, the market for Middle Eastern oil was diminished and prices dropped. The creation of OPEC did not immediately combat MOIP and Seven Sisters' pricing. It was not until the 1970s that OPEC gained international standing.

Starting in 1972, the oil-producing nations began to nationalize their petroleum industries and take greater control of global oil production and pricing. This power shift made a significant impact one year later when war erupted between Israel, supported by the United States and other European allies (e.g., the Netherlands), and the Arab nations of Syria and Egypt. In partial response to Western interference in the 1973 Arab–Israeli (or Yom Kippur) War, the Arab members of OPEC initiated a threefold-plus price increase in crude oil from October 19, 1973, to March 17, 1974, from a starting value of around $10 per barrel to more than $36 per barrel (in current U.S. dollars). Although commonly misidentified as an oil "embargo," OPEC did not generate price increases by just refusing to sell petroleum to Western nations but garnered market control by significantly reducing the supply of crude oil by five million barrels per day while petroleum demand remained high. In the United States, the increased crude oil prices were further exacerbated by the Economic Stabilization Program, implemented during the Nixon administration, which established guidelines for the allocation and consumption of petroleum-based products. The Economic Stabilization Program based the petroleum allocation on preembargo levels and did not account for a shift in fuel consumption patterns. The final result was an inequitable distribution of gasoline supplies, primarily a shortage in urban areas and a surplus in rural regions, which led to the U.S. energy crisis. The high U.S. dollar profit margins received by many oil-producing countries during this time, particularly the Persian Gulf members of OPEC, led Georgetown University economics professor Ibrahim Oweiss to coin the term "petrodollars."

Politically induced oil price fluctuations were not limited to the 1973 event. Middle East oil supply interruptions also occurred in 1979–80 following the Iranian Revolution and subsequent invasion by neighbor Iraq, which reduced worldwide production by 10 percent and more than doubled crude oil prices. Similar price fluctuations occurred as a partial consequence of the 1990–91 Persian Gulf War and the early-21st-century U.S.–Iraq conflict, which have undermined production in the region.

OPEC Today

Despite the early success of the cartel in the 1970s, OPEC failed to monopolize the petroleum energy market and exclusively control the world price of oil. At this time, crude oil prices are determined through trading on one of three chief international petroleum exchanges: the New York Mercantile Exchange, the International Petroleum Exchange in London, and the Singapore International Monetary Exchange. Dissension among OPEC members has undermined some of its successes. Iraq's production has been hampered by intermittent warfare with its neighbors and fellow OPEC members—Iran in the 1980s and Kuwait and Saudi Arabia in the 1990s. Members have also disagreed over production strategies. Countries with substantial reserves and low populations (e.g., Saudi Arabia, Kuwait) favor sustainable oil conservation policies, limiting production in favor of a long-term market share. In contrast, populous nations with a dearth of other major highly profitable exports (e.g., Nigeria, Iran) often advocate rapid production and higher prices. Technological innovations have made oil production viable in new regions (e.g., the Caspian Sea and the North Sea), increasing production in non-OPEC member nations (e.g., Russia, Georgia, Norway, Mexico) and threatening OPEC petroleum-market dominance. Ironically, much of the recent increased oil production has come from the Seven Sisters, who now trade nearly 40 percent of the global oil.

The Future

Energy conservation movements and green energy development may also jeopardize the future of OPEC. After 1973, the high price of oil stimulated the development and usage of alternative energy sources and energy reduction measures. Many Western nations have adopted daylight savings time and vehicle speed restrictions to reduce consumption. The development of new green energy technologies, such as more fuel-efficient vehicles, including hybrids and ethanol-blended fuels, and alternative power sources (e.g., geothermal, wind, solar, hydroelectric), will minimize reliance on fossil fuels (a "brown energy" source) for developed nations. Electricity-based oil consumption has also been significantly reduced by increased usage of nuclear power, natural gas, and coal. In particular, OPEC has expressed concern over "green taxes" increasingly levied on oil consumption, stating that heavy taxation imposes unnecessary limits on petroleum industry investment and creates instability in the oil industry. Despite these potential setbacks, OPEC may benefit significantly from a new set of consumers as energy-hungry nations (e.g., India and China) increasingly need petroleum to feed their expanding industrial enterprises. As such, OPEC will most likely remain a formidable participant in the world energy market even as green energy technologies become more readily available and used.

See Also: Alternative Energy; Caspian Sea; Fossil Fuels; Nonrenewable Energy Resources; Oil; Pipelines.

Further Readings

Amuzegar, Jahangir. *Managing the Oil Wealth: OPEC's Windfalls and Pitfalls.* New York: I. B. Tauris, 2001.

Falola, Toyin and Ann Genova. *The Politics of the Global Oil Industry: An Introduction.* Westport, CT: Praeger, 2005.

Organization of the Petroleum Exporting Countries. "What Is OPEC?" http://www.opec.org (Accessed January 2009).

Parra, Francisco. *Oil Politics: A Modern History of Petroleum.* New York: I. B. Tauris, 2004.

Jill S. M. Coleman
Ball State University

PETROVIOLENCE

The term *petroviolence* or *petro-violence*, popularized by University of California, Berkeley, geography professor Michael J. Watts, refers to the violence, environmental damage, and political corruption that often accompany the extraction of petroleum, particularly in developing countries. Because petroleum plays a vital role in the modern industrialized world, and also because it is a finite resource, it is a highly valued commodity. Therefore, it is not surprising that many people would want to reserve some of the profits associated with the petroleum industry for themselves, and in truth, violence, environmental devastation, and corruption have often accompanied petroleum extraction throughout the history of that industry. However, the term *petroviolence* is most commonly used in reference to Nigeria and other developing nations (from Asia to South America) with large petroleum reserves and a relatively short history of exploitation of those reserves, often by foreign companies.

Watts's Framing of the Problem

Watts's original paper identified eight properties of petroleum that he believed made the oil industry particularly prone to corruption and violence. He grounded this analysis in contemporary events taking place in Nigeria and Ecuador, while noting that they are applicable to many other countries as well. His eight points are as follows:

1. Oil is the most commercially negotiable of commodities, making it particularly attractive to those interested in profit.

2. Specific forms of social relations are required to exploit petroleum (joint ventures, concessions) that can confuse national and private interests.

3. Petroleum extraction is particularly attractive to multinational corporations and thus creates the temptation for governments to forfeit otherwise desirable qualities such as tradition, sovereignty, and independence to share in the wealth created by the oil industry.

4. Oil deposits exist in particular geographic areas (generally underground), and their economic value is literally lost to those areas as the oil is removed.

5. The overwhelming economic value of oil tends to increase state dependence on that one commodity and to make it the basis of political decisions.

6. The wealth created by oil may appear mythical or magical because it seems divorced from more familiar types of economic activity (such as agriculture or manufacturing), allowing people to become wealthy without effort while distorting local economics, and it is subject to an extreme boom-and-bust cycle, again apparently unrelated to any observable local cause (instead being most likely caused by economic activity on a global scale, which is thus outside the influence of the local worker).

7. The fact that oil exists in specific geographic areas within countries leads to conflicts between people living in those areas (e.g., the Ogoni people of the Niger delta) and the national government, including issues about who owns the oil and who should control and benefit from its extraction.

8. Oil is so valuable that it may come to dominate the national economy and reinforce certain patterns of power while weakening other parts of the economy such as agriculture that are relatively less profitable.

These factors make petroleum extraction particularly prone to the temptation to trample on the rights of individual citizens so that others may become rich. The best-known case may be that of the Ogoni people of Nigeria, a distinct ethnic group with a population of about half a million living on about 400 square miles of the Niger Delta. More than 10 major oil fields exist within their territory, with some of the largest wells being in or near the most densely settled areas. Petroleum has made Nigeria a relatively rich country in terms of gross domestic product (it is the second-largest in sub-Saharan Africa), but the benefits are distributed to only a small fraction of the population, and government corruption is rampant: Some estimate that over US$300 billion has been diverted by government officials since 1960.

Certainly the wealth created by oil does not benefit the people living in the areas where the oil exists. Although wells on Ogoni land regularly account for 40–60 percent of Nigeria's national petroleum revenues, the people living in the area remain extremely poor, receiving only a fraction (only 2 percent in the decade 1970–80) of the wealth thus created. Few households have electricity, educational levels are low, unemployment is high (few Ogoni are employed in the oil fields), and health outcomes are poor compared with the rest of Nigeria. Simultaneously, the Ogoni lands are subject to severe environmental damage from oil spills and gas flaring (the burning of natural gas that is an undesired byproduct of crude petroleum extraction; the carbon dioxide and methane thus produced make Nigeria a major contributor to global warming).

The influence of oil companies, Shell in particular, was so great in the 1990s that they were perceived almost as the local government, and requests for compensation were often addressed directly to the companies rather than to the Nigerian government. This confusion of private enterprise (as embodied by a Dutch/British multinational petroleum company) and the Nigerian national government escalated as the Ogoni become better organized and more articulate in their demands. The Nigerian military was sent to the area to support Shell, and some civil liberties, including the right to hold public gatherings and to organize for self-determination, were suspended. Some estimate that almost 2,000 Ogoni people were killed in the 1990s as a result of struggles regarding the petroleum industry in their lands.

Ken Saro-Wiwa

The individual most strongly identified with the fight against petroviolence is Ken Saro-Wiwa (Kenule Beeson Saro-Wiwa, 1941–95), a Kenyan author and environmental activist whose death in 1995 focused worldwide attention on the cause. Saro-Wiwa was a member of the Ogoni people, a minority group living in the Niger Delta whose homelands have suffered extreme environmental damage as a result of petroleum extraction by foreign companies, most notably the Dutch-British multinational Shell (aka Royal Dutch Shell), which is one of the largest companies in the world. Saro-Wiwa was a government administrator and wrote a popular television series as well as several novels and memoirs before becoming involved in the Movement for the Survival of the Ogoni People (MOSOP), an organization demanding, among other things, a fair share of the wealth created by oil extraction in their territory and restoration of lands damaged by the petroleum industry. MOSOP was also involved in prodemocracy agitation within Nigeria.

Saro-Wiwa was arrested and detained several times by the Nigerian government, beginning in 1992. In 1994, he was arrested along with eight other MOSOP leaders and was found guilty of charges of inciting violence and sentenced to death. The trial was notable for its corruption—several prosecution witnesses later admitted to accepting bribery for their testimony, and a representative of Shell Oil reportedly offered to free Saro-Wiwa if he would end protests against the company. However, his trial brought international publicity to the cause, and Saro-Wiwa was awarded the Right Livelihood Award (sometimes called the "alternative Nobel Prize") in 1994 and the Goldman Environmental Prize (for grassroots environmental activism) in 1995. Saro-Wiwa was executed by hanging in November 1995, provoking international outrage followed by sanctions against Nigeria, including expulsion from the Commonwealth of Nations (British Commonwealth).

Saro-Wiwa's family filed several lawsuits against Royal Dutch Shell in the United States under the Alien Tort Statute and Torture Victim Protection Act, charging Shell with numerous human rights violation including torture, arbitrary arrest and detention, and summary execution. A trial was scheduled in the U.S. District Court for the Southern District of New York in June 2009 but was settled out of court for US$15.5 million on June 9, 2009, a few days before the trial was to begin. Shell referred to the settlement as a humanitarian gesture to compensate Saro-Wiwa's family for their legal fees and other expenses but refused to admit any role in Saro-Wiwa's death. Some of the money would also be placed in a trust fund to benefit the Ogoni people, in recognition of the fact that the events in question took place on Ogoni land.

Central Asia and Latin America

Although Nigeria is the most famous example of the corrupting effect of petroleum extraction, there are many others. Not surprisingly, the lure of a highly valuable and finite commodity has also proven tempting to individuals within their own country who seek through intimidation, corruption, or other means to reserve the wealth gained from petroleum extraction for themselves rather than sharing it with the entire citizenry of the country. The term *petro-authoritarianism* was coined to describe the situation in which the temptation to limit a valuable commodity such as petroleum nudges a country away from democracy toward a totalitarian government structure, the better to limit the benefits from extracting those resources to just a few individuals.

Petro-authoritarianism was first used with particular reference to Russia and some central Asian nations that were formerly part of the Soviet Union. Political scientists noted that although some former Communist countries have successfully discarded their authoritarian pasts and adopted both democratic forms of government and modern, diverse economic systems, those with large reserves of petroleum and other energy resources often seem to be reverting back to totalitarian systems of government. For instance, many political economists believe that Russian policy is heavily influenced by special interests and that conglomerates such as Gazprom (the largest oil and gas company in the world and the largest company in Russia) wield undue influence in Russian matters of state. A similar relationship between a country's wealth in terms of energy resources on the one hand and political corruption and authoritarianism on the other has also been found among central Asian nations. For instance, the watchdog organization Freedom House rates Kyrgyzstan, Tajikistan, Uzbekistan, Kazakhstan, and Turkmenistan as among the least democratic and market-based among formerly socialist countries and lays the blame jointly on the availability of enormous energy wealth, the desire of a few powerful individuals to exploit this wealth for their own purposes, and the disinclination of Western democracies to protest too loudly because they require the energy resources that these countries can supply.

Several Latin American countries hold significant oil reserves, together accounting for about 8 percent of the total; the countries with the most significant resources include Venezuela, Brazil, and Mexico. As governments move to allow extraction of their petroleum reserves, often located in areas such as the Peruvian Amazon that were previously closed to development, indigenous groups living in those areas have begun to organize themselves and to protest the destruction of their homelands. Some protests have become violent, leading to the deaths of both indigenous peoples and police, and have served to organize the indigenous population as a power bloc within some countries to the degree that they have been credited for forcing governments out of office in Ecuador and Bolivia.

See Also: Environmental Stewardship; Fossil Fuels; Nonrenewable Energy Resources; Oil.

Further Readings

Ghazvinian, John. "The Curse of Oil." *The Virginia Quarterly Review* (Winter 2007). http://www.vqronline.org/articles/2007/winter/ghazvinian-curse-of-oil/ (Accessed June 2009).

Le Billon, Philippe, ed. *The Geopolitics of Resource Wars. Routledge Studies in Geopolitics.* London: Routledge, 2007.

McLuckie, Craig W. and Aubrey McPhail, eds. *Ken Saro-Wiwa: Writer and Political Activist.* Boulder, CO: Lynne Rienner, 2000.

Mouawad, Jad. "Shell to Pay $15.5 Million to Settle Nigerian Case." *The New York Times* (8 June 2009). http://www.nytimes.com/2009/06/09/business/global/09shell.html?scp=5&sq=%22ken%20saro-wiwa%22&st=cse (Accessed June 2009).

Okonta, Ike and Douglas Oronto. *Where Vultures Feast: Shell, Human Rights, and Oil in the Niger Delta.* San Francisco: Sierra Club Books, 2001.

Tynan, Deirdre. "Civil Society: Central Asia: Western Democracies Enable Petro-Authoritarianism—Report" (27 June 2008). http://www.eurasianet.org/departments/insight/articles/eav062708.shtml (Accessed June 2009).

Watts, Michael J. "Petro-Violence: Some Thoughts on Community, Extraction, and Political Ecology." Institute of International Studies, Berkeley, California, Workshop on Environmental Politica, Paper WP99-1-Watts (1999). http://repositories.cdlib.org/iis/bwep/WP99-1-Watts/ (Accessed June 2009).

Yergin, Daniel. *The Prize: The Epic Quest for Oil, Money and Power.* New York: Simon & Schuster, 1991.

Sarah Boslaugh
Washington University in St. Louis

Photovoltaics (PV)

Photovoltaics are a type of active solar energy system that transforms solar radiation into electricity through the photoelectric effect. Photovoltaics are considered a renewable energy system because their power source, the sun, is unlimited in terms of human applications. Photovoltaics are also green in the sense that they do not generate emissions during operations. Replacing electricity produced by fossil fuels with photoelectricity is a viable strategy for reducing carbon emissions, but needs to overcome high costs, the problem of intermittency (without storage its not available all the time), and transmission (its often generated away from where it would be consumed).

This close-up of a single photovoltaic cell reveals the fine wires that conduct electrical power in the cell.

Source: iStockphoto.com

Photovoltaic materials are made of several different types of semiconductor material that transmute light energy into electrical energy through a physical process called the photoelectric effect. Though this effect was noted by 19th-century researchers, it was not put into practicable use until the middle of the 20th century. Since then, an impressive number of technological advances have been made in the field, including an eight-fold increase in conversion efficiency, the minimization of requisite semiconductor material, the discovery and implementation of new types of photovoltaic material, and the incorporation of lenses and other devices to increase performance. Photovoltaics are far more versatile than any other electricity generator, and are in common use in small hand-held objects as well as utility-scale solar energy farms. Photovoltaic systems may stand alone, often tied to an energy storage system to power applications that do not coincide with diurnal production, or may be tied to a community or regional electricity distribution system, or grid (grid-tied system). Though photovoltaic manufacture and installed capacity increases each year, it has yet to achieve substantial market share with other forms of electricity production, which is dominated in much of the world by fossil fuel. This is due primarily to cost parity, though production capacity is a practical factor.

Photovoltaics are far more versatile than any other electricity generator, and are in common use in small hand-held objects as well as utility-scale solar energy farms. Photovoltaic systems may stand alone, often tied to an energy storage system to power applications that do not coincide with diurnal production, or may be tied to a community or regional electricity distribution system, or grid. Though photovoltaic manufacture and installed capacity increases each year, it has yet to achieve substantial market share with other forms of electricity production, which is dominated in much of the world by fossil fuel. This is due primarily to cost parity, though production capacity is a practical factor.

Types of Photovoltaics

Most photovoltaics are based on silicon, a semiconductor material that has the ability to propagate the photoelectric effect. Several other compounds have been implemented or identified as replacements for silicon in photovoltaic cells. Research into other materials is fueled to some degree by increasing the effectiveness of PV cells, but this exploration is primarily driven by cost and supply issues. Until recently, PV manufacturers relied on discards from the computer hardware manufacturers for high-grade silicon, but solar grade polysilicon manufacturing capacity is now growing. Reducing the quality or amount of silicon in PV cells potentially decreases the cost of photovoltaics.

Crystalline Silicon Photovoltaics (c-Si)

Most c-Si photovoltaic cells are composed of at least two layers of silicon, a semiconductor, one of which is slightly positively charged (lacking electrons) and one which is slightly negatively charged (some extra electrons). This weak charging is affected by "doping," or the introduction of small amounts of impurities into the crystalline structure of the silicon. Doping enhances the photoelectric effect—when the negatively charged (n-type, electron donor) layer is exposed to sunlight, it donates electrons to the positive (p-type, electron acceptor) side.

There are two types of c-Si photovoltaics: monocrystalline and polycrystalline. C-Si is the oldest type, or first generation, of PV. C-Si cells were produced as early as 1953; their first application was for powering spacecraft. Monocrystalline silicon is the highest quality type, in which all of the silicon is in a single lattice structure. The silicon is cast in cylindrical ingots—the ideal shape for silicon crystal. Polycrystalline PV is made from silicon that consists of many small crystalline structures rather than as a single crystal. The variation gives the appearance of these cells a noticeable texture. The main reason for the invention is to lower the material cost of the cells. Polycrystalline wafers are cut from square ingots, which slightly increase the photovoltaic surface of their panels and reduce waste during the production process. The result of this is that polycrystalline PV panels are cheaper to produce than monocrystalline cells, but have a lower conversion efficiency rate. Both types of crystalline silicon photovoltaics are composed of exceedingly thin wafers cut from two doped ingots and sandwiched together with a layer of insulative material between them. Thin wires are connected to the layers to conduct electrical flow. One such assembly is called a cell. Usually, some number of cells are connected and placed underneath glass to prevent breakage, encapsulated to protect the cells, and mounted in a frame. This is termed a *PV panel*. A number of panels may be connected together to form an array.

Amorphous Silicon (a-Si) Photovoltaics

A breakthrough in silicon processing, fueled primarily by a drive to decrease production costs of photovoltaics, led to a second generation of PV cells. There are several distinct types, all predicated on thin-film technology. Second generation cells have a thickness range of a few nanometers to tens of micrometers thick. Crystalline cells have an average thickness of 300 micrometers, so thin films use anywhere from 1/10th to 1/10,000th the light-sensitive material of first generation cells, a substantial material savings. Some thin film products also use less associated material like glass or panel bracing.

Amorphous silicon (a-Si) was the first such thin-film material, when it was discovered that non-crystalline silicon can be induced to produce electricity when exposed to light, though at a reduced efficiency. A-Si is an allotrope of crystalline silicon with no particular structure. It has slightly different light-sensitive properties. The practicable advantages of amorphous silicon are many. The thinness of the photovoltaic layer reduces the cost of the cells. Amorphous silicon is, moreover, easier to produce than crystalline wafers. A-Si cells are flexible instead of brittle, and so do not need to be stabilized by glass or metal framing. A-Si can be deposited on flexible materials such as plastic. Current photovoltaic applications include flexible cells that can be easily transported as well as modules that can be integrated into building surfaces.

Other Thin-Film Photovoltaics

Cadmium telluride and copper (Indium Gallium) selenide (CIGS) are two compounds that replace silicon as the photovoltaic material in thin-film applications. Both of these compounds are more efficient than amorphous silicon thin film, though they are still less efficient than crystalline silicon cells and, initially at least, were less expensive than silicon cells. These cells are heterogeneously composed, with several different layers of material that isolate the photovoltaic material or aid in light capture and retention.

Multi-Junction Photovoltaics

Relatively new research has revealed that PV material can be phased, or tuned, to slightly different parts of the electromagnetic spectrum. This is possible due to the varying properties of waves at the quantum level; namely, that different wavelengths contain higher or lower amounts of kinetic energy. Combining several layers of PV phased to different frequencies increases the output of the cell. A layer of a cell so tuned is called a junction; hence a cell with several such layers is called a multi-junction cell. Triple junction cells, for example, may be tuned to red, green, and blue solar wavelengths, maximizing the utilization of incoming light. Multi-junction cells are based on thin-film technology.

Though this approach does promote a marked increase in conversion efficiency, the cost of a multi-junction cell is quite a bit more expensive than single-junction cells. One solution to this problem is to couple multi-junction cells with concentrator technology (see Concentrated Photovoltaics below). The concentration of incident radiation reduces the total amount of required PV material. As of July 2007, a triple-junction cell made of thin films of a-Si and CIGS materials under concentration had an overall conversion efficiency of 42.8 percent.

Organic and Dye-Sensitized Photovoltaics

Dye-sensitized photovoltaics were first developed in the mid-1970s. These cells are based upon the semiconductor capacity of titanium dioxide (TiO_2) crystals. These crystals are saturated with a liquid electrolyte and suspended between two layers of conducting material. The photovoltaic layer is impregnated with chemical dyes of various colors (the best to date is black) designed to adsorb light. The highest efficiency of these cells as of 2001 was almost 11 percent. The main advantage of dye cells is their ease of manufacture.

Though well tested in laboratory settings, these cells have been slow to gain production capacity, due in large part to doubts about the stability of the cells in the field.

Organic cells replace silicon with complex organic compounds. Candidates for these types of cells include polymers such as polyphenelenevinylene (PPV). Organic PVs are relatively new and are based on organic LED technology; LEDs, which convert incoming electricity into photons, are essentially photovoltaics in reverse.

Organic PVs are considered a highly promising technology, as theoretically the cost of these cells would be far lower than silicon-based photovoltaics. Currently, however, organic PVs have the lowest conversion efficiency; the highest efficiency rating as of 2004 was 4.3 percent. This cell's performance was an almost twofold increase over the previous benchmark, indicating significant progress over a relatively short period of time.

Concentrated Photovoltaics

Certainly the largest impediment to the acceleration of the adoption of PVs is cost. One area of research seeks to reduce cost by increasing the conversion efficiency of the photovoltaic material through concentrating incident radiation. The idea is that limiting the amount of PV material, ostensibly the most expensive part of the assembly, will reduce the overall cost. Concentrated photovoltaics use focusing lenses or prisms to magnify incident solar radiation on the photovoltaic material. Concentrator assemblies come in a wide range of manifestations, and can magnify the sun's rays from two to several thousand times. As of May 2008, the highest concentration factor coupled with photovoltaics was 2,300 suns (times magnified). Though this did dramatically increase the amount of electrical output, it also illustrated the chief disadvantage of concentrated photovoltaics—excessive heat. The assembly noted above raised the temperature of the receiving PV material to 1,600 degrees Celsius—hot enough to melt steel. Highly concentrated PV assemblies require active heat removal systems to keep the PV material cool. PVs are subject to a degradation of performance when temperatures rise as little as 30 degrees above average ambient conditions. As a result, concentrating PV more than 10 suns begins to affect the efficiency of PV cells.

Typical PV Systems

Photovoltaic panels, or modules, are commonly grouped together into arrays, due to the fact that most panels have a relatively low maximum power output compared to typical demand. Photovoltaic panels may be connected together in a series or in parallel circuits. Due to the relative diffuseness of solar energy coupled with relatively low conversion efficiencies, PV arrays must be made as perpendicular to the sun as possible. In the Northern Hemisphere, fixed PVs usually face due south. The tilt of the array increases as latitude increases. Some PV arrays actively track the sun as it moves across the sky. PV systems require some associated electronic equipment, including charge controllers, which balance the power input of the various panels, and inverters, which are used to convert PV electricity, which is direct current (DC) to alternating current (AC), which is more often utilized. PV systems may include some type of energy storage, depending on application, or they may be tied to a secondary power source such as a regional or community power grid.

Applications

Photovoltaics are the most versatile form of electricity production available today. PV cells are used in an ever-increasing number of applications, from very small to utility-scale projects. One of the most promising aspects of PV systems is their potential application as distributed generating systems, allowing electricity consumers to operate "off-grid," meaning these consumers do not increase the demand on electricity-generating utilities. Increasing distributed energy systems is considered an effective means to decrease reliance on fossil fuel supplies.

Building-Scale Photovoltaic Arrays

These are mid-sized systems, with a rated output range of one to ten kilowatts (KW) of peak production. Building-scale systems can be sized to account for all or a portion of a building's energy needs, depending upon the amount of investment desired. Moreover, building-scale systems can be stand-alone, "off-grid" systems or be grid connected. The preponderance of PV systems are grid-tied systems, which use utility-generated electricity as a secondary source at night or during inclement weather. Stand-alone installations usually incorporate a battery storage system to provide power when demand does not coincide with production periods. Sizing storage capacity is dependent on amount of energy required.

Building-integrated photovoltaics (BIPVs) take the place of normative building materials, thereby reducing the capital cost of incorporating the system. A number of products have been developed that are infused with photovoltaic material, such as solar shingles, which are composed of flexible thin-film a-Si modules applied to a durable substrate. PV cells have also been laminated into glass and used in both vertical windows and skylights.

Very-Large-Scale Photovoltaic Systems (VLS-PV)

Though small-scale, distributed systems are a viable application of solar electricity, systems larger than 10 MW are attractive options for centralized energy providers such as governments and public and private utilities. There is some economy of scale related to associated equipment, wiring, and operation. Large-scale power plants are best suited to arid areas with a high annual number of clear days. The potential land area of deserts, equaling roughly one-third of all land on earth, is more than sufficient to provide for all energy needs; only 4 percent of desert areas would be sufficient. Several governments around the world, largely Germany, Japan, Spain, and the United States, have begun to preference photovoltaics as a utility-scale electricity provider. Approximately 20 other countries have at least one large-scale PV installation. Solar power plant initiatives in the United States are the result of action by state governments or particular utilities. As a result, the size and number of large-scale PV plants increases each year. As of September 2008, the largest PV power facility was located in Olmedilla, Spain. It had a peak production capacity of 60 MW, the rough equivalent of a medium-sized coal-fired turbine-generator. The largest PV station in the United States is located at Nellis Air Force Base in Nevada. It utilizes 70,000 PV panels to produce 14 MW of power. The Nellis facility uses solar tracking devices to maximize power production.

Cost and Effectiveness

Overall cost of electricity produced from PVs is the main impediment to widespread use. In many, but not all locations, fossil fuel–produced electricity is less expensive than PV-produced electricity. This condition is due as much to economic and political backing of fossil fuel–based utilities as absolute cost. As PVs use no fuel, the cost per watt produced is essentially amortized capital costs; the traditional period of amortization is 20 years, which is the normally ascribed warranty period for commercially available PVs. Fossil fuel power plants can amortize their construction costs over a 50-year period, and in many cases the construction of these plants is heavily subsidized. Currently, PV arrays cost US$6–$10 per watt of installed capacity. Coal power plants cost a little less than US$3 per watt of capacity, excluding fuel costs. It is anticipated that fossil fuel costs will continue to rise as supply diminishes and as transportation of fuel becomes more expensive, while the cost of photovoltaics continues to drop as new technologies mature. The National Renewable Energy Laboratory projects that the point at which these two trend lines cross will occur in the year 2015; at this point PV-produced electricity will be no more expensive than fossil-fueled electricity.

Conversion Efficiency Comparison

Historically, monocrystalline silicon cells have had the highest conversion efficiency, reaching a laboratory rating of 23.4 percent. As of January 2008, the highest efficiency PV on the market was a monocrystalline cell with a conversion efficiency of 19.3 percent. Polycrystalline cells have a laboratory-documented efficiency rating of about 18 percent. A-Si photovoltaics max out at around 11 percent efficiency, while copper-compound thin films rival polycrystalline silicon efficiency. Dye (8 percent) and organic-based cells (5 percent) are least efficient, though these are far younger technologies. The most impressive gains in conversion efficiencies are concentrated multi-junction cells. As of June 2007, a triple-junction cell under concentration achieved a laboratory tested conversion efficiency of 42.8 percent. Increasing conversion efficiency is one avenue of exploration to reduce the requisite amount of PV material required for applications, commensurately reducing the size, weight, and theoretically the cost of photovoltaic systems. By comparison, coal-fired power plants have a conversion efficiency of 32 percent; roughly 60 percent of the energy released by burning coal is lost as waste heat, and 8 percent is lost through distribution. Natural gas power plants have a higher efficiency rating, though natural gas is quite a bit more expensive fuel; in fact, PV electricity is on par in many places with natural gas–produced electricity.

See Also: Energy Storage; Green Power; Renewable Energies; Solar Concentrator; Solar Energy.

Further Readings

Borbely, A-M. and J. Kreider. *Distributed Generation: The Power Paradigm for the New Millennium*. Boca Raton, FL: CRC Press, 2001.
Hamakawa, Y., ed. *Thin-Film Solar Cells: Next Generation Photovoltaics and Its Applications*. New York: Springer Verlag, 2004.
"IBM Research Unveils Breakthrough in Solar Farm Technology." http://www-03.ibm.com/press/us/en/pressrelease/24203.wss (Accessed January 2009).
Kaltschmitt, M, W. Streicher, and A. Wiese, eds. *Renewable Energy: Technology, Economics and Environment*. New York: Springer Verlag, 2007.
Krauter, S. *Solar Electric Power Generation: Photovoltaic Energy Systems*. New York: Springer Verlag, 2006.

Kurokawa, K., ed. *Energy From the Desert: Feasibility of Very Large Scale Photovoltaic Power Generation Systems*. London: James & James Ltd., 2006.

Stein, B., J. Reynolds, W. Grondzik, and A. Kwok. *Mechanical and Electrical Equipment for Buildings,* 10th ed. New York: John Wiley and Sons, 2005.

"UD-Led Team Sets Solar Cell Record, Joins DuPont on $100 Million Project." http://www.udel.edu/PR/UDaily/2008/jul/solar072307.html (Accessed January 2009).

R. Todd Gabbard
Kansas State University

PIPELINES

Pipelines are systems for the transportation of materials. According to particular circumstances, these may be simple or complex structures. Relevant design considerations include the materials to be transported and the applicable environments and the materials selected for use in the structure of the pipeline as well as associated economics including costs and risks.

At this time, pipelines are components of utility networks, resource extraction facilities and means of delivery of products and services critical to communities. Energy pipelines are part of the infrastructure used by economic agents to transport fuel substances in gaseous or fluid bases such as natural gas or oil for various activities. Gas transported through pipelines may be either in gaseous or liquefied forms of petroleum gases. Although electricity is energy, wires or cables used to transport electricity are an acknowledged part of the infrastructure but have not usually been included in references to pipelines.

Pipelines may be short tubes within a facility or, as in the case

Welders repairing a gas pipeline. Pipeline degradation must be detected early to prevent significant losses and environmental damage.

Source: iStockphoto.com

of natural gas, oil, and water, may be hundreds of kilometers or miles long and part of a complex network of pipelines that may have diameters of up to a meter. For example, the Kern River gas transmission pipeline in the United States originates in Wyoming and travels through Colorado and Nevada before terminating in California, a distance of more than 1,600 miles (2,400 kilometers) with a capacity of two billion cubic feet of gas per day in winter. Due to the effect of temperature on the expansion of materials, warmer weather brings pipeline expansion and greater capacity. Varying seasons and pipelines of extended length, configuration, and diverse materials routed through or over land and water traversing multiple complex environments en route are necessarily also complex structures.

The designed durability of pipelines is an additional complicating factor. It is not uncommon for parts of water distribution and sewage pipelines to intentionally last 100 years or more. In parts of Europe, water distribution systems installed by the Romans are still used 2,000 years later. Changes in the physical environment over extended lifetimes can have a dramatic effect on the pipelines. Changes in political, economic, and social environments of societies that control the pipeline systems can also affect affordability, adaptability, and efficiency.

Pipeline facilities that convey utilities to human development and societies are often determinative of the distribution and characteristics of communities. According to particular circumstances, municipal utility pipelines may either precede or succeed the construction of human settlements. Water and sewage pipes are critical to the health of communities. Natural gas and oil pipelines may be required for domestic distribution of fuels as well as for international trade.

All uses of pipelines bear benefits and costs with impacts on environment and communities through which pipelines exist. As part of other facilities, pipelines may be built or modified to enable certification to regulatory or independent standards such as the ISO-14001 series. As transport systems for energy, pipelines effectively redistribute carbon and other emissions into the environment.

Growing Complexity

Pipelines as part of larger networks in a changing energy economy may be expected to get more complex. Networked pipelines may be used in a distributed dynamic energy economy to establish linkages with increasingly diverse sources of energy and varying demand and supply profiles in a "smarter" infrastructure. These systems could be used to enhance energy efficiency and comply with environmental requirements.

As part of resource extraction facilities, pipelines are included in large-scale infrastructure. They support production activities and are components of large socio-technical systems. In this function, they are essential parts of systems that contribute to the safety of people and businesses against terrorism and environmental forces. Communities and states rely on these facilities to operate and grow. Public access can also be enhanced or redistributed by adjusting the configuration and location of pipelines. The distributions achieved in this manner affect societal equity and public welfare. As managed facilities and structures, the thinking and understanding of decision makers guides the renewal, development, and operations of pipelines.

Economically, pipelines may be treated as an asset class in their own right. As large, capital-intensive assets, the return on investment is over an extended period. Over the life of the structure, there is increased risk of changes in policy, technology, and market demands that can change values. Some of this risk can be mitigated by regulatory provisions or agreements. Established pipeline systems have been economically efficient at a time of low market volatility, increasing demand, and few restrictions on a stable supply. With market volatility caused by various factors, challenges to energy security and stable fuel supplies and financing issues, affordability has become a challenge even while energy has been recognized as a necessity of life in today's society. Additional complications arise if environmental concerns are acknowledged, resulting in fundamental changes to the energy economy.

Pipelines may corrode, suffer from blockages, or leak from various causes. Rupture of the pipelines for any reason would result in emissions into the environment with economic and regulatory consequences. The composition of the transported substances may include characteristics that could increase risk of corrosion and blockages. For instance, a natural gas pipeline could suffer from a blockage as a result of the condensation of water out of the gas in the pipeline itself. Tunneling below buried pipelines can result in varying amounts of settlement-induced stress on the pipes. Cutting trenches in rock produces vibrations that could cause damage to pipeline and other structures in the vicinity.

As with much of infrastructure, pipelines play a less visible but no less central role in the daily lives of people locally and globally. In this sense, pipelines are society's vital organs in much the same sense that vessels in the human body are critical to a person's health and life. Thus the resilience of pipelines, as with vessels in the human body, is critical to the ongoing life of societies. Redundancies in the systems are necessary to accommodate the dynamics and volatility of life.

In regions with obsolete infrastructure and transportation difficulties, structural impediments to the flow of materials through pipelines may operate to control participation in the economy and related consequences. In a 2005 Infrastructure Report Card, the American Society of Civil Engineers identified a loss of 222 million gallons of drinking water per day due to leaking pipes in California alone. In the United States, the Environmental Protection Agency (EPA) estimates an increase of 45 percent for the urgent repair of sewers by 2020 and legislation includes plans for $11 billion a year for trenchless construction alone.

Detection and repair of pipelines may involve observation or other methods requiring complex instrumentation. Detection of pipeline degradation is crucial to their reliability and usefulness. Physical inspection or technological solutions as well as numerical models such as probability of detection (POD) curves may be used individually or in combination to improve efficiency of detection of such degradation in an economically viable manner. In addition to routine degradation, disasters as seen in New Orleans, Louisiana, can overwhelm pipelines.

Accessibility and Security

With global energy production, transmission, and use expected to double by 2030, accessibility and security are also increasingly relevant considerations. Modifications to the energy configuration and distribution systems imply a corresponding shift in pipeline deployment and operations. In a privatized energy market, transmission systems are identified and controlled by independent operators with attendant costs and profit interests who are subject to additional to antimonopoly or antitrust regulations.

Market inefficiencies may also affect pipelines that have been designed to be stable and durable. By design, pipelines themselves and ownership may be more suited to noncompetitive markets or natural monopolies. At the same time, in the recent volatile economic circumstances, the ability to operate infrastructure assets like pipelines under long-term agreements may be reduced. The dynamics of global warming are only likely to amplify risks and reduce stable long-term return on investments. On the other hand, to some, infrastructure investment has become highly attractive due to the monopoly-like market position and inflation-proof income potential.

Pipeline capacity and availability can influence markets by varying supply and demand ratios. Factors that affect these asset values include the continuity of linkages to markets, unplanned outages, maintenance, seasonal consumption swings, and weather-related supply disruptions.

Globalization and global warming may combine to elevate the rate of change and modify the viability of pipelines. Designing future pipelines to include flexibility to accommodate change may involve innovative approaches. Recognized options include providing for foreseeable future possibilities; applying principles of modular construction; using trenchless technologies to lay pipes inside existing conduits; designing adaptive structures; and including intergenerational compatibility by involving new architectures that minimize effort for deployment of future systems.

Alternative approaches to pipeline installations in the future could include additional complex analyses such as evaluating systems in terms of costs per unit of energy; costs resulting from co-producing multiple products or uses; and distributed production with integration of multiple forms of energy from diverse sources. Deployment of facilities that can convert or change one form of energy into another will inevitably affect pipeline configuration and installation. The introduction of carbon emissions trading will likely heighten awareness of pipelines in the overall facilities utilization, engineering structure, operations, and maintenance. When sustainability is added to the analysis, pipelines as installations and transport vehicles for energy will reasonably be subject to technological and social modification and new architectures in as yet undetermined ways.

Conceivably, future fuels may include hydrogen or other gases and may also be sent through pipelines that have yet to be devised and installed. Pipelines for the transport of hydrogen may not be commercially viable unless they can overcome vulnerabilities in an economically reasonable manner.

See Also: Biodiesel; Biogas Digester; Biomass Energy; Carbon Emissions Factors; Emissions Trading; Ethanol, Corn; Ethanol, Sugarcane; Gasohol; Green Energy Certification Schemes; Hydrogen; Internal Energy Market; Microhydro Power; Natural Gas; Nonrenewable Energy Resources; Oil; Oil Sands; Oil Shale; On-Site Renewable Energy Generation; Public Utilities; Waste Incineration.

Further Readings

Andrews, Clinton J. "Energy Conversion Goes Local: Implications for Planners." *Journal of the American Planning Association*, 74/2:231–54 (Spring 2008).

Gil, Nuno and Sara Beckman. "Introduction: Infrastructure Meets Business: Building New Bridges, Mending Old Ones." *California Management Review*, 51/2:6–29 (Winter 2009).

Klar, Assaf, Alec M. Marshall, Kenichi Soga, and Robert J. Mair. "Tunneling Effects on Jointed Pipelines." *Canadian Geotechnical Journal*, 45/1 (January 2008).

Lester de Souza
Independent Scholar

Plug-In Hybrid

The coming decline in oil supplies as well as concerns about global warming demand vehicles that use much less gasoline or diesel than conventional vehicles. Plug-in hybrids can be powered either by an internal combustion engine or by batteries feeding electric motors. The batteries can be charged either by the engine when the vehicle is moving or by an external electrical source when it is parked. Since most trips are shorter than the electric-only range of proposed hybrids, the gasoline engine would be used relatively rarely. The technology is an interim solution between the historically plentiful supply of fossil fuels and the future when oil will become increasingly scarce, expensive, and eventually unavailable. At that point, electric power for charging vehicles will have to come from renewable sources.

Automobiles have been mass-produced for about a century, since Henry Ford's Model T appeared in 1908. This first mass-produced vehicle was also the first flexible-fuel vehicle, designed to operate on ethanol, gasoline, or both. At that time gasoline, steam, and electric vehicles were on the roads, but eventually the gasoline internal combustion engine displaced all of its competitors. The engines were compact and powerful, the vehicles had a long range because of the high energy density of gasoline, and the fuel became inexpensive and widely available.

The era of inexpensive fossil fuels is coming to an end, however. After more than a century of ever-increasing consumption of oil, the world is pumping 85 million

This view of a Toyota Prius' hybrid engine shows the electric and gasoline engines side by side under the hood. A plug-in version of the Prius is in development.

Source: iStockphoto.com

barrels from the ground every day, about 1,000 barrels per second, a rate that cannot be sustained much longer. The world has already extracted about half of the recoverable oil that was originally in the ground, an energy source that ultimately came from decay of animals and plants over millions of years. Plentiful fossil fuels have enabled the Earth's population to quadruple during the last century, something that has never happened before and will never happen again. Those people depend on vast supplies of fossil fuels because highly mechanized and productive modern farming, transportation, and industry require those fuels. The Earth is of finite size, though, and it has been well explored for oil already. All of the major oil fields are in decline, but new ones are not being discovered fast enough to make up for the losses. The coming decline will be faster than the rise because the population, and therefore the demand, is so much larger. The amount of oil will decline every year until it is no longer practical to burn it as a vehicle fuel. Global warming demands that the world reduce its fossil fuel use in any case, to reduce the amount of the greenhouse gas carbon dioxide released into the atmosphere.

In this situation it is imperative that more efficient modes of transportation be developed. For the short term of a few years to a decade or so, the transportation infrastructure will not change very much because replacing the fleet of 230 million vehicles takes about

20 years. America is saddled with an infrastructure that demands huge amounts of transportation—sprawling low-density cities with widely dispersed centers of employment. Sustainable alternatives such as rebuilding cities to be compact and walkable, together with electric mass transit powered by renewable sources, are decades in the future. In the meantime, conserving the fossil fuel supply is the only alternative, and this is where more efficient vehicles come in. Conventional internal combustion engines are very inefficient, converting only about 20 percent of their fuel's energy into motive power. Then they are used in vehicles that are usually only 20 percent occupied. Diesels provide better economy because of their superior thermodynamics, but they are more expensive to manufacture and the majority of the energy that goes into them is still lost as heat.

It is important to distinguish between fuel efficiency and fuel economy. Efficiency is a measure of how much of the energy value of a vehicle's fuel reaches the wheels, while economy measures how much fuel is required to move the vehicle over a given distance. The North American transportation system, based on private vehicles, ranks poorly on both measures. Efficiency is limited by the internal combustion engine. Fuel economy is reduced by the large size of American vehicles, compared to vehicles used for private transport in the rest of the world, and it is further reduced in small trucks and SUVs (sport utility vehicles) by excessive weight and poor streamlining. Most of these vehicles are used only for the same transportation functions as standard automobiles. Hence there is a lot of room for improvements in vehicle fuel use, in smaller, sleeker vehicles using more efficient technology.

Hybrid vehicles have achieved substantially better fuel efficiency than conventional automobiles of similar size, promising to bridge the initial period of decline in oil availability. To reduce gasoline use even further, plug-in hybrids will be an interim part of the vehicle mix. The multiple sources of electric power—coal, natural gas, nuclear, and renewable—will become available to the plug-ins. As oil availability declines further in coming decades, it will be necessary to dispense with internal combustion engines altogether and to move completely to electric vehicles powered by renewable sources, if indeed private automobiles survive at all. Of course driving less also will be necessary at all stages to achieve the fuel and energy reductions that will be required.

Hybrid Technology

Current vehicles waste significant energy even beyond the low efficiency of their engines. They consume fuel even when no motive power is required, as in going downhill, slowing, or being stopped in traffic. Energy invested in accelerating the vehicles is lost as heat in braking. The engines consume excessive fuel because they must be large enough to provide peak power that is rarely required.

Hybrid vehicles use electrical power to greatly reduce these losses without reducing performance. They are more expensive to produce than conventional vehicles; in the United States, the expense is justified by reduced fuel cost, while in Europe, where vehicles are typically driven less, small cars with efficient diesel engines are generally preferred. In the hybrid vehicles currently available, all of the power driving the wheels comes eventually from an internal combustion engine. There are two basic designs: series hybrids and parallel hybrids. Both rely on internal combustion engines, electric motors, and batteries to smooth out the power requirements.

In the series arrangement, a gasoline or diesel engine runs a generator that in turn charges batteries and powers electric motors at the wheels. All of the motive power for the wheels comes from the electric motors. For peak loads the motors call on both the generator and the batteries, so that more power is available than could be obtained from the engine alone. This means that the engine can be smaller and more fuel-efficient than would be necessary to achieve the same peak power from a conventional transmission, though the motors must be strong enough to handle peak loads unassisted. There are losses in converting from mechanical power at the engine to electrical power at the generator and back to mechanical power at the wheels. Many diesel railroad locomotives use this scheme, though without the battery supplement. No series hybrid automobiles are currently manufactured.

Parallel hybrids distribute the engine's power either to batteries through a generator, to electric motors at the wheels, or directly to the wheels without going through a mechanical-electric-mechanical path. The components and controllers are more complex than in series hybrids, but the arrangement can be more efficient under many conditions, recovering more of the energy produced in the internal combustion engine. The system developed by Toyota has an internal combustion engine connected to both the vehicle's wheels and an electric motor/generator through a planetary gear arrangement that acts as a continuously variable transmission, more efficient than the fixed gear ratios of conventional transmissions because the engine can always be run at an optimum speed for current conditions. Power can be shared between the engine and the electric motor in any combination. During braking, torque (twisting power) from the wheels rotates the motor, which serves as a generator to recharge the batteries. This is called "regenerative braking," because energy stored in the vehicle's motion is regenerated as electricity. A system developed by Honda uses regenerative braking to charge batteries, which then offer some power to assist acceleration, but there is no mechanical interaction between the gasoline engine and the electric assist system.

Plug-in hybrids are an interim technology between gasoline-powered and electrically powered vehicles. The best-known conception is the proposed Chevrolet Volt, a series hybrid with large batteries that can be charged either from an external electrical power source or from a small gasoline engine on board. The series design makes sense because the electric motor must be powerful enough to power the Volt unassisted. Promised range on battery power is about 40 miles, the power in one gallon of gasoline; since the great majority of trips in the United States are less than 30 miles, most consumers would seldom engage the gasoline engine. Astronomical fuel economy figures are claimed, though they omit the plug-in electrical power that charges the batteries when the car is parked. Charging would be through an ordinary household plug, either 110 volts or 220 volts for a faster charge.

Appearance of this model in mass production depends on advances in battery technology, probably using improved lithium-ion batteries like those in cell phones and laptop computers, but of course on a larger scale. These batteries have a higher energy density than the lead-acid battery in conventional vehicles, but they are expensive; heat buildup and safety are issues. An alternative is to base a plug-in on an existing successful hybrid design such as the Toyota Prius, adding more batteries to increase the electric-only range. The Prius uses nickel-cadmium batteries, superior to the lead-acid alternative but offering less power per unit weight than lithium-ion. Toyota is developing a plug-in version of the Prius, based on the existing hybrid model. If the vehicle depends more on battery power, the gasoline tank can be made smaller to leave room for more batteries. A plug-in hybrid will always have a smaller all-electric range than a similar-sized all-electric vehicle, though.

A disadvantage of the plug-in hybrid is that the vehicle must carry both a heavy battery load and all of the equipment needed for a gasoline engine: the engine itself, radiator, gasoline tank and pump, muffler, catalytic converter, etc. The vehicle's body must also be able to handle the heat and vibration of the gasoline engine. This adds weight to the vehicle, a problem in electric vehicles because batteries have a much lower power density than gasoline or diesel fuel; a larger weight and volume of batteries is required to achieve the range of a smaller amount of fuel.

Improvements on the Chevrolet Volt design are possible. The gasoline engine might be scaled down considerably, allowing more batteries to increase the electric range somewhat. An engine designed only to charge batteries in a series hybrid could be optimized to run efficiently at one speed, because it would not need to handle other speeds. The car would normally operate in an all-electric mode, where the gasoline engine engages only when the batteries are nearly exhausted. For short trips this would never occur, as the batteries would cover the entire trip. Alternatively, to achieve the 300-mile range to which drivers are accustomed, drivers could select a long-range mode where the engine starts running as soon as a trip starts, with battery power supplementing the electricity generated by the engine. The engine need not be large enough to provide all the power needed to drive the vehicle and keep the batteries charged; they would gradually discharge over the 300-mile range of the trip, so that battery power assisted by the engine runs out at the end of that range. The engine would continue to run during a refueling stop at that point, adding further range.

Fossil Fuel and Electricity

Practicality of plug-in hybrids depends on the relative availability of fossil fuel (gasoline or diesel) and electrical power in the future. The hybrids already on the market offer improved fuel economy without compromises in range or performance. As fuel inevitably becomes scarcer, plug-in hybrids will continue to offer long range at some compromise in vehicle weight, while all-electric designs will be the only viable alternative when fossil fuels become unavailable decades in the future.

As fossil fuel becomes unavailable or prohibitively expensive, an advantageous design for plug-ins would be to assemble the gasoline engine and all its accessories, including the gasoline tank, in a module that can be removed to make room for more batteries in an all-electric mode. This would probably require government intervention, as vehicles are normally designed to meet immediate needs rather than future contingencies.

Economic and Ecological Impact

Though plug-in hybrid vehicles offer both the range of gasoline vehicles and the economy of electric drive, they will be more expensive to manufacture than conventional vehicles of comparable size. They will be viable, though, if fuel costs rise enough to make the investment in the hybrid worthwhile. This assumes that electricity remains inexpensive relative to the cost of power from gasoline. Tax policies may have to be revised so that governments can recover enough income to maintain roads and other automobile-related infrastructure, because drivers of electric vehicles do not pay fuel taxes. Higher taxes on gasoline would drive vehicle use further toward all-electric applications.

A compelling advantage of the plug-in hybrid, though, is that it allows drivers to optimize their energy use (and cost) by using the all-electric mode more or less, depending on relative energy costs and availability. This will make fossil fuel more elastic as a commodity—increasing price will tend to decrease consumption more than at present, where there is no alternative to fossil fuels for propulsion.

The ecological impact of hybrids, and especially plug-ins, depends upon the sources for the electricity that charges the batteries. Analyses of conventional hybrids conclude that the hybrid can actually use less fossil fuel per mile traveled and generate less greenhouse gas than an all-electric mode if the plug-in uses electricity generated from coal or oil. This is because the losses add up from power generation, transmission, charging batteries, holding the power, and releasing it to the electric motor. As a greater proportion of electricity originates from renewable sources, however, this balance changes in favor of the plug-in hybrid or the all-electric vehicle. Plugging in a vehicle at night for recharge uses the power grid and electrical generating plants when they are being used below their capacity, minimizing the infrastructure cost of the added electrical demand. And the vehicle can take advantage of intermittent sources such as wind by charging only when that power source is available. In the American West, about half of the electricity in the power grid already comes from non-fossil fuel sources, mostly hydroelectric and nuclear.

In summary, the plug-in hybrid promises a vehicle with range and handling similar to current vehicles, with improved operating economy and energy flexibility.

See Also: Batteries; Combustion Engine; Renewable Energies; Zero-Emission Vehicle.

Further Readings

Anderson, Curtis D. and Judy Anderson: *Electric and Hybrid Cars: A History.* Jefferson, NC: McFarland and Company, 2004.

Duvall, M. "Advanced Batteries for Electric-Drive Vehicles." Electric Power Research Institute Final Report, May 2004. http://www.evworld.com/library/EPRI_adv_batteries.pdf (Accessed January 2009).

Graham, R. "Comparing the Benefits and Impacts of Hybrid Electric Vehicle Options." Electric Power Research Institute Final Report, July 2001. http://www.evworld.com/library/EPRI_hev_options.pdf (Accessed January 2009).

Hewitt, B. "Plug-In Hybrid Electric Cars: How They'll Solve the Fuel Crunch." *Popular Mechanics*, May 2007. http://www.popularmechanics.com/automotive/new_cars/4215489.html?series=19 (Accessed January 2009).

Tannert, C. "Inside Plug-In Car Tech's Race to Production." *Popular Mechanics*, February 2008. http://www.popularmechanics.com/automotive/new_cars/4248082.html?page=1 (Accessed January 2009).

Bruce Bridgeman
University of California, Santa Cruz

POWER AND POWER PLANTS

Electric power is extremely important for the welfare of modern societies because it allows us to use a wide range of appliances that need it to operate. In fact we rely on electricity to cover our needs in an increasing manner that concerns all sectors of the economy. In the household sector, for example, electricity is dominant with applications on lighting, convenience, and cooking equipment, and often heating. In the industrial sector, pneumatic and thermal devices have been displaced by electricity-driven machinery. Finally, the transport sector sees an increased use of electric private vehicles and other electric means of public transportation such as trains and buses, mostly due to reasons related to urban environmental quality.

Power is generated at power plants, then transmitted via high-voltage power lines to the areas where it will be distributed to consumers (industrial, municipal, residential, etc.). Commercial power generation is in most cases based on the principle of electromagnetic induction, where an electrical generator uses mechanical energy to generate alternating electric power. Power plants or power stations use a variety of technologies and fuels in order to generate power. Nearly all the technical, environmental, and economic characteristics of a power plant are a result of the combination of these two factors and so is the power consumption of the plant itself; hence a reasonable way to classify power plants is by fuel and technology. Power plants that use fossil fuels such as oil, natural gas, coal, nuclear fuels, geothermal energy, or solar thermal energy are often classified as thermal power plants. A common characteristic of thermal power plants is that they convert thermal energy to electric power. Other power plants such as hydropower plants or wind turbines use the electromagnetic induction principle to generate power, but instead of thermal energy, they convert mechanical energy to electric power. Finally, solar photovoltaic (PV) panels use the photovoltaic phenomenon to generate DC electric power.

Coal-Fired Power Generation

Coal is the most important fuel of electricity generation, reaching 41 percent of the world's fuel mix and being a major component of the steel and cement industries contributes to 26 percent of the world's primary energy supply. Poland and South Africa are the countries relying the most (93 percent) on coal for their electricity generation. The United States gets about 50 percent of its primary electricity from coal. This varies by year depending on natural gas prices. It normally takes almost five years to build a coal-fired power plant and approximately one year to decommission it; and its lifespan does not exceed 30–40 years. Coal-fired power plants normally handle base load demand because they are less flexible than gas-fired power plants. There are several different types of coal based on chemical characteristics; however, the most important ones for electricity generation are brown coal (lignite) and hard coal. Brown coal is considered of lower quality, mainly due to its lower heat content and often higher content of sulfur and humidity. However, both are used extensively for large-scale electricity generation and in several countries where the reserves of high-quality hard coal are gradually being exploited, brown coal is used increasingly.

At coal-fired power plants, the chemical energy of coal is converted to electricity through a process of several steps that are described below. The first stages include breaking and drying of mined coal. Output of these stages is coal pieces of 1–2 inches in size and of significantly lower humidity than the original mined fuel. In most pulverized coal combustion (PCC) power plants, before going to the combustion chamber coal pieces are pulverized into a fine powder at coal mills. This process enables the fuel to be burned faster as it increases the surface area exposed for burning. Coal then mixes with hot air and burns, producing heat and gases that enter a boiler, producing high-pressure steam by heating purified water. At this stage, chemical energy of coal is converted to heat. Steam is then used to spin a turbine, which is an engine that converts heat into mechanical energy. It is characteristic that a turbine normally consists of a few turbines placed on the same shaft one after the other. As the steam moves from the first to the last turbine it loses heat and pressure and gains volume, so every next turbine is designed to be able to harvest energy from steam of gradually reduced quality. On the other end of the axis of the turbine is an electricity generator, a machine that converts mechanical energy to electricity. As far as electricity generation is concerned, the process is over; however, the steam exiting the turbine is then condensed to water in order to be able to return to the boiler and repeat the cycle. This is happening in heat exchangers that use cold water from a nearby lake, river, or sea or air draft in a cooling tower. The typical efficiency factor of modern power plants that use this technology ranges between 35 percent and 40 percent, but many of the older plants of this type that still operate today often have an efficiency factor as low as 30 percent.

While the previously described process concerns a simple subcritical (PCC) power plant, the need for improved efficiency or other characteristics has promoted new technologies such as fluidized bed combustion (FBC), supercritical (SC) and ultra supercritical (USC) pulverized coal combustion, and integrated gasification combined cycle (IGCC), all of which achieve better performance than a PCC plant. Specific types of FBC (pressurized and circulating) provide improved heat exchange inside the boiler that subsequently allows operation at lower temperatures, which is critical for keeping nitrogen oxides emissions lower. Another benefit of FBC systems is that their flexibility that allows burning not only coal but almost any other solid fuel makes them ideal for co-firing power plants. At FBC power plants the efficiency factor is normally higher than 40 percent. SC and USC power plants operate at significantly higher pressure and temperature than standard PCC. Further development of this type of plants is under way but heavily depends on appropriate materials for the construction of boilers that can operate in extreme conditions. Existing commercial plants using SC and USC technology are installed in Denmark, Germany, Japan, and China, and they achieve an efficiency factor of 45–47 percent. Higher efficiency means lower cost for fuel and lower CO_2 emissions—advantages that are expected to boost installation of SC and USC in the future. IGCC plants use gasification technology through which they convert coal to a mixture of H_2 and CO that is called "syngas." After being treated to remove sulfur and other solids, syngas spins a gas turbine and the waste heat produces steam and electricity in a steam turbine—what is called combined cycle.

Although coal-fired power plants are today necessary for a secure supply of electricity, they, at the same time, are responsible for serious environmental problems, such as climate change and local air pollution. In particular, coal is the most carbon-intensive fuel used in electricity generation and coal-fired electricity generation is the least efficient procedure of base-load electricity generation. These factors combined result in heavy CO_2 emissions, but unfortunately CO_2 is not the only harmful emission of coal-fired power plants. SO_2 and NO_x together with a series of heavy metals (lead, cadmium, mercury, copper) and particulate matter all contribute to significant local and regional environmental degradation. While technologies intervening before or after the combustion help to reduce emissions of almost all of these pollutants, CO_2 and its major role in climate change is not easy to control. The most promising technology available is carbon capture and storage (CCS), which basically captures CO_2 and stores it instead of allowing it to enter the atmosphere. There are currently three possible methodologies (pre-combustion, post-combustion, and oxy-fuel) that offer CCS but only demonstration projects are available—the first commercial applications for large-scale electricity generation are not expected before 2014. Demonstration has shown that although CCS is an energy-intensive process, the total CO_2 reduction can be up to 90 percent. Success of such projects will be critical for coal's role in future electricity generation.

Natural Gas–Fired Power Generation

Natural gas (NG) is found in underground reservoirs and is a mixture that consists mostly of methane, together with other heavier hydrocarbons and carbon dioxide. Most important reserves of natural gas are found in Russia, Iran, Qatar, Saudi Arabia, and the United Arab Emirates. The specific chemical characteristics of NG differ depending on the country and field, but in most cases NGs from different countries are mixed (sometimes special treatment is required to make them compatible) and burned together without a problem. Natural gas is the second most important fossil fuel for electricity generation after coal, and nearly all of the recently (last decade) installed power capacity in developed countries has been gas fired. Natural gas–fired power plants generate approximately 20 percent of the world's electricity. A gas-fired power plant can be installed quicker than any other type of fossil fuel–based plant—most often in under one year—while the plant's life span is usually in the range of 35–40 years. Most countries that use natural gas for electricity generation import it from other countries that have reserves. In comparison, a lot more countries use their own coal reserves for electricity generation or alternatively import coal that is nevertheless a much cheaper fuel than gas. This makes electricity generation with natural gas relatively expensive. At the same time, one of the greatest advantages of gas turbines is their very quick response time to demand—they can reach full power from idle in just seconds. As a result of both the high costs and the quick responsiveness, most countries are using gas-fired power plants to meet peak loads.

At gas-fired power plants, chemical energy of gas is converted into electricity. Almost all modern installations use a combined cycle gas turbine (CCGT) system. The primary cycle of this technology uses the high-temperature and high-pressure gases produced by burning natural gas to spin a turbine and generate electricity. The secondary cycle uses the waste heat of the primary cycle to produce steam, which is then used to generate more electricity. Using two cycles allows CCGT technology to overcome the usual single-cycle efficiency boundaries that limit conventional power plants to efficiency factors of approximately 35–40 percent. It is characteristic of CCGT power plants to come close to an efficiency factor of 60 percent while their typical size exceeds 500 MW, of which two-thirds are on a primary gas turbine and one-third on a secondary steam turbine.

Unlike coal, natural gas does not contain significant amounts of sulfur or other solids and as a result gas fumes from gas-fired power plants might contain only traces of sulfur dioxide and heavy metals. At the same time, methane, which is the main substance of natural gas, has the lightest possible carbon content. All that combined with the high efficiency technology used at gas-fired plants make gas-fired electricity generation approximately half as carbon intensive as the coal-fired power plant. However, nitrogen oxides emitted are not less than those from power plants that use coal, and although carbon dioxide emissions are lower than in coal-fired power plants, they still are significant and contribute to climate change. Pollution control systems are available for gas-fired power stations and they mostly change the burning process in order to prevent creation of nitrogen oxides. Their efficiency often reaches 40 percent and makes gas power plants a very attractive environmental option for electricity generation.

It is expected that in future years gas-fired power plants will continue to be one of the fastest-developing electricity generators. Accessibility to natural gas reserves and lack of fields for carbon-capture-and-storage development will give further advantage to gas as a resource for electricity generation because it is environmentally friendlier than conventional coal. New gas-fired power plants will come into operation in order to replace old nuclear and coal-fired plants and also in order to provide increased electricity demand.

Oil-Fired Power Generation

Oil used in electricity generation is mostly light oil products that come out of refineries of crude oil. Although oil is very important for the world's primary energy supply, it is the least popular fossil fuel for electricity generation today. It is used for electricity generation mainly in countries that either have their own oil reserves and/or do not have indigenous coal or access to natural gas. Traditionally the Arab Gulf States and other countries or islands (that might be countries or parts of countries) use oil in autonomous power stations).

Depending on the technology used to generate electricity from oil, different methods may be used. Oil can be burned at boilers like coal in order to generate steam and spin a steam turbine for large-scale electricity generation. However, it is more common that oil-fired power plants are equipped with exhaust gas turbines that burn oil to spin a turbine directly and generate electricity. This process is similar to natural gas turbines and may allow the operation of a secondary cycle that uses the waste heat of the primary turbine to produce steam and spin a steam turbine generating more electricity. While the two previous methods usually use heavy oil, internal combustion engines (similar to those used for cars) use a type of diesel oil to generate electricity. Internal combustion engines (ICE), though, are used only for relatively small-scale electricity generation, usually for autonomous networks on islands and wherever grid connection is not possible.

As far as emissions are concerned, oil-fired power plants are considered to be less polluting than coal-fired plants and more polluting then gas-fired plants. However, this might vary significantly depending on the technology (ICE or combined cycle have very different efficiency, hence very different emissions) and also on the fuel used. The latter has a serious impact mostly on the sulfur dioxide emissions, which are directly related to the sulfur content of the chosen oil and can be between 0.5 percent and 5 percent. Pollution control technologies to control sulfur dioxide exist; however, they are not always used, especially at small-scale power plants. Finally, oil-fired power plants emit nitrogen oxides and mercury compounds.

In most cases oil-fired power plants are used to meet peak demand (using technologies such as exhaust gas turbines or ICE in smaller systems) in grids that are based on nuclear or coal generation. However, oil-producing countries or small islands cover base load generation most often with oil-fired gas turbines (single-cycle exhaust or combined cycle) or in small systems with ICE. Despite oil's popularity, very few oil-fired power plants have been built recently, and several interconnected oil-fired power plants have been replaced with gas-fired ones once they reached retirement age.

Nuclear-Fired Power Generation

Nuclear energy mostly refers to electricity generated at power plants that use nuclear fission. Approximately 30 countries around the world rely to some extent on nuclear electricity. At the top of the list of nuclear electricity-generating countries one will find the United States, France, Japan, and Russia, while a look at the countries with nuclear power plants currently under construction reveals China, Russia, India, South Korea, and Canada at the top of the list. In total not more than 15 percent of the world's electricity is generated at nuclear power plants. Nuclear power plants have a nuclear reactor in which uranium is split, releasing energy that heats water to produce steam and spin a steam turbine that subsequently generates electricity. The procedure is not flexible and, once started, it is difficult to adjust—a characteristic that makes nuclear power plants suitable to meet only base load electricity generation. Uranium used at nuclear power plants is in the form of uranium oxide pellets that are produced from processed uranium ore. Major uranium-producing countries are Canada, Australia, Kazakhstan, Russia, and Niger.

Nuclear power plants produce no air emissions during their operation. However, during the process of electricity generation and also during preparation of uranium oxide pellets, nuclear waste is produced. No permanent way of safe disposal of nuclear waste has yet been found; nuclear waste in most cases is stored on site at the power plants where it is produced. Special measures are taken in order to prevent leakage. Significantly greater, though, is the concern not for the routine operation of nuclear plants and their nuclear waste but for any accidental nuclear event. Accidents at power plants—Three Mile Island in 1979 and Chernobyl in 1986—have both raised public opposition to nuclear energy, which has, in effect, caused significant slowdown at or even complete prevention of installation of new nuclear power plants across most developed countries.

While most of the recently installed nuclear power plants are in developing countries (particularly China and India), global warming has reignited interest in nuclear energy in the developed world. As mentioned above, nuclear power plants do not emit air pollution such as carbon dioxide and therefore are considered as carbon neutral. At the same time, several of the power plants built during the 1970s are reaching the end of their life span, which threatens electricity shortages in several developed countries. While the debate on nuclear power plants has been brought back into life, it is not expected that nuclear energy will increase its contribution to the world's electricity generation in the next decade. Even when public concern is overcome, building capacity of new reactors is limited due to the sector's underfunding that took place in the previous decades. Finally, new installations might take up to 10 years to complete and it is generally thought that almost all of them will in the best case replace the power stations that will need to be phased out during the same years.

Renewable Sources

Renewable electricity generation refers to electricity generated utilizing non-finite energy resources such as biomass, wind, hydro, solar, waves, tides, or geothermal activity. While hydropower already has been exploited adequately in most countries, the rest of the renewable energy sources have not enjoyed the same attention. In 2006 renewable energy sources met 18.4 percent of the world's electricity demand with China, Germany, the United States, Spain, and India being the leading countries in installed capacity of non-hydro renewables and China, Canada, Brazil, the United States, and Russia leading in hydropower installations.

Each of the energy resources mentioned provides different characteristics with regard to the technology needed to produce electricity, potential in various locations, and time- or season-related availability. From a technology point of view, wind, tidal, and hydro power are based on the same principle—wind or water move a turbine, which generates electricity. Biomass is mostly used in boilers (in co-firing with other fuels) for generating steam to spin a steam turbine similarly to geothermal energy. Wave power makes use of either hydro turbine or hydraulic technology to generate electricity, while solar energy concerns the photovoltaic (PV) phenomenon where PV panels generate DC electricity when exposed to solar radiation.

One of the major characteristics of renewable energy sources is that they are not available in a constant flow. Depending on the specific source, their availability is subject to the seasons (biomass, hydro), time of day (solar, tides), or is just random (wind, waves). The intermittent nature of renewable energy generation has been a major obstacle for its development on base load systems as almost always another power plant has to be in stand-by mode in order to provide backup capacity. Although a number of technical and planning solutions are available, their implementation so far has been limited mostly for reasons related to high cost, difficult spatial planning, and inadequate electricity networks.

Despite intermittency, renewable energy resources enjoy policy support in a wide range of countries as a result of their positive characteristics. These mostly relate to the increased supply security they provide by reducing energy imports and simultaneous macroeconomic benefits due to lessening of exposure to international fuel prices. Finally, what is often considered as the major advantage of renewable power generation is the relatively low environmental impact, as in most cases they generate electricity without any emissions at all. Predictions for the future show a steep increase in renewable energy installations for at least the next decade.

See Also: Alternative Energy; Alternative Fuels; Chernobyl; Coal, Clean Technology; Electricity; Fossil Fuels; Geothermal Energy; Green Power; Hydroelectric Power; Nonrenewable Energy Resources; Nuclear Power; Photovoltaic (PV); Public Utilities; Solar Energy; Three Mile Island; Tidal Power; Wave Power; Wind Power.

Further Readings

Innovation, Universities, Science and Skills Committee. *Renewable Electricity-generation Technologies*. London: House of Commons, 2008.
International Energy Agency. *Coal Information 2008*. Paris: IEA, 2008.
International Energy Agency. *Oil in Power Generation*. Paris: IEA, 1997.
Nuclear Energy Agency. *Nuclear Electricity Generation: What Are the External Costs?* Paris: OECD, 2003.
Odeh, A. Naser and T. Timothy Cockerill, "Life Cycle Analysis of UK Coal Fired Plants." *Energy Conversion and Management*, 49:212–20 (2008),
Royal Academy of Engineering. *The Costs of Generating Electricity: A Study Carried Out by PB Power for The Royal Academy of Engineering*. London: The Royal Academy of Engineering, 2004.

Konstantinos Chalvatzis
University of East Anglia

PUBLIC TRANSPORTATION

Public transportation (also known as transit or mass transit) is a transportation system designed to move large numbers of people to common destinations. It is critically important to metropolitan economies and environments. Public transportation comes in many forms, with many local eccentricities. Modes include paratransit, buses, streetcars and trams, heavy rail and metros, and commuter rails.

Paratransit

Paratransit is the smallest network of public transportation and is carried out by vehicles ranging from small jitneys to passenger vans to minibuses and shuttles. Paratransit is

These modern trams serve Budapest, Hungary—the city that introduced the electric tram in 1887. Electric trams have remained in continuous use in European cities.

Source: iStockphoto.com

highly flexible to the local demands of riders. In some locales, the vehicles may operate as collective taxis and offer door-to-door transportation at a discounted rate. In other locales, paratransit may take the form of purposely designed shuttle routes or circulator routes to serve special districts (commercial districts, college campuses, or military bases). Paratransit vehicles are typically quicker than traditional buses, as they are smaller, facilitate load-in/load-out faster, and have fewer stops. Many paratransit services are locally and privately owned and operated. This structure allows these routes to be highly adaptable to market demands, since the operators are profit driven. Furthermore, operators tend to aggressively seek out new routes and markets and will sometimes offer specialized service akin to a taxi service. These conveniences make paratransit both an alternative to buses and complementary to the entire transit system.

Paratransit is most often found in developing countries, where public investment in public transportation is lacking. While system characteristics may be similar, the vehicles vary from country to country, and even city to city. In Manila, iconic "jeep-neys" were originally built from converted World War II–era American jeeps. In Turkey, the "dolmu " network of vans and minibuses operates on semi-formal routes and is seen as the only reliable form of transit in urban areas. In Central and South America, private buses, "colectivos" (collective taxis), and an array of vans and shuttles operate as stop-gaps between established public transit routes. In the richer countries, though, paratransit tends to serve only specialized niches. "Vanpools" are used as a mass car pool, and often funded by a large employer. Most major airports are served by shuttle vans, which deliver passengers quicker than mass transit, but cheaper than a cab. In some cities, immigrant populations utilize informal, "underground" paratransit due to its convenience and familiarity. Miami and New York City are known to have these informal shared taxi networks.

The term *paratransit* has also taken on a different meaning in the American lexicon. Following the Americans with Disabilities Act (ADA) legislation in 1990, public transportation providers were required to make their vehicles and routes accessible to disabled people who reside within three-fourths mile of a fixed route. In many cities, this would mean specialized equipment and infrastructure improvements (sidewalks, bus stops, etc.). Rather than invest in these improvements, transit operators often provide "dial-a-ride" services that allow a disable person to call ahead for a minibus or van to pick them up at their residence. In the end, the bus or van operates like a shared taxi, but only for people with disabilities.

Bus Transit

Buses are typically 45–50 passenger vehicles that travel fixed routes on fixed schedules. In most public transportation networks, they are the backbone of transit. Buses come in a variety of shapes and sizes, from double-decker models to articulated buses (extra-long buses with a central pivot). Buses are usually fueled by diesel; however, smog problems in congested cities have driven transit agencies to substitute compressed natural gas buses (CNG) or modified vehicles to run on overhead electric lines. Bus service is usually managed by a public agency, affiliated with either a regional government or a municipal government, although some cities have experimented with privatized transit to offset expenses.

Compared to paratransit, buses offer opportunities to carry more passengers per unit, thus eliminating vehicle trips. However, buses are not as flexible as paratransit and are usually run on tight schedules along major commercial thoroughfares. Compared to rail services, buses offer more flexibility in routes, as there are no concerns over rights-of-way and tracks. However, buses are considered the slowest form of transit, as they are often mired in (and contribute to) inner-city traffic congestion.

While useful, buses are plagued by stigmas. The most common complaints of bus riders concern efficiency, condition of the bus, and personal comfort and safety. To address efficiency, transit agencies can designate some buses as "express," limiting the number of stops. In Los Angeles, for example, red city buses are express buses, and they include technology to trigger green lights as the bus approaches. Express routes are also becoming attractive options. These buses offer limited service from suburban areas to urban cores, with only a few stops in-between. Another alternative is to develop busways, or dedicated traffic lanes for buses. These busways relieve congestion from frequent stops, making both automobile traffic and the bus route move faster. Busways could, theoretically, double the capacity of passengers per bus route, per hour. To address concerns over bus condition, transit agencies have little choice but to upgrade equipment and replace aging fleets. New buses often operate at higher fuel efficiency and provide greater comfort to passengers.

Passenger safety and comfort is a major concern, and one of the most enduring stigmas. New buses can offer a more comfortable experience, free of loud noises, graffiti, and dirty seats. Security cameras and transit police can aid in providing a safe environment. But, on many routes, issues of overcrowding remain. As noted, express routes can alleviate some of this, but riders in some cities have vocally demanded that more routes be added to the system and more seats be made available, regardless of profitability concerns. Many buses now offer bicycle racks on the front for passengers, and amenities inside the bus like interactive route mapping, video screens, and newspapers.

Electric Trams and Streetcars

Trams, or streetcars, are vehicles that typically run in an existing street and are powered by electricity. They are smaller, and slower, than rail services, which run on dedicated railway. In their earliest form, they were considered important, primary forms of mass transit. However, they currently operate as a people mover through congested neighborhoods within a city.

The first tram dates back to 1807, in South Wales, United Kingdom. Early trams were horse-drawn carriages, which gave way to steam-powered carriages. The first electric tram was built in Budapest in 1887. In many European cities, trams never fell out of use in city centers, although inter-village tramways gave way following the advent of automobiles. In the United States, streetcars (the American term) reached their pinnacle in the mid-20th century. Cities like Los Angeles and Chicago had extensive streetcar systems, which ran from inner-city neighborhoods to suburbs. The demise of the streetcar has been well documented as a product of political conspiracy.

U.S. streetcars, at first, were often owned and/or operated by electric utility companies because of the involved technology. Over time, utility companies began to divest themselves of these enterprises, and many were de facto monopolized by a holding company representing the titans of the automobile industry. This holding company, National City Lines, was ultimately cited in federal court for its activities in Southern California, but the damage had been done. Streetcars were dismantled and replaced with General Motors buses, and tracks were either removed or paved over. Over time, freeways, land use zoning, and wide roads made automobile transit more appealing.

In recent years, streetcars have made a comeback of sorts. While retrofitting a street is very expensive, success stories like Portland, Oregon, have driven cities to examine local streetcars as an economic development strategy to circulate people throughout adjacent neighborhoods.

Heavy Rail and Metros

Heavy rail and "metro" systems are a step above streetcars and trams. The vehicles are typically larger, have a greater carrying capacity, and run on dedicated right-of-way. They are powered with electricity, and each car runs its own electric engine. In most cases, they operate with a "third rail" electrical source; for safety reasons, at-grade railways are fenced and stations have some degree of access control. "Light rail" is often used as a term for these systems; however, many of these light rails, like Los Angeles's or Phoenix's, functionally operate akin to Chicago or New York's elevated and subway lines. Metros are subject to incredible fixed expenses, but are cheaper to operate over the long term. As such, they are typically managed by some level of government or a quasi-government agency. Over time, many metros have matured into hundreds of miles of tracks. Currently, there are over 160 cities worldwide with some form of a metro system.

Metros are at their greatest effectiveness in dense environments, and as Robert Cervero has noted, the density situation is symbiotic. Without the subway, Manhattan, for example, would not be able to reach its residential and commercial density. Tokyo's and Moscow's subways are the busiest, serving upward of 2.5 billion riders annually. London, Paris, and New York have the most extensive networks.

Commuter Rail

Commuter rail is rapid rail service that links outer-ring suburbs to inner-city transportation networks. These systems exist in over 100 cities worldwide, with many proposals for new and expanded service. Commuter rails often share tracks with freight trains, and are pulled by diesel locomotives. The most effective systems run in the largest metropolitan areas, or along highly urbanized corridors. Prominent systems include MetroLink in Southern California, Metra in Chicagoland, and MetroNorth between New York City and Connecticut. Stations are sparse and typically surrounded by parking lots that allow suburban commuters to drive to/from the station.

Energy Concerns

The most immediate environmental benefit of public transportation is to remove and limit private automobiles on the road. An easement in congestion means, in theory, fewer tailpipe emissions. However, public transportation itself is dealing with energy issues. At the smallest scale, private jitneys, vans, and minibuses pollute as much as typical automobiles.

In Kolkata, for example, paratransit accounts for about 5 percent of the vehicle population, but an estimated 35 percent of the air pollution. The easiest solution is to look for fuel substitutes, like CNG. In the United States, for example, many airport shuttles and minibuses operate on CNG. However, in developing countries, owners/operators of paratransit do not have the finances or access to the infrastructure needed to convert and reliably fuel their vehicles. Diesel-operated buses are large polluters, as well. To some degree, their ozone contribution can be mitigated with the use of CNG. However, these buses are costlier to purchase and operate, and still depend on a nonrenewable fuel source.

The "cleanest" transportation options are light rail and metro. These systems are durable and run on electricity that may ultimately be produced through solar or wind energy. Commuter rails have the potential to be converted to biodiesel, a fuel source derived from renewable, biological matter. One of the first systems to be run on biodiesel was Albuquerque, New Mexico's new Railrunner commuter rail.

Conclusion

Many progressive urban planners believe that transit is the ultimate determinant of urban form, and should be planned aggressively, but carefully. Transit-oriented developments (TODs) can be located near bus, tram, and metro stations to facilitate greater ease of use. Cities that have matured in the age of sprawl must aggressively pursue at-grade metro systems, arrest the development of freeways, and create seamless connections between different transit modes (including the private automobile and bicycle). To increase ridership, however, planners may need to think outside the box. While it is easy to charge riders, arguments have been made that public transportation should be free, subsidized by taxes in the same manner that roads and freeways are. In fact, many see opportunities to actually charge for vehicle use (congestion pricing) and lower the cost of public transportation.

See Also: Automobiles; Emission Inventory; Sustainability.

Further Readings

Bansal, Monica. "Clean It Up, Don't Throw It Away: Greening Delhi's Paratransit." http://www.embarq.org/sites/default/files/Monica_Bansal_Delhi_Paratransit.pdf (Accessed May 2009).

Cervero, Robert. *The Transit Metropolis: A Global Inquiry.* Washington, DC: Island Press, 1998.

Downs, Anthony. *Still Stuck in Traffic: Coping With Peak Hour Traffic Congestion.* Washington, DC: Brookings Institution, 2004.

Marshall, Alex. *How Cities Work: Suburbs, Sprawl, and the Roads Not Taken.* Austin: University of Texas Press, 2000.

Newman, Peter and Jeffrey Kenworthy. *Sustainability and Cities: Overcoming Automobile Dependence.* Washington, DC: Island Press, 1999.

Vuchic, Vukan. *Transportation for Livable Cities.* New Brunswick, NJ: Center for Urban Policy Research, 1999.

Derek Eysenbach
University of Arizona

PUBLIC UTILITIES

Public utilities are organizations that provide public services. Public utilities may be government owned, cooperatives, or investor-owned enterprises. Private utilities are held by a single owner; as owned utilities, they are joint stock companies with stock shares sold on the open market. It is common, however, to include government, cooperative, and investor-owned utilities in the term *public utilities*. They may simply be called the "utility." In this article, the term *public utilities* refers to both types unless otherwise designated.

Cooperative utilities are owned by their customers. Cooperatives have been used to supply rural areas with water, electricity, or other services that private companies or government was unable or unwilling to provide.

Utilities provide services that are part of the infrastructure of a town, city, rural community, or at times a whole country. They number over 4,000 in the United States alone. They include transportation, water, gas, electricity, telephone, television, Internet, sewage, garbage, or other essential services. In providing these services, public utilities face similar natural, scientific, technological, economic, social, or political issues regardless of whether they are government or investor-owned organizations. However, at times there are advantages to the public, to political partisans, or to investors that accrue from the situation of having a utility owned and operated by the government or by a government-regulated investor-owned organization.

Monopolies—Better or Worse?

Historical experience has led many to conclude that the services public utilities provide are better provided by an organization with a monopoly right to be the exclusive provider of one or more of the services listed above. It would be possible for these to be provided through many companies competing; however, most of the services provided by public utilities are viewed as being best treated as a "natural monopoly" that is either government owned or publicly regulated. An example of a natural monopoly would be the water company supplying water to a city.

Those in favor of a monopoly system of public utilities have argued that public ownership or regulation is the best method for preventing wasteful competition, controlling market power, directing the utilities in ways mandated by the government (allegedly for the public good), creating a system of "fair" pricing, or for other reasons. If utility services were provided through an open market system it is likely that sooner or later the drive to achieve economies of scale would lead to the consolidation of utility companies into either an oligopoly or monopoly domination of the market. This economic condition would then allow either the one monopoly firm or one of the oligopoly firms to be able to materially affect prices and other aspects of the utility service. This monopolistic condition would very likely then lead to political demands (and has historically) for government regulation of the utility monopoly.

The electric utility business was among the first to experience the problem of trying to allow competition in a business that seems best suited to a natural monopoly. In some places where many companies vied to provide electric service in the early history of the industry, the electric wires that were hung became more than an eyesore. Electric wires in competition were loaded, in fact overloaded, onto utility poles, but without what were believed to be significant benefits to the public that owned the rights-of-way where the utility poles were placed.

The solution of allowing a single company to monopolize the utility service led to government regulation in the United States, or to government-owned utilities. The idea of a "natural monopoly" has been challenged from time to time. Deregulation of utility service has been instituted for airlines, electricity, and other utility industry sectors. The claim of opponents of monopoly utility service delivery is that competition will improve service and reduce the price of the service to consumers. The latter are virtually the same as the public.

Government ownership of utilities has been a very common solution to the politico-economic issues that utilities generate. The solution is seen by many, especially those who advocate nationalistic or socialistic ideologies, as the best way to achieve a utopian goal of benefiting the nation or the society with "justice." The problem is that government-owned utility organizations are not necessarily better than private or publicly regulated utility enterprises. In fact, government ownership does not guarantee better service or more cost-effective service. It will, however, satisfy ideological visions.

Direct government ownership of utilities is common in the United States and around the world. In some localities the electrical service provider is a local government entity. In cases where the local utility is government owned it may be able to increase its revenues over what would otherwise be the case through increased utility rates. If, for example, the city owns the electric utility in a state capital, it can make up for losses in property tax revenues due to the large number of state-owned properties with higher electric rates that are charged to the state. The profits from the higher rates can then be used for local services or projects such as swimming pools, parks, or other benefits to the public at large.

Another benefit of direct government ownership is that it is able to provide electricity or other utility service to areas that investor-owned utilities would have found unprofitable. For example, utilities such as the Tennessee Valley Authority were able to provide electricity to rural areas that would not have had electricity otherwise. In some cases, the cost of providing the service was a loss economically, but acceptable politically.

A negative consequence of government ownership of utilities has arisen in situations where the ruling elite has chosen to run the utility as a revenue stream for funding government or state activities without returning money to maintain and improve the utility's productive assets. In many countries, the utility provides poor service and is often indifferent or inefficient in responding to customer complaints such as the lack of service. The use of regulation of investor-owned utility services has often proven more beneficial to owners, customers, and the general public than many cases of government-owned utilities.

In the case of investor-owned utilities, the cost of providing service to rural areas was accomplished by charging a uniform rate sufficiently high to cover rural service costs at a profit. The uniform rate meant that rural areas were subsidized by urban and business consumers.

There are inherent similarities in utility costs regardless of whether they are provided by a government organization or an investor company. The cost of building new power plants or other facilities must be borne by taxpayers in the case of government utilities through either increased utility prices or increased taxes. In the case of increased taxes, they may not be borne equally by all. However, if a utility is privately owned, regulatory permission to increase its service capacity incurs costs that are passed on to the consumers. In either case the consumers are the taxpayers and the taxpayers are the consumers so it is hard for utility providers and taxpayers/consumers to escape the fact that utility costs have to be paid for by someone.

Regulation and Deregulation

Regulation of public utilities in the United States is usually a shared responsibility of the states and the federal government. All of the 50 states have at least one public service commission to regulate one or more utility services such as electric power, natural gas, telephone, or other services. It is not uncommon for several agencies to regulate all of the utilities in the state. The federal government regulates different aspects of public utilities through a number of different agencies.

The many public service commissions of the states have different names. For example, the Texas Railroad Commission regulates gas and oil production in Texas. The reason for its name is that it was assigned the regulatory responsibility for oil and gas in the "oil patch" during the chaotic times between 1915 and 1920.

Regulatory agencies conduct hearings and determine the rates utilities can charge. They also mandate many other rules for the public's safety or economic benefit.

In the case of energy regulation, the amount and type of regulation is often a function of the era in which the service came online. Nuclear power emerged as a source of electrical power production during the secret wartime work of building an atom bomb. Close government control has been a mark of nuclear energy ever since. On the other hand, coal production occurred during the very open free enterprise days of the 1800s. Attitudes toward utilities in the energy sector are therefore often inherited from the time of their origin as public utility providers.

Utility deregulation in the telephone industry opened the way for the entry of a large number of new providers to enter the business. Some had long wanted to impinge upon the monopoly of the largest provider—AT&T—in order to pare off its most profitable business sectors while leaving the areas that were subsidized by the uniform rate to the old monopoly to be handled at its cost.

In 1984, the AT&T monopoly was broken into a number of separate companies and regional providers such as Southern Bell became independent companies. The federal courts mandated the breakup of a monopoly that had lasted from 1877 to 1984. However, the era since the mid-1980s was one of very rapid changes in telephone service and technological development. It is very doubtful that customers saved money, but their choices in everything from the color and number of telephones in the home were greatly increased. With the advent of cellular phones, utility services were again subject to rapidly evolving competition.

Technological developments have challenged the natural monopoly status of public utilities. The cable television service providers have been challenged by satellite service providers. Electrical companies are seeing technological innovations that may challenge them with competition or with a need to accept the electrical generation that private citizens or that other organizations are able to generate.

Demands for liberalization or deregulation of natural monopolies held by public utilities have led to deregulation of parts of the postal system, the electrical retailing system, and other industries.

Around the world the poorer countries usually have utility services that are limited to wealthy neighborhoods or areas. Many rural areas or poverty-stricken neighborhoods have no access to utility supplies of water, electricity, natural gas, or other services. Exceptions that have developed since World War II include radio service because of the invention of the transistor and cellular phone access.

Of great importance to investor-owned utilities are the new technologies that are being introduced to combat global warming. These include solar, wind, geothermal, and a variety of biofuels. Many of these technologies are currently future possibilities rather than proven realities. However, many expect that a move to force companies to pay for their carbon emissions will force the advent of newer technologies much more rapidly than would have been the case through natural market forces.

See Also: Alternative Energy; Electricity; Grid-Connected System; Smart Grid.

Further Readings

Diane Publishing Company, ed. *Public Utility Holding Company Act of 1935: 1935–1992*. Darby, PA: Diane Publishing Company, 1994.

Harris, Clive and Ian Alexander. *Regulation of Investment in Utilities: Concepts and Applications*. Washington, DC: World Bank Publications, 2005.

Hayes, Hammond Vinton. *Public Utilities: Their Fair Value and Present Return*. Charleston, SC: BiblioLife, 2009.

Herzog, Thomas, ed. *Soka-Bau: Utility Sustainability Efficiency*. New York: Prestel Publishing, 2007.

Hulsink, Willem and Emiel F. M. Wubben, eds. *On Creating Competition and Strategic Restructuring: Regulatory Reform in Public Utilities*. Northampton, MA: Edward Elgar, 2003.

Palast, Greg, Jerrold Oppenheim, and Theo MacGregor. *Democracy and Regulation: How the Public Can Govern Essential Services*. London: Pluto Press, 2002.

Roess, Anne C. *Public Utilities: An Annotated Guide to Information Sources*. Lanham, MD: Rowman & Littlefield, 1991.

Rossi, Jim. *Regulatory Bargaining and Public Law*. Cambridge: Cambridge University Press, 2005.

Ruff, Matt. *Sewer, Gas and Electric: The Public Works Trilogy*. New York: Grand Central Publishing, 1998.

Andrew J. Waskey
Dalton State College

RADIATIVE FORCING

Radiative forcing (RF) is a measure of how the energy balance of the Earth–atmosphere system is influenced when factors that affect climate are altered. RF is usually quantified as the "rate of energy change per unit area of the globe as measured at the top of the atmosphere" and is expressed in units of Watts per square meter (Wm^{-2}). The influence of a factor that can cause climate change, such as a greenhouse gas, is often evaluated in terms of its RF. When RF from a factor or group of factors is positive, the energy of the Earth–atmosphere system will increase, leading to warming of the system. In contrast, negative RF causes decrease in energy, leading to cooling of the system. Under the present-day global warming and climate change conditions, RF is an important metric for assessing climate change.

The concept of RF provides a way to quantify and compare the contributions of different agents that affect surface temperature by modifying the balance between incoming and outgoing radiative energy fluxes. The term *radiative* arises because factors affecting climate do change the balance between incoming solar radiation and outgoing infrared radiation within the Earth's atmosphere. The term *forcing* is used to indicate that Earth's radiative balance is being pushed away from its normal state. In a simpler way, RF can be stated as change in the amount of energy per unit time flowing into or out of Earth's climate system. In climate science, it is defined as the change in net irradiance (difference of incoming and outgoing irradiance, radiant emittance, and radiant exitance are radiometry terms for the power of electromagnetic radiation energy) at the tropopause. The tropopause is a boundary region in the atmosphere between the troposphere and the stratosphere. The change is computed on the basis of "unperturbed" values, as defined by the Intergovernmental Panel on Climate Change (IPCC). The definition used by the IPCC is that the RF of the surface-troposphere system resulting from perturbation or introduction of an agent (say, change in greenhouse gas concentration) is the change in net (incoming minus outgoing) irradiance (solar plus longwave; in Wm^{-2}) at the tropopause, after allowing for stratospheric temperatures to readjust to radiative equilibrium, but with surface and tropospheric temperatures and state held fixed at the unperturbed values. A useful example is that of carbon dioxide (CO_2). If preindustrial CO_2 levels were increased to current CO_2 levels without changing any other aspect of the climate system, then there would be an imbalance in the radiation budget. As CO_2 is transparent to incoming radiation, but strongly absorbs outgoing longwave radiation, the outgoing longwave radiation to space would be reduced and a net heating would occur, called positive RF.

To understand radiative transfer, let us first discuss radiation balance. We know that solar energy (radiation) is the driving force for climate. The Earth and its atmosphere absorb and reflect some of the solar radiation. Earth's surface reradiates absorbed radiation back into space in the form of longwave radiation, resulting in an approximate balance between energy received and energy lost. The balance between absorbed and radiated energy, known as radiation balance, determines the mean global temperature.

Radiation balance can be altered by factors such as intensity of solar energy, reflection by clouds or gases, absorption by gases or surfaces, and emission of heat by various materials. Any such alteration is an RF and causes a new balance to be reached. Knowledge of the natural and anthropogenic processes that affect this energy balance is critical for understanding how Earth's climate will change in the future.

The anthropogenic factors contributing to RF are greenhouse gases (CO_2, nitrous oxide, methane, and halocarbons), ozone (stratospheric and tropospheric), stratospheric water vapor, surface albedo, total aerosols, and linear contrails. The forcings for all increased greenhouse gases are positive because each gas absorbs outgoing infrared radiation in the atmosphere. Among the greenhouse gases, RF of CO_2 was highest during the postindustrial period. Increases in tropospheric ozone contributed to warming, and decreases in stratospheric ozone contributed to cooling. Aerosols (small particles present in the atmosphere) influence RF directly through reflection and absorption of solar and infrared radiation in the atmosphere. Some aerosols cause positive forcing, whereas others cause negative forcing. The direct RF summed for all aerosol types is negative. Human activities since the industrial era have altered the nature of land cover and also modified the reflective properties of ice and snow. As a result, more solar radiation is reflected from Earth's surface, causing a negative forcing. Aircraft produce persistent linear trails of condensation called "contrails." These contrails reflect solar radiation and absorb infrared radiation and thereby cause small positive RF.

RF from natural changes occurs as a result of solar changes and explosive volcanic eruptions. Solar output has increased gradually in the industrial era, causing a small positive RF. Explosive volcanic eruptions can create a short-lived (two to three years) negative forcing through temporary increases in sulfate aerosols in the stratosphere.

Estimates of RF components associated with the emissions of each agent as of 2005 were made by the IPCC. The combined anthropogenic RF is estimated to be $+1.6$ (-1.0, $+0.8$) Wm^{-2}, indicating warming of the climate since 1750 as a result of human activities. Increasing concentrations of greenhouse gases CO_2, methane, nitrous oxide, halocarbons, and sulfur hexafluoride have led to a combined RF of $+2.63$ (±0.26) Wm^{-2}. Individually, RF of CO_2, methane, and nitrous oxide are $+1.66$ (±0.17), $+0.48$ (±0.05), and $+0.16$ (±0.02) Wm^{-2}, respectively. Chlorofluorocarbons, hydrochlorofluorocarbons, and chlorocarbons together contributed $+0.32$ (±0.03) Wm^{-2} to the RF. The total RF of hydrofluorocarbons and perfluorocarbons is $+0.017$ (±0.002) Wm^{-2}. The RF from increases in tropospheric ozone is estimated to be $+0.35$ [-0.1, $+0.3$] Wm^{-2} and from increases in stratospheric water vapor resulting from oxidation of methane is estimated to be $+0.07$ (± 0.05) Wm^{-2}.

RF caused by total direct aerosols is estimated to be -0.5 (±0.4) Wm^{-2}. The RF resulting from the cloud albedo effect is -0.7 (-1.1, $+0.4$) Wm^{-2}, from land cover changes is -0.2 (±0.2) Wm^{-2}, from black carbon aerosol deposited on snow is $+0.1$ (±0.1) Wm^{-2}, and from contrails from aviation is $+0.01$ (-0.007, $+0.02$) Wm^{-2}. The direct RF caused by increases in solar irradiance since 1750 is estimated to be $+0.12$ (-0.06, $+0.18$) Wm^{-2}.

The concept of RF has been central for guiding climate research and policy over the past two decades:

- It provides a simple yet fundamental index that allows us to look at how climate change is driven by the energy imbalance of the Earth system.
- It is successful in predicting change in global mean surface temperature as computed from climate models and thus allows quantitative comparison of the contributions of different agents to climate change.
- It is easy to compute and is reproducible across models, and therefore offers a convenient common metric on which policy research and recommendations can be framed.

See Also: Climate Change; Global Warming; Greenhouse Gases; Intergovernmental Panel on Climate Change.

Further Readings

Forster, P. M., et al. "Changes in Atmospheric Constituents and in Radiative Forcing." In *Climate Change 2007: The Physical Science Basis. Contribution of Working Group I to Fourth Assessment Report of IPCC.* New York: Cambridge University Press, 2007.
Radiative Forcing of Climate Change: Expanding Concept and Addressing Uncertainties. Washington, D.C.: National Academies Press, 2005.

P. Vijaya Kumar
Project Directorate on Cropping Systems Research, India

RECYCLING

Recycling is the process of taking materials that are no longer needed and turning them into new products that can be used. The benefits of recycling include reduced environmental degradation and greenhouse gas emissions, economic savings, and job creation. A variety of products can be recycled; the complexity of the process depends on the materials being recycled and those being generated. It is important to keep in mind that recycling is the last step in waste reduction: the first step is reducing consumption of products, the second step is reusing the products that are consumed, and the third step, as a last resort, is recycling products that can no longer be reused.

In the most basic terms, recycling is taking a product that is no longer useful or wanted and remanufacturing it into another product that can be used instead of sending it to the landfill. Recycling can even be as simple as reusing an item for a purpose different from its original intention, such as making old cereal boxes into magazine holders. Although numerous items can be recycled, some of the most common products include aluminum, glass, paper, steel, and plastic bottles.

Although the expansion of recycling programs will always be beneficial, already over half of U.S. aluminum cans are made from recycled aluminum, and one-quarter of the raw fibers used in the paper industry are made from recycled paper products. Some products such as glass and steel can be recycled indefinitely, whereas other items have a limited recycling life span because they break down during the recycling process. Recycling in the

Stacking waste paper in a recycling plant. Paper recycling saves energy and water and can reduce water pollutants by 35 percent compared with making paper from new wood.

Source: iStockphoto.com

United States diverts an estimated 32 percent of waste away from the landfill. Many European countries, such as Germany, Sweden, Austria, and the Netherlands, have higher rates of recycling—40–60 percent.

Recycling is by no means a new concept or invention. Historically, when goods were not made so quickly or cheaply as they are today, almost everyone recycled, and disposing of trash in landfills was rare. With the industrial age came the production of cheap goods, and in many cases it made more economic sense to throw items away than reuse or recycle them. The drawbacks of a throw-away society were realized during the Great Depression, World War I, and World War II, when people began to recognize the need to conserve, reuse, and recycle to support their own families as well as the war efforts. However, following World War II, an economic boom led to widespread consumption and a reemergence of the throw-away society. It was not until the 1970s and the emergence of the environmental movement that large-scale household recycling programs began as a response to the throw-away society.

Almost anything can be recycled, but certain things are more common including paper (colored, white, newspaper), glass, steel, plastic, tin and aluminum cans, electronics, organic wastes, foods, tires, paints, and lumber. Recycled materials, as they are converted back into raw materials before being made into new products, rarely take on exactly the same form as the original product. Products made of recycled materials that are less desirable are called down-cycling (recycled paper that contains residue), and those that become more desirable through recycling are considered to be up-cycled (e.g., artwork made from used materials).

The complexity of the process used for recycling depends on the product. For instance, paper is sorted and then put in a hot chemical and water bath to reduce it to a soupy, fibrous substance. Magnets and gravity are used to filter out materials that are mixed in with the paper, such as staples and glues. Ink from the paper is typically removed by using a chemical wash or by skimming it off. The pulp, which is sometimes bleached, is then rolled onto flat sheets, pressed, dried, and cut to size for resale. For glass, sometimes the bottles are washed, refilled, and resold; in other cases, the glass is ground into fine bits and melted down to form new glass products.

In the United States, recycling generally takes on four forms: curbside pickup, where trucks collect containers (either items placed in separate containers or items placed together in a commingled container) from households; drop-off centers, where households take their recyclables to a central location; buy-back centers, which are similar to drop-off centers except that a payment is provided to the household based on material market value; and deposit or refund programs, in which a surcharge is added to certain products (e.g., Bottle Bills) and the deposit is given back when the product is returned at a drop-off location. (A Bottle Bill is a container-deposit law in some states that requires a minimum refundable deposit on beer, soft drink, and other beverage containers to encourage a higher rate of recycling or reuse of those containers.)

In many non-industrial countries, recycling occurs in a very informal manner. For instance, in Delhi, India, there are estimated to be hundreds of thousands of poor individuals sorting through the city's waste to collect items such as plastic, paper, glass, and metal that can be resold for recycling. Although this type of informal recycling program is controversial because it is a means of survival for many slum dwellers, some cities have encouraged it because of the economic savings realized from not having to develop a formal waste and recycling program. Recycling is beneficial, as it requires no or limited extraction of new materials to make a product. Recycling also diverts items from going to landfills. Landfill space in the United States is limited; landfills are also detrimental to the environment because they can leak and pollute groundwater sources. Recycling reduces energy consumption and its associated greenhouse gas emissions:

- One glass bottle saves enough energy to light a 100-watt light bulb for four hours.
- Every ton of paper recycled saves 4,100 kilowatt-hours of energy and 7,000 gallons of water and reduces water pollutants by as much as 35 percent.
- A recycled aluminum can saves 95 percent of the energy used in making a new can, not including the energy costs of extraction, and recycling paper cuts energy usage in half.
- Every pound of steel recycled saves 5,450 BTUs of energy—enough to light a 60-watt bulb for over 26 hours.

In addition to the environmental benefits and energy savings it generates, recycling also can contribute to the economy, provide jobs, and increase state revenues. For instance, in the state of Pennsylvania, tax revenues from the recycling and reuse businesses were estimated at $305 million per year in 2008, and the payroll for employees in the recycling business was $2.9 billion.

Recycling is gaining increased acceptance worldwide, but not everyone agrees that it is the best way to deal with the environmental problems of garbage. There are several criticisms of recycling. In some cases, such as for particular plastics, it may take more energy to recycle the product than it takes to create a new product (even accounting for the entire life cycle analysis). Some critics also claim that recycling creates pollution because of the chemicals needed to create the new product. Another argument is that recycling provides a false sense of security, as people may feel that it is okay to consume larger amounts of products, such as disposable plastic water bottles, because overconsumption can be made up for by recycling.

Although the United States is recycling a higher percentage of waste than in the past, we create more waste per capita each year. In 1960, Americans averaged 2.68 pounds of waste per person per day, but this jumped to 4.4 pounds per person per day by 1997. Today, it is very easy to find one-time use items created for convenience, which only increases the amount of waste produced.

Although recycling programs continue to grow and gain acceptance—especially as the economics of recycling have improved, laws have been developed and widespread education plans have occurred throughout the United States—there has been some struggle to overcome recycled product stereotypes. Recycled products have been considered to be of lower quality and less durable. Without closing the loop (i.e., buying recycled products), recycling programs will continue to struggle.

See Also: Environmentally Preferable Purchasing; Life Cycle Analysis; Sustainability.

Further Readings

AA/CMI/ISRI. "U.S. Aluminum Can Recycling Grows in 2007: Highest Recycling Rate in Six Years." July 21, 2008. http://www.aluminum.org/AM/Template.cfm?Section=Home &TEMPLATE=/CM/ContentDisplay.cfm&CONTENTID=26715 (Accessed December 2008).

Chaturved, Bharati. "Ragpickers—The Bottom Rung in the Waste Trade Ladder." http://www.ecologycenter.org/iptf/Ragpickers/indexragpicker.html (Accessed December 2008).

Friends of the Earth. "Recycling Rates Increasing—But Fifty Percent Is Just the Start." Press release, January 23, 2006. http://www.foe.co.uk/resource/press_releases/recycling_rates_ increasing_23012006.html (Accessed December 2008).

Grabianowski, Ed. "How Recycling Works." http://science.howstuffworks.com/recycling7 .htm#McCorquodale (Accessed December 2008).

Lane County Public Works Waste Management Division. "Lane County Recycler's Handbook." 2007. http://www.lanecounty.org/PW_WMD_Recycle/documents/Brochures/ UPDATED08Recyclers_Guide2bWEB.pdf (Accessed December 2008).

Pennsylvania Department of Environmental Quality. http://www.dep.state.pa.us/dep/deputate/ airwaste/wm/recycle/Recycle.htm (Accessed December 2008).

U.S. Environmental Protection Agency. "Recycling." http://epa.gov/msw/recycle.htm (Accessed December 2008).

U.S. Environmental Protection Agency. "Wastes." http://www.epa.gov/epawaste/index.htm (Accessed December 2008).

Stacy Vynne
University of Oregon

RENEWABLE ENERGIES

Fossil fuels are being consumed at much faster rates than they are produced in the Earth's crust; hence, the world's supply of these fuels is being depleted. The depletion of these nonrenewable fuels is a major challenge facing humanity. There are alternative sources of energy that are renewable and not based on the burning of fossil fuels. Renewable energy resources are those from sources that are capable of being continually replenished after exploitation. They are not exhaustible within human timescales. They occur naturally and abundantly and provide relatively clean energy that does not produce greenhouse gases like fossil fuels do. The prospect of reducing the world's dependence on fossil fuels is, however, challenging. As global energy consumption grows each year, development of alternative energy sources becomes increasingly important. These alternative sources of energy include hydroelectric and hydrodynamics energy, solar energy, nuclear energy, geothermal energy, wind power, and biomass energy.

The criteria used in judging an energy source include its renewability, conversion into usable forms, and environmental and health issues. The cost of maintaining energy security in the world is high. Renewable energies can help to diversify energy supply and increase energy security. The potential for conflict, sabotage, and disruption of production and trade of fossil fuels cannot be dismissed. At this time, many countries depend on fossil fuels. Dependence on imported fuels leaves these countries vulnerable to disruption in supply, which might pose physical and economic hardships. Importation of crude oil, for instance, requires significant capital and infrastructure, and is often a strain on national budgets. Consequently, many countries make reducing dependence on foreign energy sources a high priority.

Hydroelectric Power

Flowing water generates energy that can be harnessed and converted to electricity. This is hydroelectricity. It comes from the damming of rivers and utilizing the potential energy stored in the water. As the water stored behind a dam is released at high pressure, its kinetic energy is transferred onto turbine blades and used to generate electricity. Hydroelectric power does not necessarily require a large dam. Some hydroelectric power plants use small canals to channel the river water from a great height through a turbine. Small hydropower plants of about 10 megawatts (MW) harness small rivers and streams. This resource has been a mainstay of rural energy development.

Hydropower is the most prominent renewable energy generator. It generates 76 percent of all electricity produced from renewable resources. About 10 percent of all power generation in the United States, for example, is through hydropower. Although this system is expensive to construct, hydroelectric power facilities have relatively low maintenance costs and are economical.

Hydroelectric power systems have many advantages and disadvantages. A major advantage of hydroelectric systems is the elimination of fuel costs. In comparison to fuel-fired generation systems, hydroelectric power facilities often have a greater longevity and are less expensive to operate and maintain. The dam also has non-energy related benefits, such as providing new recreation areas for water sports. However, there are several major disadvantages of hydroelectric systems. Hydroelectric dam construction requires the usage of nonrenewable energy resources, thus releasing significant amounts of carbon dioxide. People living where the reservoirs are planned are dislocated. Aquatic ecosystems, birds, and other wildlife are disrupted or destroyed. Dam walls also run the risk of collapse, either from mechanical failure or acts of sabotage and terrorism. Despite these negative aspects, hydroelectric power is growing, particularly in developing nations such as China. In developed nations, hydroelectric power is now more difficult to site because most suitable locations within these nations are either already being exploited or may be unavailable due to environmental and cost considerations.

Solar Energy

Solar energy manifests itself as low or high temperature solar heat, wind electricity, and photovoltaics (PV). Low temperature solar heat is produced by the absorption of sunlight by darkened surfaces that convert it into heat that can be used for warming water or other fluids. High temperature solar heat can be obtained by focusing sunlight and heating fluids to high temperature that can be used to generate electricity. Solar intensity changes coupled with topographic variability establish air movement or wind, a mechanical energy that can be converted into electricity with wind turbines, for example. Photovoltaics are a collection of semiconductors designed to convert solar energy into electricity using the photoelectric effect, a process that creates energized particles from the interaction of sunlight and electrically conducive materials. Solar technologies such as these harness the sun's energy and light to provide electricity, heat, light, hot water, and cooling for homes, businesses, and industry.

The energy from the sun is radiated in all directions in space in the form of electromagnetic radiation. Mean solar radiation (or insolation) incident upon a plane perpendicular to the Earth at the top of the atmosphere is 1367 W/m^2, a value known as the solar constant. However, the solar energy actually received at the earth's surface is considerably less and amounts depend on latitude, season, local elevation, surface type, and regional atmospheric conditions. Most radiation is absorbed at low latitudes, but the excess energy is redistributed to higher latitudes in the form of winds and ocean currents. In particular, the oceans absorb a major fraction of the incoming radiation since approximately three-quarters of the earth's surface is covered by them. Solar technologies adjust for spatial and temporal insolation variability, such as solar panels that correct for local sun angle to maximize insolation intensity.

Solar energy is a major choice in developing an affordable, feasible, decentralized global power source that can be used in nearly all climate zones. Solar energy can be applied in many ways, including: the generation of electricity using photovoltaic solar cells, concentrated solar power, solar updraft towers, and geosynchronous orbiting solar power satellites; the creation of hydrogen using photoelectrochemical cells; heating and cooling buildings through solar chimneys and passive solar building designs; and domestic heating uses, such as solar ovens for cooking and solar-thermal panels for hot water and space heating. The development of solar power systems as an alternative source of electricity could play significant complementary roles in providing isolated power systems in remote regions for lighting, water-pumping stations, communications packages, or medical equipment in hospitals. On a larger scale, solar energy could be used to run vehicles, power plants, and spaceships. Solar energy also helps create other renewable energy resources, such as the distribution of precipitation for water that is tapped by hydroelectric projects and for the growth of plants used to create biofuels. The total energy received each year from the sun is 35,000 times the total energy used by modern society, thus offering a major renewable energy resource potential. While solar energy is free, the equipment to collect and convert solar energy to electricity can be costly.

Geothermal Energy

Geothermal energy is energy obtained by tapping the heat of the earth, usually from kilometers deep into the Earth's crust. It is heat energy derived from the internal heat (molten core) of the planet and can be used to generate steam to run a steam turbine that is used to generate electricity or to provide heat. The sun also contributes to the heat gained by the Earth during daylight hours, which then radiates away during the night. Only a few meters down, the temperature tends to remain constant at about 10–16 degrees C and usually varies by a few degrees (more in the winter) from the surface and air temperature. This is the principle on which a ground-source heat pump works. The amount of temperature variance depends on the geology of the soil, climate, and season. Geothermal electricity accounts for 4 percent of all electricity derived from renewable resources.

Geothermal heat pumps constitute one of the fastest-growing applications of renewable energy in the world, with an annual growth rate of 10 percent in about 30 countries over the past 10 years. Its main advantage is that it uses normal groundwater temperatures (between 5 degrees C and 30 degrees C) that are available in all countries of the world.

Ground-source heat pumps use the relatively constant temperature of the Earth to provide heating, cooling, and domestic hot water for homes, schools, governments, and commercial buildings. A small amount of electric input is required to run their compressors that enables heat to flow from a lower to higher temperature location in compliance with the Clausius statement of the second law of thermodynamics. Geothermal energy has the advantage of being neither weather dependent or subject to fluctuating fuel prices. Three types of power plants are used to generate power from geothermal energy: dry steam, flash, and binary plants. Dry steam plants take steam out of fractures in the ground and use it to directly drive a turbine that spins a generator. Flash plants take hot water, usually at temperatures over 200 degrees C, out of the ground and allow it to boil as it rises to the surface, then separate the steam phase in steam/water separators and run the steam through a turbine. In binary plants, the hot water flows through heat exchangers, boiling an organic fluid that spins the turbine. The condensed steam and remaining geothermal fluid from all three types of plants are injected back into the hot rock for reheating.

The geothermal energy from the core of the Earth is closer to the surface in some areas than in others. Where hot underground steam or water can be tapped and brought to the surface, it may be used to generate electricity. There is also the potential to generate geothermal energy from hot dry rocks. Holes at least 3 kilometers deep are drilled into the Earth. Some of these holes pump water into the Earth, while other holes pump hot water out. The heat resource consists of hot underground radiogenic granite rocks, which heat up when there is enough sediment between the rock and the Earth's surface. The world's largest geothermal power installation is The Geysers complex in northern California, with a rated capacity of 750 MW or enough electricity to power a city the size of San Francisco. Although geothermal sites are capable of providing heat for many decades, eventually specific locations may cool down.

Wind Energy

Wind energy is used to generate electricity using wind turbines. It is another energy source that could be used without producing by-products that are harmful. Wind turbines, similar to windmills, are mounted on a tower to capture most of the wind energy at 30 meters or more above the ground. They can take advantage of the faster and less-turbulent wind. The turbines catch the wind energy with their propeller-like blades. Harnessing the wind is highly dependent on weather and location.

Wind power airflows can be used to run wind turbines. Modern wind turbines range from about 600 kW to 5 MW of rated power. The power output of a turbine is a function of the cube of the wind speed; thus, as wind speed increases, power output increases. Areas where winds are stronger and more constant, such as offshore and high altitudes, are preferred locations for wind farms. Globally, the long-term technical potential of wind energy is believed to be five times the total current global energy production, or 40 times the current electricity demand. This could require large amounts of land to be used for wind turbines, particularly in areas of higher wind resources. Wind power is growing at the rate of 30 percent annually, with a worldwide installed capacity of over 100 GW, and is widely used in several European countries and the United States. Wind generates about 3 percent of all renewable electricity globally.

Wind power is renewable and produces no greenhouse gases during operation. Wind power is one of the most environmentally friendly sources of renewable energy. A wind farm, when installed on agricultural land, has one of the lowest environmental impacts of all energy sources:

- It occupies less land area per kilowatt-hour (kWh) of electricity generated than any other energy conversion system, apart from rooftop solar energy, and is compatible with grazing and crops.
- It generates the energy used in its construction in just 3 months of operation, yet its operational lifetime is 20–25 years.
- In substituting for base-load coal power, wind power produces a net decrease in greenhouse gas emissions and air pollution and a net increase in biodiversity.
- Modern wind turbines are almost silent and rotate so slowly that they are rarely a hazard to birds.

Biomass Energy

Biomass or bio energy is the energy released principally through the burning of wood, plant materials, and organic wastes. Biomass can be defined as the biodegradable waste products and residues from agriculture (including vegetables and animal substances), forestry, as well as the biodegradable action of industrial and municipal wastes. With the exception of practices leading to deforestation, biomass energy is considered a renewable energy.

Biomass in the form of organic waste can be incinerated or processed further to produce a fuel. This is normally achieved through a number of methods including anaerobic digestion, pyrolysis, and gasification. These processes produce a biogas, generally methane, which is then burned to produce energy. Hydrodynamics is applied to the flow of liquids or to low-velocity gas flows where the gas can be considered as being essentially incompressible. This principle can be used to generate renewable energy. In many cases energy crops are also put through a gasification process and the subsequent biogas is then burned in gas-fired boilers to produce steam that in turn is used to produce electricity (steam turbines). This process involves the breakdown of hydrocarbons to a usable fuel gas by carefully controlling a heat reaction with critical amounts of oxygen.

Anaerobic digesters used in biomass energy production rank second to only to hydropower in the renewable energy production, generating 17 percent of all electricity produced from renewable resources. The process entails the organic matter being broken down by bacteria in the absence of air to produce a biogas and a residue. The biogas, normally methane, can then be used as a fuel in gas engines or boilers. The residue, called the digestate, can be returned to the land in the form of a fertilizer that is high in organic nutrients.

Biomass generates about the same amount of CO_2 as fossil fuels. All biomass needs to be grown, collected, dried, fermented, and burned. All of these steps require resources and an infrastructure. Corn, for example, is typically 66 percent starch; the remaining 33 percent is not fermented. This unfermented, otherwise-called distillers grain, is high in fats and proteins and makes good animal feed. In Brazil, where sugar cane is used, the yield is higher, and conversion to ethanol is somewhat more energy efficient than corn. Recent developments with cellulosic ethanol production may improve yields even further. Cellulosic ethanol can be made from plant matter composed primarily of inedible cellulose fibers that form the stems and branches of most plants. Crop residues (such as cornstalks, wheat straw, and rice straw), wood waste, and municipal solid waste are potential sources of cellulosic biomass. Dedicated energy crops, such as switchgrass, are also promising cellulose sources that can be sustainably produced.

Brazil has one of the largest renewable energy programs in the world, involving production of ethanol fuel from sugar cane. Ethanol provides 18 percent of the country's automotive fuel. Since the 1970s, Brazil has had an ethanol fuel program that has advanced the country to being the world's second-largest producer of ethanol (after the United States) and the world's largest exporter of the fuel. Brazil's ethanol fuel program uses modern equipment and cheap sugar cane as feedstock, and the residual cane waste (bagasse) is used to process heat and power. There are no longer light vehicles in Brazil running on pure gasoline. By the end of 2008 there were 35,000 filling stations throughout Brazil with at least one ethanol pump. Most cars on the road today in the United States can run on blends of up to 10 percent ethanol, and motor vehicle manufacturers already produce vehicles designed to run on much higher ethanol blends.

Renewable Energies Have Problems Too

In light of environmental concerns about fossil fuels, the use of renewable energy is widely viewed as a more attractive, clean, and pollution-free alternative source of energy. However, the reality is that renewable energy sources also produce adverse environmental impacts that vary with the particular energy source. Indeed, there is no completely pollution-free source of electricity. Alternative energy industries currently account for 14 percent of energy consumed worldwide. To date, alternative energy sources have been hindered by technological and environmental difficulties. For instance, although the uranium that fuels nuclear power is abundant, the risk of nuclear accidents and the difficulty associated with safe disposal of radioactive waste have led to the decline of the nuclear power industry. Radioactive materials are produced at each step in the nuclear fuel cycle. The highest levels of radioactivity are found in the spent fuel that is periodically removed from nuclear reactors when it is no longer efficient at generating heat. The safe and permanent disposal of these high-level radioactive wastes represents one of the greatest environmental challenges.

While most renewable energy sources do not produce pollution directly, the materials, industrial processes, and construction equipment used to create them may generate waste and pollution. Wind and solar energies require storage systems whose environmental impact is dependent on the technology employed. Storage batteries, for instance, are currently composed of toxic trace metals such as lead, nickel, and cadmium. There are challenges in disposing of PV solar cells in solar energy implementation. In wind power, there are visual impacts and electromagnetic interferences. Geothermal energy releases H_2S apart from causing local seismic effects and noise. In biomass energy application, there are ecological impacts of harvesting and transportation, loss of species diversity, and atmospheric emissions during harvesting and conversion. Another environmental issue, particularly with biomass and biofuels, is the large amount of land required to harvest energy, which otherwise could be used for other purposes.

Renewable energy resources also lack available and affordable transmission capacity to deliver the energy to customers. Solar and wind power are unreliable as steady sources of energy as they depend on the uncertainties of weather, thus creating an intermittent energy output. This is indeed the case for PV that requires sunshine, an energy that can be erratic in some locations due to cloud cover, pollution concentration, and seasonal sun angle variability. In the case of wind, the problem of intermittency can be solved by feeding the electricity generated into large grids. Another solution is to use electricity to compress air that can be stored and can generate electricity when needed. However, energy from geothermal, small hydroelectric, and biomass sources does not suffer from such shortcomings.

The Future of Energy

The present energy system is heavily dependent on the use of fossil fuels. Worldwide coal, oil, and gas account for 80 percent of primary energy consumption. Fossil fuel combustion is the prime source of CO_2 emissions, which are growing at the rate of 0.5 percent per year. Emissions of anthropogenic greenhouse gases are altering the atmosphere in ways that are affecting the climate. Changes have already been observed in climate patterns that correspond to scientific projections based on increasing concentrations of greenhouse gases. This is a serious challenge to sustainable development, and the main strategies to prevent it include more efficient use of energy, increased reliance on renewable energy sources, and accelerated development and deployment of new and advanced energy technologies.

Renewable energy is suitable for developing countries. In rural areas, transmission and distribution of energy generated from fossil fuels are difficult and expensive. Renewable energy projects in many developing countries have shown that they can directly contribute to poverty alleviation by providing the energy needed for creating businesses and employment. Renewable energy technologies can also make indirect contributions to alleviating poverty by providing energy for cooking, space heating, and lighting. They can also contribute to education by providing electricity to schools. The renewable energy market could increase fast enough to replace and initiate the decline of fossil fuel dominance, and the world could then avert the looming climate and peak oil crises.

Renewable energy sources are generally sustainable, with some lasting for billions of years, but the infrastructure created to exploit these resources has a more limited lifespan. Hydroelectric dams will not last indefinitely but must be repaired and eventually removed and replaced. Events like the shifting of riverbeds or changing weather patterns could potentially alter or even halt the function of hydroelectric dams, lowering the amount of time they are available to generate electricity. Biomass and geothermal energies require wise management for sustainability. The rapidly growing renewable energy industries and service sectors in many countries show clear evidence that the systematic promotion of such new technologies offers great opportunities for innovation, for the development of energy markets with locally or regionally oriented value chains, and thereby for the creation of new jobs. While the development and deployment of new state-of-the-art renewable energy technologies such as wind or photovoltaic energy require highly skilled, knowledge-intensive workforces in industrialized countries, developing countries can benefit economically from an increased use of improved biomass-based energy generation. The main beneficiaries of the adoption of renewable sources of energy will be the developing countries, where biomass will be used widely with very inefficient and wasteful technologies for cooking and heating.

The full potential and advantages of renewable energies are hindered at present because the costs of fossil fuels do not reflect their full cost. They are subsidized in several parts of the world. Removing subsidies from fossil fuels will make renewable energies competitive in many areas. Renewable energy is inexhaustible and abundant. The use of fossil fuels as sources of energy will not last forever and have proven to be one of the main sources of our environmental problems. It is clear that in time, renewable energies will dominate the world's energy system, because of their inherent advantages such as mitigation of climate change, generation of employment, and reduction of poverty, as well as increased energy security and supply.

See Also: Alternative Fuels; Biomass Energy; Electricity; Hydroelectric Power; Solar Energy; Tidal Power; Three Gorges Dam; Wind Power.

Further Readings

Da Rosa, Aldo V. *Fundamentals of Renewable Energy Processes.* London, UK: Academic Press, 2009.

Goldemberg, José. "Rationale for Renewable Energies." Thematic Background Paper of the 2004 International Conference for Renewable Energies. http://www.renewables2004.de/pdf/tbp/TBP01-rationale.pdf (Accessed August 2009).

Kemp, William H. *The Renewable Energy Handbook: A Guide to Rural Energy Independence, Off-Grid and Sustainable Living.* Tamworth, Ontario, Canada: Aztext Press, 2006.

Quaschning, Volker. *Understanding Renewable Energy Systems.* London, UK: Earth Scan Publications, 2005.

Wengenmayr, Roland and Thomas Bührke. *Renewable Energy: Sustainability Concepts for the Future.* Weinheim, Germany: Wiley-VCH, 2008.

Akan Bassey Williams
Covenant University

RENEWABLE ENERGY PORTFOLIO

We use energy generated from sources like fossil fuel and coal for our industrial and household energy needs because they are proven technologies and are easily available. Thus, we compromise with the greenhouse gas production and endanger our future through consequent global warming and climate change. However, the green energy produced through renewable energy sources (RES) would reduce the production of greenhouse gases. The United States, being a champion of advanced technology and the consumer of a large percentage of the world's energy, has taken positive steps toward the goal of reducing greenhouse gases. Although new, the technologies to produce energy from RES are proving better and cheaper with time.

In the United States, each state is required to contribute toward greenhouse gas reduction with the use of renewable energy (RE) as one of the sources of energy production. The RE portfolio (REP) standard—or renewable portfolio standard, commonly known as RPS—is probably one of the most effective ways toward this goal. REP is a state regulatory policy that requires the state electricity providers to obtain an increased and a minimum percentage of their power from RE resources by a certain date. The states have set goals to produce energy through RES, such as solar water heat, solar thermal electric, photovoltaic, landfill gas, wind, biomass, hydroelectric, geothermal electric, geothermal heat pumps, tidal energy, wave energy, ocean thermal, anaerobic digestion, fuel cells using renewable fuels, and so on.

Per RPS, the electricity supply companies are obligated to produce a certain percentage of their power through the use of RES. They earn a certificate for every unit of their electricity generated from RES and send the certificate to the RPS regulatory body to comply with the standard. Supporters of the RPS claim that the RPS implementation encourages competition, efficiency, and innovation, so that eventually energy from RES will be cheaper than the present low cost of fossil fuel.

In addition to the United States, the United Kingdom, Italy, and Belgium have adopted an RPS-type mechanism. Presently, there are 34 states plus the District of Columbia that have RPS policies in place. Together these states account for more than half of the electricity sales in the United States. Of the 34 states, five have set voluntary goals for adopting RE instead of portfolio standards with binding targets. These five states are Missouri, North Dakota, South Dakota, Utah, and Virginia. Another 17 states have yet to implement the RPS.

The amount of energy produced from RES based on RPS ranges from modest to ambitious. Some states have lofty goals of up to 27 percent of energy production coming from renewable sources by 2025 at the latest. However, all the states that have RPS or voluntary goals are determined to produce at least 10 percent of their energy from RES. Again, by state the definitions of RE vary, but their sources are from the RES mentioned above. Table 1 shows the list of states that have implemented RPS or have voluntary goals to produce energy through RES, along with a rough summary of state renewable portfolio standards and organizations that are administering these standards or assisting with the details. Percentages of states' goals and target dates are shown in the table. The date refers to when the full requirement of RPS takes effect.

Some of the states have different goals within the RPS. Xcel Energy in Minnesota has a goal of generating 30 percent of its electricity from RES by 2020, separate from the state's own goal. Many states like New Jersey and Nevada have the additional goal of generating at least some percentage of the state's total electricity production from solar sources or some other fixed sources. Many states have set staggered targets over multiple years; for example, North Carolina has a 10 percent goal by 2018 and a 12.5 percent goal by 2021.

States adopt the RPS with the goals of creating jobs, securing energy supply, reducing dependence on foreign oil, and obtaining cleaner air. The states may not have adopted the RPS with the prime motivation of addressing climate change, but RE use does deliver significant greenhouse gas reductions. For example, Texas is expected to avoid 3.3 million tons of CO_2 emissions annually with its RPS, which requires 2,000 megawatts of new renewable generation by 2009. After 2015, its greenhouse gas emissions can be reduced by almost 9 million tons annually with its RPS goal of 5,880 MW power generation from RES.

According to the Pew Center on Global Climate Change, many of these efforts toward achieving the RPS goals have been successful. For example, the state of Connecticut was able to meet its goals before the deadline, and in 2007, the state governor extended the previous RPS to all utilities in the state. The revised RPS now is 27 percent; that is, 27 percent of the state's electricity is required to come from renewable sources by 2020. The state also has separate standards for three classes of renewables, such as (1) 20 percent of the renewables must be from class 1, (2) 3 percent must be from class 1 or 2, and (3) 4 percent must be from class 3. The Connecticut standard of class 1 sources are solar, wind, new sustainable biomass, landfill gas, fuel cells (using renewable or nonrenewable

Table I Summary of State Renewable Energy Portfolio

State	Amount	Year	Organization Administering RPS
Arizona	15%	2025	Arizona Corporation Commission
California	20%	2010	California Energy Commission
Colorado	20%	2020	Colorado Public Utilities Commission
Connecticut	27%	2020	Department of Public Utility Control
District of Columbia	11%	2022	DC Public Service Commission
Delaware	20%	2019	Delaware Energy Office
Florida	RPS under development		
Hawaii	20%	2020	Hawaii Strategic Industries Division
Iowa	105 MW		Iowa Utilities Board
Illinois	25%	2025	Illinois Department of Commerce
Massachusetts	15%	2020	Massachusetts Division of Energy Resources
Maryland	20%	2022	Maryland Public Service Commission
Maine	30%; 10% more	2000; 2017	Maine Public Utilities Commission
Michigan	10%	2015	State Government**
Minnesota	25%	2025	Minnesota Department of Commerce
Missouri*	15%	2021	Missouri Public Service Commission
Montana	15%	2015	Montana Public Service Commission
New Hampshire	25%	2025	New Hampshire Office of Energy and Planning
New Jersey	22.5%	2021	New Jersey Board of Public Utilities
New Mexico	20%	2020	New Mexico Public Regulation Commission
Nevada	20%	2015	Public Utilities Commission of Nevada
New York	25%	2013	New York Public Service Commission
North Carolina	12.5%	2021	North Carolina Utilities Commission
North Dakota*	10%	2015	State Government**
Ohio	25%	2025	State Government**
Oregon	25%	2025	Oregon Energy Office
Pennsylvania	18%	2020	Pennsylvania Public Utility Commission
Rhode Island	16%	2020	Rhode Island Public Utilities Commission

(Continued)

(Continued)

South Dakota*	10%	2015	State Government**
Texas	5,880 MW	2015	Public Utility Commission of Texas
Utah*	20%	2025	Utah Department of Environmental Quality
Vermont	25%	2025	Vermont Department of Public Service
Virginia*	12% of 2007 sales	2022	Virginia Department of Mines, Minerals, and Energy
Washington	15%	2020	Washington Secretary of State
Wisconsin	10%	2015	Public Service Commission of Wisconsin

Source: PEW Center on Global Climate Change, 2009.

*State has set voluntary goals for adopting RE instead of portfolio standards with binding targets.

**State government has set the goals, but the administering authority is unknown.

fuels), ocean thermal power, wave or tidal power, low-emission advanced RE conversion technologies, and new run-of-the-river hydropower facilities with a maximum capacity of 5 megawatts. Class 2 sources are trash-to-energy facilities, biomass facilities not included in class 1, and certain hydropower facilities. Class 3 sources are customer-side power savings and state electricity savings from conservation and load management programs, and so on. Iowa has already met its RPS goal, set in 1999. Many states allow utilities to comply with the RPS through tradable RE credits. As described earlier, many states are implementing RPS voluntarily after observing the success of states that had adopted their RPS earlier.

It is noteworthy that the RPS effort has been very successful because of federal production tax credits (PTC) policies. Thus, a surge in RE capacity has been encountered in the country. However, with the withdrawal of PTC, the efficiency of energy production from RES has decreased. It is expected that in the future, without PTC, states would show new vigor in clean energy generation. However, the Edison Electric Institute, a trade association for America's investor-owned utilities, argued that a nationwide RPS would raise consumers' electricity prices and create inequities among states because of the imbalance in production costs of energy from unconventional sources and from easily available sources like fossil fuel and coal.

We cannot put a price on the catastrophe of global warming. It is practical to pay some extra money toward our energy bill now to save our lives from future peril.

See Also: Electricity; Feed-In Tariff; Fossil Fuels; Global Warming; Greenhouse Gases; Public Utilities; Renewable Energies.

Further Readings

Panda, S. S. "Energy, Renewable." In *Encyclopedia of Global Warming and Climate Change,* S. G. Philander, ed. Thousand Oaks, CA: Sage, 2008.

Pew Center on Global Climate Change. "Renewable Portfolio Standards." 2009. http://www.
    pewclimate.org/what_s_being_done/in_the_states/rps.cfm (Accessed January 2009).
U.S. Department of Energy. "States With Renewable Portfolio Standards." http://apps1.eere
    .energy.gov/states/maps/renewable_portfolio_states.cfm (Accessed January 2009).

Sudhanshu Sekhar Panda
Gainesville State College

Risk Assessment

A risk is the likelihood of harm from a hazard or any circumstances that may produce an unfavorable outcome to the natural or built environment, including human physical and mental well-being. Risk is the probability that an undesirable event (in a system of discrete outcomes) or output level (in a system of continuous outcomes) that might occur as the consequence of some action or series of actions. Risk analysis/assessment is a planning and decision support process that estimates the risk that an undesirable event—or output level—will occur and then develops and institutionally allocates that risk. A reliable analysis of risk—one that enables a confident decision about the future—requires an understanding of a system's functions and a predictability of its outputs. To the extent that a system's functions are quantitatively uncertain and that its outcomes are qualitatively undesirable, the risk associated with a particular outcome is high. In general, as the complexity of a system's structure and function increases, so does the potential uncertainty of its outcomes; conversely, as one approaches certainty, risk approaches zero.

Probability and statistics, the languages of risk analysis, were developed beginning in the mid-17th century in response to a demand to better understand games of chance. Insurance companies of the 17th century were early adopters of these mathematical principles, utilizing probability to determine the likelihood for a successful maritime trade venture. The notion of risk, however, only emerged as an object of scientific inquiry in the early 20th century, coinciding with the emergence of technologies that drove the Industrial Revolution.

The evolution of complex technologies and the growth in our understanding of their complexity underlying scientific, social, and ecological systems have revealed our increasingly limited ability to understand the inner workings of many of these systems. In the modern age, scientists have compensated for this lack of certainty by developing processes by which they strive to quantify the extent to which we do not understand a particular system. Using probabilistic reasoning, risk analysis provides us with a measure of our uncertainty.

Risk assessment enables decision makers—individuals or societies—to balance uncertain knowledge in the present against future utility. Arguably, the rapid growth of economy and wealth experienced in the industrialized nations in the 20th century could not have occurred had it not been for the development and application of modern risk assessment techniques. We use risk analysis in a Machiavellian wager that a particular future is worth the risks attendant in its attainment. On the basis of such analyses, we insure our lives against the risk of untimely death and our automobiles against an unexpected wreck. It is difficult to imagine modern life without risk assessment.

Cost-benefit analysis dominates the field of risk analysis and assessment. Derivative of neoclassical economic theory, it assumes that all relevant costs and benefits can be known and priced. Aggregate costs are compared with aggregate benefits, and if the former are greater than the latter, a decision in favor of the outcome is justified. Unfortunately, as systems become more complex and our understanding of their functions becomes more uncertain, we are unlikely to know either all the outcomes or be unable to realistically price them even if we could.

The Emergence of Precaution

The combination of systems complexity and influence has transformed the nature of risk in the late 20th and early 21st centuries. As technology has come to dominate industrialized life, so too has the complexity of systems on which it depends increased, often to the extent that no one is able to totally comprehend either their function or the total set of their consequences. Furthermore, as our world has become more integrated by communication and transportation technologies, so too has the extent to which risks derivative of these technologies are shared more widely among global communities. Our risk analyses increasingly fail to properly assess the consequences of our choices. Climate change is a particularly relevant example of the applications—and shortcomings—of industrial risk assessment methods to a fundamentally "postindustrial" problem.

Earth's climate is determined by exceedingly complex and interrelated atmospheric, oceanic, and geophysical processes that scientists still do not entirely understand. To the extent that we can explain certain atmospheric phenomena such as cold fronts, hurricanes, and temperature variability, sophisticated computer models have been created to forecast future global climate change and more regional weather phenomena. When leavened with social valuations of possible climate futures, these models help us understand the risks associated with continuing to base our economies on carbon-based fuels. The likely consequences of accepting these risks include rising global average temperatures, melting polar ice caps, rising sea levels, and widespread species dislocation and extinction.

Traditional cost-benefit analysis of the sort employed in *The Stern Review* attempts to quantify these risks and concludes that present action to mitigate climate change would be far less expensive economically than adapting to its consequences in future generations. Stern's risk analysis assumes a continuous change in climate characteristics to which human societies would be able to react. However, it fails to address the chaotic nature of climate systems. Present (risky) behavior could result in crossing an unforeseen "tipping point" that drives our climate systems into an entirely unexpected state. How does one quantify this small, but qualitatively terrifying, risk?

It is evident that an entirely different type of risk assessment is needed—one that more effectively, if more qualitatively, captures the profound uncertainty associated with some risks. An alternative to Promethean risk analysis has emerged that recognizes that prudence dictates that we refrain from action that exacerbates such risk. Cognizant of such outcomes (e.g., nonlinear climate change), the precautionary principle counsels humility in the face of uncertainty. Just because we cannot either understand or quantify such a terrible risk does not imply that it does not exist or cannot be mitigated.

The Future of Risk Assessment

By its very nature, risk assessment is future oriented. It informs democratic debate and helps decision makers—and society—decide what paths to take and which futures to desire. To the extent that the risks we face are not accurately reflected in our analyses of them, our future is precarious and uncertain, at best. To the extent that we do not understand the future consequences of our present decisions, we are unprepared as a society to cope with the future we choose.

See Also: Climate Change; Forecasting; Innovation.

Further Readings

Beck, Ulrich. *Risk Society: Towards a New Modernity.* London, UK: Sage, 1992.

Montague, Peter. "The Precautionary Principle in the Real World." *Environmental Research Foundation.* http://www.precaution.org/lib/pp_def.htm (Accessed February 2009).

Stern, Nicholas. *The Economics of Climate Change: The Stern Review.* Cambridge: Cambridge University Press, 2007.

Kent Hurst
University of Texas at Arlington

S

SMART GRID

The smart grid is an electricity transmission and distribution system that combines electric generation and distribution technologies with high-tech information and communications technologies. The benefits of a smart grid are many, although those who advocate its development are primarily concerned with developing an interactive electric transmission system that can facilitate the management, supply, demand, and distribution of electricity. Importantly, however, the development of a smart electric grid would allow for increased efficiencies in the production and consumption of electricity. It could also reduce the likelihood of transmission problems and blackouts and increase the flexibility in electricity distribution.

More than just saving energy, smart grid technologies provide the opportunity to reduce energy expenditures, particularly in the residential and commercial sectors. Whether through increased efficiencies, reductions in energy demand, or reductions in peak load demands, the development of a smart grid is expected to result in significant energy cost savings. For example, a variety of recent studies by the Electric Power Research Institute and similar organizations indicate that potential energy savings enabled by the smart grid could reduce U.S. electricity demand by an estimated 6 percent, whereas peak demand energy reductions could be as high as 27 percent. Given that power generation is already a costly endeavor for utilities and consumers, the reduction in peak demand alone would save an estimated \$175–\$332 billion over 20 years.

The current smart grid vision entails adding a unified communications and control system to the existing power delivery infrastructure so as to enable

- the transition from a radial grid system to a network of distributed and interconnected electricity generators and consumers,
- the conversion from an electro-mechanical system to one that uses digital technologies and allows for multidirectional information streams and systems automation, and
- multidirectionality in communications among the many points on the grid.

At this time, electricity is delivered over a grid or network that continues to be strongly rooted in decades-old technologies. For example, the current system cannot distinguish the path through which a quantity of electricity reaches a customer and is limited in its ability to accommodate decentralized power systems and manage intermittent energy resources

397

during peak demand periods. Current grid technologies are also unable to provide real-time pricing information, reducing the viability of market mechanisms to properly allocate scarce resources.

Of significant importance, the smart grid offers a means by which transmission-related electricity losses can be reduced. At this time, approximately two-thirds of the electricity that is generated in the United States is never used. Instead, this energy is lost in transmission. The transmission losses occur as electricity travels from source to end-user over the transmission lines. In the process, only one-third of the energy generated actually makes it to the end user. With smart grid technology, it is possible to reduce line losses through the use of intelligent technology that delivers energy as close to the end-user as possible. This capability is one of many that can be performed automatically, without the mechanical switches and manual controls currently necessary to deliver power to consumers.

The smart grid can also increase the reliability of electricity resources. As a society we are incredibly dependent on the reliable provision of electricity. As such, the electricity infrastructure must adequately and reliably adjust to large variations in energy demand across a broad spectrum of consumers to ensure that energy demands are met at all times. Unfortunately, given the limitations of the current system, decentralized and intermittent sources of power (like wind and solar) create an increased prospect of instability in electricity supply. Although renewable energy resources are great for providing low-carbon sources of electricity, these supplies tend to be less consistent than traditional fossil fuel–burning power plants. Smart grid technologies can help overcome the challenge of variable electricity supplies by providing a much more sophisticated capacity for integrating and effectively distributing decentralized and intermittent electrical inputs into the grid, including electricity generated from wind, solar, and combined head and power systems. With a smart grid, wind and solar power generators can have power storage on-site, in the form of batteries, to save the excess power that is generated on especially gusty and sunny days. Smart grids provide the capacity to dispatch stored energy from these batteries, even from resources as small as plug-in hybrid electric vehicles, during periods of peak demand when energy prices are high and electricity resources are limited.

Finally, the smart grid can provide consumers with specific and timely information about their energy consumption patterns, allowing users to better manage their electricity demands. With the current system, electric utilities and their customers are often unaware of how much electricity has been consumed until a meter is read and the monthly bill arrives. With a smart grid system, however, utilities can employ smart meters to provide real-time information on both usage and cost of electricity. As such, smart meters provide the mechanism by which utilities can allow electricity prices to fluctuate over time and for prices to reflect variations in demand. When system-wide electricity demand is low and the grid has plentiful resources to meet customers' needs, electricity prices are generally relatively low. During increased periods of electricity demand, as on a hot summer day, prices for electricity may increase to reflect the limited energy resources the grid has available to meet their needs. The smart meter also lets the customer know how much energy they are currently using: usage information alone (independent of price) has been shown to change energy consumption patterns.

Smart grid technology is relatively new, and although many pilot studies are currently under way, the technology has yet to be tested in full. Nevertheless, select portions of smart grid capabilities have been implemented in conjunction with various utility programs across the country. Such tests include efforts to develop smart grid systems that effectively integrate demand response strategies, employ smart meters, and develop systems of distributed generation including on-site solar, wind, and combined heat and power units. The nation's first smart grid is being built in Boulder, Colorado, with a $100 million investment.

Utilities and residential, commercial, and industrial customers all play a role in the smart grid. Utilities must invest in the smart grid technology associated with the generation, distribution, and management of electricity, and utility customers need to invest in smart energy devices that interact with the smart grid. Examples of such devices include programmable thermostats and home appliances that communicate with the grid so as to determine the best time to wash clothes or dishes by identifying off-peak periods, when the cost of electricity is lower and electricity is relatively plentiful.

Large utility infrastructure and consumer investments are necessary for a fully integrated smart grid to function properly, and government assistance and incentives at all levels may be required to encourage the required level of investments. One recent estimate for the installation of a nationwide smart grid projected the cost at $400 billion. Although $400 billion is indisputably a large sum of money, the alternative scenario will require $450 billion of investments in new generation capacity—that is, new power plants—to meet projected load growth under current business-as-usual scenarios. Given smart grid's projected energy, carbon, and dollar savings, in addition to its ability to provide a more stable, reliable, and integrated grid, the long-term benefits of investing in the smart grid are seen by most energy practitioners as outweighing those of maintaining the current grid system.

See Also: Electricity; Energy Payback; Grid-Connected System; Metering; Plug-In Hybrid; Power and Power Plants; Public Utilities.

Further Readings

Electric Power Research Institute (EPRI) and Global Energy Partners. "The Green Grid: Energy Savings and Carbon Emissions Reductions Enabled by a Smart Grid." EPRI Study 1016905. Palo Alto, CA: EPRI 2008.
Talbot, David. "Lifeline for Renewable Power." *Technology Review* (February 2009).
Troxell, Wade O. "Smart Grid: Transforming the US Power Grid" (2008). http://www.nrel.gov/visitors_center/pdfs/smart_grid_future.pdf (Accessed February 2009).

Karen Ehrhardt-Martinez
Vanessa McKinney
American Council for an Energy-Efficient Economy

Solar Concentrator

Concentrating solar energy technologies use optics or geometry to increase the density of energy from the sun to increase the amount of light or heat. The concentrated light or heat can be used to produce electricity. The two most prevalent uses of concentrator technology are to enhance photovoltaic output and in the operation of solar thermal electric plants. Photovoltaic electricity output is increased by utilizing optical lenses to magnify the number of photons hitting the surface of the solar cell; solar thermal plants use

The 82 mirrored panels on each of these experimental solar dishes focus light into a beam that hits receivers placed before the dishes. The energy is then transferred to a hydrogen-filled Stirling engine for conversion into electricity.

Source: U.S. Department of Energy, Sandia National Laboratories

mirrors or highly polished metal surfaces to concentrate heat from the sun's rays typically to heat a fluid.

Concentrated Photovoltaics

Much PV research is concerned with either increasing the efficiency or decreasing the costs of PV technology. The use of concentrating lenses has the capacity to address both of these aims. Concentrator lenses for PVs can increase incident light, therefore increasing overall production of photoelectricity and allowing the use of less PV material, which is the most costly portion of a normative PV cell. Several laboratory-proven concentrated PV (CPV) cells have significantly outperformed their nonconcentrated counterparts. As of June 2007, the highest-efficiency cell tested in a laboratory used concentrator technology.

Concentrators for PVs take the form of nonimaging optical lenses. Several different geometries are in use, including conventional convex lenses, compound Fresnel lenses, and prisms. Concentrator lenses are commonly grouped into three categories based on the amount of magnification the lens provides. Low-concentration PVs provide less than 10 times magnification, or 10 suns. Medium concentrators magnify sunlight to less than 100 suns, and high concentrators increase available light to 200 suns or more.

Significant issues arise when coupling concentrators with PVs—chiefly, the buildup of heat that accompanies the concentration of visible light energy. PV material loses efficiency as it is subjected to heat, and very high concentrations can raise temperatures enough to cause thermal damage. As of May 2008, the highest CPV system had a magnification of 2,300 suns, providing an output of 230 watts per square centimeter, but this magnification increased the heat to 1,600 degrees Celsius—hot enough to melt steel.

Devising economical ways to remove heat from the solar cells is the biggest challenge in full-scale production of concentrated solar cells. Some solutions to this under investigation are the incorporation of a solid- or liquid-state heat sink under the cells to remove heat, similar to computer chip cooling technology, or the incorporation of a thin infrared-reflecting film at the concentrator lens. A second issue is that as the magnification increases the optimal solar window decreases, meaning that medium and high CPVs must actively track the sun. Low CPVs, in contrast, normally increase the aperture of the solar window, allowing them to function somewhat better than normal PVs in low-light conditions. Low concentrators do not normally have issues with the buildup of heat.

Solar Thermal Concentrators

Although the increase of infrared radiation that accompanies an increase of light is a detrimental factor in terms of PV efficiency, the concentration of heat is another viable form of solar energy harvesting, termed *solar thermal electricity*. The use of concentrator technology is the foundation for this type of active solar power.

Unlike PV concentrators, thermal concentrators are opaque, reflecting light off a tributary surface area to a focal point. The temperature of the focal point can reach hundreds of degrees Celsius, which is more than sufficient to convert water into steam. Thermal concentrators can be made of mirrors or highly polished metal (usually stainless steel) and can be either flat or concave. Curved shapes ensure that all solar radiation incident on the tributary area is redirected toward the focal point, but they are more expensive to produce, especially in the case of curved mirrors. Flat surfaces are less focused but cheaper. In very large installations covering many acres of surface, the difference between curved and flat shapes is negligible. The most efficient form and type is a curved mirror; however, this is also the most expensive type of thermal concentrator to produce.

Parabolic Dishes

The shape of a parabolic dish thermal concentrator is indistinguishable from that of a satellite dish. Both are shaped to collect and direct energy to a focal point at which a receiver is suspended. A dish concentrator is relatively large in comparison, however—up to 25 meters in diameter. The ideal reflective medium is curved, high-quality mirrors to maximize the amount of energy reaching the focal point. Parabolic dish collectors may concentrate the sun's rays up to 3,000 times. Some medium of heat absorption is placed at the focal point of the dish. The fluid receives enough heat reflected from the mirrors to vaporize, which vapor, under pressure, is sufficient to turn an electrical generating turbine. Alternatively, a Stirling engine, driven by heat, may be placed at the focal point to generate electricity.

For parabolic dishes to collect heat over the course of the day, they must be able to track the sun. Practical examples of parabolic dishes are usually dual-axis tracking and are able to match both altitude (height) and azimuth (direction) of the sun to maximize heat collection.

Heliostat Concentrator Arrays

A much larger scale type of solar thermal electric facility arrays mirrors around a tower, the top of which is the focal point for an acres-large solar "farm." Instead of tracking the sun, the mirrors or polished metal surfaces are each fixed to catch the sun at a particular time of day, so that as the sun moves across the sky, a relatively constant amount of light is reflected from the mirrors to the top of the tower. The concentrators used here are usually flat, as the sheer scale of the parabolas these installations are based on virtually negate the added expense of curved shapes. An impressive amount of heat is nevertheless propagated at the focal point. Heliostat arrays can provide a concentrator factor of up to 1,000 suns.

Parabolic Troughs

Parabolic troughs take the cross-sectional shape of curved metal surfaces, polished on the inside to increase specularity, or reflectivity. As with all thermal concentrators, troughs are shaped to direct light to a specific focal point. However, unlike the types mentioned earlier, the parabolic shape is extruded for long distances, effectively turning the focal point into a line. A thin tube is usually placed at this focal point, through which a liquid—usually water—is run. At the end of the trough, the water has collected sufficient heat to be turned into steam, which can be used for heating or, most often, to produce electricity. A single trough may have a concentrating power of roughly 20 suns; in large-scale facilities, many troughs are grouped together to maximize solar heat absorption. Parabolic troughs may be fixed or may track the sun's altitude (one-axis tracking). Parabolic troughs are currently the most widely used form of solar thermal collectors for the production of electricity.

See Also: Photovoltaics (PV); Solar Energy; Solar Thermal Systems.

Further Readings

"IBM Research Unveils Breakthrough in Solar Farm Technology." http://www-03.ibm.com/press/us/en/pressrelease/24203.wss (Accessed January 2009).

Kaltschmitt, M, et al., eds. *Renewable Energy: Technology, Economics and Environment.* New York: Springer, 2007.

Lovegrove, S., et al. "Paraboloidal Dish Solar Concentrators for Multi-Megawatt Power Generation." Solar World Congress, Beijing, September 18–22, 2007.

"UD-Led Team Sets Solar Cell Record, Joins DuPont on $100 Million Project." http://www.udel.edu/PR/UDaily/2008/jul/solar072307.html (Accessed January 2009).

Winter, C.-J., et al., eds. *Solar Power Plants: Fundamentals, Technology, Systems, Economics.* New York: Springer, 1991.

R. Todd Gabbard
Kansas State University

SOLAR ENERGY

With current technology, the average home located in the U.S. southeast would require about 1,000 square feet of solar panels to cover all its electricity needs.

Source: iStockphoto.com

Solar energy refers to the energy that is created and radiated by Sol—the star at the center of the solar system—or, more significantly, the amount of that energy incident on the planet Earth. Solar energy not only is important for human technological efforts but is also the foundation for virtually all biological life on earth. Solar energy also refers to the processes by which light or heat from the sun is collected and converted to some usable form. Aside from geothermal energy, virtually every energy source commonly used by humans is directly or indirectly derived from solar energy. Historically, humans have relied on solar power for a wide number of applications. There are three general categories of solar energy systems in common use today: solar electric, solar thermal, and passive solar.

Solar Energy—The Source

The source of solar energy is the sun itself or, more specifically, the thermonuclear reactions that take place at the sun's core. Solar energy is the by-product of the nuclear fusion of hydrogen atoms into helium. Nuclear fusion is a highly exothermic reaction, producing prodigious amounts of electromagnetic radiation. This energy radiates out from the sun's core to approximately seven-tenths of the distance from the center to its surface, where it is transferred by convection to the surface of the sun. The energy is then radiated though space.

The sun emits radiation across a wide wavelength spectrum spanning from ultraviolet (approximately 200–400 nanometers) to radio waves (up to 100 meters). The sun does not emit light evenly across this spectrum, however. The highest amount of solar radiation, not coincidentally, is the portion of the spectrum visible to humans (wavelengths between 400 and 700 nanometers). A graph of the solar radiation peaks at the center of the visible light spectrum (yellow and green), falling steeply to the ultraviolet end (shorter wavelengths) and tapering more gradually on the infrared side (longer wavelengths). In short, most of the sun's radiation exists in the form of visible light and heat. Light from the sun propagates in the photosphere, a layer of space at the sun's surface where the majority of its energy is emitted to space. The amount of radiation emitted by the sun is relatively constant, varying less than 1 percent. This measure amounts to an average of 63.5×10^6 watts/meter^2 (W/m^2).

Solar Energy at Earth

A relatively small amount of this energy is radiated across space to the surface of the Earth. Incidental variations aside, the amount of energy incident on a surface, termed the radiant flux density, is inversely proportional to the square of the distance from the radiating object. Thus, the radiant flux density of solar energy incident to Earth is significantly reduced as a result of its distance (on average 1.5×10^{11} m). The average radiant flux density measured at the outer edge of Earth's atmosphere is approximately 1,370 W/m^2. This mean value is termed the solar constant.

Even though the sun's output is relatively consistent, the amount of sunlight incident on Earth is variable as a result of both astronomical and terrestrial conditions. The Earth's rotation creates the diurnal cycle of light and darkness. Seasonal variations occur because of the tilt of the Earth's axis relative to the plane of its orbit around the sun. During June, July, and August, for example, the northern hemisphere is positioned such that its surface is more perpendicular to solar radiation, creating a cyclically higher solar constant, while at the same time the southern hemisphere is more obliquely situated. This essentially reduces the density of solar radiation hitting that part of the globe, resulting in a lower solar constant for that part of the world. The geometry of the Earth—a sphere—also factors into the local variability of the solar constant. The area around the equator is most perpendicular to the sun, meaning it receives the most solar radiation. The solar equator is the line around the globe that is always perpendicular to the sun. This line is parallel to the plane of the Earth's orbit around the sun, and so is offset from the rotational equator, which at zero degrees latitude is equidistant from the Earth's poles. Generally, the higher the latitude of a geographic area, the less solar energy is available. At its poles, the surface of the Earth is close to parallel to solar radiation, meaning that very little heat is absorbed in these regions. Finally, variability is caused by the imperfectly elliptical course of the Earth's orbit around the sun, coupled with the movement of the sun in its own track around the center of the Milky Way galaxy. One position along the Earth's orbit brings it closest to the sun. This date, January 21, is termed the perihelion, which is winter in the northern hemisphere but summer in the southern hemisphere. On this date, the solar constant is roughly 1,420 W/m^2. Conversely, the lowest solar constant, 1,330 W/m^2, occurs on June 2—the aphelion, or the position at which the Earth is farthest from the sun. This inconstancy is thought to be chiefly, but not wholly, responsible for the variation in biomes when comparing areas of equal latitude in the two hemispheres.

Another local variable in regard to solar irradiation is weather and regional climate. Geographical, meteorological, and biological features may increase or decrease available solar radiation. The highest concentration of solar irradiation in the United States is in the Sonoran Desert of southeastern California and southwest Utah. This area, situated between two mountain ranges, is extremely arid, with very little precipitation or cloud cover. As a result, this area is viewed as a prime location for large-scale solar energy–harvesting installations of various types. Tracing the same latitude east across the country, there is a steady drop-off of solar radiation incident on the Earth's surface. Southern Virginia only has a third the incident radiation as southern Utah, due mainly to a marked increase in humidity, cloud cover, and precipitation.

Solar Radiation and the Earth's Atmosphere

The upper part of the Earth's atmosphere reflects 31 percent of all solar radiation, including many wavelengths that would be quite harmful to organic life. Only visible light, infrared, and a small amount of ultraviolet radiation are allowed to pass through. An additional 17 percent of the radiation is absorbed by or diffused into the atmosphere, and roughly 4 percent is reflected from the Earth's surface back into space. In all, only about 50 percent of incident radiation is received by the Earth's surface. Almost all of the solar energy received by the atmosphere and the Earth's surface is converted into longwave infrared radiation, where it eventually is re-radiated into space. Various gases in the atmosphere, however, such as methane and carbon dioxide, act as an insulating blanket, retarding the transmission of this heat.

Potential

Solar energy is a renewable source and in terrestrial terms is essentially unlimited. The sun will continue to produce energy at roughly the same output for billions of years. Fossil and nuclear fuels are in far more limited supply. The total amount of solar energy incident on the Earth can be determined by combining the mean solar constant with atmospheric and terrestrial interference and multiplying this by the surface area of the Earth. This averages 89 petawatts. The amount of energy humans use is roughly 17.8 terawatts—0.02% of available solar resources.

Solar energy is clean during generation. There are no waste products associated with solar energy collection, unlike other forms of energy such as combustion or fission. This is because solar energy systems do not produce energy but merely harvest an existing resource.

Solar energy is versatile. Humans have used energy from the sun for a wide variety of applications throughout recorded history. Current uses of the sun's energy can be divided into two types: passive and active. Passive solar strategies use the sun's rays for warmth and light. Active strategies convert solar radiation into another form, usually electricity.

Challenges

The largest issue with harvesting solar energy is that it is diffused more or less evenly across the Earth's surface. The diffuse nature of solar energy requires large collecting areas to provide enough power density for many applications. This aspect of solar energy, coupled with the relative efficiency of various harvesting strategies, is the main reason for the relatively high cost of solar power in relation to fossil fuels—another barrier to widespread solar energy applications. Currently available photovoltaics—devices that convert visible light into electricity—have nominal efficiencies between 5 and 20 percent. The diurnal cycle causes solar power availability to be periodic rather than constant—a trait common among renewable energy resources. This leaves two options: either implement an alternative energy source at night and during periods of inclement weather or adopt some type of energy storage solution. Economics and sociocultural inertia are also major impediments to the wholesale adoption of solar energy systems, though several countries and other municipalities have begun to incentivize solar energy and other renewable systems on the scale of both utilities and individuals.

Solar Energy Applications

Solar energy is the root source of many other forms of energy that we use today. The only types of energy commonly in use not directly derived from solar energy are geothermal energy, which taps into heat from deep in the Earth, and nuclear energy, which splits rare, unstable Earth elements to produce heat. Wind, for example, is generated by heat from the sun warming the Earth's oceans and land masses, causing planetary-scale convective currents. Hydroelectricity, produced mainly by damming rivers, is dependent on precipitation cycles (weather) and the seasonal melting of ice. These phenomena are linked to solar radiation. About 0.1 percent of solar radiation becomes entrained in biomass, particularly plankton and other plants, as it expires. This small amount of trapped energy is the basis for fossil fuels as well as biofuels, such as ethanol, and other forms of biopower. Fossil fuels are derived from prehistoric plant life that thrived under the sun millions of years ago and over time was converted into coal, oil, and natural gas.

Biological Life

By far the largest use of solar energy on Earth is life itself. Solar energy is the foundation on which almost all life on Earth depends. The only exception is deep-sea ecosystems that are based on the heat and nutrients released by underwater volcanoes. Plants use sunlight as the basis for photosynthesis to construct sugars out of carbon dioxide, water, and trace nutrients. This chemical energy is harvested by higher orders of life in a complex web of dependencies. Photosynthesis is also the process that produces oxygen, without which most animal life cannot exist.

Solar Electric Systems

Active energy systems convert a power source into some other usable form. Active solar energy applications turn energy from the sun into electricity. Two main types emerge: photic systems use visible light, and thermic systems use solar heat as a power source.

Photovoltaics are, as the name implies, photic. Solar thermal electric facilities, also referred to as concentrated solar power, or CSP, are thermic. Photovoltaics are made up of semiconductor material that is exposed to light from the sun. At the quantum level, individual photons from the sun hit electrons within the photovoltaic material, freeing them from their atomic bonds. These free electrons will induce electrical flow along a closed circuit. CSP systems are quite different. Using solar concentrator technology, these facilities gather heat from the sun over large areas of land. This heat is used to produce steam, which then turns a turbine-driven electrical generator—quite similar in operation to conventional centralized power plants.

Solar Electric System Advantages

Both systems produce electricity from solar energy. No other fuel source is required. No harmful environmental byproducts are created by their use. Photovoltaics are, on the whole, quite durable and are very versatile. Photovoltaics can power very small point-of-use applications and can be combined in arrays of many megawatts in size. Though not widespread, CSP systems are somewhat less expensive than large-scale photovoltaic arrays. The electrical generation technology of these facilities is virtually the same as fossil fuel power plants. The electrical output of CSP facilities is similar to that of small fossil fuel plants, in the several hundred megawatt range.

Solar Electric System Challenges

At this time, both systems produce electricity at a cost somewhat higher than the lowest fossil fuel–based utility rates. The Department of Energy estimates the average U.S. household uses roughly 27 kilowatt-hours of electricity every day. If all this energy were to be supplied by photovoltaics, the average home would require roughly 325 square feet (30 square meters) of photovoltaic cells rated at 10 percent efficiency, if the array were situated in southern Utah. If the array were situated in the southeast United States, the array would need to be almost three times as large, or 1,000 square feet (100 square meters). Other types of active solar energy collectors have higher efficiency rates.

As fossil fuels become scarcer, it is anticipated that this fossil fuel–based electrical energy will rise in cost, whereas the technology involved in photovoltaics and CSP systems will continue to fall in price. The growth of CSP facilities has been quite slow for decades, though in 2008 and 2009 a number of new facilities were announced.

Certainly the biggest challenge active solar energy systems face is the variability of solar radiation primarily resulting from the diurnal cycle and weather patterns. Both photic and thermic active solar systems work best in arid regions that are dominated by clear days.

CSP systems in particular are ill suited for areas that are largely overcast, as they rely solely on direct solar radiation. Photovoltaics have been successfully used virtually all over the world, regardless of local weather patterns. Photovoltaics can produce power in low-light and overcast conditions, though at a reduced rate.

Solar energy as the sole source of electricity requires the ability to store energy for use at night and over prolonged overcast periods. Small-scale photovoltaic systems often use chemical batteries as a storage medium, though the creation of hydrogen through solar energy–driven electrolysis has been studied and implemented to some degree. The manufactured hydrogen can be stored and used to power a fuel cell. A wide variety of large-scale energy storage systems for both CSP and utility-scale photovoltaic arrays have been proposed and, to some degree, implemented. Some examples are compressed air energy systems—in which air is compressed in underground caverns, thermal energy storage in the form of phase change material, molten salts or solid thermal mass, induced hydropower (where solar energy is used to mechanically pump water to the "upstream" side of a hydro-electric dam), flywheels, or large-scale hydrogen production.

Solar Heating

A wide and varied number of solar heating applications have been put to use by humans for millennia. There have been and continue to be a number of agricultural uses of heat from the sun, from extending the growing season through the use of greenhouses to the drying of crops. Solar radiation is used as a source of process heat for a number of industrial enterprises. Solar heat can aid quality-of-life issues, especially in economically disadvantaged regions of the world. Solar cookers have been introduced in areas where fuel for cooking is scarce. Solar distillation provides clean water in places where the local water source is contaminated or otherwise unreliable.

The most frequent use of solar heating today is for domestic purposes, either for the creation of hot water or for space heating. Solar thermal systems for heating water or air comprise, at a minimum, three components: some type of collector that absorbs light and heat from the sun, a fluid that transfers the heat, and some type of thermal storage. Solar thermal systems can be sized for a single home or for campus or district applications. Solar thermal systems of this type have relatively high conversion efficiencies (50–70 percent), though on average only about 25 percent of the absorbed energy is available for its intended use. It is rare for solar thermal systems to supply all the heat energy needed by homes or other domestic users, in large part because of seasonal or climatic variation, though this limitation can be overcome by the use of larger collector arrays, very large and/or well-insulated storage tanks, or usage prescriptions. It is commonly accepted that energy efficiency strategies are more cost-effective than active solar systems.

Building Passive Solar Systems

Solar energy can supplement or replace other energy sources for lighting and thermal comfort. Daylighting can obviate the need for electric lighting during the day. Almost all buildings have some transparent aperture for view or lighting; optimizing these strategies can reduce or eliminate electrical demand in regard to artificial lighting, especially in buildings that have diurnal occupancy patterns such as corporate or governmental offices. Passive heating strategies convert solar radiation into space heating. Direct sunlight is allowed to penetrate the building envelope through glazing; this radiation is absorbed by the interior surfaces of the building that subsequently warm the space. Several of these strategies incorporate the use of massive materials to store the collected heat for night-time use. Solar heat can also be used for cooling: A solar chimney is a tall, glazed space situated above the occupied area of a building. Sunlight penetrates the chimney, heating the air inside, which causes the air to rise through vents at the top. This creates negative air pressure inside the building, inducing air flow.

Passive solar strategies must be incorporated into the building during the design process, often requiring customization to meet the objectives of a particular project. Oversizing apertures for these systems can cause overheating during the day or excessive heat loss at night. Building shape, size, location, and orientation all affect the potential for passive solar strategies. Moreover, these strategies are region and climate specific. Low-energy or zero-energy buildings normally incorporate passive solar strategies. The highest form of passive solar buildings closely replicates natural processes. This quality is termed *biomimicry*.

See Also: Biomass Energy; Daylighting; Energy Storage; Fossil Fuels; Fuel Cells; Green Power; Hydrogen; Photovoltaics (PV); Renewable Energies; Solar Concentrator; Solar Thermal Systems.

Further Readings

Kaltschmitt, M., et al., eds. *Renewable Energy: Technology, Economics and Environment.* New York: Springer, 2007.
Stein, B., et al. *Mechanical and Electrical Equipment for Buildings,* 10th ed. New York: John Wiley & Sons, 2005.

R. Todd Gabbard
Kansas State University

SOLAR THERMAL SYSTEMS

A little more than half the solar radiation reaching the Earth's surface is in the form of heat. Solar thermal systems use the infrared radiation transmitted through space from the sun as an energy source. Humans have devised and long employed a wide number of applications that use solar thermal energy. Two main types of solar thermal systems are in use today—solar heating systems and solar thermal electric facilities, also referred to as concentrated solar power (CSP) systems.

Thermal Solar Energy

Solar radiation covers a portion of the electromagnetic spectrum ranging primarily from ultraviolet wavelengths to longwave infrared, though the distribution of energy is not uniform. The peak of the solar curve is at the middle of the visible light range of wavelengths— green and yellow. Volume-wise, roughly 9 percent of solar radiation emitted is in ultraviolet form, whereas about 38 percent is within the visible light spectrum. The remaining 53 percent is infrared radiation, or heat. This indicates that harvesting solar heat is a viable enterprise.

Two applications will be discussed here: solar heating systems and solar thermal electrical facilities. Though quite different in many ways, both applications have a similar basic requirement. These applications rely on direct rather than diffused solar radiation, and so work best in clear sky conditions. The sheer scale of solar thermal electric facilities requires that they be built in arid regions, where there are few overcast days. Solar heating systems have been installed and do work sufficiently in variable and somewhat cloudy climates, though to a lesser degree of effectiveness than in sunnier climes.

Solar Heating Systems

As the name implies, solar heating systems absorb solar thermal energy to provide heat. This heat is most commonly used to heat water for domestic use, though the same technology may be used for space heating. The basic components of solar heating systems are few: some type of solar collector—a device designed to absorb heat from the sun, a fluid-based heat exchange system, and some means to move the fluid through the absorber and to its intended point of use, often a small pump or fan. This last component is what separates solar heating systems from passive solar systems, which do not use any external power to operate.

The simplest type of solar collector is a flat plate collector. Flat plate collectors comprise a thin box with a glass or clear plastic top. All other surfaces are painted black to maximize heat absorption. A thin copper tube is routed through the box. Solar radiation (shortwave infrared) passes through the glass, is absorbed by the dark surfaces, and is reemitted as terrestrial (longwave) infrared. Glass is opaque to terrestrial radiation, and so this heat remains trapped in the collector. The heat is conducted through the copper tube to the water or other heat transfer fluid (such as ethylene glycol) inside it. This heat is then exchanged with a hot water storage tank. Solar hot water systems can provide most, if not all, domestic hot water needs. A variant of this system might include a backup heat source such as electricity or natural gas for periods of increased demand or prolonged overcast conditions.

Evacuated tube collectors are a recent technological improvement of flat plate collectors. Evacuated tubes are long, thin cylinders of glass that surround a thin tube of copper or glass containing a heat transfer fluid. The inner tubes are sealed—one end of the inner tubes connects to a manifold through which another fluid is pumped to transfer the collected heat from the evacuated tube array to the heat storage tank. The outer glass tubes are double-walled and often have a selective surface—a type of one-way mirror tinting—to aid the glass tubes' natural greenhouse effect. The space between the walls is rarefied, or evacuated, to minimize the loss of heat collected inside the tube by convection or conduction. These additional measures are quite effective in retaining absorbed heat and drastically reduce the effects of ambient conditions on the operation of the system.

As mentioned, solar heating systems are primarily used for domestic hot water purposes, but they are also used for the heating of swimming pools and sometimes for space heating, usually in the form of a radiant floor or other radiant heat system.

Another type of solar heating system uses air as its heat-transfer medium. A thermosyphon is a hybrid passive solar heating system quite similar to a flat plate collector system. Instead of water being fed through thin tubes, air is blown over the heat-collecting surface. This hot air is then drawn into a building directly or is routed through thermal mass to delay the release of the heat. The solar heat collector may be kept separate from or built into the roof or walls of a building.

Solar thermal electric facilities, also called CSP systems, are large-scale electrical generation stations. Basically, these facilities gather heat from the sun to convert water into steam. The steam is pressurized and used to turn a turbine-generator. Excepting their fuel source, these centralized facilities are quite similar to fossil fuel–based power plants. To gather a large quantity of heat, CSP systems use solar concentrator technology to focus incident solar heat to a focal point. A heat transfer fluid—often water, but sometimes oil or molten salt—is run through a tube located at the concentrator's focal point. If water is used, enough heat must be concentrated to turn the water into steam. These systems have the potential to rival the generating capacity of normative fossil fuel–based power plants. The main disadvantage of these plants is that they must use direct solar radiation. As a consequence, they must be sited in arid regions with clear skies. Devising ways to make these facilities generate electricity overnight is another issue. A wide range of potential energy storage options have been implemented or are under investigation, including heat storage in the form of molten salt or solid thermal mass, mechanical energy storage in the form of flywheels, "lifted" water for induced hydroelectricity, and compressed air energy systems, which store high-pressure air below ground. The large-scale production of hydrogen is another energy storage option. Another option is to couple a CSP system with a backup fossil fuel system such as natural gas.

Three main types of CSP systems are in use today. Parabolic trough systems use a long, parabolic reflector that concentrates sunlight along a line rather than at a point. A thin tube is placed at the focal point. As water travels down the tube, it gains heat until it evaporates into steam. The steam is collected at the far end of the trough and is channeled to a turbine-generator. Multiple rows of long east-west troughs can power a large generating station. Solar Energy Generating Systems, located in the Mojave Desert in California, is the single largest solar electricity facility worldwide, with a maximum generating capacity of 354 MW. The Solar Energy Generating Systems plant uses parabolic trough technology. Another solar thermal facility type is the heliostat or power tower. This type of system uses flat mirrors placed around a tall tower. The mirrors are oriented so that they point toward the top of the tower, similar to the geometry of a satellite dish. The mirrors do not move; instead, throughout the day the sun hits a different portion of the heliostat array. The focal-point temperature at the top of the tower is sufficient enough to instantly turn water into steam. There are currently only a few power tower stations in operation, most notably an 11-MW station in Seville, Spain, though several are under construction worldwide. The final type is the dish system. This is based on a parabolic receiving dish similar to a satellite dish or radio telescope. Heat is collected at the focal point of the dish, where it may be used directly to turn an engine or to create steam. These systems must positively track the sun and are somewhat more limited in size to roughly the 25-KW range, though multiple units can be arrayed together.

With the exception of parabolic trough systems, CSP systems have only recently been seen as a viable option for large-scale electrical power generation. A large number of CSP facilities are under construction or have been announced in the United States (about 4,000 MW of new generating capacity), Spain (400 MW under construction, almost 2,000 MW announced), and a handful of other countries (roughly 500 MW).

Other Applications

Humans have long turned solar thermal energy to their use. Greenhouses or cold frames (temporary greenhouses) are a classic solar heating system designed to extend the growing season. Products such as dried fruits and adobe bricks are cured by solar radiation. Coupled with wind, solar thermal energy is an excellent source for drying. The same solar concentrator technology used for CSP facilities can also be used for solar cooking, where heat from the sun is gathered and focused at a food item. Given a large enough surface area, solar cookers can equal or exceed the temperatures of conventional ovens. Humanitarian organizations have introduced solar cookers in less-developed regions where wood for cooking fires is scarce or nonexistent as a means to enhance the quality of life and decrease the spread of foodborne disease. Solar heat is similarly used to distill water for human consumption in places where clean water is not otherwise obtainable.

See Also: Energy Storage; Renewable Energies; Solar Concentrator; Solar Energy.

Further Readings

"CSP—How It Works." http://www.solarpaces.org/CSP_Technology/csp_technology.htm (Accessed January 2009).

Hough, Tom P., ed. *Trends in Solar Energy Research.* New York: Nova Science, 2006.

Kaltschmitt, M., et al., eds. *Renewable Energy: Technology, Economics and Environment.* New York: Springer, 2007.

Winter, C.-J., et al., eds. *Solar Power Plants: Fundamentals, Technology, Systems, Economics.* New York: Springer, 1991.

R. Todd Gabbard
Kansas State University

Sulfur Oxides (SO$_x$)

Sulfur oxides are chemical compounds of sulfur and oxygen in the form S_xO_y with lower members of the family being S_7O_2 and higher being as SO, SO_2 and SO_3. While natural sources of sulfur oxides include volcanic eruption emissions, the chemical compound is released chiefly through the combustion and processing of sulfur containing fuels and ores. Sulfur is an impurity found in most fossil fuels such as coal and oil, encompassing a content of sulfur between 0.5–3 percent and 0.5–5 percent, respectively. Large-scale energy production associated with industrialization has established sulfur oxides as one of the prime anthropogenic air pollutants.

Coal and oil contain trace impurities of sulfur. When they are burned they produce sulfur oxides, mostly sulfur monoxide SO and sulfur dioxide SO_2 and sulfur trioxide SO_3, at the same time. However, SO is an unstable molecule and quickly transforms into SO_2. Normally, SO_x are produced by the burning of sulfur following a chemical equation of the following type:

$$S_X + O_Y \rightarrow S_XO_Y$$

SO_2 is often used in the food industry, where it often appears as E220. Some of its common uses include the preservation of dried fruits, because of its antimicrobial action, which is also the reason it is found in wine with the "contains sulfites" label. However, SO_2 is used not only in the food industry and in the chemical industry as bleach but also as a refrigerant (although it has lost its popularity as a result of the emergence of Freon-type gases). In addition, SO_2 is often used for wastewater management to dechlorinate wastewater before release. Finally, SO_2 can be used for the production of sulfuric acid, but it has to be transformed to SO_3 first.

When released in the atmosphere, SO$_2$ and SO$_3$ react with atmospheric humidity and form sulfuric acid (H$_2$SO$_4$), which, when diluted in precipitation, is known as acid rain. Acid rain is harmful to man-made and natural ecosystems and also to humans. It affects the former by eroding materials such as paint or the surfaces of buildings. In particular, ancient monuments are suffering from exposure to acid rain. However serious a threat acid rain may be for man-made constructions, though, its presence in nature can be disastrous. Forests are having the tops of their trees destroyed, and at later stages whole trees are destroyed. Water in lakes becomes acidic, having a direct effect on flora and fauna and on the lake's ecosystems. Finally, humans are sensitive to the effects of SO$_x$, which usually form sulfate particles that can enter the breathing system and cause heavy breathing or, in longer exposure, respiratory diseases and eventually premature death. Skin and eyes are also sensitive to SO$_x$, especially when they occur in the form of sulfuric acid that can permanently harm the skin and the ability to see. Heavy concentration of SO$_x$ in urban environments can also lower visibility.

Acidification has been an issue of major concern in past decades when vast ecosystems in Canada, the United States, the Scandinavian countries, and central Europe have been found with high levels of acidity. International frameworks for the control of these emissions have been adopted both in the United States, with the Clean Air Act, and in Europe, with the Convention on Long-Range Transboundary Air Pollution and European Monitoring and Evaluating Programme. Although policies for reducing SO$_x$ emissions have been put into place, specific technologies have made implementation possible. Because it has been recognized that the major emitting source of SO$_x$ was coal- and oil-fired power generation, the key technology that has been developed is the flue gas desulfurization, where flue gas of a power plant is driven through lime (calcium oxide [CaO]) in a process that produces calcium sulfite, following this chemical equation:

$$CaO + SO_2 \rightarrow CaSO_3$$

When CaSO$_3$ is exposed to oxygen, it transforms into CaSO$_4$, or gypsum, which can then be used in other industrial or domestic procedures. Desulfurization units have been widely used in almost all power stations of North America, Australia, Japan, and Europe and have contributed significantly to substantial lowering of SO$_x$ emission. Gradually they are being adopted by more countries in Asia and Latin America.

See Also: CAFE Standards; Emission Inventory; Emissions Trading; Nitrogen Oxides; Power and Power Plants; Volatile Organic Compound (VOC); Waste Incineration.

Further Readings

Economic Commission for Europe. "Convention on Long-Range Transboundary Air Pollution." New York and Geneva: United Nations, 1979. http://www.unece.org/env/lrtap/full%20text/1979.CLRTAP.e.pdf (Accessed February 2009).

U.S. Environmental Protection Agency. "The Clean Air Act." U.S. Senate Committee on Environment and Public Works, 2004. http://epw.senate.gov/envlaws/cleanair.pdf (Accessed February 2009).

Konstantinos Chalvatzis
University of East Anglia

Sustainability

Sustainability has many definitions, but the core concept remains this: As we use the Earth's resources to produce economic goods and services, we must maintain the Earth's capacity to provide natural resources and absorb wastes for the benefit of future generations, and we must meet our social responsibilities to allow and enable poor countries and marginalized groups to share the Earth's bounty.

The History of Sustainability

The notion of sustainability dates back more than 40 years. It is an outgrowth of discussions among international development economists about whether constantly increasing economic growth is the best means to lift underdeveloped countries out of poverty. As Kenneth Boulding pointed out in his 1966 article "The Economics of Spaceship Earth," constantly increasing economic growth is infeasible, given the finiteness of the Earth. Sustainability was a key theme of the United Nations Conference on the Human Environment in Stockholm in 1972. Development economists such as Barbara Ward of the International Institute for Environment and Development began to focus in the early 1970s on development strategies that considered the finiteness of natural resources. Development economists coined the term *sustainable development* to express the notion that environmental protection must be integrated with development policy, because, in the long term, development could continue only if the productive and waste assimilation capacities of the environment were protected. In the past three decades, sustainability or sustainable development has become the cornerstone of international policy.

The concept of sustainability has evolved over the years through documents such as the World Conservation Strategy (1980), the Brundtland Report (1987), the Rio Declaration and Agenda 21 from the United Nations Conference on Environment and Development (1992), the Millennium Development Goals adopted at the UN Millennium Summit (2000), the Monterrey Consensus from the International Conference on Financing for Development (2002), and the Johannesburg Declaration from the World Summit on Sustainable Development (2002). Although environmental protection has remained the most prominent aspect of sustainability, poverty alleviation as a matter of social justice has become an increasingly important dimension of sustainability since the adoption of the Millennium Development Goals and the Johannesburg conference.

Defining Sustainability: The Brundtland Report

As sustainable development or sustainability became an increasingly popular understanding of the key relationship between development and the environment, the 1987 Report of the World Commission on Development and Environment, also known as the Brundtland Report, adopted a vague definition of sustainable development that remains the most cited definition of sustainable development or sustainability. It articulated the notion of intergenerational equity as "meeting the needs of the present without compromising the ability of future generations to meet their own needs."

The vagueness of the Brundtland Report definition, together with the misunderstanding that development necessarily meant constant or increasing economic growth, led many to criticize "sustainable development." As a response to that criticism, and as a way of generalizing the obligation to attend to the needs of future generations beyond developing countries, the term *sustainability* came into vogue, ultimately capturing far more popular attention in the United States than "sustainable development." In most people's minds, sustainability is probably equated with the concept of intergenerational equity expressed in the Brundtland Report definition.

Most commentators recognize that the vagueness of this definition has played a major role in popularizing sustainability in mainstream discussions of policy. An International Union for the Conservation of Nature report recently opined:

> Analysts agree that one reason for the widespread acceptance of the idea of sustainable development is precisely this looseness. It can be used to cover very divergent ideas. Environmentalists, governments, economic and political planners and business people use "sustainability" or "sustainable development" to express sometimes very diverse visions of how economy and environment should be managed. The Brundtland definition was neat but inexact. The concept is holistic, attractive, elastic but imprecise. The idea of sustainable development may bring people together but it does not necessarily help them to agree goals. In implying everything sustainable development arguably ends up meaning nothing.

Refining the Brundtland Report Definition

To actually achieve sustainability, however, we need a more concrete definition of what "sustainability" is to properly guide law and policy toward sustainability. Thus, over the last two decades, resource and ecological economists have continued their efforts to define sustainability more rigorously.

Sustainability is frequently defined by resource economists in terms of maintaining the Earth's productive and waste assimilative capacity, or natural capital, over time. Sustainability can then be expressed as four constraints on production of goods and services:

- Consumption of renewable resources must be limited to sustainable yield
- Consumption of exhaustible resources must be minimized, and the stock of renewable resources must be increased at least as fast as the stock of exhaustible resources is decreased
- Biodiversity must be preserved
- The waste assimilative capacity of the environment must be preserved

This notion of sustainability by resource and ecological economists focuses attention on the importance of the environment, as opposed to social equity and economic capacity. Thus, this concept is sometimes called "ecological sustainability."

Strong Versus Weak Sustainability

The ecological sustainability definition is considered a "strong" definition of sustainability because the ecological sustainability definition insists that future generations are entitled to at least the amount of natural capital that current generations enjoy. Other economists argue that natural capital need not be maintained and that global society should embrace "weak" sustainability. Weak sustainability would allow natural capital, that is, the Earth's productive and waste assimilative capacity, to be diminished provided that other sorts of capital (such as the traditional capital of plant and equipment, human capital, and social capital) are increased at a greater rate than natural capital is diminished and thus substitute for natural capital. The primary difficulty with strong sustainability is that it gives prefer-ence to protecting natural capital over increasing traditional capital, human capital through education, or social capital through collaborative decision making. The question becomes whether such a preference is justified in terms of the effect on intergenerational equity. The primary difficulty with weak sustainability as a definition of sustainability is that we cannot be certain that other types of capital in the long run can truly be substituted for natural capital. Without the land, natural resources, and ecosystem services provided by nature, no amount of equipment, education, or collaborative capacity will allow human and other life to continue on this planet. So, arguably, unless protecting natural capital is given preference over increasing other forms of capital, the destruction of the environment may be fatal to future generations. For example, unless global warming is controlled enough to prevent catastrophic effects such as an interruption of the ocean currents that make the northern hemisphere inhabitable, the amount of traditional capital, human capital, and social capital becomes quite irrelevant.

Three Pillars of Sustainability

The core of mainstream thinking on sustainability is that all policy and law must be made with an awareness of three separate dimensions of sustainability: economic, social, and economic. These have been drawn in a variety of ways: as "pillars," as concentric circles, or as interlocking circles.

Perhaps the most popular approach to discussing sustainability is to note that there are three so-called pillars of sustainability: environmental protection (environment), social equity (social), and economic vitality (economic). These three pillars are also called the "three Es"—for environment, equity, and economy—or the "three Ps"—for people, planet, and profits. The three pillars concept reminds us that not only is a successful and sustainable economy dependent on protecting the environment and ensuring some degree of social justice but environmental protection may be dependent on achieving a greater degree of social justice (e.g., greater gender equality, educational opportunity, and income security lead to reduced population growth). Similarly, financial resources to set aside nature reserves to protect biodiversity, to install pollution control equipment, and to restore degraded lands and clean up contaminated sites are necessary to achieve environ-mental protection.

The three pillars concept of sustainability—which suggests coequal consideration of all three pillars—produces policy dilemmas for which the concept provides no clear answers. With the three pillars approach, policy decisions simply become unprincipled pluralistic balancing of the three pillars as three factors to be considered by decision makers. So, the grand, supposedly transformative concept of sustainability is thereby reduced to an imper-ative that decision makers have knowledge of the effects of a decision on those three pillars (typically through environmental impact assessment) and consider those effects in making decisions. Indeed, the International Union for the Conservation of Nature has specifically articulated the meaning of sustainable development in terms of simply providing greater weight to the environment.

Unfortunately, historical experience suggests that in the absence of setting social equity and environmental protection as constraints on policy, the equity and environment pillars often lose out to the economic pillar, especially when those who stand to lose profits from a policy decision assert the decision will be economically inefficient or will adversely affect the economy.

Many commentators criticize the concept of sustainability or sustainable development embodied in the three pillars approach as a sound basis for law and policy because it does not insist or force decision makers to change the weight given in policy decisions to the economic pillar. Under the three pillars approach, the outcome of sustainability analysis is simply indeterminate, and the ability of the three pillars approach to produce decisions that protect the environment and promote social equity is indeed quite questionable.

The utility of the three pillars concept in actually achieving sustainability depends on at least three factors. First, how do we understand the meaning of the three pillars? For example, if economic vitality is understood as increasing the percentage of the population enjoying a reasonable standard of living, certainly the economic pillar supports the environmental pillar by creating public support and financial resources for environmental protection. However, if economic vitality is understood as constantly increasing resource use and waste production to maximize economic growth, the economic pillar clearly undercuts sustainability. We must be careful how we define the goals to be achieved as part of economic, social, and environmental sustainability.

Second, do we have the analytic tools (e.g., systems analysis, decision analysis, complexity theory, and resilience theory) to truly understand the interactions between the pillars, rather than rely on simplistic analyses of how a decision may affect the three pillars?

Finally, can we conceptualize what it means for a policy decision to be sustainable—how can decision makers integrate the three pillars into their decision making? It is mathematically impossible to maximize three different variables—one must maximize on one variable and set the other two variables as constraints. So, if we think of the three pillars as three variables in sustainable policy decisions, we can only maximize one pillar and set the other two pillars as constraints. We can maximize economic well-being subject to specified levels of social equity and environmental protection, or we can maximize social equity subject to specified levels of economic well-being and environmental protection, or we can maximize environmental protection subject to specified levels of social equity and economic well-being. But we cannot simultaneously optimize economic well-being, environmental protection, and social equity, except to the extent that there is no trade-off between the three. In general, we must choose which pillars to set at a specified level and which pillar to maximize. Paradoxically, the variable that is maximized—that is, allowed to vary—is not necessarily the pillar deemed most important. For example, in the case of global warming policy, where the survival of future generations and of the poor with limited capacity to adapt is at stake, some may want to set carbon neutrality and no increase in absolute poverty as absolute constraints and then allow the world to achieve as much economic well-being as possible within those constraints. Others may want to allow the degree of carbon reduction to vary, setting constraints that global warming policy be neutral in terms of its effects on various income groups and reduce global domestic product no more than 1 percent.

As a general matter, some scholars argue that the best way to conceptualize sustainability is as the economy being bounded or constrained by society and the environment. This explicitly recognizes that the economy must be seen as constrained by social needs and the environment.

In some cases, there is not necessarily a straightforward trade-off between the three pillars. For example, in the special case of developing countries with high levels of environmental degradation, poverty, and economic stagnation, there may be little trade-off between the pillars. Suppose the World Bank and donor countries decided to dramatically increase funds spent on reforestation in Haiti. That decision would (1) provide jobs, increasing overall economic well-being as well as the percentage of Haitians who enjoy a reasonable standard of living, and (2) decrease soil erosion, improving the productivity of lands owned by poor farmers, decreasing the frequency of floods, and decreasing the turbidity of rivers. Reducing turbidity would make water treatment more effective and less expensive, which would reduce the incidence of waterborne diseases. Improving the standard of living to the point where families could afford medicine would allow children with waterborne diseases to be treated, rather than die at an early age. Fewer child deaths and more income security would reduce the social incentive to have additional children, thus reducing population growth, which in turn would reduce natural resource consumption and reduce unemployment levels. In such a case, there is no obvious trade-off between the economic, environmental, and equity pillars. Sustainability optimists believe that this "no trade-off" case is the typical case and that so long as attention is given to all dimensions of sustainability, good decisions will be made. Sustainability pessimists believe that the "no trade-off" case is the exceptional case and that sustainability simply provides little guidance to decision makers in making difficult and critical choices.

Conclusion

Sustainability has now become the dominant paradigm of global thinking about economic development, environmental protection, and poverty alleviation. The notion of sustainability has spread to every sector of the economy, particularly with respect to ecological sustainability. We now think in terms of "green": green buildings, green cities, green industrial processes, green products, green cars, green energy, and even green law offices. No doubt our notions of what sustainability means will continue to evolve as sustainability increasingly becomes the actual basis for decision making. Only time will tell whether the concept is robust enough—and has been defined effectively enough—to provide sound guidance to those responsible for making law, government policy, and corporate policy in these critical decades of the 21st century.

See Also: Environmental Stewardship; Renewable Energies; World Commission on Environment and Development.

Further Readings

Brown, Lester R. *Plan B. 2.0: Rescuing a Planet Under Stress and a Civilization in Trouble.* New York: W. W. Norton, for the Earth Policy Institute, 2006.

Brundtland, Gro Harlem. *Our Common Future.* Oxford: Oxford University Press, for the World Commission on Environment and Development, 1987. http://www.un-documents.net/wced-ocf.htm (Accessed May 2011).

Costanza, Robert and B. C. Patten. "Defining and Predicting Sustainability." *Ecological Economics,* 15/3:193–96 (1995).

Daly, Herman. *Beyond Growth: The Economics of Sustainable Development.* Boston: Beacon, 1996.

Daly, Herman and J. Cobb. *For the Common Good: Redirecting the Economy Toward Community, the Environment and a Sustainable Future.* Boston: Beacon, 1989.

Daly, Herman and Joshua Farley. *Ecological Economics: Principles and Applications.* Washington, D.C.: Island, 2004.

Edwards, Andres R. *The Sustainability Revolution*. Gabriola Island, Canada: New Society, 2005.

Gunderson, Lance H. and Lowell Pritchard, Jr., eds. *Resilience and the Behavior of Large-Scale Systems*. Washington, D.C.: Island, 2002.

Holling, C. S. "Theories for Sustainable Futures." *Conservation Ecology*, 4/2 (2000).

International Institute for Sustainable Development. "What is Sustainable Development?" (2009). http://www.iisd.org/sd/ (Accessed May 2009).

International Union for the Conservation of Nature. *The Future of Sustainability: Re-thinking Environment and Development in the Twenty-first Century* (2006). http://cmsdata.iucn.org/downloads/iucn_future_of_sustanability.pdf (Accessed May 2009).

International Union for the Conservation of Nature. *The World Conservation Strategy*. Geneva: International Union for Conservation of Nature and Natural Resources, United Nations Environment Programme, World Wildlife Fund, 1980.

Norton, Bryan G. *Sustainability: A Philosophy of Adaptive Ecosystem Management*. Chicago: University of Chicago Press, 2005.

Pearce, David, et al. *Sustainable Development Economics and Environment in the Third World*. London: Earthscan, 2000.

Sachs, Jeffrey D. *Common Wealth*. New York: Penguin Books, 2008.

Sachs, Jeffrey D. *The End of Poverty*. New York: Penguin Books, 2005.

Smith, Susan L. "Ecologically Sustainable Development: Integrating Economics, Ecology, and Law." *Willamette Law Review*, 31/261 (1995).

UN Division of Sustainable Development. *The Johannesburg Declaration on Sustainable Development, Report of the World Summit on Sustainable Development* (2002). http://daccessdds.un.org/doc/UNDOC/GEN/N02/636/93/PDF/N0263693.pdf?OpenElement (Accessed May 2009).

UN Division of Sustainable Development. *The Johannesburg Plan of Implementation*. New York: World Summit on Sustainable Development, 2002. http://www.un.org/esa/sustdev/documents/WSSD_POI_PD/English/POIToc.htm (Accessed May 2009).

UN Division of Sustainable Development. "Major Agreements and Conventions on Sustainable Development." http://www.un.org/esa/dsd/resources/res_majoagreconvover.shtml (Accessed May 2009).

Susan L. Smith
Willamette University College of Law

T

Three Gorges Dam

The controversial Three Gorges Dam, a portion of which is shown above, supplies 10 percent of China's current electricity needs, replacing the equivalent of 40–50 million tons of coal use per year.

Source: iStockphoto.com

Highly publicized as the most monumental construction project in China since the Great Wall and the Beijing-Hangzhou Grand Canal, the Three Gorges Dam (known locally as Sanxia Ba) on the Yangtze River in central China is the largest hydroelectric power dam in the world. The concrete gravity dam spans a length of 1.4 miles (2.3 kilometers) and reaches a height of 608 feet (185 meters), creating a reservoir behind the dam site 575 feet (175 meters) deep and nearly 400 miles (650 kilometers) long (the length of Lake Superior). The Three Gorges Dam was designed primarily to minimize the devastating effects of chronic floods while creating a renewable energy source and increasing regional interior navigation abilities. The Chinese government puts the total investment at nearly $40 billion, although less-conservative estimates place the actual construction costs more than double the official figure. The project has been the subject of much international scrutiny because of its large-scale social, economic, and environmental impacts.

The Three Gorges is the more common name for the intersection of the Qutang, Wuxia, and Xiling gorges on the Yangtze River, the lower portion of the Chang Jiang (or Long River) located in the central interior Chinese province of Hubei. Floods along the Yangtze River are well known in the historic record, such as the unparalleled floods during the summer of 1954 that inundated nearly 8 million acres of agricultural land, killed over 30,000 people, and adversely affected an additional 19 million others. The 1954 event prompted Chairman Mao Zedong to order feasibility studies into controlling the Yangtze River flow through a massive dam, an idea that originated in 1919 with the founder of modern China, Dr. Sun Yat-Sen. The economic losses from the Great Leap Forward and Cultural Revolution campaigns of the 1960s and 1970s halted project development. During the 1980s, the plans were renewed. In 1992, the National People's Congress narrowly approved the project, amid record-number abstentions, and construction officially began on December 14, 1994. Although the main wall was completed in May 2006, the dam did not become fully operational until early 2009.

The Three Gorges has the largest total hydroelectric capacity of any dam, being capable of generating 22,500 megawatts (MW) of power when all 34 generators are in place; in comparison, the U.S. Hoover Dam on the Colorado River produces a little over 2,000 MW. The Three Gorges Power Plant generates approximately 85 billion kilowatt-hours of energy per year, second only to the Brazilian Itaipu Dam on the Parana River, with an annual output of nearly 94 billion kilowatt-hours. This output supplies 10 percent of China's electricity requirements, especially to the central Chinese provinces of Anhui, Henen, Hubei, Hunan, and Jiangsu. The power produced is expected to alleviate the energy requirements of China's burgeoning industrial sector.

Hydropower provides a green energy alternative to the more common coal-burning power plants. China is the largest coal consumer in the world, using 1.5–2 billion tons per year to meet three-fourths of its energy needs, primarily for electricity generation and residential heating. Coal releases harmful derivatives when burned, including particulate matter (e.g., fly ash), greenhouse gases (e.g., carbon dioxide), acid rain agents (e.g., sulfur dioxide), and biophysical toxins (e.g., arsenic). The hydroelectric power produced by the Three Gorges Dam affords an annual energy equivalent to 40–50 million tons of coal without the harmful atmospheric by-products.

In addition to energy production, the Three Gorges Dam allows for greater control of the Yangtze River water flow and channel depth. The dam provides flood protection, greatly reducing the socioeconomic risk for those downstream in central and eastern China. The reservoir created behind the dam provides water storage for irrigation, minimizing the drought impacts on crop yield. Another benefit is increased navigability along the Chang Jiang, as the once volatile river currents are tamed. An extensive system of locks also enables ocean-bound commercial ships to travel much farther into the interior, extending cargo traffic 1,500 miles (2,400 kilometers) along its course from coastal Shanghai to Chongqing, one of the busiest inland ports in the world.

Although the dam has been hailed as an engineering success, critics suggest the negative aspects of the megaproject far outweigh the gains. The Three Gorges Dam has reduced the surface river current from 13 feet (4 meters) per second to 1 foot (0.3 meters) per second. The stagnation of water behind the dam has resulted in sediment accumulation into the reservoir, thereby slowly diminishing its water storage capacity, commercial shipping depth, and hydroelectric capabilities. For those downstream of the reservoir, stagnant water means a decline in soil quality as nutrient-rich minerals settle behind the dam. Although termed a green energy source, hydroelectric dams may actually promote water pollution by concentrating human, animal, and chemical waste in areas where formerly rapidly moving waters would quickly dilute and disperse waste material. This polluted water threatens residential water supplies and the local ecology, including many endangered species (e.g., the river dolphin) native to central China. The rising waters have also put pressure on the physical landscape, making landslides more common and, some have suggested, increasing earthquake potential.

The creation of the reservoir has inundated over 1,500 cities, towns, and villages and over 100,000 acres of fertile agricultural lands, displacing an estimated 1.5 million people and eradicating cultural histories. Farmers forced from their ancestral homelands are rebuilding in marginally productive areas without the aid of their established socioeconomic support structures. Others find themselves having to adjust to a new urban lifestyle and crafting an existence in a more industrialized, modern China. Financial compensation for those displaced has been largely insufficient because the costs involved were grossly underestimated, the alternative land dispersed was poorer in quality, and corrupt local officials were not properly allocating funds.

Since its inception, environmental activists and nongovernmental organizations have been outspoken about these potential detrimental effects of the Three Gorges Dam to the physical and cultural landscape. Journalist Dai Qing has written and spoken extensively about the environmental degradation and widespread governmental corruption associated with the Three Gorges project—a criticism that earned her a 10-month prison sentence in 1989. Ironically, the most recent disparaging remarks have come from Chinese government officials who note that water pollution, landslides, and erosion have already become problematic along the Three Gorges and caution that careful environmental stewardship is needed to prevent ecological disaster. The absence of President Hu Jintao and Prime Minister Wen Jiabao at the 2006 Three Gorges completion ceremony was suggested by some to be an acknowledgement of the concerns raised and that the administration was perhaps distancing itself from such projects. However, new megadam development continues, such as the Xiluodu Dam on the Jinsha River, a Yangtze River tributary.

See Also: Environmental Stewardship; Hydroelectric Power; Power and Power Plants; Renewable Energies.

Further Readings

Cech, Thomas V. *Principals of Water Resources: History, Development, Management, and Policy.* Hoboken, NJ: John Wiley & Sons, 2005.
Hvistendahl, Mara. "China's Three Gorges Dam: An Environmental Catastrophe?" *Scientific American.* http://www.sciam.com (Accessed March 2008).
Qing, Dai. *The River Dragon Has Come!: The Three Gorges Dam and the Fate of China's Yangtze River and Its People.* Amonk, NY: M. E. Sharp, 1998.
Yardley, James. "Chinese Dam Projects Criticized for Their Human Costs." *The New York Times* (19 November 2007).

Jill S. M. Coleman
Ball State University

THREE MILE ISLAND

The Three Mile Island Unit 2 (TMI-2) nuclear power plant operated by the Metropolitan Edison utility company near Harrisburg, Pennsylvania, is best known as the site of a core meltdown that began on March 28, 1979. Like the majority of U.S. nuclear power plants, TMI-2 was a pressurized-water nuclear reactor. Water is pressurized to 150 times that of

The accident at the Three Mile Island nuclear plant drastically increased public suspicion of nuclear power and opposition to nuclear waste sites in the United States. Nevertheless, part of the plant continues to operate.

Source: iStockphoto.com

atmospheric pressure and then heated to 600 degrees Fahrenheit by passing through the nuclear core. The high pressure prevents the water from boiling, which enables it to cool the radioactive core and transfer heat to an area outside of the reactor where steam can be used to power a turbine for the generation of electricity.

The TMI-2 plant suffered a severe core meltdown when either mechanical or electrical failure caused the water pumps, turbine, and reactor to shut down. A valve designed to relieve the resulting high pressure remained stuck in an open position, which released cooling water and caused the core of the reactor to overheat. The metal tubes holding the nuclear fuel ruptured, and the fuel pellets began to melt. Instrument failure in the control room complicated the response to the incident. The reactor containment vessel failed to isolate radioactive water, which was improperly vented as steam. A hydrogen bubble formed on top of the reactor as the temperature of the water rose so high that the water molecules were split into oxygen and hydrogen.

By March 30, Pennsylvania governor Richard L. Thornburgh advised pregnant women and pre-school-age children within five miles of the plant to relocate, and thousands of people evacuated the area. The greatest fear was that the containment structure at the plant would be breached and release massive amounts of radiation into the surrounding area as a result of either the melting nuclear fuel or a hydrogen bubble explosion. Neither of these fears were realized. About half of the nuclear fuel melted, and by the end of the first day the core had adequately cooled to ensure containment. By the third day of the incident, plant operators were eventually successful in shrinking the hydrogen bubble.

The incident did not result in any deaths or injuries, radiation doses from the steam released were found to be minimal compared to natural background levels, and cancer deaths among nearby residents have not been elevated. In the immediate aftermath of the TMI-2 incident, both TMI-2 and the nearby TMI-1 reactors were shut down. TMI-1 reopened in October 1985 and continues to operate, whereas TMI-2 is permanently shut down.

The incident had a wide-ranging effect on the policies and law governing nuclear power. It was the most dramatic event in a parade of embarrassing nuclear blunders in the 1960s and 1970s. After World War II, the federal government extended nuclear policy from weapons production to energy production with the Atomic Energy Act of 1946 and its amendments in 1954, which established the Atomic Energy Commission (AEC) to at once promote private investment in nuclear power generation and regulate this new industry. For three decades the AEC successfully promoted nuclear power with research and development funding, as well as subsidized uranium enrichment and prototype reactors, contributing more than twice the amount of private investment in nuclear power. Private investment in nuclear power plants had increased dramatically in the 1960s and reached a peak in the early 1970s. Utilities ordered 136 new nuclear plants between 1971 and 1976. However, the regulatory function of the AEC did not keep pace with the rapid development of a nuclear power industry. This fact was widely recognized in the 1970s through highly visible examples of mismanagement and dissension among nuclear experts on public health and safety. Radioactive waste was found to be leaking from existing disposal sites, and state leaders began to challenge new AEC waste disposal plans, revealing poor site selection—such as a cavern just 500 feet from the Tuscaloosa Aquifer and salt mines in Kansas compromised by water leakage. In 1966, the Fermi-I nuclear plant experienced a partial core meltdown. In 1975, 2,300 scientists, many of whom had worked to develop the first atomic bomb, signed a petition urging the United States to reduce the construction of nuclear power plants. That same year Congress abolished the AEC, assigning the agency's regulatory responsibilities to the independent Nuclear Regulatory Commission. This did little to relieve the growing distrust of nuclear power.

By 1979, press coverage of nuclear power was predominantly negative, as were Congressional hearings, and the stock market value of nuclear utilities had fallen. Just 12 days before the meltdown at TMI-2 the film *The China Syndrome*, starring Jack Lemmon, Jane Fonda, and Michael Douglas, reflected popular suspicions associated with nuclear power. The film depicted an arrogant disregard for public safety on the part of utility companies and government regulators, leading to a near meltdown at a California nuclear plant. In the film, the "China syndrome" is described as the disaster that occurs when the nuclear core of a power plant melts down into the ground until it hits groundwater and then explodes back through the ground as radioactive steam. Lemmon's character, the power plant manager, explains that such a scenario "renders an area the size of Pennsylvania permanently uninhabitable." After the real-life flirtation with the China syndrome at TMI-2, new orders for nuclear power plants evaporated entirely. The utilities also canceled many of the orders they had placed in earlier years.

The national political climate was highly critical of federal efforts to regulate nuclear energy and radioactive waste and was sympathetic to state demands for more responsibility over disposal. Many members of Congress shared Senator Gary Hart's sentiment that nuclear policy was a "continual encroachment on vital state prerogatives." In this context Richard W. Riley, the governor of South Carolina, ordered a convoy of trucks carrying radioactive waste from the Three Mile Island accident bound for the Barnwell, South Carolina, radioactive waste site to turn back toward Pennsylvania. The governors of two other radioactive waste sites in Nevada and Washington also closed access to waste shipments in an effort to force Congress to draft major legislation on radioactive waste disposal. Environmentalists and state governments lobbied for a strong state veto provision over siting, while the nuclear industry opposed such measures. In 1980, Congress passed the low-level Radioactive Waste Policy Act, which devolved responsibility for the siting of low-level radioactive waste sites to states and regional compacts. In 1982, Congress passed the Nuclear Waste Policy Act governing the disposal of high-level radioactive waste and spent fuel at two permanent, underground repositories. Under this law states could veto a federal decision to build a waste disposal facility within their borders, but the veto could be overridden by both houses of Congress. Neither of these laws has yet resulted in the operation of new radioactive waste disposal sites. State, local, and/or community opposition to siting decisions, inspired in part by fears resulting from accidents such as TMI-2, has stymied new waste sites, which presents a significant barrier to the expansion of nuclear power in the United States.

See Also: Chernobyl; Nuclear Power; Uranium; Yucca Mountain.

Further Readings

Osif, Bonnie A., et al. *TMI 25 Years Later: The Three Mile Island Nuclear Power Plant Accident and Its Impact.* University Park: Penn State University Press, 2004.

U.S. Nuclear Regulatory Commission. "Three Mile Island Accident Fact Sheet." http://www .nrc.gov/reading-rm/doc-collections/fact-sheets/3mile-isle.html (Accessed February 2009).

Walker, J. Samuel. *Three Mile Island: A Nuclear Crisis in Historical Perspective.* Berkeley: University of California Press, 2004.

Daniel J. Sherman
Independent Scholar

Tidal Power

Tides are the cyclic rising and falling of the ocean surface caused by the gravitational attraction of the moon and the sun acting on the ocean waters of the rotating earth. Tides cause changes in the depth of marine and estuarine water bodies and produce oscillating currents known as tidal streams. Moreover, tidal power can be used to mean the generation of electricity using the mechanical force created by the rise and fall of ocean surge. It is a renewable, largely abundant, nondepleting, and clean source of energy. Tides are very powerful, with the sea moving very quickly, producing an immense force of moving water that can be adequately harnessed for gainful work. Tidal energy is produced by using special energy generators purposely designed to convert the mechanical power of ocean currents to electricity. These generators are generally called tidal "stream" generators. They contain turbines mounted on gearbox shafts and operate in a manner similar to wind turbines, but they produce more electrical energy at low tidal velocities compared with the same amount of wind speeds for wind turbines. The generators are placed underwater in places or areas where high tides are common and recurrent. They are designed to capture the kinetic energy of moving water and subsequently convert it to electrical energy. A key feature of most tidal stream turbines is that the turbines use technology already developed from the wind industry. The only difference is the support structure.

Apart from the use of tidal stream generators, another approach to creating tidal power is the use of dams or barrage systems. The dams are based on using embankments at a bay or estuary with a large tidal range. The tidal range must usually be above 5 meters. Power is generated at ebb tides as the barrage system creates a significant head of water. When the tide rises, the barrage opens and allows water to flow inward; it closes when the tide stops. Water is then trapped within the barrage basin and thereby creates a hydrostatic head. Because of this hydrostatic head, the water flows through the passage leading through turbines, driving the turbines and consequently generating electricity as it flows in and out. Tidal barrage systems work more like hydropower systems. The disadvantages of using the barrage system include the very high civil infrastructural cost, the inadequacy of suitable sites for construction, and the daily intermittent nature of the ocean tides.

Tidal variation is a common phenomenon. It is caused by the moon's movement, its interaction with the sun, and the gravitational pull between the moon and Earth. Tides may either be diurnal (producing one tidal cycle of high and low water levels per day) or semidiurnal (having two high and two low water levels per day). This diurnal difference produces daily inequality in the water gradients, which change with time and are generally low when the moon is over the equator. The history of tide discovery and its relation to the sun's and moon's activities has come a long way. For example, the link between tides and the moon was first theorized by Seleucus of Seleucia in the second century. He wrote that the rising and falling of tides were the result of the attraction of the moon and that the height of the tides depends on the location of the moon relative to the sun. In Europe around 730 C.E., the Venerable Bede described how the rise of the tide on one coast of the British Isles coincided with the fall on the other side and described the progression in times of the same high water along the Northumbrian coast.

The use of tidal power provides some advantages. Once a tidal power plant has been built, its operation and maintenance is relatively cheap. It does not produce greenhouse gases, as no burning of fuels is involved. It only uses natural energy from flowing water and tidal currents to power turbine installations. It produces no air pollutants and provides clean and reliable energy capable of powering industries and communities. It is renewable because the tides will continue to ebb and flow, and the energy harvest can keep going on and on (because as long as the Earth and moon continue to move and interact with each other and with the sun, and the ocean exists on the surface of the earth, there will always be tidal movements and variations, and so tidal power will always be available). The tides, though intermittent, are predictable and thus encourage energy planning. The use of tidal power also removes foreign exchange expenditures that would have been channeled to purchasing petroleum fuels and creates employment opportunities. It can adequately be relied on where high tides are harvestable when adequate energy storage capacities are provided or alternatives are available during the absence of tides. A major disadvantage of tidal power stations is that they can only generate power when the tide is flowing in or out, meaning that the power plants can only generate electrical energy for about 10 to 12 hours per day. However, tides are totally predictable, so other power stations can generate power at those times when the tidal station is out of action. As for the environmental impact assessment of tidal power plants, the barrage plants have been found to affect the salinity and circulation of the seawater in which they are located. Tidal stream generators also can affect fish—the fast-moving turbine blades of these generators can kill fish that flow through the systems.

See Also: Hydroelectric Power; Renewable Energies; Wind Power; Wind Turbine.

Further Readings

Alternative Energy News. "Tidal Power: News and Information About Tidal Energy Technology." http://www.alternative-energy-news.info/technology/hydro/tidal-power (Accessed January 2009).
Renewable Energy Center. "Wave and Tidal Power." http://www.therenewableenergycentre .co.uk/wave-and-tidal-power (Accessed January 2009).

Ajayi Oluseyi Olanrewaju
Covenant University, Nigeria

Total Primary Energy Supply

Total primary energy supply (TPES) is a measure of the energy inputs to an economy. The calculation methodology allows diverse energy resources including fossil fuels, renewables, nuclear energy, and heat to be aggregated and compared. TPES has risen by 2 percent per year on average since 1973 and is forecast to continue rising at 1.5 percent per year, largely as a result of increased demand in Asia. In combination with measures of population, economic wealth, or greenhouse gas emissions, TPES can be used to create economic efficiency indicators; however, such metrics must be carefully interpreted in light of local social, economic, and geographic factors.

TPES is formally defined by the International Energy Agency (IEA) as indigenous production + imports − exports − international marine bunkers ± stock changes. Therefore, in any given year, the TPES of a country will depend on the energy resources extracted from its own territories, the net of energy resource trades with other nations, and any changes in bulk resource storage (e.g., the U.S. Strategic Petroleum Reserve). International marine bunkers, that is, those fuels supplied to ocean-going vessels, are excluded from national TPES calculations but added to global figures. TPES is typically reported in units of millions of tonnes of oil equivalent (Mtoe); in SI units, 1 Mtoe equals 41.868 picojoules ($\times 10^{15}$ Joules).

Energy resources included in the calculation of TPES include fossil fuels (such as coal, crude oil, and natural gas), combustible renewables and wastes (e.g., biomass), primary electricity (i.e., from renewable and nuclear sources), and heat. For fossil fuels, combustible renewables, or other energy commodities, standard tables of calorific values enable the conversion of volumes or weights to tonnes of oil equivalent. However, care must be taken when converting primary electricity sources to TPES values. For hydropower, wind, solar, and other renewable electricity sources, primary energy is assumed to be equivalent to the amount of generated electricity, not the energy content of the captured wind, water, or solar flows. For nuclear or geothermal electricity, heat is considered to be the primary energy source, and therefore the conversion losses of the generating turbines must be accounted for.

The IEA recommends using a conversion factor of 33 percent for nuclear electricity and 10 percent for geothermal energy, unless better values for local turbine technologies are known.

In 2006, global total primary supply was 11,741 Mtoe. Of this amount, 81 percent comes from fossil fuels like coal, oil, and gas; 6.2 percent comes from nuclear power; and the rest comes largely from renewable sources. Regionally, 47 percent of global TPES is used by the Organisation for Economic Co-operation and Development (OECD) nations (of which 50 percent is used in North America, 34 percent in Europe, and 16 percent in Asia-Pacific); 16 percent by China; approximately 10 percent in each of eastern Europe, the former Soviet Union, and Asia excluding China; and approximately 5 percent in each of Latin America, the Middle East, and Africa (marine bunkers account for 1.6 percent). In 1973, global TPES was 6,115 Mtoe, implying an annual average growth rate of 2 percent over the past three decades. By 2030, the IEA forecasts a global TPES of between 15,800 and 17,700 Mtoe (annual average growth of approximately 1.2–1.7 percent), with growth largest in China and Asia.

TPES is frequently used to compare the relative efficiencies of national economies. The energy use of countries can be analyzed by presenting TPES on a per capita or per unit gross domestic product basis (sometimes adjusted for purchase power parity). On a per capita basis, the OECD nations and the former Soviet Union are the highest consumers at 4.7 and 3.6 toe per capita, respectively; Asia (0.63 toe/capita) and Africa (0.66 toe/capita) are the lowest users. On a wealth basis (i.e., toe per '000$ gross domestic product purchase power parity), the most efficient regions are Latin America (0.15), Asia (0.17), and the OECD (0.18); the least efficient are the former Soviet Union (0.45) and the Middle East (0.36). Similarly, the environmental impacts of energy use can be assessed using measures of carbon dioxide (CO_2) emissions per unit TPES. By this metric, the lowest emissions factors can be found in Latin America and Africa (1.8 and 1.4 tonnes CO_2/toe, respectively) and the highest found in China (2.98), the former Soviet Union, eastern Europe, the Middle East, and OECD nations (2.32–2.51).

Strong correlations have been found between TPES and national gross domestic product. However, although such trends are useful for initial analyses, the influence of local factors must be accounted for. For example, a nation's TPES can be significantly affected by its climate (i.e., demand for heating or air conditioning services) or its economic structure (e.g., whether an economy is based in manufacturing or services). Geography can also be important: Luxembourg has a high TPES per capita (10 toe/cap), but this is partly a result of low taxes, which attracts the citizens of neighboring Belgium, France, and Germany for fuel purchases. Similarly, Iceland also has a high TPES per capita (14 toe/cap), but it has cheap and abundant supplies of geothermal energy, resulting in a low emissions factor of 0.50 tonnes CO_2/toe. Total primary energy supply is therefore an important metric in energy analysis, but one that needs to be carefully interpreted.

See Also: Nonrenewable Energy Resources; Renewable Energies.

Further Readings

Department for Business, Enterprise & Regulatory Reform. "Digest of United Kingdom Energy Statistics. Annex A: Energy and Commodity Balances, Conversion Factors and Calorific Values." 2008. http://stats.berr.gov.uk/energystats/dukes08_aa.pdf (Accessed January 2009).

International Energy Agency. "Key World Energy Statistics." 2008. http://www.iea.org/textbase/nppdf/free/2008/key_stats_2008.pdf (Accessed January 2009).

International Energy Agency. "World Total Primary Energy Supply 2005." 2007. http://www.iea.org/textbase/country/maps/world/tpes.htm (Accessed January 2009).

James Keirstead
Imperial College London

URANIUM

Uranium is the heaviest naturally occurring element found on Earth. It is naturally radioactive (i.e., it decays into different elements over time) and consists of two primary forms: uranium-238 (^{238}U) and uranium-235 (^{235}U), which make up about 99.3 percent and 0.7 percent of natural uranium, respectively. Each isotope (form) has slightly different radioactive decay sequences, ending in a nonradioactive (stable) isotope of lead. The intermediate decay products include highly radioactive elements such as radon, radium, and polonium.

The key property of uranium that makes it a potential energy source is its ability to absorb neutrons into its atomic nucleus. When ^{235}U absorbs a high-speed neutron, this can lead to the atom splitting, or "fissioning," and releasing significant energy and several neutrons at the same time, which in turn can fission more atoms, and so on, in a chain reaction. This process means ^{235}U is called "fissile." If this chain reaction is uncontrolled, it will lead to a nuclear explosion. Alternatively, if the fission reaction is moderated by limiting neutron speed and the rate of fission, the energy release can be captured as heat and used for driving a steam turbine for electricity.

When ^{238}U absorbs a neutron, it rapidly undergoes radioactive decay to plutonium-239 (^{239}Pu). This process means ^{238}U is called "fertile." Plutonium did not naturally exist before the dawn of the nuclear age. Similar to ^{235}U, ^{239}Pu is fissile and can be used either in nuclear weapons or nuclear reactors.

To make use of uranium in nuclear power reactors, it is common to use enrichment to increase the percentage of ^{235}U and improve the characteristics of the fission process in a reactor. To achieve this, centrifuge or gaseous diffusion technologies are used. Uranium is converted to uranium hexafluoride, and the percentage of ^{235}U is increased to between 3 and 5 percent (for comparison, a nuclear weapon generally requires >90 percent ^{235}U). Enriched uranium is then converted to uranium dioxide and manufactured into fuel rods. After consumption in a nuclear reactor, the spent fuel rods are intensely radioactive (because of the large number of intensely radioactive fission products) but still contain residual ^{235}U and ^{238}U, as well as ^{239}Pu (as a result of conversion from ^{238}U). It is therefore possible to reprocess the spent fuel rods to extract the ^{235}U and ^{239}Pu for the production of new fuel (or nuclear weapons).

The overall process for nuclear power could be described as a "cycle," as the process includes mining, enrichment, fuel manufacture, reactor use, reprocessing, and new fuel from reprocessing. Although this is theoretically possible, no regional nuclear industry yet operates a fully closed cycle, and the more appropriate description is "nuclear chain." A critical issue with nuclear power is the long-term management of the intensely radioactive spent fuel rods, called high-level nuclear waste. Despite some 60 years of research, there is still no operating repository for such waste.

History of Uranium

Uranium was first identified as a unique element in 1789 by German chemist Martin Klaproth, based on samples from the metal mines of Bohemia (now northwestern Czech Republic). The main uses for uranium were limited to glass and ceramics, often for glazes, and scientific curiosity. In 1896, French physicist Henri Becquerel discovered the phenomenon of radioactivity from uranium, leading to a rapid evolution in the field and the subsequent discovery of numerous radioactive elements such as thorium, polonium, radon, and radium (among many others).

One intensely radioactive element, radium-226 (^{226}Ra, formed from the decay of ^{238}U), was almost immediately heralded as a great cure for cancer. Although radium was extremely difficult to procure because of its low concentration in uranium ore, a minor boom in uranium mining occurred to extract radium. The main countries involved in radium mining were the Czech Republic, the United States, the Belgian Congo, and Canada. By the 1920s, however, significant cancer rates were being noted in radium workers (such as the radium dial painters) and uranium mine workers (following on from health effects first noted from the 16th century among Bohemian miners). The radium industry declined, though it had reached a price of some $100,000 per gram during this time.

In 1939, German physicists proved the concept of fission—the splitting of uranium atoms into new, smaller elements—paving the way for the atomic bomb and nuclear power. The United States was the first country to successfully build, test, and use nuclear bombs, achieved through the secret Manhattan Project during World War II. Russia, Britain, France, and China had all developed nuclear weapons by the mid-1960s—the Cold War was in full motion, and it fostered a new global uranium mining boom. The issue of nuclear weapons remains pivotal to the debate about uranium as a source of energy.

By the mid-1960s, using nuclear reactors for electricity generation was made a reality, providing a major nonmilitary use for uranium, and gave rise to yet another uranium boom. Although nuclear power underwent rapid growth until the 1980s, despite optimistic projections, it has remained a relatively minor source of electricity. The major demand for uranium is now for nuclear power, as nuclear weapons states do not need new uranium.

There are broadly considered three main historical phases for uranium—the radium era (early 1900s), the nuclear weapons age (1940s to present), and the civilian nuclear power period (1970s to present).

Geology of Uranium

Uranium is a relatively versatile element in the surface environment compared with many heavy metals. Uranium can be easily dissolved into water under the right chemical conditions (e.g., acidic and oxygen-rich waters) but will also precipitate under different settings (e.g., oxygen-depleted waters). It is commonly found in minerals with oxygen, such as uraninite (uranium dioxide) or pitchblende (a uranium dioxide–uranium trioxide mix, giving rise to U_3O_8), or in combination with iron and titanium (e.g., brannerite, davidite), silicate minerals (e.g., coffinite), or potassium and vanadium (e.g., carnotite). The major minerals present in a uranium deposit will depend on how the rocks were formed and on any alterations resulting from ongoing chemical weathering.

There are numerous types of uranium deposits currently known, including unconformity, surficial calcrete, sandstone, conglomerate, breccia complex, volcanic, metasomatite and vein deposits, and other minor types. Uranium can also be found at low concentrations in phosphate rocks, leading to an important by-product from phosphate mining for fertilizer. During the weapons era, sandstone deposits were dominant; however, since the 1970s, unconformity deposits are now the major source of global uranium production. It is common for countries to have uranium resources grouped in a few deposit types. For example, Canada is dominated by unconformity deposits with lesser amounts in conglomerate resources, whereas Australia is dominated by the supergiant Olympic Dam breccia complex (which contains copper, uranium, gold, and silver) and has several major unconformity deposits.

Uranium Mining

Uranium can be mined and produced in a conventional manner, through open-cut or underground mining and processing the ore, or unconventional methods such as in situ leach (ISL) mining.

The major conventional uranium mines of the world range in grade (concentration) from about 0.04 percent to 15 percent U_3O_8, with most deposits between 0.1 to 3 percent U_3O_8. The contained uranium can vary from 1,000 to 100,00 t U_3O_8 or more. Mining and processing involves the generation of significant quantities of solid wastes. For ore grading 0.1 percent U_3O_8, some 99.9 percent is therefore waste or "tailings." In addition, open-cut mines generate large amounts of nonmineralized rock and low-grade waste (low levels of uranium). The management of solid mine wastes is a major long-term environmental issue.

Unconventional uranium mines, based on ISL mining, use boreholes and cyclical pumping of chemical solutions through the ore zones to leach the uranium in situ, from where it is extracted in a surface facility. ISL mining is only viable for certain sandstone deposits, and the geology, ideally, should include low-permeability layers above and below the deposit to prevent the migration of mining solutions and the contamination of surrounding groundwater. Either acid or alkaline solutions can be used, depending on the deposit's characteristics and environmental concerns. In general, alkaline is seen as less damaging to groundwater compared with acid. The ISL method is widespread in Kazakhstan, with minor use in Australia, China, and the United States, though all regions have significant concerns regarding effects on groundwater resources.

Health Effects From Uranium Mining

A unique aspect of uranium mining, compared to, say, gold mining, is the health risk associated with radiation exposure. Although there is natural background radiation, the additional radiation exposure for uranium miners leads to an increase in risks such as cancer and other health impacts. The radium era gave rise to early awareness about these impacts (e.g., radium dial painters), but it was not until the mid-20th century that the major risk for miners was identified as radon gas. Radon is formed from radium decay and is intensely radioactive, as are its subsequent decay products. For uranium miners, in situations where radon gas can accumulate (e.g., underground mines), this exposure can lead to a significant increase in health risk. As knowledge has evolved, mining practices have improved so that workers receive less radiation exposure in modern mines. The extent of residual radiation exposure risks in uranium mining remains a highly contentious topic.

Sustainability Aspects of Uranium Mining

The application of sustainability principles to uranium mining is challenging. One of the key aspects is the extent of economic resources—that is, the amount of uranium extractable under various economic assumptions (e.g., demand, technology, exploration, price, and so on). In addition, a critical issue at present is the extent of greenhouse gas emissions associated with all methods of energy production.

Uranium, as a metallic resource, can be considered to be effectively finite. That is, it is not being renewed or regenerated in Earth's natural processes, so all extraction and consumption are leading to a gradual decline in the amount available for future generations. At each major phase of uranium mining (radium, weapons, power), uranium was perceived to be in short supply, and extensive exploration work went on to discover significant new resources. By 2007, annual demand was about 78,000 t U_3O_8, and cumulative global uranium production was 2 million t U_3O_8. The estimated global economic resources of uranium in 2006 were 10.4 million t U_3O_8—a figure that has grown gradually over time and does not account for all known resources (e.g., phosphate), nor does this account for the plutonium economy (converting ^{238}U to ^{239}Pu to create an additional energy resource).

A major area of concern for uranium is the environmental impacts during and after mining. In general, each phase has improved its environmental and radiological management, but the extent of effects from the Cold War era is a major environmental legacy. For example, in the United States and the former German Democratic Republic, billions of dollars have had to be spent on the remediation and rehabilitation of former Cold War–era mines. The long-term success of these programs remains to be seen, with early evidence for both success and failure at many sites.

Furthermore, a critical aspect of the potential for uranium and nuclear power to be perceived as green is the relatively lower greenhouse gas emissions associated with nuclear power compared with fossil fuels. Although life cycle assessments of embodied energy and greenhouse gas emissions for nuclear power remain contested in the scientific literature, it is clear that uranium mining is a major contributor of greenhouse emissions in nuclear power. In general, uranium production requires about 200 gigajoules (GJ) of energy and leads to about 27 metric tons of greenhouse emissions (as CO_{2-e}) per ton of uranium oxide. In addition, nearly one million liters of water are required. These environmental costs, however, are inversely dependent on ore grade. Given that ore grades are in gradual decline, this indicates a slow increase in environmental costs such as greenhouse emissions in the future.

Key Aspects of Uranium

In summary, uranium can provide a significant source of energy through use in nuclear reactors based on ^{235}U and/or ^{239}Pu. The major concerns related to uranium mining include the extent of remaining economic resources, nuclear weapons, and the environmental impacts during and after mining—especially the extent of greenhouse gas emissions associated with uranium mining and the nuclear chain. Overall, the future of uranium will be constrained by the extent to which society continues with nuclear power, and especially whether nuclear power achieves broad acceptance as an electricity source in comparison to conventional renewable energy sources (e.g., solar, wind, geothermal).

See Also: Chernobyl; Nuclear Power; Nuclear Proliferation; Three Mile Island; Yucca Mountain.

Further Readings

Makhijani, Hu, et al., eds. *Nuclear Wastelands—A Global Guide to Nuclear Weapons Production and Its Health and Environmental Effects.* Cambridge, MA: MIT Press, International Physicians for the Prevention of Nuclear War (IPPNW), and Institute for Energy & Environmental Research (IEER), 1995.

Mudd, G. M. and M. Diesendorf. "Sustainability of Uranium Mining: Towards Quantifying Resources and Eco-Efficiency." *Environmental Science & Technology,* 42/7 (2008).

OECD Nuclear Energy Agency and International Atomic Energy Agency. *Uranium 2007: Resources, Production and Demand.* Paris: OECD, 2008.

Standing Committee on Industry and Resources. *Australia's Uranium—Greenhouse Friendly Fuel for an Energy Hungry World.* Canberra: Australian Parliament, November 2006.

Gavin M. Mudd
Monash University

Volatile Organic Compound (VOC)

Volatile organic compounds (VOCs) refer to a broad group of chemicals that have a high-enough vapor pressure to evaporate and enter the atmosphere at ambient temperatures. This definition includes a number of compounds that are natural, as well as synthetic. The materials emitting the VOCs may be solid or liquid. Some VOCs exist only as gas at ambient temperatures. VOCs may have short- or long-term adverse health effects. In environmental legislation, the term is often used to suggest air pollutants. Beyond the direct effects of VOCs, the production of ozone is often affected by VOC levels, adding to air pollution.

VOCs are generated by a variety of thermal and nonthermal processes. Burning organic or synthetic materials can release VOCs. Off-gassing from a variety of industrial or consumer products can create VOCs. Paints are a major source. Various solvents evaporate, releasing VOCs into the environment. VOCs are produced by many processes used to make other materials. VOCs are not only the product of the industrial world, however: Plants produce VOCs, and some estimates put this production at half the overall production of VOCs.

Often overlooked in discussions about VOCs is the fact that methane is a VOC. At ambient temperatures, it is already a gas. Significant production of methane occurs in combustion, but there are natural sources including decay of biomass and from animals. Some discussions of VOCs will differentiate methane and "non-methane volatile organic compounds." Methane is considered a greenhouse gas, along with carbon dioxide, affecting the climate.

Green energy and VOCs intersect in a number of settings. These include combustion for energy production, various energy-producing processes, increased efficiency of use of energy, and productions of a variety of items used in the field of green energy.

Energy production by burning will produce VOCs. Ideally, a green energy process would minimize the amount of pollutants, including VOCs. However, green energy also includes use of sustainable sources including wood and other biomass materials. Such use may produce VOCs, potentially in greater amounts than traditional, "nongreen" sources such as oil or coal if pollution controls are not in place.

Gasoline for vehicle use is a major source of VOCs. In the refining of oil to gasoline, production of VOCs occurs. Then in the combustion of that gasoline in driving, further

VOC production occurs. Such production is a major source of man-facilitated VOC production.

In the production of ethanol, which is touted as a renewable source of energy, VOCs are produced in several ways. Ethanol production is very energy-intensive, hence VOCs are produced. The ethanol itself also then becomes a source of VOCs.

On the positive side, VOCs are able to be used in fuel cells that can produce energy without VOC production. Indeed, some industries are capturing the VOCs produced in the plant and feeding them into fuel cells to produce electricity used in the plant. This promises to be a good source of green energy.

Biofiltration can remove VOCs that are generated in a variety of processes. This type of scrubbing can reduce the emissions of VOCs where the production process creates them. Pulsed microwaves are sometimes used in scrubbers to fully oxidize the VOCs. Capture techniques can be used to supply fuel cells.

To improve the effectiveness of energy used, thereby reducing the need for energy production, many modern buildings are designed to be highly efficient in terms of insulation and air exchanges. This results in potential for increased VOCs in the building environment. Sources might include off-gassing from products such as paints, carpets, furniture, and accessories. Because of reduced air exchanges, this may result in the buildup of VOCs to the point of the building's inhabitants developing symptoms—the so-called sick building syndrome. This effect can be magnified if the building's energy—typically heating—is produced by combustion of biomass products.

Such increases in VOCs can also occur in buildings used for commercial purposes, where VOC-generating materials are used and the building is a high-efficiency one. For example, solvents used in production of various products can build up significant levels of VOCs when the building is "tight" and/or ventilation and air exchanges are reduced to improve heating or cooling efficiency.

Green energy often involves alternative forms of energy production that include wind power, geothermal energy, tidal energy, or solar energy. Such forms of energy production do not in themselves produce VOCs. Nuclear energy, although controversial, is another form of energy production that does not create VOCs. However, there are often a variety of electronics that are involved in these various forms of energy production.

Electronic devices are produced in a variety of ways. Many involve processes that themselves create VOCs. The cleaning of various electronic parts may include the use of solvents that produce VOCs. Thus, although VOCs are not produced by the electronics used in alternate energy methods, they may be produced in the processes used to create the electronics that such alternative energy methods use. Such production of VOCs may be amiable to reduction by use of alternative chemicals. Removal and recovery of VOCs may be possible through biofiltration or other scrubbing methods.

In summary, volatile organic compounds are the by-product of various forms of energy production and use. These compounds have a variety of negative effects on life and climate. Green energy offers the opportunity to reduce such compounds. However, such green energy processes may also result in the indirect production of volatile organic compounds. This must be factored into decisions about various forms of energy. The potential for capturing and using VOCs in green energy production through techniques such as fuel cells offers an exciting opportunity for cleaner energy production.

See Also: Alternative Energy; Biomass Energy; Combustion Engine; Fuel Cells; Greenhouse Gases; Waste Incineration.

Further Readings

Chang, Meng-Wen and Jia-Ming Chern. "Stripping of Organic Compounds from Wastewater as an Auxiliary Fuel of Regenerative Thermal Oxidizer." *Journal of Hazardous Materials*, http://dx.doi.org/10.1016/j.jhazmat.2009.01.025 (Accessed September 2009).

Goldman, Lynn and Christine Coussens, eds. *Implications of Nanotechnology for Environmental Health Research*. Washington, D.C.: Board on Health Sciences Policy and National Academies Press, 2005.

Helm, John L., ed. *Energy Production, Consumption, and Consequences, National Academy of Engineering*. Washington, D.C.: National Academies Press, 1990.

Nikiema, Josiane, P.-A. Dastous, and Michèle Heitz. "Elimination of Volatile Organic Compounds by Biofiltration: A Review." *Review of Environmental Health*, v.4 (2007).

Organisation for Economic Co-operation and Development (OECD). *Environmental Outlook*. Paris, France: OECD Publishing, 2001.

Wu Lin, Fenq and Lo LoYang. "Measurement of Toxic Volatile Organic Compounds in Indoor Air of Semiconductor Foundries Using Multisorbent Adsorption/Thermal Desorption Coupled With Gas Chromatography-Mass Spectrometry." *Journal of Chromatography*, 996:225–31 (2003).

Karl Auerbach
University of Rochester

Waste Incineration

The practice of waste incineration is not new. Disposing of remnants or waste from human activities by incineration has been practiced since fire has been known to human beings. Changes in the environment have made waste incineration emissions and residues a sensitive matter. Technologies for incineration are being developed to take local environments into account to improve the efficiencies and the end products of the processes. Combustion leading to vaporization of substances, atmospheric emissions, and residues is complicated when feedstock used includes inorganic and synthetic substances. Integrating the reuse and recycling of emissions and residues in the design and operation of incineration, the processes can give new meaning to the phrase "garbage in, garbage out." Depending on the mix of compounds introduced in the system, the resulting end substances can be reused or recycled in other processes. The costs of treatment can impose economic imperatives for suppliers to reduce waste while accounting more completely for human activities.

The change in terminology from garbage to waste reflects a shift in perspective and permits analysis of the processes by which the waste is generated. In this context, waste recognized as the inefficiency of production processes can be an incentive for improvements and accords waste management significance on the management agenda. Incineration may be an option in waste management. In some contexts, incineration may be either an alternative or a means to reduce landfill. Another option is to include incineration in combination with other waste-processing technologies. Geographical and social realities may also determine the use of incineration technologies. Availability of waste from a number of facilities can also support a stand-alone incineration facility. Depending on the technology, there would also be energy inputs that can be accounted for in determining the economic viability of the incineration processes.

Emissions and residues from incineration may include volatile organic compounds as well as heavy metals in fly ash and bottom ash residues. Some substances that are regulated include cadmium, carbon monoxide, dioxins/furans, hydrogen chloride, lead, mercury, opacity, nitrogen oxides, particulate matter, and sulfur dioxide. Technologies used, including oxidizing or reducing processes, can also result in different emissions and residues.

Historical uses for incineration continue to be relevant today. For instance, in farming, incineration has been used to reclaim nutrients in the form of ash residue that can be plowed back to enrich the soil. Similar considerations can be raised in determining the value of incineration. Cultural and religious reasons may prompt the consideration of incineration independently of recognizing it as an economically viable alternative to dumping waste in landfills.

Circumstances have contributed to the research and development of innovative incineration technologies. Emergent incineration technologies must be compliant with regulatory regimes that take the environment into consideration. In some jurisdictions, waste treatment including incineration has been deployed to reduce existing levels of greenhouse gas emissions and meeting Kyoto Protocol obligations. In contrast, unlike recycling of plastic or landfill sites, incineration can result in an increase of carbon dioxide emissions.

Landfill gas collection systems can be used to reduce methane emissions into the atmosphere and can even be redirected to generate power. Anaerobic digestion is used to generate usable methane and carbon dioxide. Composting can be included to generate or capture nutrients for inclusion in soil conditioners. An example of an innovative incineration process involves heating waste in two stages to produce a hydrogen and carbon monoxide mixture called synthesis gas or "syngas." The heating process leaves inorganic materials that are deposited on molten glass that can be recycled or further treated. The syngas is treated to produce ethanol and methanol, which may be used as gasoline and biodiesel additives, respectively.

In another design, incinerated domestic and industrial solid waste is used to supply heat to local communities and in the production of electricity. Steam and gas can also be used in driving gas turbines to produce electricity. Pyrolysis is used to generate carbon monoxide, hydrogen, and methane, which can be used in exothermic reactions to produce usable energy. Each design satisfies different parameters and may be appropriate in various environments.

The distinction between waste destined for disposal or recovery can sometimes be problematic. Regulations have been developed to assist in the distinction that can determine whether the waste is a problem or a solution. More accurately, the distinction is less clear and can only be a difference of emphasis on whether the process leads to the waste being a problem or solution.

The installation of maximum achievable control technologies is sometimes required by regulations for commercial and industrial waste incinerators. For instance, in the United States, the Environmental Protection Agency included maximum achievable control technologies–based emission performance standards for new sources commencing construction after November 30, 1999, pursuant to Clean Air Act section 129.

Whether integrated as an addition to a facility or as a stand-alone facility, there are similar issues for incineration. Feedstock for incineration needs to be evaluated for suitability and can determine the selection of technology. Where appropriate feedstock and appropriate incineration technology are available, the environmental contexts for emissions and residues as well as location-specific considerations apply. Biomedical waste, for instance, can be a source of organic and inorganic emissions and residues that reflect all the inputs into the medical facility from patients and treatment services. Incineration in this context can offer sanitizing and disinfectant features while also becoming a source of compounds that are at least as complex as the pharmaceuticals used in treatment.

Heterogeneous feedstock such as biomedical waste can include a wider variety of materials involving everything from stainless steel to cotton and radioactive waste compared with single-product feedstock such as paper or methane. In addition, a high-efficiency particulate air filter may be necessary to process airborne substances and separate out particulate matter. Pathogen destruction can require additional nonincineration processes including microwave heating, autoclave, and hot air or steam systems. The issue in this disinfection process involves heating to sterilize without the release of volatile organic compounds from plastic items in the feedstock. Scrubbers can be used to remove acids, heavy metals, dioxins, and other organic compounds, and the waste stream can be further processed for aesthetics before discharging into the air.

Commercial and industrial waste cannot be treated in the same manner as medical waste or municipal waste. Other than in nuclear fusion processes, nuclear waste such as that intended for landfill in Yucca Mountain may conceivably never become appropriate for incineration.

Regulatory regimes for waste incineration and related activities exist in some but not all jurisdictions. Where they exist, such regulations include environmental, land use planning, energy, and business regulatory systems. Records must be kept to demonstrate actual operations and compliance with guidelines. Periodic reports are required.

Waste incineration can be used as a source of energy and can be integrated into a facility either as a cogeneration operation or a stand-alone facility contributing to the energy grid. Here, the appreciation of the feedstock as required input rather than as unwanted "garbage" materials presents an opportunity and resource to satisfy the energy market. Waste-to-energy technology includes the combustion of methane gas from existing landfill sites to generate usable energy. Plastics can sometimes be used instead of heavy oil as a fuel in blast furnaces in an energy recovery process.

Despite initially high capital costs, in some studies it was determined that incineration was cheaper than feedstock recycling with about the same environmental advantages. Additional operating costs may be imposed by regulatory requirements for operator training and qualification and for annual refresher courses.

See Also: Alternative Energy; Alternative Fuels; Biogas Digester; Biomass Energy; Combined Heat and Power; Emission Inventory; Flaring; Gasification; Green Power; Landfill Methane; Life Cycle Analysis; Power and Power Plants; Volatile Organic Compound (VOC).

Further Readings

Bullis, Kevin. "Garbage Power." *Technology Review*, 110/2:25 (March 2007).
Stehlik, Petr. "Heat Exchangers as Equipment and Integrated Items in Waste and Biomass Processing." *Heat Transfer Engineering*, 28/5:383–97 (May 2007).
Valenti, Michael. "Rx for Medical Waste." *Mechanical Engineering*, 122/9 (September 2000).

Lester de Souza
Independent Scholar

Wave Power

As people and nations are becoming more aware of the effect our lifestyles have on the planet, more efforts are being made on local and global scales to find what are known as renewable energy sources. Renewable energy sources are naturally consistent sources, such as the waves of the ocean, and can be used to generate energy such as electricity and heat while not having a negative effect on the environment. One such form of renewable energy is wave power.

This Pelamis Wave Power machine is part of the Aguçadoura Wave Park, the world's first commercial wave farm, which is generating power for 1,500 homes in Portugal.

Source: Wikipedia

Wave power, not to be confused with tidal power, is the energy of the ocean's surface waves, captured in the up-and-down movement of long snake-like machines that float on the ocean surface and converted into usable energy forms such as electricity without producing any greenhouse gases such as carbon dioxide. Waves are generated by the changing of ocean tides, and tides are generated by the gravitational pull of the Earth's moon on its waters. Therefore, wave power is constantly generated, and tapping into this source of power will not deplete its own source in turn, unlike traditional energy sources such as coal and oil.

A major benefit of wave power is that it is generally more reliable as a source of energy than other "clean energy" sources such as air power. This enhanced reliability is because periods without wind are more common than periods with weak waves. In addition, wave power can be harnessed close to the shoreline, offshore, or anywhere in-between. Downsides to wave energy include the difficulty in maintaining a wave farm during storm seasons, protecting the equipment from damage caused by the ocean's salt water, and the potential negative effect on the surrounding marine environment and marine economy. Nevertheless, engineers are optimistic that the technological challenges can be overcome, as the field of wave power and its mechanics is still relatively new.

Although the idea of capturing the sea's energy was first mentioned over a century ago, the inspiration for contemporary wave farms came from a Scottish scientist in 1974. Stephen Salter (b. 1938), then a professor at the University of Edinburgh, demonstrated how the majority of wave motion could be stopped against the body of a device and how the majority of this resulting harnessed power could subsequently be converted into energy. Salter's device was named the Edinburgh Duck, but the name did not stick. Today it is known as Salter's Duck.

Locations where wave power is harnessed are called wave farms. The first commercial wave farm opened on September 23, 2008, off the coast of Portugal. It is called the Aguçadoura Wave Park. Three machines convert the energy into electricity that powers approximately 1,500 homes in Portugal. There are plans to expand the Aguçadoura Wave Park to nearly 10 times its current generation capacity.

This development of wave power has been an international effort. The machines of the Portuguese Aguçadoura Wave Park were developed by a Scottish firm, Pelamis Wave Power, based in Edinburgh. Initial funding for the wave farm came from Babcock & Brown, a global investment firm founded in San Francisco, although it is currently headquartered in Sydney, Australia. Portugal is a nation eager to develop renewable energy sources, and it quickly signed on to wave power while its chief scientific developer, Great Britain, lagged slightly behind.

The latter nation has plans to open its own wave farm off Cornwall's north shore, to be called Wave Hub. Wave Hub is projected to open in 2010 and, at peak capacity, to power roughly 7,500 homes. Pelamis Wave Power will develop Wave Hub's power converters as well. Several other nations including Australia and the United States are also considering wave farms to generate clean energy.

See Also: Alternative Energy; Greenhouse Gases; Nonrenewable Energy Sources; Renewable Energies; Tidal Power.

Further Readings

Boyle, G. *Renewable Energy.* New York: Oxford University Press, 2004.
Cruz, J. *Ocean Wave Energy: Current Status and Future Perspectives.* Green Energy and Technology Series. New York: Springer, 2008.
Falnes, J. *Ocean Waves and Oscillating Systems: Linear Interactions Including Wave-Energy Extraction.* New York: Cambridge University Press, 2005.
Jha, A. "Making Waves: UK Firm Harnesses Power of the Sea . . . in Portugal." *The Guardian* (25 September 2008).
McCormick, M. *Ocean Wave Energy Conversion.* Mineola, NY: Dover Publications, 2007.

Claudia Winograd
University of Illinois at Urbana-Champaign

WIND POWER

The movement of rushing air creates wind. When land air that has been heated by the sun rises, it leaves a space to be filled by cooler surrounding air. The filling takes place immediately as the space is created in a way that could be described and felt as a rush. This fast movement of air is called wind. Wind power (or energy) can be described as the process by which the mechanical energy in the air movement is employed to produce electricity or do meaningful work. Wind power as an energy source is a green energy source. It is a nondepleting, naturally available, nontoxic, and environmentally friendly

Windmills have been used for power for centuries and were once widespread on U.S. farms. This 1885 diagram shows a large windmill installation on a farm built by the Halladay Standard Company of Batavia, Illinois.

Source: National Oceanic and Atmospheric Administration

source of valuable and usable energy. Wind energy has historically been used directly to power ships, pump water, or grind grain, but the principal application of wind power today is to generate electricity. Large-scale wind farms are typically connected to the local electric power transmission network; smaller turbines are used to provide electricity to isolated locations.

The secret behind the fast-growing economy in recent times can be linked to adequate infrastructural development, which invariably is not unconnected to constant and sufficient power supply. A lack of energy or its inadequacy in an economy had variously led to social and economic poverty, underdevelopment, unemployment, underutilization of rural human resources, economic stagnation, underperformance of industry and industrial sectors, and low turnover. People in the rural areas of developing countries and some very poor nations around the world lack access to modern energy supply, with so many depending basically on traditional biomass to meet their energy needs. The severity of their poverty levels, however, is contributing to the reluctance of such nations to embrace the globally accepted sources of energy. Although still depending on burning coals, woods, and so on, for energy, the environmental contributions of such practices is negative, as it causes pollution and the emission of harmful gases and by-products. Moreover, other sources like fossil fuels and nuclear reactors, which have been relied on in the past and are still used in many industrialized nations, also have introduced some undesirable effects to the atmosphere, creating serious environmental concerns. The depletion of these sources points to the fact that other more permanent means and sources need to be sought.

The challenge of providing adequate and sufficient amounts of energy for the populace is a global issue; the depth of concern may vary from developed to underdeveloped nations, but the exercise of making energy scarcity a foregone issue is a major phenomenon throughout the world. Thus, ways of adequately meeting the growing energy demands of the over 6 billion population of the world is an issue that must be met with utmost seriousness. Overreliance on the majority of sources termed the nonrenewable sources of energy is being discouraged throughout the world because of their unsustainability and, at times, their environmental unfriendliness. Green energy sources are a way to resolve the energy challenge of the world, but in using such sources, there are criteria that must be satisfied. These criteria include being affordable (i.e., it must be cheap), easily accessible, widely available, and environmentally friendly. A source of energy that satisfies these definitions is the wind.

History of Wind Energy

The history of using wind power as a source of energy is a long one, dating back to the early times of Egypt's civilization, when it was used to drive boats from shore to shore. However, the first true windmill is recorded to have been built as early as 2000 B.C.E. in ancient Babylon; it was later used by the Babylonians to grind grains in the 10th century. By the 12th century, the Western world began accessing the windmill for the same purpose of milling grain, and later for pumping water. In the 19th century, the United States began creating farm windmills, and by 1889, there were 77 windmill factories. Between the 1930s and 1940s, many wind turbines for electricity purposes were built in the United States to supply power to farms located farther away from power lines. After these periods, rapid developments of wind technology were marred by intermittent stoppages probably resulting from the discovery of oil and gas and partly to the instability of the policies of the regulating bodies. Today, the hiking oil prices and concerns for the environment have made the wind one of the viable sources of energy for the ever-growing world population. Developed countries are fast embracing it as a better and sustainable means of power generation and a suitable alternative fuel for oil and gas. Its friendliness to the environment and availability are also contributing to its acceptance as a better solution to the lingering energy needs of developing countries. Around the world, wind power as a source of energy is gaining prominence, and although it is backed by its long history, the technology is still new, unlike that for solar energy. The availability of wind power is undoubted, but many countries have yet to embrace it: In 2006, only 13 countries has installed wind capacity of over 1,000 megawatts, with Germany having the greatest amount followed by Spain and the United States. By 2007, global nameplate capacity of wind-powered generators was 94.1 gigawatts. Although wind power produces only about 1 percent of the world's electricity, it is growing rapidly—increasing more than fivefold globally between 2000 and 2007. Wind power generating systems are one of the most useful energy resources using the natural environment; their use is encouraged because of the reduced losses in power system transmission and distribution equipment.

Technology of Wind Power

Turning wind power into electricity is very common around the world. In 2006 alone, wind turbines were used to generate a total of 26.6 billion kilowatt-hours per year of electricity in the United States, representing about 0.4 percent of the nation's total electricity generation. The technology of turning wind power into electricity works with a windmill (a device used to harness wind power) or wind turbines (or wind generator). With the windmill, the blades pick up the mechanical energy from the moving wind and turn a driveshaft connected to an electric generator, which in turn produces electricity. The wind turbines, in contrast, directly convert the kinetic energy of the wind into electrical energy. The production of electricity from wind involves creating wind farms—areas of land, usually large flat, containing clusters of dozens of wind machines used for the production of electricity. The location of these farms depends on factors ranging from the availability of wind, wind speed, and the absence or presence of windbreaks. Moreover, the availability and speeds of wind varies throughout a country and from season to season. Areas located between deserts and large water bodies, such as seas or oceans, enjoy a rich supply of wind during summer. This is because the heating of the deserts by the sun makes hot air rise; then cooler air from the surrounding large water bodies moves in to fill the space created by the rising hot air. The movement of these land and sea/ocean breezes generates enormous wind power that can be used to do useful work. The magnitude of wind power available in a location can be determined from the mean wind speed of the location. The wind power is proportional to the cube of the mean wind speed. A mathematical relationship that can be used for the computation is given as

$$P = \tfrac{1}{2}\,\rho V^3 C_p = K \bullet V^3.$$

Where P = wind power flux,

ρ = air density,

V = mean wind speeds,

C_p = Coefficient of power, and

K = constant of proportionality and equal to $\tfrac{1}{2}\,\rho C_p$.

Thus, with this mathematical relationship, the amount of wind power available in an area over a period of time can simply be deduced from the knowledge of the mean wind speed over the area for a certain period. The period for accurate reading and forecasting is usually taken to be 10 years or more. In addition, knowing the average wind power for a place can lead to an understanding of the average wind energy flux density over an entire area.

Advantages of Using Wind Power

Wind power is one of the lowest-priced renewable energy technologies available today, costing 4–6 cents per kilowatt-hour, depending on the wind resource base and the financing of the particular project. The construction time for wind energy technology is less than for other energy technologies, it uses cost-free fuel, the operation and maintenance cost is very low, and capacity addition can be in modular form, making it adaptable to increasing demand. In addition, it has the ability to generate high amounts of electricity to meet world consumption; it is a clean, environmentally friendly, readily available, usable, and nondepleting energy that can reduce dependence on hydropower, fossil fuels, biomass, and other environmentally unfriendly sources of energy; and its installations are very durable. Remote areas that are not connected to the national grid can rely on wind power to produce electricity, meaning that wind power has a potential to hasten development of rural areas, enhance technology and awareness, and reduce poverty because electric power promotes many businesses, and some businesses that depend on electricity will go into such rural areas, leading to employment.

See Also: Green Power; Renewable Energies; Wind Turbine.

Further Readings

Dincer, Ibrahim. "Renewable Energy and Sustainable Development: A Critical Review." *Renewable and Sustainable Energy*, v.4 (2000).
Han, S. G., I. K. Yu, and M. Park. "PSCAD/EMTDC-based Simulation of Wind Power Generation System." *Renewable Energy*, v.32/1 (2007).

Ridao, Ángel R., Ernesto H. Garcia, Begoña M. Escobar, and Montserrat Z. Toro. "Solar Energy in Andalusia (Spain): Present State and Prospects for the Future." *Renew Sustain Energy*, v.11/1 (2007).

Scheer, Hermann. *A Solar Manifesto*. London: James and James, 2001.

Sorenson, Harry A. 1983. *Energy Conversion Systems*. New York: Wiley, 1983.

United Nations. "The Energy Challenge for Achieving the Millennium Development Goals." http://www.undp.org/energy/docs2/UN-ENRG%20paper.pdf (Accessed January 2009).

U.S. Department of Energy. "Energy Information Administration, International Energy Outlook 2005." http://www.eia.doe.gov/oiaf/pdf/0484(2005).pdf (Accessed January 2009).

Voorspools, Kris R. and William D. D'Haeseleer. "Critical Evaluation of Methods for Wind Power Appraisal." *Renewable and Sustainable Energy Reviews*, v.11/1 (2007).

Wind Energy Manual. 2004. http://www.energy.iastate.edu/renewable/wind/wem/wem-04_history.html (Accessed January 2009).

World Bank. "Human Resources Development and Operations Policy. Acute Respiratory Infections." 1993. http://www.worldbank.org/html/extdr/hnp/hddflash/hcnote/hrn004.html (Accessed January 2009).

Ajayi Oluseyi Olanrewaju
Covenant University, Nigeria

WIND TURBINE

The photo shows a wind turbine during final assembly as a crane lifts the rotor toward the generator housing.

Source: iStockphoto.com

Wind turbines are mechanical instruments used for converting mechanical power from the wind to electrical power. It is a rotating machine that can be used to do meaningful work or to generate electricity from the widely available clean fuel called wind power. It comprises four basic components: rotor, tower, generator, and nacelle. The rotor picks up mechanical energy from the wind through attached blades and in turn rotates; the rotation of the rotor is then picked up by the generator to convert the motion to electricity. The nacelle is the compartment that houses the generator and other equipment; the tower is the base on which the turbine is mounted. Other equipment that makes up a wind turbine includes the gearbox, used to increase the speed of the shaft between the rotor hub and generator; the brakes, used to stop shaft movement during system failure or power overload; the control unit, which monitors and shuts down the system in case of malfunction and controls the lurch mechanism; the yaw controller, which moves rotors to align with wind direction; and the electrical parts that carry electricity from the generator.

The history of using wind machines is ancient. The first known practical wind machine was a windmill built in the seventh century in Iran. It was used mainly for grinding and milling and also for fetching water from wells. By the late 19th century, the first known electrical generating wind machine was constructed and operated. Today, hundreds of wind machines (turbines) exist in different parts of the world, from Spain to the United States; there are numerous wind farms employing wind turbines for the production of electrical energy.

Types of Wind Turbines

There are basically two types of wind turbines based on the axis of rotation of the turbines. These are the vertical axis wind turbines (turbines that rotate about a vertical axis) and the horizontal axis wind turbines (turbines that rotate about a horizontal axis). These machines can be massive structures with very long blades. The horizontal axis wind turbines produce energy at higher efficiency than the vertical axis wind turbines and are commonly in use, whereas the vertical axis wind turbines have the advantage of their rotor shaft arrangement, because they do not need to be pointed into the wind to be effective on sites where the wind direction is unpredictable. The size of wind turbines also varies from small to medium to large. Small wind turbines are capable of generating from 50 watts to 60 kilowatts of electricity, and their rotors are of various diameters ranging from less than 1 meter to 15 meters. These kinds of wind turbines can be used as stand-alone power sources in remote areas where connection to the national grid is not feasible. Medium-sized wind turbines are capable of generating electricity from 50 kilowatts up to 1,500 kilowatts, and their rotor sizes range in diameter from 15 meters to 60 meters. On a commercial scale, most medium-sized turbines can generate electricity within the range of 500 to 750 kilowatts. Large wind turbines are huge machines capable of generating 2 to 3 megawatts of electricity. They have proven to be less economical and less reliable on a commercial scale because of their high cost of installation and maintenance.

To generate electricity using wind turbines, wind velocity above 5 meters/second is useful; however, wind speeds above a certain limit are not allowed to drive the rotor shaft because they make the turbine spin too fast, adding excessive torque, which can damage the turbine because the blades get ripped off by the winds. The alternator, too, can become excessively heated and get damaged. Very strong winds also affect the turbine tower a great deal. Thus, to prevent the turbines from damage resulting from strong winds, a mechanical braking system (manual or automatic) is used. It works in such a way as to turn the blades away from the wind direction, thereby disallowing the blade's response to the strong wind and preventing excessive spinning of the rotor shaft, which has a tendency to damage the turbine parts.

Using wind turbines to generate electricity has some advantages. Environmentally, operating wind turbines does not lead to the emission of greenhouse gases. Economically, in addition, the foreign exchange involved in the use of fossil fuel–driven turbines is conserved. However, there are some disadvantages posed by the use of wind turbines. These include the variability in the amount of power generated resulting from the intermittence in the availability of wind. This variability makes precise energy planning and forecasting difficult, as there is no sure evidence that such an amount of wind power will be consistently available in a particular year or period. The economy of managing wind farms is another major consideration: Because quality wind resources are located away from places where people could easily inhabit, the provision of transmission capacity and logistics can be quite challenging. Danger to wildlife has often been the complaint of wind farms, because some birds and bats that run into the blades are killed. Noise from the blades' rotation is another complaint, but when the farms are located away from inhabited areas, the noise does not constitute a concern. However, turbines located offshore can produce noise that can interact with ocean currents and can affect sea animal distribution. Operating wind turbines has also presented some safety issues. Workers sustaining injuries—and some dying—in the course of installation or maintenance have been reported; there are also accidents from transport vehicles and motorists distracted by the sight and shadow flicker of wind turbines along highways. Turbine blade failures could also account for some injuries and deaths, because when the blades fail they fall and could hit or kill someone.

See Also: Fossil Fuels; Green Power; Greenhouse Gases; Renewable Energies; Wind Power.

Further Readings

Ajayi, Oluseyi O. "Assessment of Utilization of Wind Energy Resources in Nigeria." *Energy Policy,* 37:750–53 (2009).
Burton, Tony, et al. *Wind Energy Handbook*, 1st ed. New York: John Wiley & Sons, 2001.
"Wind Energy and the Environment." http://www.ewea.org/fileadmin/ewea_documents/documents/press_releases/factsheet_environment2.pdf (Accessed January 2009).

Ajayi Oluseyi Olanrewaju
Covenant University, Nigeria

WOOD ENERGY

Wood is a sustainable source of energy, provided we apply sustainable forest management practices. It offers significant advantages over traditional, carbon-based fuels: reduced carbon dioxide (CO_2) and other pollutant emissions; reduced transportation costs if available locally; reduced dependence on foreign energy sources; reduced environmental damage from extraction, spills, and toxins; convenience; and increased energy efficiency, and it is clean-burning when burned in modern stoves or boilers.

Wood is a biomass energy source; grasses, corn, and other agricultural crops are also considered biomass energy. The term *biomass* can refer to any crop; alternatively, it can be used to distinguish grasses and other crops from chunkwood.

Traditional wood fuels include chunkwood, chips, and pellets. Chunkwood (also called firewood) can be obtained in long lengths, green cut and split, or seasoned cut and split. Wood chips can be paper grade, bole-tree, wood waste, or bole chips. Wood pellets come in three grades: premium, standard, and industrial. They are made from densified wood waste material, typically from logging, sawmill, or packaging residues. Wet sawdust is compressed into pellets under high heat and pressure. There are no additives in wood pellets, so they burn cleanly. Chemically, wood is composed of approximately 50 percent carbon, 42 percent oxygen, and 6 percent hydrogen. In terms of fiber, it has three major components: it is 52 percent cellulose, 22 percent semicellulose, and 25 percent lignin. Wood has a high heat value range of 8,000–12,200 British thermal units (Btus)/oven dry pound. Moisture content in wood varies from 5 to 65 percent.

Replacing fossil fuels with wood (or another biomass energy source) has the environmental benefit of reducing our contribution to global warming. This is because wood combustion cycles carbon that was already in the natural carbon cycle. The net effect is that no new CO_2 is added to the atmosphere, provided the forests from which the wood was extracted were managed sustainably. In contrast, burning fossil fuels takes carbon that was locked away underground (as crude oil, gas, and coal) and transfers it to the atmosphere as CO_2. CO_2 buildup in the atmosphere is one of the leading causes of global warming. Furthermore, significant quantities of CO_2 are absorbed by plants through photosynthesis and then released through plant decay. Removing biomass fuel from forests using sustainable forestry practices stimulates the growth of replacement wood. This replacement growth absorbs approximately the same amount of CO_2 as was released during combustion. Thus, the U.S. Environmental Protection Agency does not count wood or any biomass energy as putting off greenhouse gas emissions because it is considered part of the short-term CO_2 cycle of the biosphere. Finally, a large percentage of the biomass that is burned to generate energy is waste from the forest products industry, such as sawmill waste. This waste would release CO_2 and methane while decomposing in landfills and waste piles. Thus, using wood waste to produce energy minimizes the emissions of both of these greenhouse gases.

In addition to forest products industry waste, another prevalent source of wood energy is low-grade wood grown in forests, such as cut tree tops from timber harvests or land-clearing operations, that is not of a quality and size sufficient to produce sawlogs, lumber, or furniture veneers. Low-grade bole, top, and limb wood can be used to make pulp (for paper), firewood, and biomass (chips and pellets).

Firewood from standing trees, called stumpage, can be obtained from timber harvests, land-clearing, or timber stand improvement operations. If the wood is clean, cut to the appropriate length to fit into a wood stove, and dry, it is considered "seasoned." If it has not been dried, it is called "green." Green wood is difficult to burn and harmful to the environment, so is should be stacked and stored for a year to season before being used.

How Wood Burns

The combustion in a wood stove results in CO_2, water vapor, ash, and heat. Incomplete combustion also produces carbon monoxide, various hydrocarbons, and other gases considered pollutants. The gases that are vaporized from burning wood contain about 50–60 percent of the heat value of the wood and can only be burned at very high temperatures—around 1,100 degrees Fahrenheit. The way the wood burns, the condition of the stove, and the moisture content of the wood determine the amount of smoke the stove emits and the amount of creosote that remains in the chimney.

The greatest heating efficiency and least smoke (and pollution) occur when the stove is burning at high temperatures. The addition of a catalytic converter/combustor to the stove allows the volatile gases emitted in the combustion process to burn at a much lower temperature. This reduces smoke and creosote formation by 90 percent and increases heat output by 30 percent.

The standard unit of measure for firewood is the cord—a stack four feet high by four feet deep and eight feet long. A cord contains 128 cubic feet of solid space, or about 80–90 cubic feet of solid wood, the rest being air space between the stacked pieces of wood. Sometimes firewood is sold on a weight basis, which can be confusing, because weight will vary by species and dryness. Different types of wood have different properties in terms of heat value, ease of burning, ease of splitting, and aroma. Ask a local supplier or forester about locally available species.

The per unit energy content of a fuel is measured in Btus. One Btu is equal to the amount of heat required to raise the temperature of one pound of water by 1 degree Fahrenheit. Different fuels, and even different wood species, produce different amounts of heat energy when burned. Also, all heating systems lose some heat to start up, to cool down, and from heat escaping up the chimney with combustion gases. Look for an annual fuel use efficiency of 85 percent or higher with use of a wood stove.

As with any heating system, it is critical that a wood burning system be installed and maintained properly for safety and performance reasons. Local fire departments, building inspectors, and insurance companies can provide advice on proper materials and installation methods, safe venting systems, distance from combustible materials, and installation of smoke detectors.

Wood energy is best purchased locally for several reasons. By purchasing locally, one reduces the amount of fuel needed for transporting the energy, thus also reducing its cost and environmental impact of unwanted emissions. Second, because of the potential for the dispersion of invasive species, it is better not to take wood more than 50 miles from its source. Third, locally produced energy contributes value to the local economy. And fourth, the buyer can more likely verify that the wood is extracted sustainably.

Advantages and Disadvantages

The advantages of heating a home with firewood in a wood stove or fireplace stove include the comfort of a wood fire on a cold night; the self-reliance that comes with cutting, splitting, and stacking a cord of wood; independence from utilities; and a reduced conventional fuel bill. In addition, no electricity is needed to burn firewood, and it is a renewable and potentially locally available fuel source. Neither wood nor its ash residue is toxic. Some disadvantages include the need to gather, handle, and store large volumes of wood fuel (unless someone delivers it to the woodshed and stacks it); the dirt, dust, and bugs brought into the house with the wood; the weekly cleaning of residual ashes, flue, and chimney; and the need to feed the wood into the wood stove every eight hours. Conventional fireplaces (without a stove inside) are essentially aesthetically pleasing but are not useful as a heat source.

Masonry heaters, such as a Russian fireplace, Finnish contraflow, and German and Austrian tiled heaters, rely on very hot fires—in excess of 2,000 degrees Fahrenheit—of short duration. Fires lasting an hour or two heat a ton or more of masonry mass by directing the gases through a series of channels, and the stored heat radiates for 12–24 hours. The extremely hot fires result in very clean burns, as the tars and organic compounds are consumed in the firebox. These large, heavy heaters act more like central heating, radiating an even, long-term heat. They are the most efficient wood heaters and require the least maintenance. They are the most costly to build initially but have the highest payback as a result of reduced fuel and cleaning requirements and lack of parts to wear out. They are best located in a large central living area, where the radiant heat can reach most of the living space. Faced with stone, brick, or tile, they are beautiful centerpieces. Environmentally, they produce the smallest quantity of emissions and use a quarter of the wood that a stove would require to heat the same space.

Wood furnaces and boilers, typically located in basements, are larger units that circulate the heated air or water (a more efficient heat-transfer medium). They are appropriate for large or sprawling houses. They are more convenient to use than woodstoves but are considerably more expensive. They require additional ductwork, pumps, blowers, and electricity. Outdoor boilers are best used for heating a group of buildings. Because they are located outside the home or other building, they do not require expensive chimneys, the mess associated with wood burning is left outside, and there is little risk of fire in the home should the chimney catch fire, except for perhaps airborne burning ash. Because they tend to be oversized, they are run less full and therefore burn at lower temperatures.

Firewood Versus Wood Pellets

Firewood needs to be stored outside, under cover, with enough opportunity for air to circulate to keep it dry and prevent mold. Newer stoves are extremely airtight and durable, can heat an entire house efficiently, are very attractive, and are easy to operate. Wood is often used as a secondary fuel source—to provide a warm, constant heat when convenient, coupled with another type of furnace or heat source (e.g., oil, propane, or solar or geothermal radiant heat). When sizing a wood stove, consider the amount of heat wished to be produced in the space it will serve. Do not oversize: A partially full stove will not burn at a high enough temperature to burn off the gases.

Pellet stoves have several advantages over (fire) wood stoves: they are self-feeding from a hopper, they can have an automatic ignition system and be operated by thermostat (like traditional fuel burners), and they have longer burning time and consistently even heat output. They produce lower emissions than firewood as a result of technology within the stove that ensures almost complete combustion of the fuel. Pellet stoves can sometimes be vented with an inexpensive double-walled pipe, rather than a full chimney. Furthermore, they not only use renewable fuel but it is often produced from waste wood that would otherwise have to be disposed of. The disadvantages are that pellets need to be manufactured (not cut from a woodlot) and kept very dry during storage (up to one year). Wet pellets turn back into sawdust that cannot be used in a pellet stove. In addition, the pellet stove costs more than a traditional wood stove, has electric-powered forced air intake (a problem during power outages, unless it has a backup battery), and more things can go wrong with its two blowers, motors, and controls. Pellets can be in short supply and more costly than firewood. Pellet stoves are classified by the amount of heat they generate—high versus low output—and the size of the hopper. Because they are power vented, they can be installed almost anywhere in the home. There are pellet stoves that can burn pellets with corn as well. For convenience, the pellet stove wins out; for self-reliance and cost savings, the wood-burning stove prevails.

A wood pellet stove is sized for residences; a wood pellet boiler or furnace, in contrast, is sized for larger commercial and institutional heating loads—generally for heating 10,000–25,000 square feet of heated space. (Furnaces circulate air; boilers circulate water.) Commercial furnaces and boilers require 15–30 minutes of operator attention daily to empty the ash bin (once a week) and check boiler settings. All moving parts and motors need to be greased annually and checked for wear. They run as quietly as any other furnace or boiler and have minimum dust and wood debris, and the smell of burning wood is usually undetectable as a result of their highly efficient combustion. The boiler can be adapted to use wood pellets, chips, tablets, and/or smaller chunkwood. It can be economical to use locally available, partially chipped chunkwood, but boilers need additional watching, as some of the pieces may be too large to feed easily into the boiler and may block the feed conveyor.

Wood boilers are becoming popular for schools and eco-friendly businesses. Communities are increasingly using community forests to supply biomass for their public buildings, saving taxpayers money and modeling renewable energy technology. Some communities permit their residents to extract firewood from their forests under certain conditions.

See Also: Alternative Energy; Biomass Energy; Carbon Sequestration; Climate Change; Greenhouse Gases; Renewable Energies; Sustainability.

Further Readings

Biomass Energy Resource Center. "Carbon Dioxide & Biomass Energy." http://www .biomasscenter.org (Accessed July 2009).

MacKay, Susan, et al. *Burning Wood and Coal.* Ithaca, NY: Northeast Regional Agricultural Engineering Service Publication No. NRAES-23, 1985.

Shelton, J. W. *Wood Heat Safety.* Charlotte, VT: Garden Way Publishing, 1979.

Thomas, Dirk. *The Woodburner's Companion.* Chambersburg, PA: Alan C. Hood & Co., 2006.

Elizabeth L. Golden
University of Cincinnati

WORLD COMMISSION ON ENVIRONMENT AND DEVELOPMENT

The World Commission on Environment and Development (WCED) is a special commission established in 1983 by the United Nations General Assembly with the aim of preparing a report on global environment and development problematic to 2000 and beyond.

The establishment of the WCED was a response to the concerns about global environmental degradation that first emerged in the late 1960s and then were put forcefully on the international agenda in Stockholm in 1972 at the United Nations Conference on the Human Environment. That it took a decade from the Stockholm Conference to the establishment of the WCED was the consequence of the oil crises of the 1970s, which caused a loss of momentum for the environmental agenda.

The commission was set up as an independent body outside the control of governments and the United Nations. Its mandate was to include, in particular, proposed strategies for environmentally sustainable development. The commission—also known as the Brundtland Commission after its chairman, Gro Harlem Brundtland—was tasked with delineating critical environment and development challenges threatening the planet and with outlining strategies to meet those challenges, particularly in terms of international cooperation and of raising awareness and understanding of these issues on the part of individuals, nongovernmental organizations, businesses, and governments: A new global agenda for change was to be envisioned within the emerging principles of environmentally sustainable development.

One of the goals of the commission was in fact to capture and further define shared perspectives on long-term environmental issues, as well as to outline a strategy for tackling those problems up to 2000 and beyond. The detailed objectives of the commission were articulated in the General Assembly resolution in the following way:

- preparing long-term environmental strategies for achieving sustainable development to 2000 and beyond
- recommending ways in which environmental concerns may be translated into greater cooperation among developing countries and between countries at different stages of socioeconomic development and allow achievement of common objectives formulated on the basis of the recognized interrelationships among people, resources, environment, and development
- considering how the international community can more effectively deal with global environmental concerns
- helping to define a shared set of perceptions of long-term issues of environmental protection and enhancement, as well as a long-term agenda for action for the coming decades
- outlining aspirational goals for the entire world community in light of the overall objectives of a long-term environmentally sustainable global development

The WCED set out to fulfill its mandate by identifying, at its inaugural meeting, eight key issues for analysis during the course of its work: Perspectives on Population, Environment, and Sustainable Development; Energy: Environment and Development; Industry: Environment and Development; Food Security, Agriculture, Forestry, Environment, and Development; Human Settlements: Environment and Development; International Economic Relations, Environment, and Development; Decision Support Systems for Environmental Management; and International Cooperation.

The commission proceeded to hold public hearings in many regions of the world where senior government representatives, scientists, research institutes, industry and business members, nongovernmental organizations, and the general public could openly submit their views to the commission. These public hearings became a "trademark" feature of the commission and were a testimony to its intended goal of enabling the broadest possible participation and the demonstration of the true global nature of the issues it had mandate to investigate. In addition to the public hearings, the commission appointed expert special advisors to assist its work on each of the eight key issues.

The outcome of WCED was a report titled "Our Common Future," in which the commission laid out the global challenges to be faced in common and the new political, legal, and policy framework necessary to achieve sustainable development.

The report was delivered to the United Nations General Assembly at its 42nd session in 1987, and it became one of the seminal environmental documents of the 20th century. In particular, it completed a process of convergence of environmental and developmental concerns and problems, leading to an new, integrated policy approach.

The report offered what became the most used definition of sustainable development, famously defined in the report as development that meets the needs of the present without compromising the ability of future generations to meet their own needs. In this definition, the Commission highlighted two key elements: the element of needs and that of limits. While the element of needs brought forth forcefully the issue of poverty as a major concern at the crossroad of international developmental and environmental politics, the element of limits highlighted the need to recognize the ecological limits to growth, as well as the fact that however technology can expand those limits, there exist absolute limits to the Earth's carrying capacity.

The report is a global call to action, encouraging all nations of the world to join in a common effort to address problems of environmental degradation and economic development, recognizing explicitly the interlinkages between the two. It emphasizes the necessity of an integrative approach to find solutions to the ongoing global tragedy of the commons and outlines principles of shared vision and shared responsibility.

This last aspect was an important novelty of the report, as it approached the key areas of examination in an integrated, holistic manner, recognizing and illustrating the fundamental interconnections between poverty and population, poverty and environment, and economic development and environment. In this manner, the report developed a common and integrative approach to peace, security, development, and the environment with cascading effects in terms of institutional and legal change.

Importantly, the commission laid the foundation for some concepts that would become key ideas and principles of international politics and law in the field of sustainable development, such as the fundamental link between economy and environment, the need to prioritize meeting the needs of the poor, the relationship between poverty and environmental degradation, the focus on intergenerational responsibility, and the need of broad participation of citizenry to decision making processes. It also pointed toward a restructuring of international economic relations toward equity and sustainability.

In this respect, the report recognized the importance of realigning international law and institutions to the expanding scale of the effects of development on the global environment, and it emphasized the need to incorporate at the national and international levels consideration for the rights of present and future generations of living in an environment adequate for their health and well-being.

Although the report's predictions did not materialize, it built important momentum toward international cooperation and the establishment of a common framework of multilateral action, which was concretely brought to bear at the United Nations Conference on Environment and Development in 1992, also known as the Earth Summit. There, the concept of sustainable development developed in the WCED report was enshrined in the Rio Declaration, in Agenda 21, and in the International Agreements on climate and biodiversity, the Framework Convention on Climate Change, and the Convention on Biological Diversity.

See Also: Climate Change; Sustainability.

Further Readings

Sands, Philip. *Principles of International Environmental Law*, 2nd ed. Cambridge, MA: Cambridge University Press, 2003.
United Nations General Assembly, A/RES/38/161, 19 December 1983.
World Commission on Environment and Development. *Our Common Future*. Oxford: Oxford University Press, 1987.

Vito De Lucia
Independent Scholar

Y

YUCCA MOUNTAIN

Workers beginning the excavation of a tunnel at Yucca Mountain in 1993. Engineers planned for the site to accommodate 77,700 tons of high-level radioactive waste 1,000 feet below the surface.

Source: U.S. Department of Energy, Office of Civilian Radioactive Waste Management

Ever since the first atomic bomb was exploded at Trinity, New Mexico, the United States has been producing low- to high-level radioactive liquids and solids, mostly by-products of manufacturing nuclear warheads during the Cold War, generation of nuclear energy, and engineering and medical technologies. The territory of the United States contains tens of thousands of tons, hundreds of millions of cubic feet, and millions of gallons of radioactive waste—spent fuel, acids, solvents, tools, metal scraps, clothing, and tailings—made of or imbued with uranium, plutonium, neptunium, cesium, or strontium, all of which are to some degree poisonous to human and nonhuman life. In fact, the amount of enervated fuel from defense and commercial reactors has been increasing over the last several years at a rate of 2,000 tons per year. Since the first man-made nuclear reaction, governments and scientists have been trying to figure out how to dispose of this waste so that it does not threaten, or at least minimally threatens, the health and welfare of people in the United States now and for years to come. One solution, laid out under the Nuclear Waste Policy Act of 1982, was to build a single repository. In 1987, Yucca Mountain, located in the southwestern corner of Nevada in an area where the government has tested nuclear bombs for years, was selected as the site.

The issue of nuclear waste storage would not be so pressing if it were not for the poisonous nature of the materials produced. The radioactive elements mentioned above all have atomic structures that are highly unstable and hyperenergized. Some emit particles and waves, such as gamma rays, billions of times each second, some for a fraction of a second, and others for billions of years. When these rays penetrate the human body, they can alter the atomic structure of our cells, causing them to grow and decay in unhealthy ways. The term *millirem* is a measure of radiation received by human beings. A yearly average of 350 millirems of radiation emanating from various sources within our daily surroundings penetrates our bodies. A single dose of 400 millirems would kill 50 percent of people. Human skin has no barrier to the detrimental effects of high doses of radiation. Due to the effects to humans, radioactive waste resting in 131 locations in 39 states, and most actively produced in the 103 nuclear power plants that supply 20 percent of the electricity in the United States, must be thickly encased and/or deeply entombed.

During the 1950s and 1960s, steel barrels filled with radioactive waste were placed in shallow pits and trenches and covered with a layer of graded earth. Much of what was covered came from nuclear reactors generating electricity—a process that begins with the extraction of uranium ore—a nonrenewable energy resource—from deposits located in a few areas of the world. About 2.2 pounds of uranium can be wrested from 1,000 tons of ore. A single pound of uranium-235 contains the energy equivalent of nearly 5,500 barrels of oil. In power plants, uranium pellets fill fuel rods grouped into fuel assemblies. These assemblies promote the controlled splitting of uranium's atomic nucleus—nuclear fission—which generates heat used to produce steam, which then turns turbines to make electricity. When the fuel is spent, water cools the rods. Today spent rods are usually encased in steel-and-concrete casks and shelved on-site, a method that contains possible seepage of radiation for decades. The U.S. military deals with its radioactive waste in a similar manner.

Yet, for some time, scientists, engineers, and government officials have viewed this kind of waste storage as inefficient and dangerous, therefore constituting a nuclear disposal problem. Twenty-five years ago, the U.S. government, with the backing of the nuclear power industry, mandated the construction of a permanent repository. Yucca Mountain—a 1,500-foot-high spine of compressed volcanic ash 90 miles northwest of Las Vegas in an isolated desert region with high temperatures and minimal rainfall—was chosen as a repository because of its fitting physical geography.

The Mechanics of Waste Storage

A large tunnel is bored into the mountain about 1,000 feet below its peak. Extending from this main thoroughfare are smaller passageways, reaching into chambers plated with titanium, which is used to prevent rainwater that running down the surface of the mountain from reaching the 77,700 tons of high-level radioactive waste, set in 10-inch-thick stainless steel and nickel alloy canisters. The canisters will eventually come from 31 different sites throughout the United States, and when this transportation is fully realized, the waste will travel through 43 states via railway and/or highway. Once the canisters reach the gates of Yucca, they will be loaded onto electric trains, and taken to robots in the chambers, ready to unload and position the canisters into rows. The repository will operate for 150 years at a cost of $96.2 billion, which includes building and studying the site. Yucca will be monitored for up to 300 years, then the gates will be closed and the waste finally entombed.

Safety Concerns

The U.S. government estimates that those living near the site 10,000 years from now will receive 15 millirems of radiation annually, about as much as an X-ray; and in a million years, 100 millirems. After that, exposure amounts would continue to rise as the energetic, resolute particles and waves begin to substantially breach all barriers.

Human exposure could also occur if an underground stream of water moving through subterranean rock makes its way into the chambers, corrodes the layers of protective metal shields, and carries radiation to the surface. This possibility, along with that of earthquakes, volcanic eruptions, climate change, and accidents while transporting nuclear waste through U.S. towns and cities, concerns many experts. Legal challenges and scientific debate may eventually lead to abandoning using Yucca Mountain as a single repository for radioactive waste produced in the United States.

Alternatives to Yucca Mountain and similar geologic repository methods have been proposed, each not without risks. Monitored Retrievable Storage (MRS) facilities (such as the one proposed for the Skull Valley Goshute Indian Reservation in Utah) could store spent fuel from commercial nuclear power reactors for decades until other long-term management solutions are developed. However, some experts believe that these interim storage sites also pose serious risks associated with nuclear waste transportation (e.g., potential multiple movements and extensive distances through higher populated areas).

Others have suggested placing nuclear waste canisters deep into the oceanic crust, an option that limits the potential of radioactive material penetrating the drinking water supply. Less practical and more costly alternatives also include releasing the material into outer space. Until a universally adopted nuclear waste solution is established, Yucca Mountain and other proposed long-term storage sites will continue to be evaluated as viable choices.

See Also: Chernobyl; Nuclear Power; Nuclear Proliferation.

Further Readings

Herszenhorn, David M. "Yucca Mountain Plan for Nuclear Waste Dies." *The New York Times* (31 March 2009). http://thecaucus.blogs.nytimes.com/2009/03/31/yucca-mountain-plan-for-nuclear-waste-dies/?scp=2&sq=yucca%20mountain%20&st=cse (Accessed June 2009).

Long, Michael E. "Half-Life: The Legal Legacy of America's Nuclear Waste." *National Geographic* (v. 202/1, 2002).

Shrader-Frechette, Kristin S. *Burying Uncertainty: Risk and the Case Against Geological Disposal of Nuclear Waste.* Berkeley, CA: University of California Press, 1993.

Zapler, Mike. "Yucca Mountain." *Planning* (February 2003).

Ken Whalen
American University of Afghanistan

Z

Zero-Emission Vehicle

The U.S. Department of Energy's Argonne National Laboratory tested this BMW Hydrogen 7 prototype in early 2008 and found it to be a near-zero emission vehicle, with some of the lowest emissions of any combustion engine vehicle built to that date.

Source: U.S. Department of Energy, Argonne National Laboratory

A zero-emission vehicle (ZEV) is a vehicle that does not emit pollutants from the onboard source of power. By this definition, an electric car could qualify as a ZEV despite the fact that pollutants might have been produced at the electrical generation plant (e.g., from burning coal) that produced the electricity to charge the battery, because no pollutants are emitted by the battery when it is powering the car. However, a typical gasoline-powered car would not qualify because the gasoline burned in the process of internal combustion creates pollutants including carbon monoxide, nitrogen oxides, and sulfur that are emitted through the tailpipe when the car is in operation. Closely related to ZEVs are partial-zero-emission vehicles (PZEVs) and advanced-technology-partial-zero-emission vehicles (AT PZEVs), which have near-zero tailpipe emissions.

The primary impetus for developing ZEVs is to reduce the amount of greenhouse gases and other air pollutants emitted by gasoline or petroleum-based diesel-powered vehicles that currently meet much of the world's transportation needs. Development and promotion of ZEVs are named as a means to achieve one of four priority goals (reducing pollution, noise, and greenhouse gas production) in the 2009 Amsterdam Declaration of the Transport Health and Environment Pan-European Programme of the World Health Organization.

Types of Emission-Reducing Technology

Although the ultimate goal may be zero emissions, governments recognize that intermediate steps that gradually reduce pollution may be necessary. Many different technologies have been used to reduce vehicle-related pollution. One is simply to modify gasoline-powered vehicles to reduce or eliminate tailpipe emissions (caused by incomplete combustion) and evaporative emissions (caused by gasoline vapors escaping from the fuel system). These modifications include advanced controls on engines and sealed fuel systems. As of 2004, 37 gas-powered vehicle models were certified as PZEVs in California, making them substantially less polluting than the average new automobile.

Electric cars are powered by battery packs and do not include an internal combustion engine; they are sometimes referred to as battery electric vehicles and are the most common type of ZEV available. At this time, the most common type of electric car is the REVA, a small car produced in India and intended for city trips and short commutes. It is used in several European countries, including the United Kingdom, Belgium, Germany, Spain, and Norway. The Tesla Roadster, a sports car developed by the Silicon Valley (California) company Tesla Motors, came on the market in 2008 and is available in Europe (right-drive countries only), Canada, and the United States. Several other manufacturers have reported that they will produce electric cars in the near future, including Mitsubishi, Subaru, and Toyota.

Electric buses may be powered by batteries or may draw power from electrical wires; trolley buses drawing power from overhead wires are commonly seen in cities today. The technology for the electric trolley was demonstrated in the late 19th century and became very common in major cities including New York and London. It remains common in Europe, China, and Russia today, although less so in the United States. Battery-powered buses (which have the advantage of being able to go anywhere, rather than just where the trolley wires exist) are used in a few American cities, including Chattanooga, Tennessee; Miami Beach, Florida; and Los Angeles, California.

Hybrid electric vehicles combine a conventional fuel system such as gasoline or diesel with a rechargeable energy storage system (battery). They are thus not ZEVs but emit less pollution, as they burn less gasoline than a standard vehicle. Although hybrid vehicles were produced in the early 20th century, the first hybrid vehicle available to a large market was the Toyota Prius, introduced in 1997. In the United States, hybrids are the most widely adopted of the various alternatives to the ordinary internal combustion engine automobile. In 2003, over 40,000 hybrid cars were registered in the United States, with over a quarter of those registered in California. Several manufacturers produce hybrid vehicles, including Honda, Toyota, Audi, Ford, and Lexus.

Vehicles that run on a fuel other than gasoline or petroleum-based diesel are often referred to as alternative-fueled vehicles (AFVs). Possible types of fuel include compressed natural gas (CNG), biodiesel, propane, methanol, hydrogen, ethanol, and electricity. Some cars also run on a combination of gasoline and another fuel; for instance, flex fuel vehicles are powered by a combination of ethanol or other alcohol fuel and gasoline, and bi-fuel vehicles have two separate fuel systems, one for gasoline or diesel and a second for an alternative fuel such as compressed natural gas or propane. Of these, compressed natural gas may be the most widely adopted: One estimate is that over seven million CNG vehicles were on the road as of 2008, with the largest numbers used in South America. CNG buses are used in several countries, including the United States, India, Australia, and Germany.

Hydrogen fuel cell vehicles have been recommended by some as a potential replacement for the internal combustion engine, but although prototypes have been developed, there are no hydrogen fuel cell cars on the market at this time. Right now, fuel cells are fragile and expensive to produce and will not operate at cold temperatures. In addition, there is the problem that although hydrogen is a natural element, as a practical matter it must be created and stored—a process that creates pollution. Most hydrogen today is produced using fossil fuels, particularly natural gas, and at this time, more greenhouse gases are created by generating the hydrogen to run a fuel cell car than by running an ordinary gasoline-powered internal-combustion-engine car.

Although most discussion of ZEVs has focused on variants of the automobile, there are other types of vehicles that could be considered in this category as well. These include human-powered vehicles such as the bicycle (on land) and canoes and rowboats (on the water), and animal-powered vehicles such as horse carts. Segway personal transporters operate on batteries, as do some brands of golf carts, and would also qualify as ZEVs.

California Automobile Standards

In the United States, California has been the leader in the promotion and development of ZEVs: California's efforts began with the Low Emissions Vehicle program adopted in 1990. As of 2008, the state's Air Resources Board estimates that over 750,000 Californians were driving PZEVs or AT PZEVs, as well as 100,000 Californians who were driving gas-electric hybrids. All three types of vehicle emit far fewer pollutants than conventional gasoline cars.

All new vehicles sold in California must be certified as meeting one of the six categories of emissions ratings as published by the California Air Resources Board. The emissions rating of a car is indicated on the Vehicle Emissions Control Information Label applied to each car. Details of what is necessary to meet each standard vary with the size and type of vehicle, although since 2008 sport utility vehicles have been held to the same standard as smaller cars. The categories are:

- LEV: low-emissions vehicle
- ULEV: ultra-low-emissions vehicle (50 percent cleaner than the average new 2003 model vehicle)
- SULEV: super-ultra-low-emissions vehicle (90 percent cleaner than the average new 2003 model vehicle)
- PZEV: partial-zero-emissions vehicle (meets SULEV tailpipe standards, has zero evaporative emissions and a 15-year, 150,000-mile warranty)
- AT PZEV: advanced-technology-partial-zero-emissions vehicle (meets PZEV standards and includes ZEV-enabling technology)
- ZEV: zero-emissions vehicle (zero tailpipe emissions, 98 percent cleaner than the average new 2003 model vehicle)

Although relatively few ZEVs are on the road at present, the inclusion of the other categories is significant because the overall goal of the program is to reduce emissions and fuel consumption. Incentives for buying a vehicle within the California ZEV program include rebates, tax credits, insurance discounts, exemptions from the high occupancy vehicle lane restrictions, and free or reduced-rate parking in some cities. California also has a zero emission bus program to encourage the use of ZEVs in bus fleets.

See Also: Alternative Fuels; Bicycles; California; Carbon Footprint and Neutrality; Electric Vehicle; Fuel Cells; Natural Gas; Plug-In Hybrid.

Further Readings

California Environmental Protection Agency Air Resources Board. "Zero Emission Vehicle (ZEV) Program." http://www.arb.ca.gov/msprog/zevprog/zevprog.htm (Accessed June 2009).

Dixon, Lloyd, et al. *Driving Emissions to Zero: Are the Benefits of California's Zero Emission Vehicle Program Worth the Cost?* Santa Monica, CA: RAND Institute, 2002. http://www.rand.org/pubs/monograph_reports/2007/MR1578.pdf (Accessed June 2009).

Hasegawa, Yozo. *Clean Car Wars: How Honda and Toyota Are Winning the Battle of the Eco-Friendly Autos.* Singapore: John Wiley & Sons, 2008.

Heywood, John B. "Fuelling our Transportation Future." *Scientific American* (September 2006).

"Revenge of the Electric Car" http://revengeoftheelectriccar.com/ (Accessed June 2009).

Who Killed the Electric Car? Dir. Chris Paine; Sony Pictures. DVD release date November 14, 2006.

World Health Organization. "Making the Link: Transport Choices for our Health, Environment and Prosperity" (The Amsterdam Declaration). Transport Health and Environment Pan-European Programme (January 2009). http://www.euro.who.int/Document/E92356.pdf (Accessed June 2009).

Sarah Boslaugh
Washington University in St. Louis

Green Energy Glossary

A

Absorber: The component of a solar thermal collector that absorbs solar radiation and converts it to heat, or, as in a solar photovoltaic device, the material that readily absorbs photons to generate charge (free electrons or holes).

Absorption: The passing of a substance or force into the body of another substance.

Absorption Coefficient: In reference to a solar energy conversion devices, the degree to which a substance will absorb solar energy.

Adiabatic: Without loss or gain of heat to a system.

Aerobic Bacteria: Microorganisms that require free oxygen, or air, to live, and contribute to the decomposition of organic material in soil or composting systems.

Air: The mixture of gases that surrounds the Earth and forms its atmosphere, composed of, by volume, 21 percent oxygen, 78 percent nitrogen, and 1 percent other elements.

Air Pollution: The presence of contaminants in the air in concentrations that prevent the normal dispersive ability of the air, and interfere with biological processes and human economics.

Air Quality Standards: The prescribed level of pollutants allowed in outside or indoor air as established by legislation.

Albedo: The ratio of light reflected by a surface to the light falling on it.

Alternative Fuels: A popular term for nonconventional transportation fuels derived from natural gas or biomass materials.

Ambient Air: The air external to a building or device.

Anaerobic Bacteria: Microorganisms that live in oxygen-deprived environments.

Anaerobic Digestion: The process by which organic matter is decomposed by anaerobic bacteria.

Anemometer: An instrument for measuring the force or velocity of wind; a wind gauge.

Annual Load Fraction: The fraction of annual energy demand supplied by a solar system.

Annual Solar Savings: The annual solar savings of a solar building is the energy savings attributable to a solar feature relative to the energy requirements of a non-solar building.

Appliance: A device for converting one form of energy or fuel into useful energy or work.

Appliance Standards: Standards established by the U.S. Congress for energy-consuming appliances in the National Appliance Energy Conservation Act (NAECA) of 1987.

Argon: A colorless, odorless inert gas sometimes used in the spaces between the panes in energy-efficient windows.

Available Heat: The amount of heat energy that may be converted into useful energy from a fuel.

B

Balance-of-System: In a renewable energy system, refers to all components other than the mechanism used to harvest the resource (such as photovoltaic panels or a wind turbine).

Batch Heater: This simple passive solar hot water system consists of one or more storage tanks placed in an insulated box that has a glazed side facing the sun.

Bioconversion: The conversion of one form of energy into another by the action of plants or microorganisms, such as the conversion of biomass to ethanol, methanol, or methane.

Biogas: A combustible gas created by anaerobic decomposition of organic material, composed primarily of methane, carbon dioxide, and hydrogen sulfide.

Biomass: Any organic matter that is available on a renewable basis, including agricultural crops and agricultural wastes and residues, wood and wood wastes and residues, animal wastes, municipal wastes, and aquatic plants.

Blackbody: An ideal substance that absorbs all radiation falling on it, reflecting nothing.

Bottoming-Cycle: A means of increasing the thermal efficiency of a steam electric–generating system by converting some waste heat from the condenser into electricity.

Bread Box System: This simple passive solar hot water system consists of one or more storage tanks placed in an insulated box that has a glazed side facing the sun.

C

Carbon Monoxide: A colorless, odorless, but poisonous combustible gas with the formula CO.

Caulking: A material used to seal areas of potential air leakage into or out of a building.

Cellulose: The fundamental constituent of all vegetative tissue; the most abundant material in the world.

Central Receiver Solar Power Plants: Also known as power towers, these use fields of two-axis tracking mirrors known as heliostats. Each heliostat is individually positioned by a computer control system to reflect the sun's rays to a tower-mounted thermal receiver. The effect of many heliostats reflecting to a common point creates the combined energy of thousands of suns, which produces high-temperature thermal energy.

Clean Power Generator: A company or other organizational unit that produces electricity from sources that are thought to be environmentally cleaner than traditional sources. Clean or green power is usually defined as power from renewable energy such as wind, solar, and biomass energy.

Climate Change: A term used to describe short and long-term affects on Earth's climate as a result of human activities such as fossil fuel combustion and vegetation clearing and burning.

Cogeneration: The generation of electricity or shaft power by an energy conversion system, and the concurrent use of rejected thermal energy from the conversion system as an auxiliary energy source.

Collector Tilt: The angle that a solar collector is positioned from horizontal.

Combustion: The process of burning; the oxidation of a material by applying heat, which unites oxygen with a material or fuel.

Composting: The process of degrading organic material (biomass) by microorganisms in aerobic conditions.

Concentrating (Solar) Collector: A solar collector that uses reflective surfaces to concentrate sunlight onto a small area, where it is absorbed and converted to heat or, in the case of solar photovoltaic (PV) devices, into electricity.

Conventional Fuel: The fossil fuels: coal, oil, and natural gas.

Coproducts: The potentially useful byproducts of the ethanol fermentation process.

Crystalline Silicon Photovoltaic Cell: A type of photovoltaic cell made from a single crystal or a polycrystalline slice of silicon. Crystalline silicon cells can be joined together to form a module (or panel).

Cube Law: In reference to wind energy, for any given instant, the power available in the wind is proportional to the cube of the wind velocity; when wind speed doubles, the power availability increases eight times.

D

Dam: A structure for impeding and controlling the flow of water in a water course, which increases the water elevation to create the hydraulic head. The reservoir creates, in effect, stored energy.

Decomposition: The process of breaking down organic material; reduction of the net energy level and change in physical and chemical composition of organic material.

Dimmer: A light control device that allows light levels to be manually adjusted. A dimmer can save energy by reducing the amount of power delivered to the light.

Direct Gain: The process by which sunlight directly enters a building through the windows and is absorbed and stored in massive floors or walls.

Drag: Resistance caused by friction in the direction opposite to that of movement of components such as wind turbine blades.

E

Economizer: A heat exchanger for recovering heat from flue gases for heating water or air.

Effective Capacity: The maximum load that a device is capable of carrying.

Efficiency (Appliance) Ratings: A measure of the efficiency of an appliance's energy efficiency.

Electrical Charge: A condition that results from an imbalance between the number of protons and the number of electrons in a substance.

Electric System: The physically connected generation, transmission, and distribution facilities and components operated as a unit.

Electromagnetic Energy: Energy generated from an electromagnetic field produced by an electric current flowing through a superconducting wire kept at a specific low temperature.

Emissions: A substances or pollutants emitted as a result of a process.

Emissivity: The ratio of the radiant energy (heat) leaving (being emitted by) a surface to that of a black body at the same temperature and with the same area; expressed as a number between zero and one.

Energy: The capability of doing work; different forms of energy can be converted to other forms, but the total amount of energy remains the same.

Energy Crops: Crops grown specifically for their fuel value.

Entropy: A measure of the unavailable or unusable energy in a system; energy that cannot be converted to another form.

Environment: All the natural and living things in the world. The earth, air, weather, plants, and animals all make up the environment.

Exothermic: A reaction or process that produces heat; a combustion reaction.

F

Fermentation: The decomposition of organic material to alcohol, methane, and other substances by organisms such as yeast or bacteria, usually in the absence of oxygen.

First Law of Thermodynamics: States that energy cannot be created or destroyed, but only changed from one form to another.

Fossil Fuels: Fuels formed in the ground from the remains of dead plants and animals. It takes millions of years to form fossil fuels. Oil, natural gas, and coal are fossil fuels.

Fuel: Any material that can be burned to make energy.

G

Geothermal Energy: Any and all energy produced by the internal heat of the Earth.

Global Warming: The increase in average global temperatures due to the greenhouse effect.

Green Certificates: Green certificates represent the environmental attributes of power produced from renewable resources.

Greenhouse Effect: The heating effect due to the trapping of long wave (length) radiation by greenhouse gases produced from natural and human sources.

Greenhouse Gases: Gases such as water vapor, carbon dioxide, tropospheric ozone, methane, and low level ozone that are transparent to solar radiation, but opaque to long wave radiation, and contribute to the greenhouse effect.

Green Power: A popular term for energy produced from clean, renewable energy resources.

Grid: A common term referring to an electricity transmission and distribution system.

H

Heat: A form of thermal energy resulting from combustion, chemical reaction, friction, or movement of electricity.

Heat-Absorbing Window Glass: A type of window glass that contains special tints that cause the window to absorb as much as 45 percent of incoming solar energy, to reduce heat gain in an interior space.

Heliochemical Process: The utilization of solar energy through photosynthesis.

Heliothermal: Any process that uses solar radiation to produce useful heat.

Home Energy Rating Systems (HERS): A nationally recognized energy rating program that gives builders, mortgage lenders, secondary lending markets, homeowners, sellers, and buyers a precise evaluation of energy losing deficiencies in homes.

Hybrid System: A renewable energy system that includes two different types of technologies that produce the same type of energy.

Hydroelectric Power Plant: A power plant that produces electricity by the force of water falling through a hydro turbine that spins a generator.

I

Incident Solar Radiation: The amount of solar radiation striking a surface per unit of time and area.

Insolated Solar Gain System: A type of passive solar heating system where heat is collected in one area for use in another.

Insulation: Materials that prevent or slow down the movement of heat.

Integrated Heating Systems: A type of heating appliance that performs more than one function, for example space and water heating.

Inverter: A device that converts direct current electricity (from, for example, a solar photovoltaic module or array) to alternating current for use directly to operate appliances or to supply power to an electricity grid.

K

Kinetic Energy: Energy available as a result of motion that varies directly in proportion to an object's mass and the square of its velocity.

L

Lift: The force that pulls a wind turbine blade, as opposed to drag.

Load: The power required to run a defined circuit or system, such as a refrigerator, building, or an entire electricity distribution system.

M

Megawatt (MW): One thousand kilowatts, or one million watts; standard measure of electric power plant generating capacity.

Methane: The main constituent of "natural gas" that is formed naturally by methanogenic anaerobic bacteria, or can be manufactured, and is used as a fuel and for manufacturing chemicals.

Module: The smallest self-contained, environmentally protected structure housing interconnected photovoltaic cells and providing a single dc electrical output; also called a panel.

N

Natural Gas: A hydrocarbon gas obtained from underground sources, often in association with petroleum and coal deposits.

Net Energy Production (or Balance): The amount of useful energy produced by a system less the amount of energy required to produce the fuel.

Nonrenewable Fuels: Fuels that cannot be easily made or renewed, such as oil, natural gas, and coal.

Nuclear Energy: Energy that comes from splitting atoms of radioactive materials, such as uranium, and producing radioactive wastes.

O

Ocean Energy Systems: Energy conversion technologies that harness the energy in tides, waves, and thermal gradients in the oceans.

Oil (Fuel): A product of crude oil that is used for space heating, gasoline engines, and electrical generation.

P

Panel (Solar): A term generally applied to individual solar collectors, and typically to solar photovoltaic collectors or modules.

Panemone: A drag-type wind machine that can react to wind from any direction.

Passive Solar (Building) Design: A building design that uses structural elements of a building to heat and cool a building, without the use of mechanical equipment.

Penstock: A component of a hydropower plant; a pipe that delivers water to the turbine.

Phantom Load: Any appliance that consumes power even when it is turned off.

Phase Change: The process of changing from one physical state (solid, liquid, or gas) to another, with a necessary or coincidental input or release of energy.

Photovoltaic (Solar) Cell: Treated semiconductor material that converts solar irradiance to electricity. When grouped, they are called solar arrays, modules, or panels.

Pitch Control: A method of controlling a wind turbine's speed by varying the orientation, or pitch of the blades, altering its aerodynamics and efficiency.

Power: Energy that is capable or available for doing work; the time rate at which work is performed.

Power (Solar) Tower: A term used to describe solar thermal, central receiver power systems, where an array of reflectors focus sunlight onto a central receiver and absorber mounted on a tower.

Process Heat: Thermal energy that is used in agricultural and industrial operations.

Propeller (Hydro) Turbine: A turbine that has a runner with attached blades similar to a propeller used to drive a ship. As water passes over the curved propeller blades, it causes rotation of the shaft.

Pyrolysis: The transformation of a compound or material into one or more substances by heat alone (without oxidation). Often called destructive distillation.

R

Radiant Energy: Energy that transmits away from its source in all directions.

Radiation: The transfer of heat through matter or space by means of electromagnetic waves.

Recycling: The process of converting materials that are no longer useful as designed or intended into a new product.

Refraction: The change in direction of a ray of light when it passes through one media to another with differing optical densities.

Renewable Energy: Energy derived from resources that are regenerative or for all practical purposes cannot be depleted. Types of renewable energy resources include moving water (hydro, tidal, and wave power), thermal gradients in ocean water, biomass, geothermal energy, solar energy, and wind energy. Municipal solid waste (garbage) is also considered a renewable energy resource.

Resistance: The inherent characteristic of a material to inhibit the transfer of energy.

Rotor: An electric generator consists of an armature and a field structure. In wind energy it is the blades and rotating components.

Run-of-River Hydropower: A type of hydroelectric facility that uses the river flow with very little alteration and little or no impoundment of the water.

S

Semiconductor: Any material that has a limited capacity for conducting an electric current.

Silicon: A chemical element, of atomic number 14, that is semi-metallic, and an excellent semiconductor material used in solar photovoltaic devices; commonly found in sand.

Single-Crystal Material: In reference to solar photovoltaic devices, a material that is composed of a single crystal or a few large crystals.

Skylight: A window located on the roof of a structure to provide interior building spaces with natural daylight, warmth, and ventilation.

Smart Window: A term used to describe a technologically advanced window system containing glazing that can change or switch its optical qualities when a low voltage electrical signal is applied to it, or in response to changes in heat or light.

Solar Collector: A device used to collect, absorb, and transfer solar energy to a working fluid.

Solar Cooling: The use of solar thermal energy or solar electricity to power a cooling appliance.

Solar Energy: Electromagnetic energy transmitted from the sun (solar radiation).

Solar Fraction: The percentage of a building's seasonal energy requirements that can be met by solar energy devices or systems.

Solar Irradiation: The amount of solar radiation, both direct and diffuse, received at any location.

Solar Mass: A term used for materials used to absorb and store solar energy.

Solar Spectrum: The total distribution of electromagnetic radiation emanating from the sun.

Stall: In reference to a wind turbine, a condition when the rotor stops turning.

Steam Turbine: A windmill that is powered by steam, usually to produce electricity.

Sunspace: A room that faces south (in the northern hemisphere), or a small structure attached to the south side of a house.

Superconducting Magnetic Energy Storage (SMES): SMES technology uses the superconducting characteristics of low-temperature materials to produce intense magnetic fields to store energy.

System Mix: The proportion of electricity distributed by a power provider that is generated from available sources such as coal, natural gas, petroleum, nuclear, hydropower, wind, or geothermal.

T

Temperature Coefficient (of a Solar Photovoltaic Cell): The amount that the voltage, current, and/or power output of a solar cell changes due to a change in the cell temperature.

Thermal Capacitance: The ability of a material to absorb and store heat for later use.

Thermal Energy: The energy developed through the use of heat energy.

Thermal Mass: Materials that store heat.

Thermal Storage Walls: A south-facing wall that is glazed on the outside. Solar heat strikes the glazing and is absorbed into the wall, which conducts the heat into the room over time.

Thermodynamic Cycle: An idealized process in which a working fluid successively changes its state (from a liquid to a gas and back to a liquid) for the purpose of producing useful work or energy, or transferring energy.

Thermodynamics: The study of the transformation of energy from one form to another, and its practical application.

Thermosiphon: The natural, convective movement of air or water due to differences in temperature.

Tidal Power: The power available from the rise and fall of ocean tides.

Total Heat: The sum of the sensible and latent heat in a substance or fluid above a base point, usually 32 degrees F.

Total Incident Radiation: The total radiation incident on a specific surface area over a time interval.

Tracking Solar Array: A solar energy array that follows the path of the sun to maximize the solar radiation incident on the cell's surface.

Transmission: The process of sending or moving electricity from one point to another.

Turbine: A device for converting the flow of a fluid (air, steam, water, or hot gases) into mechanical motion.

U

Unglazed Solar Collector: A solar thermal collector that has an absorber without a glazed covering, like those used to heat swimming pools.

Useful Heat: Heat stored above room temperature (in a solar heating system).

V

Volt (V): A unit of electrical force equal to that amount of electromotive force that will cause a steady current of one ampere to flow through a resistance of one ohm.

Voltage: The amount of electromotive force, measured in volts, that exists between two points.

W

Water Turbine: A turbine that uses water pressure to rotate its blades, usually for generating electricity.

Water Wheel: A wheel that is designed to use the weight and/or force of moving water to turn it, primarily to operate machinery or grind grain.

Wave Power: The concept of capturing and converting the energy available in the motion of ocean waves to energy.

Weatherization: Caulking and weather stripping (sealing gaps around windows and doors) to reduce air infiltration and exfiltration into/out of a building.

Wind Energy: Energy available from the movement of the wind across a landscape caused by the heating of the atmosphere, earth, and oceans by the sun.

Wind Energy Conversion System (WECS) or Device: An apparatus for converting the energy available in the wind to mechanical energy that can be used to power machinery.

Wind Generator: A WECS designed to produce electricity.

Windmill: A WECS that is used to grind grain and typically has a high-solidity rotor; commonly used to refer to all types of WECS.

Wind Power Plant: A group of wind turbines interconnected to a common power provider system.

Wind Turbine: A term used for a wind energy conversion device that produces electricity; typically having one, two, or three blades.

Wind Velocity: The wind speed and direction in an undisturbed flow.

Working Fluid: A fluid used to absorb and transfer heat energy.

Y

Yaw: The rotation of a horizontal axis wind turbine around its tower or vertical axis.

Source: U.S. Energy Information Administration (http://www.eia.doe.gov/tools/glossary)

Green Energy
Resource Guide

Books

Alliance to Save Energy. *New York State Energy Efficiency Industry Directory: A Networking and Buying Guide.* Albany, NY: New York State Energy Research and Development Authority, 1997.

ASHRAE. *The ASHRAE GreenGuide.* Amsterdam, The Netherlands: Elsevier, 2005.

Barnham, Kay. 2008. *Save Energy.* New York: Crabtree Publishing, 2008.

Berger, John J. *Charging Ahead: The Business of Renewable Energy and What it Means for America.* Berkeley, CA: University of California Press, 1998.

Berinstein, Paula. *Alternative Energy: Facts, Statistics, and Issue.* Westport, CT: Oryx Press, 2001.

Bloom, Fred. *Photovoltaic Technology in the USDA Forest Service.* Albuquerque, NM: Sandia National Laboratories, 1996.

Bromley, Marianne. *Wildlife Management Implications of Petroleum Exploration and Development in Wildland Environments.* Ogden, UT: U.S. Dept. of Agriculture, Forest Service, Intermountain Research Station, 1985.

Brooke, John. *Wave Energy Conversion.* Amsterdam, The Netherlands: Elsevier, 2003.

Brown, Lester Russell. *Six Steps to a Sustainable Society.* Washington, D.C.: Worldwatch Institute, 1982.

California Department of Education. *Environmental Education: Compendium for Energy Resources.* Sacramento, CA: California Dept. of Education, 1992.

California Office of Appropriate Technology. *Present Value: Constructing a Sustainable Future.* Sacramento, CA: 1979.

Castaner, L. and T. Markvart. *Practical Handbook of Photovoltaics: Fundamentals and Applications.* Amsterdam, The Netherlands: Elsevier, 2003.

Chandler, William U. *Energy Productivity: Key to Environmental Protection and Economic Progress.* Washington, D.C.: Worldwatch Institute, 1985.

Cleveland, C. and C. Morris. *Dictionary of Energy.* Escondido, CA: Morris Books, 2005.

Communications Consortium Media Center. *A Sustainable Energy Blueprint.* Washington, D.C.: Communications Consortium Media Center, 1992.

da Rosa, Aldo. *Fundamentals of Renewable Energy Renewable Energy Processes.* Amsterdam, The Netherlands: Elsevier, 2008.

Department of Energy. *National Energy Strategy: Powerful Ideas for America, One Year Later.* Washington, D.C.: Department of Energy, 1992.

Devins, D.W. *Energy, Its Physical Impact on the Environment.* New York: Wiley, 1982.

Droege, Peter. *Urban Energy Transition.* Newcastle, Australia: World Council for Renewable Energy, 2008.

Duncan, Trent. *Renew the Pub Lands: Photovoltaic Technology in the Bureau of Land Management.* Albuquerque, NM: Sandia National Laboratories, 1996.

Federal Emergency Management Agency. *Dispersed, Decentralized and Renewable Energy Sources: Alternatives to National Vulnerability and War, Final Report.* Washington, D.C.: Federal Emergency Management Agency, 1980.

Flavin, Christopher. *Electricity's Future: The Shift to Efficiency and Small-Scale Power.* Washington, D.C.: Worldwatch Institute, 1984.

Flavin, Christopher. *Energy and Architecture: The Solar and Conservation Potential.* Washington, D.C.: Worldwatch Institute, 1980.

Flavin, Christopher. *Power Surge: Guide to the Coming Energy Revolution.* New York: W. W. Norton, 1994.

Foley, Gerald. *The Energy Question.* New York: Viking Penguin, 1987.

Fusaro, Peter and Marion Yuen. *Green Trading Market: Developing the Second Wave.* Amsterdam, The Netherlands: Elsevier, 2005.

Gershon, David. *Household Ecoteam Workbook.* Woodstock, NY: Global Action Plan for the Earth, 1992.

Holing, Dwight. *Coastal Alert: Ecosystems, Energy, and Offshore Oil Drilling.* Washington, D.C.: Island Press, 1990.

Hore-Lacy, Ian. *Nuclear Energy in the 21st Century.* Amsterdam, The Netherlands: Elsevier, 2006.

Inhaber, Herbert. *Energy Risk Assessment.* New York: Gordon and Breach, 1982.

Interstate Renewable Energy Council. *Procurement Guide for Renewable Energy Systems.* Washington, D.C.: U.S. Government Printing Office, 1993.

Javna, John. *30 Simple Energy Things You Can Do to Save the Earth.* Berkeley, CA: Earth Works Group, 1990.

Komp, Richard J. *Practical Photovoltaics; Electricity From Solar Cells.* Ann Arbor, MI: AATEC Publications, 1995.

Kozloff, Keith. *Rethinking Development Assistance for Renewable Electricity.* Washington, D.C.: World Resources Institute, 1994.

Letcher, Trevor. *Future Energy.* Amsterdam, The Netherlands: Elsevier, 2006.

Marion, William and Stephen Wilcox. *Solar Radiation Data Manual for Flat-Plate and Concentrating Collectors.* Golden, CO: National Renewable Energy Laboratory, 1995.

McCracken, Garry and Peter Stott. *Fusion: The Energy of the Universe.* Amsterdam, The Netherlands: Elsevier, 2005.

McKie, Robin. *Energy.* New York: Hampstead Press, 1989.

Miami International Conference on Alternative Energy Sources. *Solar Collectors Storage.* Ann Arbor, MI: Ann Arbor Sciences, 1982.

Miller, Alan S. *Growing Power: Bioenergy for Development and Industry.* Washington, D.C.: World Resources Institute, 1986.

National Conference on Renewable Natural Resources. *Working Group Background Papers.* Washington, D.C.: American Forestry Association, 1980.

National Technical Information Service. *Stand-Alone Photovoltaic Systems; A Handbook of Recommended Design Practices*. Springfield, VA: Sandia National Laboratories, National Technical Information Service, U.S. Department of Commerce, 1995.

Niele, Fank. *Energy: The Engine of Evolution*. Amsterdam, The Netherlands: Elsevier, 2005.

Oguti, Takasi. *Sun-Earth Energy Transfer*. Oslo, Norway: Norwegian Academy of Science and Letters, 1994.

Rooney, Anne. *Solar Power*. Pleasantville, NY: Gareth Stevens Publications, 2008.

Silveira, Semida. *Bioenergy–Realizing the Potential*. Amsterdam, The Netherlands: Elsevier, 2005.

Solar Design Associates. *Photovoltaics in the Built Environment*. Springfield, VA: National Technical Information Service, U.S. Department of Commerce, 1997.

Sorensen, B. *Renewable Energy*. Amsterdam, The Netherlands: Elsevier, 2004.

Sorensen, B. *Renewable Energy Conversion, Transmission and Storage*. Amsterdam, The Netherlands, 2008.

Sorensen, B. *Renewable Energy Focus Handbook*. Amsterdam, The Netherlands: Elsevier, 2007.

Sperling, Daniel and James Cannon. *The Hydrogen Energy Transition: Cutting Carbon From Transportation*. The Netherlands, Amsterdam: Elsevier, 2004.

Suppes, Galen and Truman Storvick. *Sustainable Nuclear Power*. Amsterdam, The Netherlands: Elsevier, 2006.

Sweet, William. *Kicking the Carbon Habit*. New York: Columbia University Press, 2006.

Thomas, Isabel. *The Pros and Cons of Solar Power*. New York: Rosen Central, 2008.

Union of Concerned Scientists. *America's Energy Choices: Investing in a Strong Economy and Clean Environment*. Cambridge, MA: Union of Concerned Scientists, 1991.

U.S. Department of Energy. *A Place in the Sun; Solar Buildings*. Merrifield, VA: Energy Efficiency and Renewable Energy Clearinghouse, 1995.

Watson, Robert K. *Looking for Oil in All the Wrong Places: Facts About Oil, Natural Gas and Efficiency Resources*. Washington, D.C.: Natural Resources Defense Council, 1991.

Williams, J. Richard. *Solar Energy: Technology and Applications*. Ann Arbor, MI: Ann Arbor Science Publishers, 1974.

Williams, L. O. *An End to Global Warming*. Amsterdam, The Netherlands: Elsevier, 2002.

Journals

Alternatives (Alternatives Inc.)
American Naturalist (Thomson Corporation)
Annual Review of Ecology and Systematics (Annual Reviews)

BioScience (American Institute and Biological Sciences)

Energy (Elsevier)
Energy and Buildings (Elsevier)
Energy and Environment (Multi-Science Publishing)
Energy Conversion and Management (Elsevier)
Energy Policy (Elsevier)
Environmental Management (Academic Press)

EPA Journal (Environmental Protection Agency)

Fuel (Elsevier)

Geothermics (Elsevier)
Global Environmental Change (Royal Society of Chemistry)

International Journal of Energy Research (John Wiley & Sons)
International Journal of Greenhouse Gas Control (Elsevier)
International Journal of Hydrogen Energy (Elsevier)
International Journal of Sustainable Development & World Ecology (Taylor & Francis)

Journal of Environmental Management (Academic Press)
Journal of Environment and Development (SAGE Publications)

Natural Resources Forum (Blackwell Publishing)
Natural Resources Journal (University of New Mexico)

Refocus (Elsevier)
Renewable & Sustainable Energy Reviews (Elsevier)
Renewable Energy (Elsevier)
Renewable Energy Focus (Elsevier)
Renewable Energy World (Earthscan Publications)

Solar Energy (Elsevier)

Whole Earth Review (Point Foundation)

Websites

Alliance for Energy and Economic Growth
www.yourenergyfuture.org

Alternative Energy Institute, Inc.
www.altenergy.org

Alternative Technology Association
www.ata.org.au

American Society of Mechanical Engineers
www.asme.org

American Solar Energy Society
www.ases.org

American Wind Energy Association
www.awea.org

Building Green, LLC
www.buildinggreen.com

California Energy Commission Home Page
www.energy.ca.gov

Calpirg
www.calpirg.org

Database of State Incentives for Renewables & Efficiency
www.dsireusa.org

Environmental Literacy Council—Energy
www.enviroliteracy.org/category.php/4.html

Hydrogen Energy Center
www.hydrogenenergycenter.org

International Green Energy Council
greenenergycouncil.com

International Solar Energy Society
www.ises.org

National Energy Education Development Project
www.need.org

National Renewable Energy Laboratory
www.nrel.gov

Renewable Energy Policy Project
www.repp.org

Solar Energy International
www.solarenergy.org

U.S. Department of Energy
www.doe.gov

U.S. Environmental Protection Agency
www.epa.gov

Worldwatch Institute
www.worldwatch.org

Green Energy Appendix

Biz/ed

http://www.bized.co.uk

Biz/ed is a prize-winning online service that is an excellent source of information on the business and economic aspects of many topics relevant to green energy, including packaging and pollution, road congestion, fair trade, and alternative energy production. The site's target market is university and college students and professors but it will also be useful to journalists, policy makers, and the general public. The site includes a glossary, acronym finder, diagram bank, and search interface, while individual topics include an overview of the subject including current European Union (EU) legislation, and links to further information both within the website and externally. It also includes a regularly updated section of information about topics in the news and numerous data sets, which can be freely downloaded, as well as a guide to locating other data on the web. Six "virtual worlds," including a developing country, a factory, and a farm can be accessed through the website: these "worlds" integrate information about economic theory with current information and data that allow users to place information about topics (e.g., the externalities of pollution) in context.

The Carbon Trust

http://www.carbontrust.co.uk/default.ct

The Carbon Trust is an organization located in the United Kingdom whose purpose is to cut current and future carbon emissions and develop commercial low carbon technologies. This website includes information about climate change and carbon reduction, including basic discussions of the topic, news items, strategies to reduce carbon, case studies, videos, teaching resources, information about research grants and financing available, and links to relevant reports, press releases, and speeches. It also includes information about the Carbon Reduction Label, which shows a product's carbon footprint, i.e., the amount of greenhouse gases emitted during its life cycle (manufacture, distribution, use, and disposal), and a list of brands and retailers using the Carbon Reduction Label. It also includes two interactive calculators to calculate carbon footprints (one for individuals and one for businesses creating a product or service) and offers advice (with links to relevant information sources) on reducing the carbon footprint.

The Chicago Climate Exchange

http://www.chicagoclimatex.com

This is the website of the Chicago Climate Exchange (CCX), the only cap and trade system in North America that deals in all six greenhouse gases. The exchange facilitates the trading of Carbon Financial Instrument (CFI) contracts, each of which represents the equivalent of 100 metric tons of carbon dioxide. Members of the exchange, which includes companies such as Ford, DuPont, Rolls-Royce, and Eastman Kodak as well as municipalities including King County, Washington, the State of Illinois, and Boulder, Colorado, and universities including Michigan State and Tufts, enter into a voluntary but legally binding commitment to meet emissions targets (calculated as a reduction from their baseline emissions). Those below the targets can sell or bank their allowances while those above can stay in compliance by purchasing allowances or participating in offset projects. The website contains background information about climate exchanges in general and the CCX in particular, news about the exchange, a calendar of events, and links to information about other exchanges including the European Climate Exchange and the Montreal Climate Exchange.

The European Environment Agency

http://www.eea.europa.eu

The European Environment Agency (EEA) is an agency appointed by the European Union (EU) to provide independent information on the environment to those involved in creating and implementing policy as well as to the general public in order to facilitate the exchange of information and promote sound decision making. Information on the EEA's website is organized into five categories: air pollution, biodiversity, climate change, land use, and water. Within each section the site provides an overview of the situation in the EU (for instance, whether a type of pollution has increased or declined in recent years), general information about the topic, information about relevant EEA activities, policy information, news articles, publications and multimedia, maps and graphs, data sets, a calendar of upcoming events, and external links to other organizations dealing with the topic. The site also includes a glossary of terms relevant to the environment with translations of the term into the other EU languages, a link to other networks concerned with the European environment, and press releases.

The Green Power Network

http://apps3.eere.energy.gov/greenpower

This website, created and maintained by the U.S. Department of Energy, provides news and information on green power (energy) products, providers, consumer protection issues and policy discussion. Most of the website is devoted to green power generation within the United States, although there is also a section devoted to international issues with links to news articles and programs. The website includes links to published reports on green power (from governmental, academic and private sources), an archive of news stories about green power and listings of green energy products available in each state as well as greenhouse gas offsets, which are available nationally with a state-by-state listing of policies. There is also a section for renewable energy RFPs (requests for proposals), which lists grants and business opportunities for sustainable energy projects and research.

The U.S. Green Building Council

https://www.usgbc.org/DisplayPage.aspx?CMSPageID=124

The U.S. Green Building Council (USGBC) is a nonprofit organization headquartered in Washington, D.C., that promotes environmentally sustainable building practices and developed the Leadership in Energy and Environmental Design (LEED) system, which certifies buildings as "green" if they meet certain standards concerning conservation, waste reduction, reduction of greenhouse gas emissions, and promotion of the occupant's health and safety. The website includes basic information about the impact of the built environment on the natural environment and the benefits of green building as well as specific information about the LEED program, education and training, educational resources and news about green building. In addition, it includes information about USGBC membership, chapters, and committees.

Vehicle Technologies Program

http://www1.eere.energy.gov/vehiclesandfuels

This website, created and maintained by the Office of Energy Efficiency and Renewable Energy of the U.S. Department of Energy, disseminates information about transportation-related programs and technologies to reduce petroleum use and reduce environmental impact. Information is organized into 10 categories: hybrid and vehicle systems, energy storage (including batteries), power electronics & electrical machines, advanced combustion engines, fuels & lubricants, materials technologies, analysis & tools (including data, downloadable software tools and links to reports from several federal agencies), EPAct transportation regulatory activities, clean cities, and research partnerships. The emphasis is on automobile and truck technologies (i.e., not rail or other forms of mass transit) and includes basic information and downloadable fact sheets, news items, information about federal and state incentives and laws relating to alternative fuels, as well as on related topics such as fuel efficiency. For potential purchasers of new cars there is information about fuel efficiency, air pollution ratings, and safety data, and an interface to search for alternative fueling stations within the United States.

Sarah Boslaugh
Washington University in St. Louis

Index

Article titles and their page numbers are in **bold**.

Abandoned Mine Land fund, 82, 295
Absorption natural gas process
    technique, 300
Acetogenic bacteria, 219
Acetylene, 178
Acidification of ocean water, 75
Acid rain
   biomass energy production and, 40
   coal burning and, 83, 100, 420
   from fossil fuel combustion, 82, 180–181,
     309, 336
   gas flaring and, 169
   responses to, 215
   from sulfur dioxide, 109, 180–181, 412
Active solar power, 145, 220
Adsorption natural gas process
    technique, 300
Advanced Energy Initiative (AEI, Department
   of Energy), 99–100
Advanced-technology-partial-zero-emission
   vehicles (AT PZEVs), 461
Aerosols
   CFCs and, 71
   greenhouse effect of, 72–73, 74
Africa
   biomass energy produced in, 39
   gas flaring in, 169
   total primary energy supply (TPES)
    used by, 427
Agenda 21, 413, 455
Agent Orange, 102, 103
Agricultural Nonpoint Source (AGNPS) water
   pollution management, 307
Agriculture/agroecosystems
   carbon releases and, 54–55
   entropy and, 137–139
   environmental stewardship and, 146

   natural entropy cycle and, 138
   sustainability and, 138
   unsustainability of maize agroecosystem
    and, 139
Agrofuels, 219
Air conditioning automatic cycling, 288
Alaska
   Alaska National Interest Lands
    Conservation Act (1980), 14
   Alaska Native Claims Settlement Act, 15
   Arctic National Wildlife Refuge, 13–18
   Athabascan people of, 15
   Brooks Range, 15
   Exxon Valdez oil spill and, 160–162
   Gwich'in people of, 15, 17
   Inupiat peoples of, 15, 17
   Kaktovik Indian Corporation in, 15, 16
   National Petroleum Reserve—Alaska, 15
   North Slope, 15–16
   "1002 Area," 13–18
   Prudhoe Bay, 15–16
   Trans-Alaska Pipeline, 16
   See also **Exxon Valdez**
Albania, 263
Albedo global warming factor, 204, 207, 378
Algae
   as biofuel source, 21, 33, 34, 154, 219, 327
   as oil and oil products remediation, 326
Algeria, 169
Alkaline fuel cells, 184
Alternating current (AC), 107,
   112–113, 184, 293
Alternative energy, 1–3
   alternative-fueled vehicles and, 462
   biodiesel, 1–2
   daylighting tools and, 95–97
   definition of, 1

dioxin production in production of, 104
ethanols, 1–2
geothermal energy, 3
hydroelectricity, 2
OPEC affected by, 341
photovoltaic (PV) electric generators,
    2–3
renewable heating fuels, 3
solar energy, 1, 2
solar thermal systems, 3
wind turbines, 1 (photo), 2
See also **Alternative fuels; Innovation;**
    Landfill methane; *specific alternative*
    energy source
Alternative fuels, 4–8
biofuels, 5–6
cellulose ethanol, 6
coal, 4–5
in developing countries, 4–5
diesel engine oil-based fuel sources, 6
environmental stewardship and, 148
fossil fuel energy crisis and, 4
gasoline-electric hybrid vehicles and, 7
hydroelectricity generation, 7
hydrogen, 6–7
liquid biofuels gasoline substitutes, 5
oil consumption decline and, 4
oil reserves and prices issues and, 4–5
solar energy, 7–8
solar-powered vehicles, 7
source materials for, 4
sugarcane ethanol, 6, 20
unconventional fossil fuel extraction
    methods and, 5
wind power, 8
See also specific fuel alternative
Aluminum Company of America, 86
American Competitiveness Initiative (ACI,
    Department of Energy), 99–100
American Council for an Energy-Efficient
    Economy, 12
American Institute of Architects, 96
American National Science Foundation, 313
Americans with Disabilities Act
    (ADA, 1990), 368
Ammonia, 178
Amorphous silicon (a-Si) photovoltaics,
    348–349
Amsterdam Declaration of the Transport
    Health and Environment Pan-European
    Programme of WHO, 461
Anadromous fish species, 87
Anaerobic digestion, 36, 219, 275, 386, 440

Angola
    gas flaring in, 169
    OPEC membership of, 338, 339
Aniline, 178
An Inconvenient Truth film, 75
Anode (negative electrode) of fuel cells, 183
Antarctica
    glacial melt, sea level rise and, 205
    global warming data and, 202, 203 (fig.)
    ozone layer hole above, 215
Anthracene, 178
Anthracite coal, 4–5, 50, 79, 80, 130, 178
ANWR. *See* **Arctic National**
    Wildlife Refuge (ANWR)
Aphelion position of Earth, 404
Appalachian Mountains, 81
Appalachian Voices, 295
Appliance standards, 8–13
    Collaborative Labeling and Appliance
        Standards Program and, 10
    Energy Policy Act (2005) and, 10
    federal appliance regulations and, 8
    future trends in, 12
    goals of, 9–10
    history of, 10
    household and commercial energy
        consumption focus of, 8–9
    minimum efficiency standards and, 10
    National Appliance Energy Conservation
        Act (NAECA, 1987), 10, 11 (table)
    players in, 12
    refrigerators and, 9
    residential electricity consumption by end
        use and, 10 (table)
    savings from, 11 (table), 12
    U.S. carbon emissions savings and, 9
    voluntary standards and, 10
Arab oil embargo (1973), 43, 47, 92,
    98, 329, 340
Arctic
    glacial melt, sea level rise and, 205
    global warming data and, 202
Arctic National Wildlife Refuge (ANWR),
    13–18
    Alaska National Interest Lands
        Conservation Act (1980) and, 14
    biological diversity within, 14–15
    creation of, 13–14
    drilling debates and, 16–17
    ecological zones within, 14–15
    Exxon Valdez oil spill and, 16
    history of, 13–14
    "minimal management" classification of, 14

natural resource management and, 13–14
prospective *vs.* proven oil reserves in,
    17–18
Section 1002 of, 13, 15–17
settlements and land rights issues and, 15
AREVA nuclear conglomerate, 318
Argentina
    International Renewable Energy Agency
        (IRENA) and, 266
    natural gas vehicles in, 301
Arizona
    appliance efficiency standards in, 10
    Black Mesa coal slurry pipeline in, 82
    renewable energy portfolio of, 391 (table)
Arrhenius, S., 73, 202
Ash, 181, 439
Asia
    electric vehicles in, 7
    feed-in tariffs policy in, 167
    flue gas desulfurization in, 412
    petroviolence in, 345–346
    total primary energy supply (TPES)
        used by, 427
Associated natural gas, 298–299
AT&T, 374
Athabascan peoples, Alaska, 13
Athabasca oil sand fields, Alberta,
    Canada, 5
Atmospheric heat island effect, 234
Atomic Energy Act (1946), 423
Atomic Energy Commission (AEC),
    98, 423
Atoms for Peace, 311, 318
Audi, 462
Austin Energy green pricing utility, 225
Australia
    compressed natural gas buses in, 462
    electric trolleys and buses in, 115
    electric vehicle infrastructure in, 114
    emissions reductions in, Berlin Mandate
        and, 26
    flue gas desulfurization in, 412
    geothermal energy produced in,
        200 (table)
    green energy certificates in, 212
    House Energy Rating used in, 236
    International Renewable Energy Agency
        (IRENA) and, 266
    Kyoto Protocol and, 270
    National Pollutant Inventory in, 118
    oil shale in, 334
    Tindo electric bus system in, 115
    uranium resources in, 313, 366, 431

Austria
    electric trolleys and buses in, 115
    geothermal energy produced in, 200 (table)
    recycling in, 380
Automobiles, 18–21
    biofuels issue and, 20–21
    compressed natural gas fuel and, 21, 462
    electric vehicles and, 19
    environmental effects of, 19
    environmental protection measures
        and, 143
    ethanol *vs.* gasoline in, 155
    fuel economy environmental
        standards and, 144
    gas-electric hybrids and, 20
    hydrogen fuel cells and, 21
    internal combustion engines and, 182
    propane fuel source and, 21
    in U.S., 18 (photo)
    See also **CAFE (Corporate Average Fuel
        Economy) standards; Combustion
        engine; Electric vehicle; Flex fuel
        vehicles (FFVs); Internal combustion
        engine; Oil; Plug-in hybrid;
        Zero-emission vehicle (ZEV)**
Azeotrope, 154
Azerbaijan, 66, 169

Bacterial hydrolysis, 219
Bagasse, as ethanol source, 155, 156, 195
Bangladesh, 219
Bank of America green lending initiative, 210
Banqiao Dam disaster, China, 244
Batch biogas digester, 36–37
Batteries, 23–25
    alkaline battery, 23–24
    capacitors and ultracapacitors and, 135
    chemical batteries, 135–136
    disposal of, 23, 24
    electrical energy storage in, 107
    electric vehicle batteries, 113–114
    Eveready Battery Company (Energizer
        Battery Company), 23–34
    fuel cells *vs.*, 183
    history of, 23–24
    lithium batteries, electric cars, 24, 114
    lithium-ion battery, 23 (photo), 24, 114
    microhydro power stored in, 293
    plug-in hybrid and, 356–361
    rechargeable, 23, 24, 107, 183
    recycling of, 24
    sodium-sulfur units and, 135
    structure of, 24

valve-regulated batteries, 135
vanadium batteries, 135
voltaic batteries, 23, 24, 107
wet cell lead acid batteries, 135
See also **Electric vehicle; Photovoltaics (PV)**
Batt insulation, 255–256
Bay of Fundy, Nova Scotia, 261
Beaver Lake Watershed Management
    system, 307
Becquerel, Henri, 430
Belgian Congo, 430
Belgium
    renewable energy certificates in,
        168, 212, 390
    REVA electric car used in, 462
    total primary energy supply (TPES)
        measure of, 427
Benchmarking, 28
Benzene, 169, 178
Berlin Mandate, 25–27
    Conference of the Parties to the UN
        Framework Convention on Climate
        Change (UNFCCC) and, 25–26
    greenhouse gas emissions issue and, 25–26
    International Panel on Climate Change
        (IPCC) and, 25–26
    joint implementation issue of, 26
    Kyoto Protocol and, 25, 26
    Second Assessment Report (IPCC) and, 26
Bernstein, Lenny, 204 (fig.)
Best management practices, 27–29
    best practices, good practices terms
        and, 29
    continuous improvement concept and, 28
    definition of, 27
    electricity industry and, 28
    "ends" of work and, 27
    goals accomplishment and, 27
    ISO 14000 and, 29
    of nonpoint source water pollution,
        305–307
    operations *vs.* project management and,
        27–28
    pioneers in, 28
    routines *vs.*, 28
    Frederick Winslow Taylor's work in, 28
    techniques of, 28
    work breakdown structure method of, 28
    See also **Emission inventory; Energy audit;**
        Environmentally preferable purchasing
        (EPP); Exergy
Better Assessment Science Integrating Point
    and Nonpoint Sources (BASINS), 307

Better Place, 114
Bhutan, 219
Bicycles, 29–32
    bicycle commuters, Beijing, China,
        30 (photo)
    efficiency and functions of, 29–31
    green public policy and, 31
    history of, 30
    limitations of, 31
    rights-of-way and safety issues of, 31–32
    as urban transportation, 29, 30, 31–32
    as zero-emission vehicle, 463
Binary cycle geothermal applications,
    198, 199, 221, 385
Biochar gasification product, 191, 193
Biodiesel, 32–35
    algae-derived biodiesel and, 21, 33, 34
    as biofuel, 32–35
    biomass production of, 40
    CAFE standards and, 44
    costs and benefits of, 33–34, 40
    diesel engines and, 20
    feedstocks to produce, 2, 6, 20, 32–33,
        34–35, 39
    food security issues and, 33, 34
    future trends in, 34
    GHGs emissions of, 33
    increased access to, 33 (photo)
    jatropha feedstock source of, 34
    life cycle energy reduction of, 2
    performance capacity of, 2
    rapeseed and soybean to produce, 33
    scaling issue of, 6
    sun and carbon roles in, 1
    U.S. "splash and dash" subsidies and, 34
    wet biomass feedsource of, 34
    See also **Combustion engine**
Biofiltration, 436
Biofuels
    agrofuels as, 154, 219
    biofuel vehicles and, 20–21
    biogas, 193, 217, 275, 386
    biomass energy and, 5, 39–40
    "cloud points" factor in, 20
    definition of, 153
    as energy innovation, 252
    energy policies regarding, 132
    environmental stewardship and, 148
    ethanol, 5, 20
    ethyl alcohol ethanol fuel source, 20
    as green power, renewable energy source,
        218–219
    hydrogen extraction from, 248–249

IPCC studies of, 260
life cycle production approach to, 148
liquid biofuels gasoline substitutes, 5
net energy value of, 240
second-generation biofuel generation
    and, 144
solar energy element in, 405
See also Biogas digester; Biomass energy;
    Ethanol; Ethanol, corn; Ethanol,
    sugarcane; Gasohol; Methane;
    Wood energy
Biogas, 193, 217, 275, 386
See also Biogas digester
Biogas digester, 35–37
advantages of, 37
anaerobic bacteria role in,
    35–36, 219, 386
batch biogas digester, 36–37
methane gas product of, 36
plant designs of, 36
See also Biomass energy
Biological life, solar energy and,
    402, 405
Biomass energy, 37–41
agroecosystem entropy and, 139
agrofuels and, 219
bioelectricity in U.S. and, 38–39
from biofuels, 5–6, 109,
    218–219, 386–387
Biomass Energy and Alcohol
    Fuels Act and, 100
carbon-based feed stock to produce, 1–2
carbon dioxide emissions of, 38, 386
carbon neutrality status of,
    153, 193, 219, 450
carbon sequestration process and, 154
carbon source from, 64
definition of, 37–38, 386
domestic and industrial waste
    managed by, 40
environment affected by, 39–41, 148, 387
gasification and, 191
as green power renewable energy source,
    83, 108, 109, 217, 218–219, 223, 252,
    386–387, 388
hydrogen extraction from, 248–249
lignocellulosic feedstocks and, 195, 196,
    326, 386–387
methane gas byproduct of, 40, 153, 275,
    386, 435
on-site renewable energy and, 337
solar energy element in, 405
sources of, 1–2, 37, 38, 219

sulfur dioxide and nitrogen oxide
    byproducts of, 40–41, 109
sun's role in, 1
thermal and chemical types of, 38
transportation industry use of, 39–40
wood waste source of, 38 (photo)
See also Biofuels; Biogas digester; Ethanol,
    corn; Ethanol, sugarcane; Flex fuel
    vehicles (FFVs); Wood energy
Biomass Energy and Alcohol Fuels Act, 100
Bio-oil gasification product, 193
Bitumen
carbon source from, 5, 64
as heavy oil form of petroleum, 324
in tar sands, 180, 324
See also Oil sands
Bituminous coal, 5, 50, 79, 80, 130, 178
Bleve (boiling liquid expanding vapor
    explosion) explosions, 164
Blown-in insulation, 256
BMW Hydrogen 7 zero-emission prototype an,
    461 (photo)
Borohydride, 248
Bosnia, 263
Bottle Bills, 381
Boulding, Kenneth, 413
Brannerite, 431
Brayton engines, 110
Brazil
biodiesel produced in, 33, 34
flex fuel vehicles in, 171
gas flaring in, 169
hydroelectric power produced in, 223
International Renewable Energy Agency
    (IRENA), 266
Itaipu Dam, Parana River, 420
natural gas vehicles in, 301
oil reserves in, 346
oil shale in, 334
Petrobras oil company in, 329
renewable energy-fired power plants
    in, 366
sugarcane-based ethanol produced in,
    6, 155, 156, 171, 172, 194,
    223, 386, 387
British Columbia, Canada, 249
British Nuclear Fuels Limited, 318
British Petroleum (BP), 4, 329, 330, 339
British thermal unit (Btu),
    106, 302, 333, 451
Brown, Jerry, 47
Brown coal, 79, 362
See also Lignite coal

Brownfields, 279
Brundland, Gro Harlem, 453
Bruntland Report, 413–414
Buckminsterfullerene carbon molecule, 135
Buckypaper energy storage material, 135
Building-scale photovoltaic arrays, 351
Bulgaria, 263
Bureau of Land Management, 28
Bush, George W.
  Alaska oil drilling policy of, 16
  California emission standards policy
    of, 44
  Coal-to-Liquids Fuel Promotion Act (2007)
    and, 84, 100
  energy legislation of, 84, 99–100
Bus public transportation, 368–369
Butane natural gas component, 298, 308

Cadillac Desert (Reisner), 86
Cadmium, 439
Cadmium batteries, 24
Cadmium telluride photovoltaics, 348–349
CAFE (Corporate Average Fuel Economy)
  standards, 43–45
  California emission standards and, 44–45
  Energy Independence and Security Act
    (2007) and, 34, 44
  flex fuel, and biodiesel-ready
    vehicles and, 44
  "footprint" approach to, 43–44
  GHG emissions, global warming
    issue and, 44
  miles per gallon element of, 43
  National Highway Traffic Safety
    Administration (NHTSA) and,
    43–44
  smaller car safety issue and, 44
  technology factor in, 43
California, 45–49
  appliance efficiency standards in, 10, 12
  automobile standards in, 463
  California Air Resources Board, 463–464
  California Climate Action Registry, 118
  California Energy Commission, 10, 47, 48
  California Environmental Air Quality Act
    (1970), 47
  Chevron in, 330
  clean air and water acts of, 46, 47
  Coastal Zone Conservation Initiative
    (1972), 47
  CO_2 emission standards of, 44–45
  drinking water pipelines in, 355

electric tram public transportation in,
  369–370
electric vehicle infrastructure in, 114
energy consumption and efficiency measured
  in, 259–260
energy technologies in, 46
environment activism in, 46–47
Geysers geothermal area, 197, 198, 199,
  223, 385
GHG emissions of, 45
Global Warming Solutions Act (2006), 59
green energy certificates in, 212
green energy efforts in, 46, 47–48
Green For All movement, 48
Hetch Hetchy Valley, 47
hybrid vehicles in, 462
"hydrogen highways" in, 249
Kenneth Hahn State Recreation Area,
  Los Angles, 326
Kern River gas transmission pipline in, 353
Los Alamos National Laboratory Hot Dry
  Rock geothermal program, 198
Low-Carbon Fuel Standard (CLFS), 34
Low Emissions Vehicle program, 463–464
Mammoth Lakes binary cycle geothermal
  power plant, 198
methanol vehicles and, 171
MetroLink public transportation, 370
partial-zero-emission vehicles
  (PZEVs) in, 462
population growth in, 46
Public Utility Regulatory Policies Act
  (1978), 166–167
Rancho Seco Power Plant closure, 47
renewable energy portfolio of, 391 (table)
Renewable Portfolio Standard of, 48
smart meters used in, 288
Solar Energy Generating Systems, Mojave
  Desert, 410
solar technology in, 46 (photo), 47, 110
Sonoran Desert, 404
water infrastructure of, 46
wind facilities in, 47
zero-emission vehicles (ZEVs) in,
  463–464
Cambodia, 219
Campbell, Colin J., 238
Canada
  acid rain in, 412
  Athabasca oil sand fields, Alberta, 5
  Climate Exchange PLC in, 62
  coal reserves in, 80

Columbia River dams in, 86
emissions reductions in, Berlin Mandate
    and, 26
flex fuel vehicles in, 171
hydroelectric power produced in, 223
international energy distribution and,
    261–262
National Pollutant Release Inventory in,
    118, 119
nuclear power reactors in, 365
oil consumption decline in, 4
oil sands, Fort McMurray, Alberta, Canada,
    332, 332 (photo)
oil (tar) sands in, 5
oil shale in, 334
radium mining in, 430
renewable energy-fired power plants
    in, 366
Telsa electric sports car used in, 462
uranium resources in, 313, 366, 431
Cannibalism of energy, 128
Capacitors, 135
Cap-and-trade system climate change
    approach, 57–58, 59–60, 84, 133, 144,
    207, 216, 269, 270
Capped vacuum well of landfill, 275 (photo)
Carbon capture and storage technologies,
    77, 84, 144, 363, 364
Carbon dioxide (CO₂)
    anaerobic digestion and, 38, 219, 386
    Antarctic ice core data on, 202, 203 (fig.)
    Assigned Amount Unit (Kyoto Protocol)
        and, 271
    carbon capture and storage technologies
        and, 77, 78, 84, 144, 363, 364
    carbon dioxide equivalence value and, 78
    Certified Emission Reduction (CER) credits
        and, 272
    climate forcing effect of, 214
    coal-fired power plants and, 363
    of dry steam geothermal energy
        systems, 199
    embodied carbon and, 78–79, 178
    flaring and, 169
    gasification process and, 191
    geo-engineering technology and, 207
    as greenhouse gas, global warming and,
        73, 109, 181, 202–203, 203 (fig.),
        204 (fig.), 214–215, 216, 269, 289,
        290 (table), 315
    hydroelectric dam construction and, 383
    in landfill gas (LFG), 275

in natural gas, 179, 298, 300, 364
as natural GHG, 213–214, 404
from oil refining, 326
oil sands and, 240, 325
radiative forcing and, 377–378
reforming of hydrocarbons and, 248
from waste incineration, 440
wood energy and, 449–450
See also **Carbon emissions factors** (CEF);
    Carbon footprint and neutrality;
    Carbon sequestration; Carbon tax;
    Carbon trading and offsetting; Metric
    tons of carbon equivalent (MTCE)
Carbon emissions factors (CEF), 49–51
    appliance-related energy consumption
        and, 9
    biomass energy production and, 39–41
    calculation of, 50
    CEFs by units of energy (U.S.) and, 49
    coal types and, 50
    direct vs. indirect CO₂ emissions and, 49
    electricity generated from coal plants
        and, 50
    fossil fuels consumption and, 50
    GHGs and global climate issues and, 49
    pipelines and, 356
    standards using, 49
    trade policies and carbon credits and, 51
    uses of, 50–51
    See also **Carbon footprint and neutrality;**
        Emission inventory; Kyoto Protocol
Carbon footprint and neutrality, 52–54
    appliance standards, carbon reductions of,
        11 (table), 12
    biomass energy and, 153, 218–219
    calculation of, 52–53
    carbon footprint term meaning and, 52
    carbon neutrality term meaning and, 53
    carbon offset concept and, 53
    climate changes and, 52
    electric vehicle reduction of, 112
    of fossil fuels, 259
    greenhouse gases and, 52
    net zero carbon emission concept
        and, 53
    nuclear power plants and, 77, 366
    primary vs. secondary types of, 52
    renewable energy credits, green power
        and, 226
    See also **Climate neutrality; Combustion**
        engine; Food miles; Life cycle analysis
Carbon monoxide, 248, 439, 461

Carbon neutrality, 144
  See also **Carbon footprint and neutrality**
Carbon sequestration, 54–57
  agricultural practices and, 55–56
  biomass carbon and, 54, 56, 153–154, 193
  carbon cycle and, 55
  carbon sinks and, 54
  CO_2 emissions reduced by, 54, 56–57
  as energy storage, 135
  explanation of, 54
  forestry practices and, 55–56
  gasification and, 191
  hydrogen production and, 248, 249
  offshore drilling and, 322
  as opportunity and option for mitigation,
      56–57
  peat bogs, Ireland, 54 (photo)
  terrestrial carbon in peat lands and, 57
  tropical deforestation reduction and,
      55–56
Carbon sinks, 54, 62, 73
Carbon tax, 57–61
  advantages of, 59–60
  mechanism of, 58
  negative externalities concept and, 58–59
  political manipulation issue and, 60
  price variations issue and, 60
  rationale for, 58–59, 168
Carbon trading and offsetting, 61–66
  carbon emissions factors (CEFs) and,
      51, 78
  carbon emissions markets and, 61
  carbon in products *vs.* production processes
      and, 64
  carbon offsetting and, 53, 63–64, 78
  carbon trading and, 62–63, 78
  Chicago Climate Exchange and,
      53, 62, 122
  Clean Development Mechanism strategy
      (Kyoto Protocol) of, 56–57, 61, 63, 207
  emissions certificates and, 53
  gate cap on carbon-emitting fuels and, 78
  GHG emissions increase and, 61–62
  issues regarding, 64–65
  redistribution and rebalance of emissions
      goals of, 62, 65
  "smart meters" used in, 64
  technological advances and, 65
  tradable credits concept and, 62–63
  trading in futures and, 62
  See also **Climate neutrality; Emissions
      trading**

Carnegie Mellon University, 285
Carnotite, 431
Cartagenesis process of fossil fuel
    formation, 177
Cartels, 339
Carter, Jimmy
  energy policies of, 98
  pro-solar policies of, 47
Casing-head natural gas, 299
Caspian Sea, 66–69
  Baku "Black Gold Capital" of,
      66, 67 (photo)
  Bibi-Heybat Bay, 66
  Energy Community Treaty of EU
      and, 263
  fossil fuels of, 66, 67 (photo)
  living conditions and, 68
  post-Soviet Caspian region and, 68
  sovereignty issues regarding, 68–69
  Soviet Caspian region and, 67–68
  sturgeon caviar and, 69
  United Nations Conventions on the
      Laws of the Sea and, 69
Cathode (positive electrode) of fuel cells, 183
Caviar, 69
CEF. *See* **Carbon emissions factors** (CEF)
Cellulosic ethanol, 2, 6, 386–387
  See also Lignocellulosic feedstocks
Central Asia, 345–346
Certified Emission Reduction (CER) credits.
      See **Kyoto Protocol**
Cervero, Robert, 370
Cesium, 70, 457
CFB. *See* **Compact fluorescent bulb** (CFB)
Chamberlain, Thomas, 73
Char, 191, 193
Chemisorption natural gas process
    technique, 300
Chernobyl, 69–71
  damage resulting from, 70, 313
  events of, 70, 313
  See also **Nuclear power**
Chevrolet Volt plug-in hybrid vehicle,
      359, 360
Chevron, 329, 330, 339
"Chicago Pile 1," first atomic cell, 311
China
  anaerobic digestion system used in, 219
  Banqiao Dam disaster in, 244
  bicycle commuters, Beijing, China,
      30 (photo)
  biomass energy production in, 132

China National Petroleum state-owned oil company in, 329
Chinese Electricity Regulatory Commission (SERC) and, 167
Chinese National Development and Reform Committee and, 167
Climate Exchange PLC in, 62
coal-fired electrical power plants in, 308, 420
coal reserves in, 80
CO_2 emissions produced in, 50
electric trolleys and buses in, 115, 462
energy market in, 325
feed-in tariffs policy in, 167
geothermal energy produced in, 200 (table)
hydroelectric power dams in, 7, 223, 383
Kyoto Protocol and, 207, 216, 270
microhydropower used in, 292, 293
natural gas in, 298
nuclear power reduction in, 313, 366
nuclear weapons developed in, 317, 430
oil consumption increase in, 4, 341
oil shale–fired power plants in, 334
oil shale in, 334
renewable energy-fired power plants in, 366
solar power use in, 223
Three Gorges Dam, 7, 419–421
total primary energy supply (TPES) used by, 427
uranium resources in, 431
Xiluodo Dam, Jinsha River, 421
Yangtze River, 419–420, 421
The China Syndrome film, 423
Chlorofluorocarbons (CFCs), 71–72
carbon presence in, 64
climate forcing effect of, 214
as greenhouse gas, global warming and, 73, 213, 289–290, 290 (table)
haloalkanes chemical group and, 71
heat pumps use of, 71
Montreal Protocol and, 71–72, 215
radiative forcing and, 378
uses of, 71
CHP. *See* **Combined heat and power** (CHP)
Chu, Steven, 101
Clean Air Act, 44, 295, 412, 440
Clean air and water acts, California, 47
Clean Development Mechanism. *See* **Kyoto Protocol**
Clean Water Act, 161, 162, 305, 306
Climate change, 72–76
CAFE standards and, 44–45

CO_2 and, 73, 74, 75, 180–181
definitions regarding, 72
developing countries affected by, 75
evidence regarding, 75–76
fossil fuels and, 181
global gas flaring activities and, 169–170
global issue of, 74–75
Global Warming Solutions Act (2006, California) and, 59
"greenhouse effect" on, 73
Kyoto Protocol and, 72
modeling techniques projections of, 75–76
rate of, 72
State of Fear (Crichton) and, 73
United Nations Framework Convention on Climate Change (UNFCCC) and, 74–75
See also **Berlin Mandate; Carbon emissions factors (CEF); Carbon footprint and neutrality; Carbon tax; Chlorofluorocarbons (CFCs); Climate neutrality; Global warming; Greenhouse gases (GHG); Intergovernmental Panel on Climate Change (IPCC); Kyoto Protocol;** United Nations Environment Programme; United Nations Framework Convention on Climate Change (UNFCCC)
Climate Exchange PLC, 53, 62, 122
Climate neutrality, 76–79
"carbon capture and storage" technologies and, 77, 84, 144, 363, 364
carbon trading and, 78
climate forcings term and, 76
CO_2 equivalence and, 78
definition of, 76
direct *vs.* indirect emissions and, 78–79
"feedback mechanisms" and "tipping point" concepts and, 77, 394
geo-engineering technologies and, 78
multiscalar nature of, 77
nuclear power production issue and, 77
practical achievement of, 77
reasons for, 76–77
technologies of, 77
zero-carbon solutions term and, 76, 77
Closed-cell spray foam insulation, 256
Closed energy storage systems, 134–135
Coal. *See* **Coal, clean technology; Fossil fuels**
Coal, clean technology, 79–85
Abandoned Mine Land fund and, 82, 295
acid rain and, 83, 100, 420
anthracite coal formation, 5, 80

bituminous coal formation, 79–80
cannel coal formation, 80
carbon footprint of, 259
carbonization and, 178
carbon tax and, 58
"clean coal" power plant, Germany
    and, 84
coal-bed methane and, 239, 297–298, 299
coal cleaning and, 82
coal creation process and, 4–5
coal-fired electricity plants and, 82–83, 84,
    100, 178–179, 308, 362–363, 420
coal gas and, 83, 178
coalification process of, 178
Coal-to-Liquids Fuel Promotion Act (2007)
    and, 84
CO_2 capture and, 84
co-firing power plants and, 308
coke in steel production and, 83
components of coal and, 178, 308
costs associated with, 243
demand and regulation issues and, 81–82
electrical energy infrastructure and, 108
environmental pollutants from, 181, 363
environmental protection measures
    and, 143
ethanol (corn) production using, 151
as fossil fuel, 308
gasifying coal and, 83, 192–193
hydrogen production and, 248
hydrogen tetraoxosulfate acid (acid mine
    drainage) and, 181
increase in use of, 83
lignite coal formation and, 5, 50, 79, 84,
    130, 362
liquified coal, 5
methane and, 275
mining policies regarding, 130
mining techniques of, 80–81, 130
montaintop mining, 81
new clean technologies and, 83–84,
    100–101, 363
nitrogen-oxygen compounds and, 82–83,
    100, 308, 363
run-of-mine coal and, 82
slurry pipelines and, 82
steam for electricity generation and,
    178–179, 362–363
subbituminous coal formation, 79
Surface Mining Control and Reclamation
    Act (SMCRA, 1977) and, 81–82
synthesis gas and, 308

types of coal, 79–80
types of coal, CEFs calculation and, 50
undersea beds and mining of, 80
See also Fossil fuels; Mountaintop removal
Coal tar, 178
Coastal Zone Conservation Initiative
    (1972), 47
Coffinite, 431
Co-firing power plants, 308
Cogeneration
    cogeneration energy storage, 136
    cogeneration thermal power plants, 110
    methane as energy source in, 276
Cohen, Bernard, 224
Coke in steel production, 83, 178
Cold fusion, 188
Collaborative Labeling and Appliance
    Standards Program, 10
Colorado
    heating degree day measures in, 232
    Hoover Dam, 241 (photo), 292, 420
    Kern River gas transmission pipline in, 353
    oil shale in, 334
    renewable energy portfolio of, 391 (table)
    smart grid in, 399
Columbia River, 85–87
    Cadillac Desert (Reisner) and, 86
    Columbia River Risk Assessment Project
        and, 86
    Elwha Ecosystem Restoration Project
        and, 87
    Grand Coulee Dam on, 86, 86 (photo)
    human settlement history of, 85
    hydroelectric capacity of, 86
    industrial electricity provided by, 86
    Northern Pacific Railroad and, 85
    The Organic Machine (White) and, 85
    river renewal through dam removal
        and, 87
    salmon spawning issues and, 87
Combined cycle gas turbine, 110,
    300, 364, 365
Combined heat and power (CHP), 87–89
    applications using, 88–89
    barriers faced by, 89
    Edison Pearl Street station,
        New York City, 88
    efficiencies of, 88
    emissions reduced by, 88
    integrated system feature of, 88
    Public Utilities Regulatory Policies Act
        (1978) and, 88

site of demand element in, 87–88
technology advances and, 88
thermal energy recovery of, 88
U.S. policy regarding, 89
Combustion engine, 90–91
BMW Hydrogen 7 zero-emission prototype, 461 (photo)
diesel engine and, 90
ethanol from corn and, 149–150
external combustion engine and, 91
four-stroke reciprocating internal combustion engine and, 90
HEMI engine and, 90–91
hydrogen fuel and, 91
internal combustion engine and, 90–91
rotary engine, 91
volatile organic compounds (VOCs) and, 435–436
See also **Hydrogen;** Internal combustion engine
Common Agricultural Policy (Europe), 33
Commuter rail public transportation, 370
Compact fluorescent bulb (CFB), **91–93**
advantages of, 91–92
carbon emission factors and, 51
Energy Star data regarding, 93
GHGs produced by, 91, 92
mercury component in, 93
nonintegrated *vs.* integrated CFBs and, 92
prices of, 92
savings from use of, 93
spiral design of, 92, 92 (photo)
ultraviolet light emitted by, 93
Compressed air energy storage, 135
Compressed air vehicles, 21
Compressed gas energy storage, 135, 462
Concentrated photovoltaics, 350, 400, 406
Condensed matter nuclear science, 188
Confirmed home energy rating, 236
Congo, 169, 314
Connecticut
appliance efficiency standards in, 10
commuter rail transportation in, 370
The Glass House, New Caanan, 96
renewable energy portfolio of, 390, 391 (table), 392
smart meters used in, 288
Conoco-Phillips, 329, 330, 339
Conservation movement, 147
Continuous improvement concept, best management practices, 28
Contour mining, 81

Convention on Biological Diversity, 455
Convention on Long-Range Transboundary Air Pollution (Europe), 412
Cooking oil, as biofuel source, 33
Cooling degree day, 232
Cooperative utilities, 372
Copper (Indium Gallium) selenide (CIGS) photovoltaics, 349
Coppicing, sustainable management practice, 39
Corn ethanol. *See* **Ethanol, corn**
Corporate Average Fuel Economy regulations, 144
See also **CAFE** (Corporate Average Fuel Economy) **standards**
Coskata, 171
Costa Rica, 200 (table)
Cost-benefit analysis, 394
Coulomb barrier, in nuclear fusion reaction, 187
Coulombs (C) units of electrical charge, 106
Council for Solid Waste Solutions, 284
Cow manure, biomass energy from, 36, 37, 219
Cradle-to-grave concept, 116, 283
Crichton, Michael, 73
Croatia, 263
Crystalline silicon (c-Si) photovoltaics, 348
Crystal Palace, 95–96
Cumulative Average Temperature, 232
Curie, Marie, 311
Curie, Pierre, 311
Czech Republic, 430

Daily mean temperature, 231–232
Daimler Chrysler, 171
Dai Qing, 421
Dairyland Power Cooperative, La Crosse, Wisconsin, 37, 277
Database of State Incentives for Renewable Energy, 338
Davidite, 431
Daylighting, 95–97
benefits of, 96
greener building design tool of, 96–97
heating and cooling systems adjunct of, 95
historic examples of, 95
modern architectural design examples of, 96
natural daylight illumination use and, 95
in premodern architecture, 95
Delaware, 391 (table)

Demonstration power plant (DEMO) fusion device, 188
Denaturing process of ethanol, 154–155
Denmark
  coal-fired electrical power plants in, 308
  electric vehicle infrastructure in, 114
  feed-in tariffs policy in, 167
  International Renewable Energy Agency (IRENA) and, 266
  wind power produced in, 222
Department of Energy, U.S. (DOE), 98–102
  administrative organization of, 99
  Advanced Energy Initiative (AEI) and, 99–100
  agencies of, 98, 99
  American Competitiveness Initiative (ACI) and, 99–100
  Arab oil embargo, energy crisis (1973) and, 98
  biodiesel and, 34
  Canadian oil sands and, 332
  Department of Energy Organization Act (1977) created by, 98
  Energy Information Administration and, 17, 68, 101, 105
  energy security issues and, 100
  Energy Star data and, 110
  Federal Energy Regulatory Commission of, 99
  Handford plutonium production plant, Superfund site and, 86
  Home Energy Rating reports issued by, 236
  hydroelectric power incentives of, 243
  hydroelectric power potential capacity data of, 243
  hydrogen energy storage system and, 134 (photo)
  inertial confinement fusion research of, 188
  international nuclear programs and, 318
  Lawrence Livermore Laboratory, National Ignition Facility of, 188
  mission and goals of, 98
  National Policy Plans of, 99–100
  renewable energy technologies and, 100–101
  responsibilities of, 99–100
  U.S. electrical use data of, 406

  Yucca Mountain tunnel excavation and, 457 (photo)
  zero-net energy commercial buildings and, 101
  See also **Energy policy; Federal Energy Regulatory Commission (FERC)**
Deuterium-tritium nuclear reaction, 188
Developing countries
  anaerobic digestion system used in, 219
  Berlin Mandate and, 26
  biodiesel produced in, 33
  carbon leakage concept and, 117
  Clean Development Mechanism strategy (Kyoto Protocol) and, 56, 207
  coal energy source in, 4–5
  environmental measures priorities of, 143
  hydroelectric power dams in, 7, 383
  Kyoto Protocol and, 75, 207, 270–273
  on-site renewable energy used in, 336–337
  paratransit in, 368
  petroviolence in, 343–346
  public utilities in, 374
  renewable energy sources use in, 388, 444
  sustainability issues and, 416–417, 444
  See also specific country
Devil's Tower offshore drilling platform, Gulf of Mexico, 321 (photo)
Diagenesis process of fossil fuel formation, 177
Diesel, 247 (table), 324, 358, 369
Diesel, Rudolph, 20, 90
Diesel oil-based fuel sources, 6
Dioxin, 102–104
  Agent Orange and, 102, 103
  endocrine disruptors function of, 103
  health effects of, 103–104
  sources of, 102–103
  toxicity variations of, 103
  from waste incineration, 439
Direct carbon conversion fuel cell, 185
Direct current, 184
Direct energy, 115
Direct methanol fuel cells, 184
Dispersed generation grid-connected systems, 228
Display Energy Certificates (U.K.), 237
Disruptive innovation, 254
Dissolved natural gas, 299
Distributed generation of energy, grid-connected systems, 228, 398, 399
Distribution domain, of electric power system, 227–228

District of Columbia. *See* Washington, D.C.
Diurnal cycle
    of solar energy, 405, 406, 407
    of tidal power, 425
Diversion hydroelectric power, 242
Dominion Virginia Power, 110–111
Dry natural gas, 298, 299
Dry-sensitized photovoltaics, 349–350
Dry steam resources geothermal applications,
    197–198, 199, 220, 385

Earth Summit, Rio de Janeiro
    Berlin Mandate signed at, 25
    objective of, 25, 269
    United Nations Framework Convention
        on Climate Change (UNFCCC) and,
        74–75, 269
East Kentucky Power Cooperative, 277
Economic Stabilization Program, Nixon
    administration, 340
Eco Power Program, green energy
    certification, 212
Ecosystems, 54–55, 87, 137–139, 146
    See also Agriculture/agroecosystems
Ecuador, 338, 343
Edinburgh Duck wave power model, 442
Edison Electric Institute trade
    association, 392
Edison Pearl Street station,
    New York City, 88
Egypt, 340
Eisenhower, Dwight D.
    Arctic National Wildlife Range
        created by, 13
    Atoms for Peace discourse and, 311, 318
Electricity, 105–111
    alternating current (AC) and, 107
    battery storage of, 107
    best management practices in electricity
        industry and, 28
    British thermal unit (Btu) and, 106
    coal-fired plants production of, 105 (photo),
        178–179, 362–363
    consumer conservation and, 110–111
    coulombs (C) units of charge and, 106
    electrical energy *vs.* electrical power and,
        106–107
    electric trams and streetcars public
        transportation, 368 (photo), 369–370
    electric utility monopoly and, 372–373
    electromagnetic induction and, 107
    feed-in tariffs and, 166–167

FERC regulation of, 164
    fossil fuels reliance and, 106, 109
    from fuel cells, 182–186
    generation alternatives and efficiency of,
        109–110
    GHGs from generation of, 109
    green certificates and, 110–111
    hydroelectric power plant and, 107
    infrastructure and policy regarding,
        108–109
    joule (J), watts (W), and kilowatt hours
        (kWh) units of, 106
    kinetic *vs.* potential energy and, 106
    lighting application of, 108
    natural gas–fired production of, 364
    net-metering policy and, 166
    renewable energy certificates and,
        167–168
    renewable power stations and, 108
    transmission of, 106–108, 337
    usage of, 105–106, 107–108
    volts measure (V) of, 107
    See also **Appliance standards; Combined
        heat and power (CHP); Electric vehicle;
        Fuel cells; Hydroelectric power;
        Metering; Power and power plants;
        Tidal power**
Electric Power Research Institute, 397
Electric trams and streetcars, 368 (photo),
    369–370
Electric vehicle, 112–115
    as alternative-fuel vehicle, 19
    battery issues and, 113–114
    Better Place, battery charging spots
        and, 114
    carbon footprint reduction and, 112
    components of, 112–113
    environmental stewardship and, 148
    hybrid vehicles and, 113, 462
    hydrogen vehicles and, 114
    infrastructure for, 112 (photo), 114–115
    life cycle production approach to, 148
    lithium batteries and, 24, 114
    tax incentives for, 114–115
    Tesla Motors prototypes of, 19
    zero-emissions goal of, 113, 461, 462
    See also **Plug-in hybrid**
Electrochemical energy storage systems, 134
Electrolysis hydrogen production, 6–7
Electrolyte (ionic conductor) of
    fuel cells, 183
Electromechanical induction meter, 287

El Salvador, 200 (table)
Elwha Dam, Columbia River, 87
Embedded energy, 115
Embodied energy, 115–117
    applications of, 116–117
    boundaries controversy regarding, 116
    direct *vs.* indirect energy and, 115
    embodied carbon and, 78–79
    food consumption and, 117
    household energy requirements and,
        116–117
    hybrid analysis of, 116
    input-output models of, 116
    international implications of, 117
    methodologies for calculation of, 115–116
    process analysis model of, 116
Emerald People's Utility District,
        Oregon, 277
Emerson, Ralph Waldo, 147
Emission inventory, 118–120
    business efficiencies and, 119
    carbon trading use of, 119
    compliance and transparency issues of,
        119–120
    "criteria" air pollutants and, 118
    definitions regarding, 118
    Department of Energy fuel use data
        and, 118
    Emission Tracking System/Continuous
        Emissions Monitoring data and, 118
    *Massachusetts et al. v. Environmental
        Protection Agency et al.* and, 118, 123
    Maximum Achievable Control Technology
        program and, 118
    National Emission Inventory (NEI) database
        of EPA and, 118
    pollutant release and transfer registers
        and, 118
    regional approaches to, 119
    Toxic Release Inventory database and, 118
    validation of data and, 119
    See also **Emissions trading**
Emissions trading, 120–123
    Climate Exchange PLC example of,
        53, 62, 122
    command and control regulatory regimes
        and, 121
    market or economic approach to, 121
    *Massachusetts et al. v. Environmental
        Protection Agency et al.* and, 123
    net results for emission rights holder
        and, 121

    offset strategy and, 121
    phasing-out of leaded fuel example
        of, 122
    pipelines and, 356
    property right to emissions and,
        120–121, 123
    reduction responsibilities and, 121
    renewable energy promoted by, 168
    results of, 122–123
    trading system design issues and,
        121–122
    See also **Carbon emissions factors** (CEF);
        **Climate neutrality; Green energy
        certification schemes; Kyoto Protocol**
Energy
    Earth's core heat and sunlight sources of,
        130–131
    energy-efficiency incentives and standards
        and, 144
    environmental stewardship and,
        147–148
    feed-in tariffs and, 166
    kinetic *vs.* potential, 106
    See also **Alternative energy; Electricity;
        Energy audit; Energy payback; Energy
        policy; Energy storage; Federal Energy
        Regulatory Commission** (FERC);
        Green pricing
Energy audit, 124–126
    elements of, 124
    household audit, 124
    industrial firm audit, 124–126
    See also **Appliance standards; Emission
        inventory; Federal Energy Regulatory
        Commission** (FERC); **Metering**
Energy Improvement and Extension Act
        (2008), 243
Energy Independence and Security Act
        (2007), 34, 44
Energy Information Administration, 17, 68,
        101, 105, 241, 243
Energy payback, 126–129
    cannibalistic energy formula and, 128
    embodied energy and, 117
    formula to calculate, 126
    life cycle analysis used in, 127
    net energy gain term and, 126
    net negative effect on GHG emissions and,
        127–128
    policy making use of, 127
Energy Performance Certificates (U.K),
        236, 237

Energy policy, 129–133
alternative energy investments and, 131
biofuels alternative fuel policies and, 132
in California, 46–48
cap-and-trade approach to, 57–58, 59–60,
84, 133, 144, 207, 216, 269, 270
characteristics of, 129
electricity infrastructure and, 108–109
energy payback calculation and, 127
forecasting and, 175–177
importance of, 129–130
new policies, 131–133
nuclear power issues and, 131–132
political factors and, 130
private *vs.* public policies and, 129
production, distribution, and consumption
issues and, 130–131
renewable power stations and, 108–109
solar power and, 132
taxing energy consumption and, 131
traditional policies, 130–131
of U.S., 132–133
See also **Appliance standards; Biodiesel;
Carbon tax; Department of Energy,
U.S. (DOE); Environmental
stewardship; International Renewable
Energy Agency (IRENA);** *specific
legislation*
Energy Policy Act (1992), 236
Energy Policy Act (2005), 10, 34, 84, 229, 243
Energy Policy and Conservation Act
(EPCA), 43
Energy Rated Homes of America, 236–237
Energy Security Act (1980), 100
Energy Star data, 93, 110
Energy storage, 133–137
Buckminsterfullerene carbon molecule
and, 135
Buckypaper material and, 135
chemical batteries, 135–136
closed *vs.* open storage systems and,
134–135
cogeneration designs and, 136
energy grid evolution and, 136
energy harvester storage systems, 134
environmental measures regarding,
144–145
generators role in, 134
hydrogen energy storage system and,
134 (photo), 136
hydrogen fuel cells, 136
inefficiencies in, 136

mechanical energy storage systems, 135
natural *vs.* human design energy storage
examples, 133–134
pumped hydroelectric storage (PHS) plants
and, 135
renewable energy, stability of energy
supplies and, 135
stability issues of, 135
supply, demand, and technology parameters
of, 134
thermal energy storage systems, 134–135
See also **Photovoltaics (PV)**
Enhanced geothermal applications, 198
Eni S.p.A. oil company, Italy, 329
Entropy, 137–140
agroecosystems and, 138
definitions regarding, 137
entropy pump formula and, 137
natural ecosystems and, 137
"sustainability" of agroecosystems measure
and, 137
technosphere disequilibrium and, 138
thermodynamic equilibrium process
and, 137
See also **Exergy**
Environmental impact statements
(EIS), 165
**Environmentally preferable purchasing (EPP),
140–142**
benefits of, 140
definition of, 140
elements of, 140, 141
EPA EPP program of, 141–142
internal organizational guidelines of,
140–141
recycling and, 379–382
U.S. government spending data and, 141
Environmental measures, 142–145
affordability, availability, and acceptability
elements of, 142
in California, 46–48
Exxon Valdez oil spill and, 160–162
FERC environmental measures effectiveness
database and, 165
green buildings and, 143 (photo)
of nonpoint source water pollution,
305–307
policies and technologies of, 143–144
profitability of energy production issue
and, 143
World Energy Council (WEC) objectives
regarding, 142–145

Environmental Protection Agency (EPA), 43
    California emission standards waiver
        and, 44
    Emission Factor and Inventory
        Group of, 118
    Emission Trading Program of, 121, 123
    Environmentally Preferable Purchasing
        (EPP) program of, 141
    environmental stewardship
        definition of, 146
    green power as renewable energy and,
        217–218
    landfill gas energy projects data of, 276
    Landfill Methane Outreach Program
        of, 277
    life cycle analysis tools of, 285
    *Massachusetts et al. v. Environmental
        Protection Agency et al.* and,
        118, 123
    maximum achievable control technology
        and, 440
    mountaintop removal opposed by, 295
    National Emission Inventory (NEI) database
        of, 118
    nonpoint source water pollution
        data of, 305
    sewer pipelines data of, 355
    wood energy evaluated by, 450
Environmental Resources Trust, 212
Environmental stewardship, 145–149
    in California, 46–48
    care and maintenance themes of, 146
    central idea of, 145–146
    conservation and sustainability concepts
        and, 146
    conservation movement *vs.,* 147
    ecological "systems view" of nature
        and, 146
    economic factors related to, 147
    Exxon Valdez oil spill and, 160–162
    global energy issues and, 147–148
    green energy choices and, 146
    land ethics and, 146
    life cycle of product approach to,
        147, 148, 284–285
    mountaintop removal coal mining technique
        and, 294–296
    multiscale decision making and, 146
    origins of, 146–147
    ownership claims motivations
        approach to, 147
    preservationism *vs.,* 146–147

A Sand County Almanac (Leopold) and, 146
    spiritual motivations for, 147
    See also **Best management practices;
        Environmentally preferable purchasing
        (EPP); Environmental Protection
        Agency (EPA); Life cycle analysis**
Enviro Watts program of East Kentucky
    Power Cooperative, 277
EPP. *See* **Environmentally preferable
    purchasing (EPP)**
Equatorial Guinea, 169
Equilibrium potential, of fuel cells,
    182–183
Estonia, 334
Ethane natural gas component, 298, 308
Ethanol
    biomass energy production of, 39, 40, 219
    carbon source from, 64
    cellulose biomass sources of, 6
    ethyl alcohol ethanol fuel source, 20
    flex fuels and, 150, 171
    gasoline *vs.,* 194–195
    hydrogen extraction and, 248–249
    net energy value of, 33, 151–152,
        195–196, 240
    sugarcane ethanol, 6, 20
    syngas production and, 440
    volatile organic compounds (VOCs)
        production from, 436
    See also **Ethanol, corn; Ethanol, sugarcane;
        Flex fuel vehicles (FFVs)**
Ethanol, corn, 149–153
    carbon-based feed stock to produce, 1–2
    combustion engine compatibility with,
        149–150
    corn for food and feed issue and,
        152, 156, 172
    dry mill *vs.* wet mill process and, 195
    embodied energy in, 117
    ethanol-gas blend and, 5, 20
    gasoline substitutes need and,
        149–150, 171
    life cycle approach to production of,
        149, 150–152
    net energy benefit measures and,
        151–152, 195–196
    political factors regarding, 152
    popularity of, 152, 171
    production inefficiencies of, 5
    production process of, 150–151
    sugarcane ethanol *vs.,* 155–156
    sun and carbon roles in, 1

tetraethyl lead fuel and, 150
unsustainability of maize agroecosystem
    and, 139
viability data regarding, 150
See also **Gasohol**
Ethanol, sugarcane, 153–157
    as alternative fuel, 6, 20, 171
    biofuels, biomass sources of,
        153–154, 386, 387
    carbon sequestration process and, 154
    embodied energy in, 117
    environmental issues regarding, 172
    gasoline *vs.*, 155, 194–195
    production process of, 154–155
    sources issues and, 155–156
    sugarbeets and sweet sorghum and,
        154, 156
    viability of, 156
    world production of, 155
    yeast fermentation element of,
        153 (photo), 156
    See also **Gasohol**
Ethene, 324
Ethyl alcohol ethanol fuel source, 20
EU. *See* European Union (EU)
Europe
    acid rain in, 412
    appliance standards in, 10
    bicycle transportation in, 30, 31
    biodiesel produced in, 6, 33–34
    Climate Exchange PLC in, 62
    coal reserves in, 80
    Convention on Long-Range Transboundary
        Air Pollution and, 412
    ecobalance analysis in, 283
    electric tram public transportation in,
        368 (photo)
    electric trolley in, 462
    electric vehicles in, 7
    European Atomic Energy Community, 318
    European Commission Joint Research
        Center, 285
    European Environment Agency, 166
    European Monitoring and Evaluating
        Programme in, 412
    feed-in tariffs in, 166, 167, 211
    flue gas desulfurization in, 412
    forest expansion in, 56
    gas flaring in, 169
    oil consumption decline in, 4
    recycling in, 380
    REVA electric car used in, 462

solar power use in, 223
supermarket purchasing policies, food miles
    and, 174
Telsa electric sports car used in, 462
total primary energy supply (TPES)
    used by, 427
tram public transportation in, 369
weather contracts based on heating degree
    days in, 232
wind power in, 8, 108, 222, 385
See also European Union (EU); **International
    Renewable Energy Agency** (IRENA)
European Union (EU)
    Caspian Sea oil, Russia, 69
    emissions inventories of, 119
    emissions reductions in, Berlin Mandate
        and, 26
    emissions trading in, 168, 207, 216
    Energy Community Treaty of, 263
    EUGENE green energy standard, 212
    EU Reform Treaty (2007), 264–265
    European Union Emissions Trading Scheme,
        62, 271
    flex fuel vehicles in, 171
    green energy certificates in, 212–213
    Greenhouse Gas Emissions Trading System
        of, 58, 59, 60
    Green Paper on energy sources in, 325–326
    internal energy market of, 263
    International Renewable Energy Agency
        (IRENA), 266
    Kyoto Protocol and, 270
    oil fossil fuel value in, 325
    See also Europe; **International Renewable
        Energy Agency** (IRENA)
European Wind Energy Association, 108
EUROSOLAR, 266
Eutrophication of water bodies, 305
Evacuated tube solar collector, 409
Exclusive economic zone (EEZ), 322
Exergy, 157–160
    definition of, 157–158, 159
    entropy and, 157
    exergy analysis formulas and, 158
    exergy cost of production process
        and, 159
    exergy of a mineral deposit and, 159
    exergy replacement cost and, 159
    quality deterioration and, 157
    resource accounting and, 158–149
    thermodynamics laws and, 157, 158–159,
        239–240

External combustion engine, 91
Exxon Mobil, 58, 160–162, 329–330, 339
Exxon Valdez, **160–162**
  the accident, 161
  Alaskan oil drilling issue and, 16
  litigation regarding, 161–162
  Prince William Sound, Alaska, 161 (photo)

Factory-built insulated wall systems, 256
Fagerman, Jan, 251
Farm Security and Rural Investment Act
    ("Farm Bill," 2002), 34
Federal Energy Regulatory Commission
    (FERC), 163–166
  authority limits of, 164–165
  Department of Energy and, 163
  electricity sales regulated by, 164
  energy audits managed by, 164
  energy markets monitored by, 164
  environmental impact statements
      (EIS) and, 165
  environmental measures effectiveness
      database and, 165
  environmental protection efforts of, 164
  funding of, 163
  goals and responsibilities of, 163–164
  natural gas pipelines and electrical
      transmission lines regulated by, 163
  structure of, 163
  See also **Department of Energy,**
      U.S. (DOE)
Federal Highway Administration, 151
Federal Power Act, 163
Feed-in tariff, 166–168
  carbon taxing *vs.,* 168
  emissions trading *vs.,* 168
  green energy certification *vs.,* 211
  policy structure incentives and, 166
  renewable energy certificates *vs.,*
      167–168
  renewable energy sources and, 167
FERC. *See* **Federal Energy Regulatory**
    Commission (FERC)
Fermentation, 36
Fermi, Enrico, 311
FFVs. *See* **Flex fuel vehicles** (FFVs)
Final Protocol for In Situ Bioremediation of
    Chlorinated Solvents Using Edible Oil
    (USAF), 326–327
Finland, 312
Fischer-Tropsch process, 192, 193
Fish & Wildlife Service, U.S., 14, 16

Fission reactions, 131–132, 187,
    221–222, 458
  See also **Nuclear power**
Flaring, 168–170
  environment affected by, 169, 344
  human health affected by, 169
  methane emissions and, 168, 169, 215, 275
  purposes of, 168–169
Flash steam geothermal applications,
    198, 199, 385
Flat plate solar collector, 409
Flex fuel vehicles (FFVs), **170–172**
  benefits and drawbacks of, 171–172
  CAFE standards and, 44
  definition of, 170
  examples of, 171
  food *vs.* fuel debate regarding, 172
  gas-electric hybrids, 20, 462
  Model T Ford and, 150, 171, 357
  natural gas vehicles and, 301
  oil embargo crisis and, 92
  OPEC affected by, 341
  petrol-ethanol fuel blend and, 20, 462
  smart electric meters and, 288
  zinc-air fuel cell vehicles and, 185
  See also **Electric vehicle**
Florida
  Florida Power and Light green power
      program in, 226
  renewable energy portfolio of, 391 (table)
  solar feed-in tariff in, 166
Flowing how water geothermal
    applications, 197
Flue gas desulfurization technology, 412
Fluidized bed combustion coal boilers, 83, 363
Food miles, 172–175
  contextualization of, 174
  distributive and social justice issues and, 174
  environmental impacts of food issues and,
      172–173
  food production costs and, 173
  "local trap" concept and, 174
  policy solutions regarding, 174
  processed foods issue in, 174
  supermarket hub-and-spoke distribution
      systems and, 173
  tomatoes example of, 174
  transportation factors in, 173
  "Weighted Average Source Distance"
      calculation method of, 173
Food security, 33, 34, 152, 156, 172
Ford, Gerald, 98

Ford, Henry, 150, 171, 357
Ford Motor Company, 462
Forecasting, 175–177
conditional forecasts and, 175
decision making and planning and, 175
definition of, 175
discontinuous path element in, 176–177
energy examples of, 177
error measures in, 176
future values of a time-series and, 175
incremental path element in, 176
motivations for, 175
one-off events and, 175
technology's future forecasting and, 176
timescales element of, 175
See also **Heating degree day**
Forests
carbon cycle and, 54–56
deforestation and, 54
Forestry Clean Development Mechanism
    projects and, 56–57
Good Practices Guidance for Land-Use
    Change and Forestry project and, 258
tropical deforestation and, 55–56
wood energy and, 449, 450
Fossil fuels, 177–182
ash issues and, 181
biomass energy *vs.*, 39–41
carbon and hydrogen atomic bonds
    and, 180
carbon element in, 177–178
climate neutrality and, 77
coal, 178–179
electrical energy generation from,
    106, 109
energy audits and, 125–126
energy crisis and, 4
energy policies regarding, 131
environmental exploitation results of, 148
ethanol (corn) production using, 151
flaring and, 168–170
formation process of, 177
fuel efficiency *vs.* fuel economy and, 358
GHGs and global warming and, 181
heating values and, 180
Hubbert's Peak and, 238–240
hydrocarbon composition of, 308
hydrogen and carbon ratio differences in,
    177–178
hydrogen tetraoxosulfate acid (acid mine
    drainage) and, 181
natural gas, 179

negative externalities associated with, 58–59
nonrenewable resources as, 177
oil and natural gas creation process and, 4
oil consumption decline and, 4
oil consumption to oil availability
    ratio and, 4
oil sand extraction methods and, 5
oil shales and liquefied coal, 5
ozone layer depletion and, 181
petroleum, 179–181, 308
reserves *vs.* resources and, 182
solar energy factor in, 405
unconventional extraction methods of, 5
World Energy Council (WEC) perspectives
    on, 141–145, 153
See also **Carbon emissions factors (CEF);
    Carbon footprint and neutrality;
    Carbon tax; Carbon trading and
    offsetting; Caspian Sea; Climate change;
    Global warming; Greenhouse gases
    (GHG); Landfill methane; Metric
    tons of carbon equivalent (MTCE);
    Nonrenewable energy resources; Oil;**
    specific fuel source
Fourier, Jean Baptiste, 202
Framatome organization, 312
Framework Convention on Climate
    Change, 455
France
appliance standards in, 10
AREVA multinational industrial
    conglomerate and, 312
Bibliothèque Nationale de France,
    Paris, 96
European pressurized water nuclear
    reactor in, 312
geothermal energy produced in, 200 (table)
International Thermonuclear Experimental
    Reactor (ITER) in, 188
Kyoto Protocol and, 273
metro systems in, 370
nuclear fission policies in, 131
nuclear power reactors in, 222, 311–312,
    314, 365
nuclear weapons developed in, 317, 430
Royal Dutch Shell oil company and,
    328, 339, 344–345
total primary energy supply (TPES) measure
    of, 427
uranium dependence of, 313–314
Francis reaction microhydro turbine, 293
Freon, 71

Fuel cells, 182–186
  applications of, 185–186
  batteries *vs.*, 182–183
  components of, 183
  direct carbon conversion fuel cell, 185, 252
  electricity conversion from chemical energy
    by, 182
  "equilibrium potential" element of,
    182–183
  how fuel cells work and, 182–183
  hydrogen fuel cells, 184, 222, 248, 249,
    250, 337
  internal combustion engines *vs.*, 183–184
  microbial fuel cell, 185
  natural gas and, 300
  on-site renewable energy and, 337
  portable fuel cells, 186
  reservoir of rechargeable electrolytes
    and, 184
  stationary power from, 186
  types of, 184
  volatile organic compounds (VOCs)
    and, 436
  zinc-air fuel cells (ZAFC), 185
  See also **Electric vehicle; Hydrogen**
Fusion, 187–189
  cold fusion controversy and, 188
  condensed matter nuclear science and, 188
  controlled fusion and, 187
  Coulomb barrier in, 187
  demonstration power plant (DEMO) device
    and, 188
  fission *vs.*, 187
  as green power source, 224
  inertial confinement fusion and, 188
  International Thermonuclear Experimental
    Reactor (ITER) and, 188
  magnetic confinement approach to, 187
  nuclear reactions and, 187
  proton-proton chain reaction in, 187
  technological challenges of, 187
  "tokamaks" fusion reactors and, 187–188
  See also **Nuclear power**

Gabon
  gas flaring in, 169
  OPEC membership of, 338
Galloway, J. N., 303
**Garbage in, garbage out recycling
    concept, 439**
Gas cap of natural gas, 29, 297
Gas-electric hybrids, 7

Gasification, 191–194
  applications of, 192
  of biomass sources, 193, 386
  in electricity generation, 192–193, 363
  Fischer-Tropsch fuels and, 192, 193
  hydrogen source through, 248
  integrated gasification combined cycle
    (IGCC) power plants and, 192, 363
  liquid fuels from, 193
  polluting compounds removal by, 191
  process description and, 191
  pyrolysis element in, 191
  syngas product of, 191, 192, 195, 363
  See also **Gasohol**
Gasohol, 194–196
  gasoline *vs.*, 194–195
  lignocellulosic feedstocks and, 195, 196,
    326, 386–387
  new technologies of, 195–196
  See also **Ethanol, corn; Ethanol, sugarcane;
    Gasification**
Gasoline
  ethanol compared to, 149–150, 155, 171,
    194–195
  as fossil fuel, 308
  gasoline-electric hybrid vehicles and, 7
  gasoline tax issues and, 360
  leaded gasoline and, 122
  liquid biofuels gasoline substitutes, 5
  volatile organic compound and, 435–436
Gate cap on carbon-emitting fuels, 78
Gazprom, 329, 345–346
General Electric, 92
General Motors, 71
  ethanol research of, 171
  hydrogen energy storage system and,
    134 (photo)
**Generation domain, of electric
    power system, 227**
**Generation IV International Forum
    (GIF), 313**
Geo-engineering technologies, 78
Georgia (Russia), 340
Geothermal energy, 196–201
  advantages *vs.* disadvantages of, 199–200
  "baseload" running requirement of, 199
  binary cycle, 198, 199, 221, 385
  carbon negative feature of, 221
  costs associated with, 221, 243
  dry steam resources, 197–198,
    199, 220, 385
  electrical energy infrastructure and, 108

as energy innovation, 252
energy policies regarding, 130
energy storage systems and, 135
environmental impacts of, 387
flash steam, 198, 199, 385
flowing hot water, 197
geologic settings and, 197
Geothermal Energy Act and, 100
Geysers geothermal area, California,
    197, 198, 199, 223, 385
as green power renewable energy source,
    207, 217, 220–221, 366, 384–385, 388
ground source heat pumps (GSHP),
    198–199, 384–385
heat pumps, 3, 384–385
hot dry rock, 198, 200–201, 385
hydropower use with, 199
IPCC studies of, 260
Larderello power generator, Italy,
    197–198, 220
Ocean Thermal Energy Conservation Act
    and, 100
Public Utilities Regulatory Policies Act
    (1978) and, 197
site-specific resource factor in, 199
space requirement of, 199
volcanic and magmatic activity and,
    196–197, 260
worldwide power production from,
    200–201, 200 (table)
See also Gasification
Geothermal Energy Act, 100
Germany
    agroecosystems, entropy in, 138–139
    Bauhaus in, 96
    bicycle transportation in, 31
    "clean coal" power plant in, 84
    coal-fired electrical power plants in, 308
    compressed natural gas buses in, 462
    feed-in tariffs policy in, 167
    geothermal energy produced in,
        200 (table)
    International Renewable Energy Agency
        (IRENA) and, 266
    Kyoto Protocol and, 273
    oil shale–fired power plants in, 334
    photovoltaics green power in, 222
    recycling in, 380
    renewable energy-fired power plants
        in, 366
    REVA electric car used in, 462
    syngas used in, 192

total primary energy supply (TPES) measure
    of, 427
tram public transportation in, 369
uranium mining remediation in, 432
wind power use in, 445
GHG. See Greenhouse gases (GHG)
Gibbs Free energy, 158
Gilbreth, Frank and Lillian, 28
Glines Canyon Dam, 87
Global Nuclear Energy Partnership, 100, 318
Global warming, 201–208
    albedo factor in, 204, 207
    carbon dioxide and radiative forcing data
        and, 203, 204 (fig.)
    consequences of, 204–207
    disease vector changes and, 206
    Earth's cyclic cooling and warming phases
        and, 202
    ecological habitat shifts and, 206
    feedback mechanism limitations and, 207
    geo-engineering technologies and, 207
    GHGs, fossil fuels and, 180, 181, 201–208
    Global Warming Solutions Act (2006,
        California) and, 59
    mitigation of, 207–208
    oceanic thermohaline circulation and,
        205, 206 (fig.)
    polar and glacial ice core data and,
        202, 202 (photo)
    population relocation and, 206
    precipitation pattern changes and, 206
    radiative forcing and, 203–204, 205 (fig.),
        378–379
    the science regarding, 201–204
    sea level changes and, 205
    See also Berlin Mandate; Carbon emissions
        factors (CEF); Carbon footprint and
        neutrality; Chlorofluorocarbons
        (CFCs); Climate change; Greenhouse
        gases (GHG); Kyoto Protocol
Global Warming Solutions Act
    (2006, California), 59
Good practices, 29
Gore, Al, 75
Grand Coulee Dam, 86 (photo)
Greece, 167, 298
Green banking, 208–211
    checks from recycled paper, 209
    eco-banking practices, 209–210
    electronic banking cost savings and, 209
    examples of, 208–209
    green credit and debit cards, 210

greenwashing practices and, 209
paperless transactions, 209
security issues and, 209
Green buildings, 143 (photo), 144
Green-collar employment concepts, 48
Green–e certificate program (California),
212, 226
Green energy certification schemes, 211–213
attributes of, 211
daylighting tools and, 95–97
Dominion Virginia Power certificates,
110–111
double counting issue in, 212
feed-in tariffs *vs.*, 211
"greeness" of energy element of, 211
hydroelectric power and, 243
mandatory market *vs.* voluntary market
types of, 211–212, 226
market compatibility issues and, 213
renewable energy certificates (RECs) term
and, 211
See also **Emissions trading; Green pricing**
Green For All movement, 48
Greenhouse gases (GHG), 213–217
anthropogenic GHGs, global warming and,
109–110, 213, 215–216
automobile carbon dioxide and, 19
background regarding, 213
biodiesel production and, 33
biomass energy production and, 39–41
CAFE standards and, 44–45
carbon equivalency and, 289–290
compact fluorescent bulbs and, 91, 92
"criteria" air pollutants and, 118
of dry steam geothermal energy
systems, 199
end-of-pipe process modifications
and, 215
Energy Star ratings and, 110
food miles and, 172–173
"forcing" concept and, 214
Fourth Assessment Report of IPCC
and, 216
gas flaring and, 169
global policy response to, 216–217
"greenhouse effect" explained and, 73
health and property losses from, 118
increase in, 61–62
IPCC measures of, 259
Kyoto Protocol and, 216
Low-Carbon Fuel Standard (CLFS,
California) and, 34

*Massachusetts et al. v. Environmental
Protection Agency et al.* and, 118, 123
pollutants definition and, 118
radiative forcing and, 378–379
reductions of, 390
renewable power stations and, 108
responding to, 215–216
steam production of hydrogen and, 6
See also **Berlin Mandate; Carbon
emissions factors (CEF); Carbon
footprint and neutrality; Carbon tax;
Chlorofluorocarbons (CFCs); Climate
change; Climate neutrality; Coal,
clean technology; Emission inventory;
Emissions trading; Energy payback;
Global warming; Kyoto Protocol;
Landfill methane; Methane; Metric tons
of carbon equivalent (MTCE); Nitrogen
oxides; Sulfur oxides (SO$_x$);** *specific
substance*
Greenland, 202, 205
Green power, 217–225
anaerobic digestion, 219, 386
biomass, 217, 218–219, 223
ethanol fuel, 223
future sources of, 224
geothermal power, 217, 220–221, 223
hydrogen energy, 223
hydropower, 221, 223
nuclear power, 217, 221–222
OPEC affected by, 341
potential of, 222–223
renewable energy term and, 217–218
in rural areas, 223
solar power, 217, 220, 223
sources of, 217–218
switchgrass, 218 (photo), 387
tidal power, 221, 223
uranium, 224
wind power, 217, 219–220, 222, 223
See also **Green energy certification schemes;
International Renewable Energy
Agency (IRENA); Nuclear power;
On-site renewable energy generation;
Photovoltaics (PV); Renewable energies;**
specific power source
Green pricing, 225–227
carbon neutrality offsets and, 226
customer costs of, 225–226
fixed-price contracts of, 225–226
green tags strategy and, 226
popularity of, 226

renewable energy credits (RECs) and, 225, 226–227
third-party certification and, 226
See also **Internal energy market**
Green tags, 226
Greenwashing, 209
Grid-connected system, 227–229
distributed or dispersed generation terms and, 228
electric power system structure and, 227–228
environmental issues of, 228
features of, 227
national regulatory frameworks and, 228
net metering issue and, 228–229
technical and economic issues regarding, 228–229
technical components of, 228
See also **Smart grid**
Ground source heat pumps (GSHP) geothermal applications, 198–199, 384–385
Guatemala, 200 (table)
Gulf Stream, 205, 206 (fig.)

Haber process of fertilizer production, 301
Halladay Standard Company, Batavia, Illinois, 444 (photo)
Hammer, Edward E., 92
Hampton, Brett, 218 (photo)
Hanford nuclear site, 86
Hansen, James, 204, 214, 216
Hart, Gary, 423
Hart, William, 298
Hawaii
electric vehicle infrastructure in, 114
Hilo geothermal power plant in, 198
renewable energy portfolio of, 391 (table)
Hazelwood, Joseph, 161, 162
Heating degree day, 231–233
base temperatures of, 231
calculation process of, 231
cooling degree day *vs.*, 232
Cumulative Average Temperature and, 232
energy use and atmospheric conditions index and, 231
heating contracts based on, 232
latitude and altitude factors in, 232–233

potential energy demand predictions and, 233
variables in, 231–232
weather derivatives financial instrument and, 232
Heat island effect, 233–235
absorbed solar radiation and, 233–234
built environments and, 233–234
daily maximum and minimum temperatures factor in, 234
energy consumption affected by, 235
factors in, 233
longwave thermal infrared energy and, 234
mitigation of, 235
population and temperature relationship formula and, 233
regional weather patterns factor in, 234
seasons and meteorological condition factors in, 234–235
sensible heat sources and, 234
surface *vs.* atmospheric heat islands and, 234
urban *vs.* rural temperatures and, 233
Heat pumps
chlorofluorocarbons used in, 71
"dumping" heat cooling function of, 199
geoexchange technology of, 3
as green energy application, 145
ground source heat pumps (GSHP) geothermal application and, 198–199, 384–385
Heavy rail and metro public transportation, 370
Heliostat concentrator arrays, 401, 410
Helium natural gas component of, 298
Helliostat or power tower solar thermal facility, 410
HEMI engine, 90–91
Hendryx, Michael, 295
HERS. *See* **Home Energy Rating Systems (HERS)**
Herzegovina, 263
Hetch Hetchy Valley, California, 47
Hiroshima nuclear bomb, 311, 317
Hobhouse, Hermione, 95
Holland, 115
Home Energy Rating Systems (HERS), 235–238
energy mortgages and, 236, 237
Energy Rated Homes of America and, 236–237

energy-use index of "American Standard Building" and, 235–236
history of, 236–237
Home Energy Rating report and, 236
National Shelter Industry Energy Advisory Council and, 236
projected *vs.* confirmed rating and, 236
rating and assessment and, 236
standards used in, 236
U.K. Energy Performance Certificates Scheme and, 236, 237
See also **Insulation**
Honda, 462
Hoover Dam, 241 (photo), 292, 420
Horizontal axis wind turbines, 448
Horizontal nuclear proliferation, 316
Horse manure, biomass energy from, 36
Hot dry rock geothermal applications, 198, 200–201, 385
Hot fractured rock geothermal applications, 198
Household energy audit, 124
Hubbert, Marion King, 238
Hubbert's peak, 238–241
alternative energy sources development and, 239, 259, 331
economic and geopolitical implications of, 238, 239–240
energy innovation and, 252
explanation of, 238
natural gas production and, 238–239
net energy concept and, 239–240
oil production and, 238, 239, 357
oil shale and oil sands and, 240
renewable energy sources and, 240
Hummer, 171
Hungary, 139
Kyoto Protocol and, 273
trams in Budapest, 368 (photo)
Hurricane Ivan, 322
Hurricane Katrina, 101, 322
Hurricane Rita, 322
HVAC (heating, venting, and air conditioning) ductwork, 257
Hybrid analysis of embodied energy, 116
Hybrid vehicles, 113, 462
Hydrocarbons, 177, 179, 180, 247–248, 249, 308, 324, 326
Hydroelectric power, 241–245
of Columbia River, 85–87
costs associated with, 221, 242–243, 383
economic advantages of, 242–243, 383

efficiency of, 242, 244
electrical energy infrastructure and, 108
as energy innovation, 252
energy policies regarding, 130, 131
energy production process of, 2, 7, 241–242, 383
environmental and social advantages of, 243–244
FERC regulation of, 164
fish affected by, 244
flood control, irrigation, and recreation benefits of, 244
geothermal energy use with, 199
as green power renewable energy source, 83, 217, 221, 223, 244, 366, 383, 420
grid power storage of, 242, 244
habitat displacements and, 244, 383
Hoover Dam example of, 241 (photo), 292, 420
IPCC studies of, 260
Kyoto Protocol and, 272
life span of dams and, 131, 388
methane release and, 245
potential capacity data of, 243
pumped hydroelectric storage (PHS) plants and, 135
river quality maintenance and, 244–245
social and environmental concerns regarding, 221, 244–245, 260, 383
solar energy factor in, 405
Three Gorges Dam, 410–421
types of, 242
water as clean fuel source and, 244
wind turbines and, 220
See also **Microhydro power; Three Gorges Dam**
Hydrofluorocarbons, 52, 78, 269, 289
radiative forcing and, 378
Hydrogen, 245–250
as alternative-fueled vehicle, 462
biomass extraction source of, 248–249
California hydrogen refueling stations and, 249
carbon sequestration and, 248, 249
clean combustion of, 249
coal gasification and, 248, 363
concentrated solar power systems and, 410
containment and transport methods of, 247, 252
efficiency issues of, 250
electrolysis production of, 6–7
energy density of, 247

future trends regarding, 249–250, 463
gasification process and, 191, 363
hydrogen embrittlement and, 246
hydrogen energy storage system and, 134 (photo), 136
hydrogen fuel cells and, 184, 222, 246, 246 (photo), 248, 249, 250, 462–463
"Hydrogen on Demand™" system and, 184, 248–249
hydrolysis source of, 248
moderate *vs.* high temperature fuel cell types and, 184
in natural gas, 239, 249, 298
net energy value of, 239–240
pipeline transmission of, 356
properties of, 246–247, 247 (table)
reforming of hydrocarbons source of, 247–248, 248 (fig.), 249
renewable energy technology and, 100, 223
sources of, 247–249
steam production of, 6
storage methods and, 184
as syngas component, 249
U.S. space program and, 246
uses of, 249
as vehicle fuel source, 21, 91, 114
volatility of, 7
Hydrogen chloride, 178, 439
Hydrogen cyanide, 178
Hydrogen sulfide, 169, 178, 179, 181, 298, 300
Hydrologic Simulation Program-Fortran (HSPF), 307

Iceland
geothermal energy produced in, 196 (photo), 197, 199, 200 (table), 221, 223, 427
Krafla volcano region of, 196 (photo)
Kyoto Protocol and, 269, 270
Idaho, 86, 288
Illinois
commuter rail transportation in, 370
Coskata ethanol technology and, 171
Crown Hall, Chicago, 96
renewable energy portfolio of, 391 (table)
smart meters used in, 288
windmills, Halladay Standard Company, Batavia, 444 (photo)
Impoundment hydroelectric power, 242, 244–245

Impulse microhydro power turbine, 293
Incremental innovation, 254
India
anaerobic digestion system used in, 219
CO_2 emissions produced in, 50
compressed natural gas buses in, 462
cow manure burned in, 37
energy market in, 325
feed-in tariff policy in, 167
hydroelectric power produced in, 223
International Renewable Energy Agency (IRENA) and, 266
Kyoto Protocol and, 207, 216, 270
natural gas in, 298
natural gas vehicles in, 301
nuclear power reactors in, 365, 366
nuclear weapons in, 317
oil consumption increase in, 4, 341
recycling in, 381
renewable energy-fired power plants in, 366
REVA electric car in, 462
solar power use in, 223
Indirect energy, 115
Indonesia
gas flaring in, 169
geothermal energy produced in, 200 (table)
International Renewable Energy Agency (IRENA) and, 266
OPEC membership of, 339
Royal Dutch oil company in, 328
Industrial firm energy audit, 124–126
Industrial Revolution
carbon dioxide levels and, 52
coal as energy source and, 80, 83
GHGs and global warming since, 72, 203–204, 214
off shore drilling and, 66
oil energy source of, 238
risk analysis and, 393
steam engines and, 130
Inertial confinement fusion, 188
Innovation, 251–255
classes of, 251
concept of, 251
end-user innovation and, 254
energy centralization *vs.* distribution, 252
in energy distribution, 252
in energy generation, 252
in energy use, 253
fields of, 252
forecasting and, 175–177
incremental *vs.* disruptive innovation, 254

invention *vs.*, 251
linear model of, 253
open innovation and, 254
Peak Resources theory and, 252
sustainable energy systems and, 251
technology adoption lifecycle model of,
    253–254
See also **Best management practices**
Input-output models of embodied energy, 116
In situ leach (ISL) uranium mining, 431
In situ retorting oil shale technique, 334
Insulated concrete forms, 256
Insulating paint additives, 256
Insulation, 255–258
    batt insulation, 255–256
    blown in insulation, 256
    closed-cell spray foam insulation, 256
    factory-built insulated wall systems, 256
    HVAC (heating, venting, and air
        conditioning) ductwork and, 257
    insulated concrete forms, 256
    insulating paint additives, 256
    open-cell spray foam insulation, 256
    proper installation of, 255, 257
    R-value of, 256–257
    special attention areas and, 257
    spray foam insulation, 255 (photo), 256
    structural insulated panels, 256
    vapor barriers and, 255
    wet-spray cellulose insulation, 256
Integrated compact fluorescent bulb, 92
Integrated gasification combined cycle (IGCC)
    power plants, 192, 363
Intergovernmental Negotiating Committee
    for a Framework Convention on Climate
    Change, 74
Intergovernmental Panel on Climate Change
    (IPCC), 258–262, 259
    First Assessment Report of, 74, 269
    Second Assessment Report of, 74–75, 259
    Third Assessment Report of, 75, 205,
        258–259
    Fourth Assessment Report of, 75, 203–204,
        204 (fig.), 205, 216, 261
    Fifth Assessment Report of, 261
    Berlin Mandate and, 25–26
    California's energy consumption measured
        by, 259–260
    carbon emissions factors used by, 49
    carbon footprint issues and, 259
    climate change reports of, 260
    "Connectivity Week" of, 261

constituency of, 258
energy budget for Earth study of, 259
energy economic subsystems and, 259
formation and mandate of, 74
functions of, 258
GHG emissions measured by, 259
global warming potential (GWP) of GHGs
    and, 290, 290 (table)
Good Practices Guidance for Land-Use
    Change and Forestry project of, 258
Industrial Energies Studies report of, 259
Intergovernmental Negotiating Committee
    for a Framework Convention on
    Climate Change and, 74
International Energy Studies Group (IESG)
    of, 258–259
Kyoto Protocol and, 74–75
Modeling Energy and Climate Change
    Features report of, 259
radiative forcing (RF) defined by, 377
Renewable Energy Sources and Climate
    Change Mitigation workshop of, 260
Second World Climate Conference of, 74
solar energy reports of, 260
Special Report on Emissions
    Scenarios of, 258
Special Report on Methodological and
    Technical Issues in Technology
    Transfers of, 258
Special Reports series of, 259
structure of, 74, 258
United Nations Framework Convention on
    Climate Change (UNFCCC) and,
    74–75
Working Groups of, 74, 258
See also **International Renewable Energy**
    Agency (IRENA)
Internal combustion engine
    ethanol used in, 194
    fossil fuels, GHGs, and global warming and,
        90–91, 182, 183–184, 214, 357, 461
    inefficiency of, 186, 358
    landfill methane gas use in, 276
    oil-fired electrical power plants and, 365
    plug-in hybrid and, 356–357
    series *vs.* parallel hybrid vehicles and,
        358–359
    See also **Combustion engine**
Internal energy market, 262–265
    of Canada and U.S., 262–263
    distribution system operators and, 264
    electricity *vs.* gas differentiation and, 263

emissions trading market element of, 263
Energy Community Treaty of EU and, 263
energy consumer protection and, 264
of European Union, 263, 264–265
GHS emission allowance and, 264
Kyoto Protocol and, 264
privatizing the market and, 263–264
retail user and home user concerns and,
    264–265
security of supply issues and, 264
taxation element of, 264
See also Emissions trading
International Atomic Energy Agency (IAEA),
    266, 311–312, 317, 318
International Conference on Financing for
    Development, 413
International Energy Agency (IEA), 218, 222,
    265, 266, 426–427
International Energy and Green Building
    Programs, 278
International Energy Studies Group (IESG), of
    IPCC, 258–259
International Institute for Environment and
    Development, 413
International Organization for Standardization
    14064 registry, 118
International Petroleum Exchange, 340
International Renewable Energy Agency
    (IRENA), 265–267
    creation of, 265–266
    equality of all members of, 267
    goals and organization of, 266–267
    issues addressed by, 266
    structure of, 267
International Thermonuclear Experimental
    Reactor (ITER), 188
International Union for the Conservation of
    Nature, 414, 415
Inupiat peoples, Alaska, 13
Iodine-131, 70
Iowa, 391 (table)
IPCC. See Intergovernmental Panel on Climate
    Change (IPCC)
Iran
    Caspian Sea border country of, 66, 67,
        68–69
    gas flaring in, 169
    National Iranian Oil Company in, 329
    natural gas produced in, 299, 364
    natural gas vehicles in, 301
    nuclear power issues and, 315
    nuclear weapons in, 317, 318

oil supply from, 325
OPEC membership of, 338, 339
uranium resources in, 310
Iraq
    gas flaring in, 169
    oil supply from, 325
    OPEC membership of, 338, 339
Ireland, 54 (photo)
IRENA. See International Renewable Energy
    Agency (IRENA)
Iron pyrite ("fool's gold"), 82
ISO 14000, 14001, 29, 354
Israel
    Arab–Israeli War and, 340
    nuclear weapons in, 317
    oil shale–fired power plants in, 334
Isuzu, 171
Italy
    carbon tax in, 108
    Eni S.p.A. oil company of, 329
    geothermal energy produced in, 200 (table)
    Larderello geothermal power in,
        197–198, 220
    mandatory market green certificates in, 212
    natural gas vehicles in, 301
    oil shale in, 334
    renewable energy certificates in, 168, 390

Jacobs, J., 156
Japan
    coal-fired electrical power plants in, 308
    emissions reductions in, Berlin Mandate
        and, 26
    flue gas desulfurization in, 412
    geothermal energy produced in, 200 (table)
    hydrogen fuel cells in, 186
    metro systems in, 370
    nuclear power reactors in, 222, 365
    photovoltaics (PV) green power in, 222
    See also Kyoto Protocol
Jatropha, as biofuel source, 34, 154, 219
Johannesburg Declaration, 413
Joint Implementation. See Kyoto Protocol
Joliot-Curie, Frédéric, 311
Jonas, Hans, 315

Kahn, A. Q., 318
Kansas, 288, 423
Kaplan reaction microhydro turbine, 293
Kazakhstan
    Caspian Sea border country of, 66, 67
    gas flaring in, 169

petroviolence in, 346
uranium resources in, 313, 366, 431
Keeling, Charles, 73
Kentucky
    coal mining in, 294, 295, 296
    East Kentucky Power Cooperative in, 277
Kenya
    geothermal energy produced in,
        200 (table)
    microhydropower used in, 293
    Ken Saro-Wiwa environmental activism and,
        344–345
    solar power use in, 223
Kern River gas transmission pipeline, 353
Kerogen, 177, 180, 298, 334
Kinetic energy
    kinetic hydroelectric power and, 2, 7, 129,
        135, 241, 242, 313, 383, 425
    of nuclear power, 222
    from potential energy, mechanical energy
        and, 106, 129, 135
    of windmills, wind power, 8, 445
Klaproth, Martin, 430
Kosovo, 263
Kowarski, Lew, 311
Kuwait
    gas flaring in, 169
    OPEC membership of, 338, 340
Kyoto Protocol, 269–273
    Annex B countries responsibilities and,
        270, 271, 272
    Annex I countries (industrialized) and,
        270, 272
    cap-and-trade approach of,
        57–58, 269, 270
    carbon emissions standards and, 216
    carbon equivalency values and,
        78, 289–290
    carbon leakage concept, 117
    carbon tax and, 108–109
    carbon trading and offsetting provisions of,
        53, 61, 62, 123, 207, 270–271, 272
    Certified Emission Reduction (CER) credits
        of, 272
    Clean Development Mechanism (CDM)
        strategy in, 56, 61, 63, 65, 207, 216,
        270, 271–272, 273
    climate change focus of, 72
    Designated Operational Entity (DOE)
        functions and, 272
    emission reduction units (ERUs) and, 272
    "55 percent" clause of, 269

flexibility mechanisms of, 270–271
GHG emissions focus of, 52, 207,
    216, 269, 270
hydropower plants and dam issues
    and, 272
internal energy market and, 264
Joint Implementation mechanisms of,
    26, 61, 63, 270, 272–273
legal provisions of, 270–271
"linking directive" of, 271
successor to, 207, 216, 329
track 1 or track 2 Joint Implementation and,
    272–273
United Nations Framework Convention
    on Climate Change (UNFCCC) and,
    74–75, 207, 216
U.S. rejection of, 123, 269, 270
See also Berlin Mandate; Emissions trading;
    International Renewable Energy
    Agency (IRENA); Metric tons of carbon
    equivalent (MTCE)
Kyrgyzstan, 346

Labbé, Marie-Hèléne, 314
Land ethics, 146
Landfill methane, 275–277
    capped vacuum well of, 275 (photo)
    cogeneration projects and, 276
    electrical and transportation systems usage
        of, 276
    Enviro Watts program, East Kentucky
        Power Cooperative, 277
    Evergreen Renewable Energy Program,
        Wisconsin, 277
    Exelon Power, Pennsylvania, 277
    as GHG, global warming and, 276
    industrial use of, 276
    landfill gas (LFG) and, 275, 298, 440
    Landfill Methane Outreach Program of EPA
        and, 277
    LFG energy projects and, 276
    natural gas pipeline system and,
        276–277
    New London, Texas methane accident
        and, 275
    Short Mountain Methane Power Plant,
        Eugene, Oregon, 277
    technology for extraction of, 276
    U.S. energy grid and, 276
Laos, 219
Larderello geothermal generator, Italy,
    197–198, 220

Latin America
  flue gas desulfurization in, 412
  petroviolence in, 345–346
  state-owned oil company in, 329
  total primary energy supply (TPES)
    used by, 427
Lead, 439
Leaded fuel, 150
Leadership in Energy and Environmental
  Design (LEED)-certified laboratory,
  Albuquerque, New Mexico, 143 (photo)
LEED standards (Leadership in Energy and
  Environmental Design), 277–283
  accreditation and, 281
  energy and atmosphere criteria of, 280
  future trends in, 282–283
  "green" environmentally sustainable
    construction standards and, 277–278
  history of, 278
  indoor environmental quality criteria
    of, 281
  innovation and design process criteria
    of, 281
  limitations of, 282
  materials and resources criteria of, 280
  Phases of, 278, 279 (table)
  point rating system for project certification
    by, 278–281, 279 (table)
  sustainable sites criteria of, 278–279
  U.S. Green Building Council (USGBC)
    and, 278
  water efficiency criteria of, 280
Leontief, Wassily, 116
Leopold, Aldo, 146
Lexus, 462
Libya
  gas flaring in, 169
  nuclear weapons in, 317
  OPEC membership of, 338
Life cycle analysis, 283–286
  applications of, 284–285
  of biodiesel, 33
  "cradle-to-grave" analysis and, 283
  data collection phase of, 283
  ecobalance term and, 283
  energy audits and global modeling studies
    and, 283
  energy payback calculation and, 127
  environmental stewardship and,
    148, 284–285
  of ethanol from corn, 149, 150–152
  life cycle assessment phase of, 283–284

of nuclear power, 432
recycling considerations and, 284
resource and environmental profile analysis
  phrase and, 283
stages in, 283–284
tools used in, 285
See also Carbon footprint and neutrality;
  Embodied energy; Environmentally
  preferable purchasing (EPP)
Lighting. See Compact fluorescent bulb (CFB);
  Daylighting; Electricity
Light rail public transportation, 370
Lignite coal, 5, 50, 79, 84, 130, 178, 362
Lignocellulosic feedstocks, 195, 196, 326,
  386–387
See also Cellulosic ethanol
Linear contrails, 378
Linear model of innovation, 253
Liquified coal, 5
Liquified petroleum gas, 50, 130, 164, 179
Lithium iodide battery, 24
Lithium-ion battery, 23 (photo), 114, 359
Longwave thermal infrared energy, 234
Los Alamos National Laboratory Hot Dry
  Rock geothermal program, 198
Love Canal, Niagara Falls, New York, 104
Lovejoy, Thomas, 75
Low-Carbon Fuel Standard (LCFS,
  California), 34
Lund, John W., 200 (table)
Luxembourg, 427

Macedonia, 263
Magnetic confinement controlled fusion, 187
Magnetorestrictive energy storage
  systems, 134
Maine, 391 (table)
Malaysia, 329
Mammoth Lakes binary cycle geothermal
  power plant, California, 198
Mandatory market green certificates,
  211–212
Mandatory Oil Import Quota (MOIP)
  program (1959, U.S.), 339
Manhattan Project, 86, 311, 430
Maryland, 10, 391 (table)
Massachusetts
  appliance efficiency standards in, 10
  mandatory market green certificates in, 212
  Massachusetts et al. v. Environmental
    Protection Agency et al. and, 118, 123
  renewable energy portfolio of, 391 (table)

Maximum achievable control technology, 118, 440

Mazda, 171

Mellenium Development Goals, 413

Mercedes, 171

Mercury
in compact fluorescent bulbs, 93
from waste incineration, 439

Mercury automobile company, 171

Metal hydride batteries, 24

Metcalf, Gilbert, 59

Metering, 287–289
air conditioning cycling programs and, 288
broadband over power line grid communication of, 288–289
electromechanical induction meter and, 287
high *vs.* low system demand and, 287, 288
plug-in hybrid electric vehicles and, 288
real time utility use information and, 288
remote reading of, 287
"smart meters" and, 287–289
variable electricity rates and, 288

Methane
alternative-fueled vehicles and, 462
anaerobic digestion plants and, 36, 219, 275, 386, 440
from animal digestion, 215
Antarctic ice core data on, 202, 203 (fig.)
as biogas digester product, 36
biomass energy and, 38–39, 386, 435
carbon tax and, 58
climate change (forcing) affected by, 78, 214
coal-bed methane and, 239, 297–298, 299
electricity generated from, 40
energy density of, 247 (table)
flaring and, 168, 169, 215, 275
gasification process and, 191, 386
as GHG emission, global warming and, 40, 52, 64, 73, 207, 214–215, 269, 289–290, 290 (table), 378
hydroelectric power release of, 245
as jet engine fuel, 301
in melting permafrost, 214
natural gas and, 179, 275, 298–300, 301, 308, 309, 364
as natural GHG, 213–214, 404
oil component of, 324
radiative forcing and, 378
syngas production and, 440
as a volatile organic compound, 435
See also **Landfill methane; Natural gas**

Methanol, 61, 64, 247 (table), 462

Methyl tertiary butyl ether (MTBE), 152

Metric tons of carbon equivalent (MTCE), 289–291
calculation of, 291
GHGs, global warming potential (GWP) and, 289–290, 290 (table)
radiative forcing values of GHGs and, 289–290

Mexico
gas flaring in, 169
geothermal energy produced in, 200 (table)
National Emission Inventory program in, 118
oil produced in, 339, 340, 346

Michigan, 391 (table)

Microbial fuel cell, 185

Microhydro power, 291–294
advantages and disadvantages of, 293
carbon neutrality of, 292
environmental, social, and economic advantages of, 292
as hydroelectric power subcategory, 291
impulse *vs.* reaction turbines used in, 293
large *vs.* small scale generation of, 292
on-site renewable energy and, 337
penstock feature of, 292
picohydro power, 292, 293
run-of-river, 292, 293
storage of, 293
technical considerations regarding, 292–293
water pressure factors and, 292
See also **Columbia River; Hydroelectric power**

Microscale hydroelectric power, 242

Middle East
oil price fluctuations and, 340
state-owned oil company in, 329
total primary energy supply (TPES) used by, 427

Midgley, Thomas, Jr., 71

Midwestern Regional Greenhouse Gas Reduction Accord (U.S.), 58

Millennium Cell Inc., 184, 248

Millirem radiation measure, 458

Minnesota
heating degree day measures in, 232
renewable energy portfolio of, 391 (table)
Xcel Energy wind power facility in, 107, 390

Missouri
    renewable energy portfolio of,
        390, 391 (table)
    Times Beach dioxin exposure in, 104
Mitsubishi, 462
Mobil Corporation, 329
Molasses, as ethanol source, 154, 156
Molina, Frank Sherwood, 71
Molton carbonate (direct fuel cells), 184
Monitored Retrievable Storage (MRS)
    facilities, 459
Montana
    bituminous coal in, 79
    renewable energy portfolio of,
        391 (table)
Montenegro, 263
Montreal Protocol, 71–72, 215
Mountaintop removal, 294–296
    Abandoned Mine Land Reclamation Fund,
        82, 295
    advantages *vs*. disadvantages of, 295–296
    drinking water contaminated by, 295–296
    environmental impact of, 294 (photo)
    process of, 81, 294
    Samples Mine, West Virginia, 294 (photo)
    Surface Mining Control and Reclamation
        Act and, 294–295
    valley fills or "hollow fills" and, 81, 294
    See also **Coal, clean technology**
Movement for the Survival of the Ogoni
    People (MOSOP), 345
MTCE. *See* **Metric tons of carbon equivalent
    (MTCE)**
Muir, John, 47, 147
Multi-junction photovoltaics, 349
Multiple equation forecasting methods, 176
Mustard seed, as biofuel source, 33

Nacelle wind turbine component, 447
Nadel, Steven, 10
Nagasaki nuclear bomb, 311, 317
Napthalene, 178
National Appliance Energy Conservation Act
    (NAECA, 1987), 10, 11 (table)
National Association of Regulatory Utility
    Commissioners, 12
National Center for Climatic Data, 233
National Emission Inventory (NEI) database
    of EPA, 118
National Energy Act, 98
National Environmental Policy Act (1969),
    44, 165

National Highway Traffic Safety
    Administration (NHTSA), 43–44
National Ignition Facility (U.S.), 188
National Iranian Oil Company, 329
National Oceanic and Atmospheric
    Administration, 202 (photo), 215
National Pollutant Discharge Elimination
    System, 306
National Renewable Energy Laboratory, 34
National Resource defense Council, 278
National Shelter Industry Energy Advisory
    Council, 236
Natural gas, 297–302
    absorption *vs*. adsorption process technique
        and, 300
    advantages and disadvantages of, 309
    alkanes in, 298
    bleve explosions and, 164
    cap rock, gas cap and, 297
    carbon emission factors of, 50, 364
    carbon footprint of, 259
    carbon tax on, 58
    Caspian Sea reserves of, 68
    challenges in use of, 301–302
    chemisorption process technique and, 300
    classification systems of, 298–298
    coal-bed methane and, 239, 297–298, 299
    combined cycle gas turbine and, 300, 364
    components of, 298–299, 308
    compressed natural gas vehicles and, 462
    costs associated with, 243
    diluents and contaminants of, 298
    as direct energy example, 115
    electricity generated using, 239, 309, 364
    energy policy regarding, 130
    ethanol (corn) production using, 151
    flaring and, 169–170, 169 (photo)
    geothermal energy use with, 199
    GHGs from electricity production and, 58
    heating efficiency of, 111, 300–301
    Hubbert's Peak and, 238–240
    hydrogen and, 239, 249, 301
    incomplete combustion issues of, 179
    Kern River gas transmission pipeline
        and, 353
    landfill methane and, 276–277, 298
    liquified ship transport of, 297 (photo), 298,
        299–300, 302
    methane component of, 179, 275, 298–300,
        299–300, 301, 309, 364
    natural gas vehicles and, 300–301, 308–309,
        369, 462

nitrous oxides and, 298, 309, 364
nonfuel use of, 179, 300–301, 308–309
nonhydrocarbon components of, 179
oil component of, 324
petroleum and, 297
plankton, algae, and protozoans and, 297
production process of, 299–300
storage limitations of, 179, 301–302
sulfur oxides release from burning of,
    180–181, 309
synthetic agricultural fertilizers and, 239
therms or cubic feet (ccfs) measures
    of, 106
transportation issues of, 179
See also **Fossil fuels; Methane;**
    Offshore drilling
Natural monopoly, 372–373
Negative radiative forcing, 204
Nepal
    anaerobic digestion system used in, 219
    microhydropower used in, 292, 293
Neptunium, 457
Net energy value
    of biofuels production, 33, 240
    energy payback and, 126
    of ethanol, 33, 151–152, 195–196, 240
    of hydrogen, 239–240
    net zero carbon emission and, 53
    of oil shale and oil sands, 240
    zero-net energy commercial buildings
        and, 101
Netherlands, the
    bicycle transportation in, 31
    carbon tax in, 108
    Rabobank green banking policies in, 210
    recycling in, 380
Net irradiance, radiative forcing, 377
Net metering electricity policy, 166
Net metering of grid-connected systems,
    228–229
Net zero carbon emission, 53
Nevada
    Black Mesa coal slurry pipeline in, 82
    Kern River gas transmission pipeline
        in, 353
    radioactive waste disposal in, 424
    renewable energy portfolio of,
        390, 391 (table)
    Steamboat Springs geothermal power plant
        in, 198
    See also **Yucca Mountain**
New Hampshire, 391 (table)

New Jersey
    appliance efficiency standards in, 10
    renewable energy portfolio of,
        390, 391 (table)
New Mexico
    "green" buildings in, 143 (photo)
    Permaculture Credit Union eco-banking
        practices in, 210
    renewable energy portfolio of, 391 (table)
    Waste Isolation Pilot Plant in, 315
New York (state), 10, 391 (table)
New York City
    commuter rail transportation in, 370
    Edison Pearl Street station,
        New York City, 88
    New York Mercantile Exchange, 340
    Pennsylvania Station, 96
New Zealand
    emissions reductions in, Berlin Mandate
        and, 26
    fruits exported from, 174
    geothermal energy produced in, 200 (table)
    Kyoto Protocol and, 273
    Tararua Wind Farm projects in, 273
Nicaragua, 200 (table)
Nickel cadmium batteries, 24, 359
Niger, 366
Nigeria
    gas flaring in, 169, 344
    Movement for the Survival of the Ogoni
        People (MOSOP) and, 345
    natural gas–fired turbines in, 309
    OPEC membership of, 338, 340
    petroviolence in, 343, 344–345
NIMBY (Not In My Backyard), 315
Nissan, 171
Nitrogen oxides, 302–305
    from biomass energy production, 40–41
    climate forcing effect of, 214
    from coal burning, 82–83, 100, 308, 363
    denitrification process and, 304
    from electrical power plants, 109
    environmental problems associated with,
        303–304
    fertilization process and, 302, 303
    flaring and, 169
    from fossil fuel burning, 180, 461
    as greenhouse gas, global warming and,
        40–41, 73, 78, 213–214, 215, 269,
        289–290, 290 (table), 378
    human health affected by, 304
    natural gas component of, 298, 309, 364

radiative forcing and, 378
reforming of hydrocarbons and, 248
sources of, 302–303
from waste incineration, 439
Nitrogen trifluoride, 290 (table)
Nixon, Richard, 98, 340
Nonassociated natural gas, 298–299
Nongovernmental organizations (NGOs),
    62–63
Nonintegrated compact fluorescent
    bulb, 92
Nonpoint source (NPS), 305–307
    Clean Water Act and, 306
    eutrophication of water bodies and, 305
    examples of, 305–306, 306 (table)
    National Pollutant Discharge Elimination
        System and, 306
    regulation of, 305, 306–307
    water pollution from, 305
    Water Quality Act (1987) and, 306
Nonproliferation Treaty (NPT) and, 317
Nonrenewable energy resources, 307–310
    definition of, 308
    fossil fuels, 308–309
    nuclear fuels, 309–310
    renewable energy resources vs., 308
    total primary energy supply (TPES)
        calculations and, 426–427
    See also Fossil fuels; Nuclear power;
        Uranium; specific energy resource
North Carolina, 390, 391 (table)
North Dakota, 232, 390, 391 (table)
North Korea, 317, 318
Norway
    oil produced in, 340
    REVA electric car used in, 462
Not In My Backyard (NIMBY), 315
Nova Scotia, Canada, 261
NPS. See Nonpoint source (NPS)
Nuclear power, 310–316
    antinuclear movements and, 314–315
    Atomic Energy Commission and, 98, 423
    Atoms for Peace discourse and, 311, 318
    carbon neutrality issue regarding, 77, 366
    Chernobyl accident and, 313, 366
    The China Syndrome film and, 423
    controversy regarding, 110, 310, 313–315
    cool water element in, 312
    costs associated with, 243
    Department of Energy responsibilities
        and, 101
    electrical energy infrastructure and, 108

electricity power plants generated by,
    108, 109, 365–366
energy policy regarding, 130
energy storage element in, 134
environmental protection measures and, 143
European pressurized water reactor,
    France, 312
federal regulation of, 374
first atomic cell and, 311
Framatome organization and, 312
Global Nuclear Energy Partnership and,
    100, 318
as green, sustainable, or renewable energy,
    83, 217, 310, 311, 315, 317
history of, 311–312
International Atomic Energy Agency
    Incident (IAEA) data on, 311–312
kinetic energy produced in, 312
life cycle production approach to, 148
nuclear energy for peaceful purposes
    and, 317
nuclear fission process and, 131–132,
    187, 221–222, 309–310, 311,
    312–313, 365, 458
nuclear fusion process and, 132, 145, 224,
    312–313
Nuclear Regulatory Commission and,
    98, 164, 423
Nuclear Waste Policy Act (1982) and,
    424, 457
power capacity and operating cycle of,
    312–313
power station electrical conversion
    and, 109
radioactive waste issues and, 310, 311,
    314–315, 316, 366, 387, 423,
    424, 441, 458
Radioactive Waste Policy Act (1980)
    and, 424
radium discovery and, 311
Rancho Seco Power Plant closure,
    California, 47
safety issues of, 132, 311, 314, 366, 387
second-generation reactors and,
    311 (photo), 312
terrorism security issues and, 132, 314
thermal energy from, 110
third and fourth-generation reactors and,
    145, 312–313
Three Mile Island accident and, 313, 366
"tokamak" technology and, 187–188,
    312–313

Torness Nuclear Power Station, Dunbar,
    Scotland, 311 (photo)
uranium development and, 132, 309–310,
    313, 315–316, 365–366
uranium mining, GHGs and, 432–433
Waste Isolation Pilot Plant,
    New Mexico, 315
See also Chernobyl; Nuclear proliferation;
    Three Mile Island; Yucca Mountain
Nuclear proliferation, 316–319
    controls and abatements of, 322
    Department of Energy responsibilities
        and, 101
    fissile materials acquisition issue and,
        317–318
    full nuclear fuel cycle issues and, 318
    globalized nuclear economy issues
        and, 318
    horizontal vs. vertical proliferation
        and, 316
    International Atomic Energy Agency (IAEA)
        and, 317, 318
    mixed oxide reactor fuels issue and, 318
    Nonproliferation Treaty (NPT) and, 317
    nuclear energy for peaceful purposes
        and, 317
    quantitative vs. qualitative proliferation and,
        316–317
    term explanation and, 316–317
    terrorist threat and, 317
    uncontrolled fission chain reaction in, 312
    UN "nuclear club" and, 317
    uranium acquisition and, 318
    U.S. "stockpile stewardship" program
        and, 188
    See also Nuclear power; Yucca Mountain

Obama, Barack
    CAFE standards policies of, 44
    California environmental policies and,
        44, 48
    coal policies of, 84, 296
    energy efficiency and conservation policies
        of, 101, 133
    Yucca Mountain funding policy of, 315
Oceanic thermohaline circulation,
    205, 206 (fig.)
Ocean iron fertilization, 78
Ocean Thermal Energy Conservation
    Act, 100
Office of Surface Mining Reclamation and
    Enforcement agency, 294–295

Offsetting. See Carbon trading and offsetting
Offshore drilling, 321–323
    carbon dioxide sequestration and, 322
    Caspian Sea and, 66–69
    Coastal Zone Conservation Initiative (1972)
        and, 47
    Devil's Tower, Gulf of Mexico,
        321 (photo)
    environmental protection issues and, 322
    exclusive economic zone (EEZ) and, 322
    hurricanes and, 322
    oil and gas deposits associated
        with, 321
    platform variations and, 321–322
    spills and leaks issues and, 322
    United Nations Convention on the Law of
        the Sea (UNCLOS) and, 322–323
    waste produced by, 322
    water depth issues and, 321
    See also Arctic National Wildlife Refuge
        (ANWR); Caspian Sea
Ogoni people of Nigeria, 344–345
Ohio, 391 (table)
Oil, 323–328
    Arab oil embargo (1973) and,
        43, 47, 92, 98, 329, 340
    best management practices in electricity
        industry and, 28
    carbon dioxide in, 325
    Caspian Sea resources of, 66–69
    crude petroleum and, 324
    economic implications of, 325
    electricity power plants generated
        by, 365
    energy to output ratio of, 333
    environmental implications of, 326
    Exxon Valdez oil spill and, 160–162
    flaring and, 169
    formation of, 324
    gaseous, liquid, and solid hydrocarbons
        components of, 324
    Hubbert's Peak and, 238–240
    as nonrenewable energy source,
        324, 325–326
    oil drilling and refining waste products
        and, 326
    polycyclic aromatic hydrocarbons (PAHs)
        and, 326
    properties of, 323–324
    remediation techniques and, 326–327
    remote sensing techniques and, 325
    "signatures" for oils and, 324–325

social justice issues and, 326
in vivo vs. in vitro carbon-based oils and,
    324, 327–328
See also **Arctic National Wildlife Refuge
    (ANWR); Caspian Sea; Fossil fuels;
    Offshore drilling; Oil majors; Oil sands;
    Oil shale**
Oil majors, 328–331
    alternative energy concerns of, 331
    "Big Oil" *vs.* consumers and, 330–331
    British Petroleum (BP), 329, 330, 339
    Chevron, 329, 330, 339
    Conoco-Phillips, 329, 330, 339
    Exxon Mobil, 329–330, 339
    GHG emission tax and, 330
    "hold up" problem and, 329
    Mobil, 329, 339
    Royal Dutch Shell, 328, 329, 330, 339,
        344–345
    "Seven New Sisters" and, 329
    "Seven Sisters" and, 329, 339
    Shell Oil Company, 328–329, 344–345
    Standard Oil, 328, 329
    state-owned oil companies *vs.*, 329
    subsidy companies of, 329
    Texaco, 329, 339
    Total S. A., 330
    vertical integration feature of, 328, 329
    vertical monopoly and, 329
Oil Pollution Act (1999), 162
Oil sands, 331–334
    bitumen in, 324, 332
    carbon source from, 64
    components of, 332
    downstream remediation and, 333
    energy to output ratio of, 333
    environmental impact issues and, 333
    extraction methods for, 332
    open pit mining issue and, 333
    sample of, Fort McMurray, Alberta,
        Canada, 332 (photo)
Oil shale, 334–336
    bitumen in sedimentary rock and, 80, 334
    environmental impact issues and, 334
    as green power renewable energy
        source, 83
    hydrocarbon deposits in, 180
    kerogen in, 180, 334
    mining methods of, 334
    as natural gas substitute, 334
    oil shale–fired power plants and, 334
    in situ retorting technique and, 334

Oil spills, 181
    See also ***Exxon Valdez***
Omnibus Budget Reconciliation
    Act (1986), 163
**On-site renewable energy generation,
    336–338**
    benefits of, 336–337
    biomass combustion, 337
    costs and pollution reduced by, 336
    Database of State Incentives for Renewable
        Energy and, 338
    green power and sustainability
        elements of, 337
    hydrogen fuel cells, 337
    low power outputs feature of, 337
    microhydro technology, 337
    solar (photovoltaic panels), 337
    tax and funding incentives for, 338
    wind turbines, 330
    See also **Hydroelectric power**
Ontario, Canada, 262–263
Opacity, 439
OPEC. *See* **Organization of Petroleum
    Exporting Countries** (OPEC)
Open-cell spray foam insulation, 256
Open energy storage systems, 134–135
Operations management, 27–28
Oregon
    appliance efficiency standards in, 10
    Columbia River and, 85
    electric tram public transportation in, 370
    electric vehicle recharging station in,
        112 (photo)
    renewable energy portfolio of, 391 (table)
    ShoreBank Pacific eco-banking practices
        in, 210
    Short Mountain Methane Power Plant,
        Eugene, 277
The Organic Machine (White), 85
Organic photovoltaics, 349–350
Organisation for Economic Co-operation and
    Development (OECD), 427
**Organization of Petroleum Exporting
    Countries** (OPEC), 338–341
    Arab-Israeli War and, 340
    Arab oil embargo (1973) and,
        43, 47, 92, 98, 329, 340
    cartel designation of, 339
    disagreements among, 340
    economic and political power of, 339
    energy conservation and green energy
        influence and, 341

energy policies of, 131
international petroleum exchanges and, 340
Mandatory Oil Import Quota (MOIP)
    program (1959) and, 339
membership of, 338–339
nationalization of petroleum industries
    and, 340
oil reserves of, 339
petrodollars term and, 340
"Seven Sisters" oil companies and, 329, 339
See also **Oil**
Otto, Nikolaus, 90
Oweiss, Ibrahim, 340
Ozone layer
chlorofluorocarbons (CFCs) depletion of,
    71, 215
fossil fuel combustion depletion of,
    180, 181
GHGs, radiative forcing and, 378
volatile organic compounds (VOCs)
    and, 435

Pakistan
natural gas vehicles in, 301
nuclear power issues and, 315
nuclear weapons in, 317, 318
Palm oil, as biofuel source, 33, 219, 327
Panda, S. S., 218
Papua New Guinea, 200 (table)
Parabolic dishes, 400 (photo), 401, 410
Parabolic troughs, 401–402, 410
Parallel hybrid engines, 358–359
Paratransit public transportation, 367–368
Partial-zero-emission vehicles (PZEVs),
    461–462
Passive solar power, 2, 145, 220,
    404, 407–408, 409
Peanut oil, diesel fuel from, 20
Peat lands, 54 (photo), 57
Pelamis Wave Power machine, Aguçadora
    Wave Park, Portugal, 442 (photo), 443
Pellet stoves, 452
Pennsylvania
Exelon Power, Pennsylvania and, 277
recycling in, 381
renewable energy portfolio of, 391 (table)
smart meters used in, 288
Three Mile Island nuclear power plant in,
    421–424
See also **Three Mile Island**
Penstock hydropower feature, 242, 244, 292
Pentane natural gas component, 298, 308

Perfluorocarbons, 52, 78, 269, 289, 378
Perihelion position of Earth, 404
Peru, 346
Petro-authoritarianism term, 345–346
Petrobras oil company, Brazil, 329
Petrodollars term, 340
Petroleos de Venezuela S. A. oil company, 329
Petroleum
components of, 179
as domestic and industrial energy
    source, 180
as fossil fuel, 179–181, 308
petroleum products and, 308
refining of, 179–180
See also **Fossil fuels; Oil**
Petroviolence, 343–346
in Central Asia and Latin America,
    345–346
definition of, 343
factors in, Watt's work and, 343–344
petro-authoritarianism term and, 345–346
Ken Saro-Wiwa's work and, 344–345
Pew Center on Global Climate Change, 390
Phenol, 178
Philippines
biogas digesters in, 37
geothermal energy produced in,
    199, 200 (table), 221, 223
microhydropower used in, 292
paratransit in, 368
Phosphoric acid fuel cells, 184
Photosynthesis
biomass resource formation and,
    1, 38, 137, 324
carbon sequestration and, 54, 77, 450
plant growth and, 39, 54, 55, 131, 405
Photovoltaics (PV), 347–353
amorphous silicon (a-Si), 348–349
building-scale photovoltaic arrays, 351
cadmium telluride, 348–349
concentrated photovoltaics, 350, 400, 406
conversion efficiency comparison of, 352
copper (Indium Gallium) selenide
    (CIGS), 349
crystalline silicon (c-Si), 348
embodied energy in, 117
energy payback calculation and, 127
energy production process of, 2–3, 7–8
as energy storage systems, 134, 136
IPCC studies of, 260
life cycle production approach to, 148
multi-junction, 349

on-site renewable energy generation and, 337

organic and dry-sensitized, 349–350

as renewable energy green power, 222–223, 347

semiconductor material of, 347, 347 (photo)

solar power and, 7–8, 136, 220, 337, 347, 383–384, 387

solar-powered vehicles and, 7

technology advances in, 347

versatility of, 347

very-large-scale photovoltaic systems (VLS-PV), 351–352

See also **Solar concentrator**

Picohydro power, 292, 293

Piezos energy harvester system, 134

Pig manure, biomass energy from, 36, 37

Pimental, David, 138

Pipelines, 353–356

accessibility and security issues of, 355–356

accidental spills and, 181

as an asset class, 354

of Caspian Sea, 69

complexity of, 354–355

degradation and repair of, 353 (photo), 354–355

design considerations of, 353–354

durability of, 354

energy infrastructure and, 353, 355

FERC regulation of, 163, 164, 165

global warming factor and, 355–356

infrastructure features of, 354

of liquified natural gas (LNG), 300, 302, 353

networked pipelines and, 354

Pitchblende, 431

Plass, Gilbert, 73

Plug-in hybrid, 356–361

battery power of, 356–367, 357 (photo)

battery technology and, 359

Chevrolet Volt, 359, 360

disadvantages of, 359–360

economic and ecological impact of, 360–361

fossil fuel and electricity and, 360

fossil fuel decline and, 357

fuel efficiency *vs.* fuel economy and, 358

full electric vehicles *vs.*, 113

gasoline tax issues and, 360

hybrid technology and, 358–360

regenerative braking technology and, 359

series *vs.* parallel hybrids and, 358–359

Toyota Prius, 7, 186, 357 (photo), 359, 462

Plutonium, 70, 86, 318, 429, 457

Poland, 168, 362

Pollutant release and transfer registers (PRATRs), 118

Polo, Marco, 66

Polonium, 429, 430

Polycyclic aromatic hydrocarbons (PAHs), oil remediation, 326

Polymer electrolyte fuel cells, 184

Pongamia pinnata, as biofuel source, 154, 219

Portugal

geothermal energy produced in, 200 (table)

Pelamis Wave Power machine, Aguçadora Wave Park, 442 (photo), 443

wave farm of, 221

Positive radiative forcing, 204

Potential energy, 106

Power and power plants, 361–367

coal-fired, 362–363

co-firing power plants and, 308

electric power importance and, 361–362

fuel *vs.* technology power plant classification and, 362

natural gas–fired, 364

nuclear-fired, 365–366

oil-fired, 365

renewable sources power generation and, 366–367

Three Gorges Dam, 419–421

See also **Chernobyl; Combined heat and power (CHP); Nuclear power; Smart grid**

Preservationist environmental agenda, 146–147

Process analysis model of embodied energy, 116

Production tax credits (PTC), 392

Projected home energy rating, 236

Project management, 27–28

Propane

as natural gas component, 298, 308

as vehicle fuel, 21, 61, 247 (table), 462

Propeller reaction microhydro turbine, 293

Propylene, 324

Proton exchange membrane (PEM) fuel cells, 184

Public transportation, 367–371

bus transit, 368–369

commuter rail, 370
electric trams and streetcars, 368 (photo),
    369–370
energy concerns regarding, 370–371
heavy rail and metros, 370
paratransit, 367–368
Public utilities, 372–375
cooperative utilities and, 372
Energy Policy Act (2005) and,
    10, 34, 84, 229
global warming issues and, 374–375
government ownership issues and,
    372, 373
investor owned utilities and, 372, 373
monopolies issue and, 372–373
Public Utility Regulatory Policies Act (1978)
    and, 88, 166–167, 197, 229
regulation vs. deregulation of,
    373, 374–375
structure variations of, 370
technology advances and, 374
See also Combined heat and power (CHP);
    Energy policy; Internal energy market;
    Metering; Smart grid
Pumped hydroelectric storage (PHS) plants,
    135, 242
Pumped storage hydroelectric power, 242
PV. See Photovoltaics (PV)
Pyridine, 178
Pyrolysis, 191
Pyrolysis technologies, 249

Qatar
gas flaring in, 169
natural gas produced in, 299, 364
OPEC membership of, 338
Qualitative forecasting methods, 176
Qualitative nuclear proliferation, 316–317
Quantitative nuclear proliferation, 316–317
Quebec, Canada, 262

Radiant flux density of solar energy
    at Earth, 403
Radiative forcing (RF), 377–379
albedo factor in, 204, 207, 378
anthropogenic factors of, 378
carbon dioxide (CO_2) example of,
    377–378
climate research and policy and, 379
components' estimation and, 378
definition of, 377
from direct aerosols, 378–379

GHGs, global warming and, 202–203,
    203 (fig.), 207, 214, 378–379
natural solar changes and volcanic activity
    and, 378
natural vs. anthropogenic forcings and,
    204, 205 (fig.)
negative vs. positive RF and, 204, 377
net irradiance element in, 377
radiative balance, radiative transfer
    and, 378
tropopause boundary region of atmosphere
    and, 377
unperturbed values in, 377
See also Metric tons of carbon equivalent
    (MTCE)
Radioactive Waste Policy Act (1980), 424
Radium, 429, 430, 431–432
Radon, 429, 430, 432
Rahmsdorf, Stefan, 206 (fig.)
Rancho Seco Power Plant closure
    (California), 47
Rankine engines, 110
Rapseed, as biofuel source, 20, 33, 327
Reaction microhydro power turbine, 293
Reagan, Ronald, 47
Recycling, 379–382
Bottle Bills and, 381
criticisms of, 381
definition of, 379
down- vs. up-cycling and, 379
economic, energy, and environmental
    benefits of, 381
examples of, 379–380
paper recycling, 380 (photo)
product stereotypes and, 382
throwaway society and, 380
See also Environmentally preferable
    purchasing (EPP); Life cycle analysis
Reforming of hydrocarbons as hydrogen
    source, 247–248, 248 (fig.), 249
Regional Greenhouse Gas Initiative
    (U.S.), 58
Regression forecasting methods, 176
Reisner, Marc, 86
Renewable energies, 382–389
affordability, availability, and acceptability
    of, 142, 367
biogas, 35–37
biomass energy, 37–41
in California, 46–48
challenges faced by, 387–388
criteria used to judge, 382–383

definition of, 308
development of, 382–383
DOE development of, 100–101
electrical energy infrastructure and, 108–109
embodied energy in, 117
energy payback calculations and, 127
environmental stewardship and, 143–144, 148, 367, 387
Evergreen Renewable Energy Program, Wisconsin, 277
feed-in tariffs and, 166, 167
fossil fuel depletion and, 382–383
future trends in, 388–389
geothermal energy, 384–385
green certificates and, 111
hydroelectric power, 383
intermittent nature of, 366, 387, 398
nonrenewable energy resources *vs.*, 308
for plug-in hybrid vehicles, 356–361
Renewable Energy and Energy Efficient Partnership and, 265
Renewable energy credits (RECs) and, 167–168, 225, 226–227
Renewable Energy Policy Network for the 21st Century and, 265
Renewable Energy Resources Act and, 100
"Renewable Energy Roadmap" (European Commission), 34
Renewable Energy Sources and Climate Change Mitigation workshop of IPCC and, 260
renewable heating fuels, 3
solar energy, 383–384
sources of, 218
stability of energy supplies and, 135
Three Gorges Dam and, 419–421
total primary energy supply (TPES) calculations and, 426–427
transmission issues of, 387–388
wind energy, 1 (photo), 385–386
See also **Alternative energy; Biodiesel; Biomass energy; Energy policy; Flex fuel vehicles (FFVs); Geothermal energy; Green energy certification schemes; Green power; Grid-connected system; Hydroelectric power; Innovation; International Renewable Energy Agency (IRENA); Microhydro power; Nuclear power; On-site renewable energy generation; Photovoltaics (PV); Renewable energy**

portfolio; **Tidal power; Wave power; Wind power; Wind turbines; Wood energy;** *specific energy source*
Renewable energy portfolio, 389–393
federal production tax credits (PTC) and, 392
GHG reductions and, 390
state renewable energy portfolios and, 389–390, 391–392 (table)
Xcel Energy, Minnesota, 390
Residue natural gas, 299
REVA electric car, 462
Revised Universal Soil Loss Equation (RUSLE), 307
RF. *See* **Radiative forcing (RF)**
Rhode Island
appliance efficiency standards in, 10
Citizens Bank, green banking and, 209
renewable energy portfolio of, 391 (table)
Rifkin, Jeremy, 249
Riley, Richard W., 423–424
Rio Declaration, 413, 455
Rio de Janeiro. *See* Earth Summit, Rio de Janeiro
Risk assessment, 393–395
climate change risk assessment and, 394
cost-benefit analysis and, 394
definitions regarding, 393
forecasting and, 175–177
future emphasis nature of, 394–395
modern economic growth and, 393
precautionary principle and, 394
probability and statistics tools of, 393
technology advances and, 394
Rockefeller, John D., 150, 329
Romania, 263, 273
Romm, Joseph, 249
Room-and-pillar mining system, 80
Rotary engine, 91
Rotor wind turbine component, 447
Rowland, Mario, 71
Royal Dutch oil company, 328, 329, 330, 339, 344–345
Run-of-mine coal, 82
Run-of-river hydroelectric/microhydro power, 242, 292, 293
Russia
Caspian Sea border country of, 66, 67–68, 69
coal reserves in, 80
electric trolley in, 462
gas flaring in, 169

Gazprom state-owned oil company in, 329, 345–346
geothermal energy produced in, 200 (table)
Kyoto Protocol and, 269
metro systems in, 370
natural gas produced in, 299, 364
nuclear power reactors in, 365
nuclear weapons developed in, 430
oil supply from, 325, 340
petro-authoritarianism in, 345–346
renewable energy-fired power plants in, 366
Russian Federation Ministry for Atomic Energy and, 318
state-owned oil company in, 329
uranium resources in, 366
See also Soviet Union
R-value of insulation, 256–257

Salter, Stephen, 442
Salter's Duck wave power model, 442
A Sand County Almanac (Leopold), 146
Saro-Wiwa, Ken, 344–345
Saturn EV-1 electric vehicle, 113
Saudi Arabia
    natural gas produced in, 299, 364
    OPEC membership of, 338, 339, 340
    Saudi Aramco state-owned oil company in, 329
Schumpeter, Joseph, 251
Scotland, 311 (photo)
Seaton, Fred Andrew, 13
Second-generation biofuel generation, 144
Seleucus of Seleucia, 425
Serbia, 263
Series hybrid engines, 358–359
"Seven New Sisters" oil companies, 329
"Seven Sisters" oil companies, 329, 339
Shale natural gas, 299
Shell Oil Company, 328–329, 344–345
Short Mountain Methane Power Plant, Eugene, Oregon, 277
Sick building syndrome, 436
Silicon oils, 327
Singapore International Monetary Exchange, 340
Skull Valley Goshute Indian Reservation, Utah, 459
Smart grid, 397–399
    benefits of, 397–398
    broadband over power line meter communication and, 288–289
    "Connectivity Week" of IPCC and, 261

energy consumption patterns information and, 398
energy cost savings of, 397
energy distribution innovation and, 252
peak energy demands reduction and, 397
reliability of electricity resources increased by, 398
smart meters and, 287–289, 398
stored energy and, 398
transmission electricity losses reduced by, 398
unified communications and control system of, 397
utility infrastructure and, 399
See also Energy policy; Metering
Smart meters in carbon offsetting, 64
Smart utility meters, 287, 398
Smog, 40, 109, 309, 369
Snake River, 86
Sodium borate, 248
Sodium/bromine fuel cells, 184
Soil and Water Assessment Tool (SWAT), 307
Solar concentrator, 399–402
    concentrated photovoltaics, 400, 406
    heliostat concentrator arrays, 401, 410
    nonimaging optical lenses, 400
    parabolic dishes, 400 (photo), 401, 410
    parabolic troughs, 401–402, 410
    photovoltaic electricity output and, 399–400
    removing heat from solar cells and, 400
    solar thermal plants and, 399–400
    sun's energy density increased by, 399–400
    See also Solar energy; Solar thermal systems
Solar energy, 402–408
    active and passive solar heating and, 145, 220, 404, 407–408
    agroecosystem entropy and, 139
    applications of, 220, 384
    astronomical and terrestrial conditions factor in, 403
    biogas vs., 36
    biological life and, 402, 405
    biomass and, 405
    biomimicry concept and, 408
    California solar panels field and, 46 (photo)
    challenges of, 405
    diffuse nature of, 405
    diurnal cycle limitation of, 387, 405, 406, 407
    Earth's geometry and orbit factors in, 403–404
    electrical energy infrastructure and, 108

electromagnetic radiation and, 384
as energy innovation, 252
energy payback calculation and, 127
energy policies regarding, 132
energy storage, smart grid and, 398
environmental stewardship and, 148, 387
feed-in tariff and, 166
as green power renewable energy source,
    1, 83, 207, 217, 218, 220, 330, 366,
    383–384, 404–405
hydroelectricity and, 405
intermittent nature of, 387
IPCC studies of, 260
nuclear fusion reactions at sun's core
    and, 403
off-grid applications of, 220
on-site renewable energy generation
    and, 337
passive solar power and, 2, 404,
    407–408, 409
perhihelion vs. aphelion positions of Earth
    and, 404
photovoltaic (PV) technology and, 7–8, 136,
    220, 260, 330, 383–384, 387, 406–407
potential solar energy, 404
radiant flux density of solar energy and, 403
residential solar panels and, 402 (photo)
solar electric systems and, 405–408
Solar Energy and Energy Conservation Act
    and, 100
Solar Energy and Energy Conservation Bank
    Act and, 100
solar heating, 407, 409–410
solar-powered vehicles, 7
solar radiation and Earth's atmosphere
    and, 404
solar radiation management and, 78
solar thermal vs. solar electric types of, 220
sun's energy and, 384, 402–405
weather and regional climate factors
    in, 404
wind and ocean currents and, 384, 405
See also Photovoltaics (PV); Solar
    concentrator; Solar thermal systems
Solar thermal systems, 408–411
applications of, 410–411
concentrated solar power (CSP) systems,
    400–401, 408, 409–410
costs associated with, 243
electricity generated from, 3, 110
evacuated tube solar collector and, 409
flat plate solar collector and, 409

greenhouses, 410
helliostat or power tower, 410
hydronic in-floor heating, 3
infrared radiation from sun and, 408
solar cooking, 410–411
solar electric systems vs., 220
Solar Energy Generating Systems, Mojave
    Desert, 410
solar heating systems and, 145, 220, 404,
    408, 409–410
solar radiation, 410
solar thermal electric facilities, 408
thermal solar energy, 408
thermosyphon passive system, 409
ultraviolet solar radiation, 408
See also Solar concentrator; Solar energy
Solid acid fuel cells, 184
Solid oxide fuel cells, 184
Solomon, Susan, 215
Sorghum (corn) ethanol source, 20, 219
Sour natural gas, 298, 299
South Africa, 317, 362
South Asia, 80
See also specific country
South Carolina, 423–424
South Dakota, 390, 392 (table)
Southeast Asia, 34, 292
See also specific country
South Korea, 365
Soviet Union
Chernobyl nuclear incident and, 69–71
Cold War and, 98–99
gas flaring in, 169
nuclear power reactors in, 311
nuclear weapons in, 317
petro-authoritarianism in, 345–346
total primary energy supply (TPES)
    used by, 427
See also Chernobyl; Russia
SO_x. See Sulfur oxides (SO_x)
Soybeans, as biofuel source, 6, 20, 33, 34,
    154, 219, 327
Spain
concentrated solar power (CSP) systems
    in, 410
feed-in tariffs policy in, 167
International Renewable Energy Agency
    (IRENA) and, 266
power tower solar thermal systems in, 410
renewable energy-fired power plants in, 366
REVA electric car used in, 462
wind power use in, 445, 448

Spray foam insulation, 255 (photo), 256
Standard Oil Trust, 150, 329
State of Fear (Crichton), 73
Steam engines, 109–110, 130
Steam hydrogen production, 6
Steinborn, Wolf, 138
Stirling engine, 91, 401
Streetcars and trams, 368 (photo), 369–370
Strip mining, 80–81, 130
Strontium-90, 70, 457
Structural insulated panels, 256
Stumpage firewood, 450
Subaru, 462
Subbituminous coal, 5, 50, 79, 178
Sugarbeets
    as biofuel source, 219
    as ethanol source, 155, 156
Sugarcane, as biofuel source, 219
    See also **Ethanol, sugarcane**
Sulfur hexafluoride, 52, 269
Sulfur oxides (SO$_x$), 411–412
    acid rain and, 109, 180–181, 412
    from biomass energy production,
        40–41, 109
    from coal burning, 100, 363, 411
    of dry steam geothermal energy
        systems, 199
    from electrical power plants, 109
    flaring and, 169
    flue gas desulfurization technology
        and, 412
    as greenhouse gas, global warming and,
        40–41, 73, 78, 180, 289–290,
        289 (table), 461
    gympsum formation and, 412
    human health affected by, 109
    natural gas and, 180–181, 309
    oil-fired electrical power plants and,
        365, 411
    from oil refining, 326
    radiative forcing and, 378
    sulfur containing fuels and ores and, 411
    sulfur dioxide "acid rain" and, 82, 100, 309
    uses of, 411
    from waste incineration, 439
Sunflower crops, as biofuel source,
        154, 219, 327
Supercritical pulverized coal combustion, 363
Superfund sites, 86
Surface heat island effect, 234
Surface Mining Control and Reclamation Act
    (SMCRA, 1977), 81–82, 294–295

Sustainability, 413–418
    agroecosystem entropy and, 138–139
    bicycle transportation and, 29–32
    the Brundtland Report and, 413–414
    core concept of, 413
    daylighting tools and, 95–97
    definitions regarding, 413–414
    economic vitality pillar of, 415–416
    environmental protection pillar of, 415–416
    history of, 413
    innovation element in, 251–255
    international policy, documents and,
        413, 415
    pipelines and, 356
    policy decision-making and, 416–417
    recycling and, 379–382
    social equity pillar of, 413, 415–416
    strong *vs.* weak sustainability and,
        414–415
    sustainable development term *vs.*,
        413–414
    uranium mining and, 432
    See also **Best management practices;
        Embodied energy; Environmental
        stewardship; Food miles; Life cycle
        analysis; Nuclear power; Wood energy**
Sustainable Agriculture, Food, and the
    Environment report (U.K.), 173
Svirejeva-Hopkins, Anastasia, 139
Svirezhev, Uri, 138
Sweden
    carbon tax in, 108
    flex fuel vehicles in, 171
    mandatory market green certificates in, 212
    recycling in, 380
Sweet natural gas, 298, 299
Sweet sorghum, as biofuel source, 219
Switchgrass renewable agrofuel,
        218 (photo), 387
Switzerland, 115
Syngas
    biomass energy form of, 40, 84
    coal and, 308
    as gasification product, 191, 192, 195, 363
    hydrogen component of, 249
    natural gas and, 300
    from waste incineration, 440
Synthetic gas, 178, 191
Syria
    Arab–Israeli War and, 340
    nuclear weapons in, 317
Szilard, Lea, 311

Taiwan, 200 (table)
Tajikistan, 346
Tar sands, 83, 180, 324, 332
    See also **Oil sands**
Taylor, Frederick Winslow, 28
Taylorism, 28
Technology adoption lifecycle model of
    innovation, 253–254
Technosphere, 138
Telsa Roadster electric sports car, 462
Temperature inversion systems, 83
Tennessee, 294
Tennessee Valley Authority, 164, 373
Terrestrial albedo, 204
Tesla Roadster electric vehicle, 114
Texaco oil company, 329, 339
Texas
    Austin Energy green pricing utility in, 225
    mandatory market green certificates
        in, 212
    New London, Texas methane accident, 275
    renewable energy portfolio of,
        390, 392 (table)
    Texas Railroad Commission, 374
Thailand, 200 (table)
Thermal bypass of heat, 255, 257
Thermal power
    Brayton cycle engines and, 110
    as coal competition, 83
    cogeneration plants and, 110
    combined cycle system power stations
        and, 110
    power station electrical conversion
        and, 109
    Rankine engines and, 110
    thermal energy storage systems and,
        134–135
    See also **Gasification; Geothermal energy**
Thermodynamic laws, 157–159, 239–240
    See also **Exergy**
Thermohaline circulation of oceans,
    205, 206 (fig.)
Thermosyphon passive solar heating
    system, 409
Thomson, Elihu, 287
Thoreau, Henry David, 147
Thorium, 430
Three Gorges Dam, 419–421
    China's electricity needs and, 419 (photo)
    flood management by, 419–420
    functions of, 419
    green energy alternative from, 420

hydroelectric capacity of, 7, 420
locks system of, 420
negative impacts of, 420–421
Three Mile Island, 421–424
    core meltdown incident at, 421–423,
        422 (photo)
    nuclear policy impacted by, 423–424
    See also **Nuclear power**
Tidal power, 424–426
    advantages *vs.* disadvantages of, 425–426
    diurnal cycle of, 425
    as green power renewable energy source,
        83, 218, 221, 223, 366, 424–426
    IPCC studies of, 261
    kinetic energy of moving water and, 425
    Portugal's wave farm and, 221
    "stream" generators and, 424–425
    tidal barrage systems use and, 425
    tidal streams produced by, 424
    tidal variation element in, 425
    underwater windmills as, 221
Tight natural gas, 299
Tillerson, Rex, 58
Time series forecasting methods, 176
Tokamak nuclear fusion technology, 187–188,
    312–313
Toluene, 169, 178
Torness Nuclear Power Station, Dunbar,
    Scotland, 311 (photo)
Total primary energy supply (TPES), 426–428
    calculation methodology of, 426
    definition of, 426
    efficiencies of national economies compared
        by, 427
    factors in, 426
    fuel types in calculation of, 426
    geography factor in, 427
    global TPES and, 427
    national gross domestic product correlated
        with, 427
Total S. A., 329, 330
Toyota, 462
Toyota Prius, 7, 186, 357 (photo), 359, 462
Toyota Tundra, 171
TPES. *See* **Total primary energy supply** (TPES)
Tragedy of the commons, 455
Trams and streetcars public transportation,
    368 (photo), 369–370
Trans-Alaska Pipeline, 16
Transmission domain, of electric power
    system, 227
Tritium, 188, 246, 315, 318

Tropical deforestation, 55–56
Tropopause boundary region of atmosphere, 377, 378
Turkey
  geothermal energy produced in, 200 (table)
  IPCC Session in, 261
  paratransit in, 368
Turkmenistan
  Caspian Sea border country of, 66, 67
  petroviolence in, 346
Tyndall, John, 202

Ukraine, 310, 325
  See also Chernobyl
Ultracapacitors, 135
Ultra supercritical pulverized coal combustion, 363
Underwater windmills tidal power, 221
Uninterruptible power supplies, 135
United Arab Emirates
  International Renewable Energy Agency (IRENA) and, 266
  natural gas produced in, 299, 364
  OPEC membership of, 338
United Kingdom (U.K.)
  British Nuclear Fuels Limited, 318
  cap-and-trade system in, 59–60
  Display Energy Certificates, 237
  Domestic Energy Assessor, 237
  Energy Performance Certificates, 236, 237
  food miles concept in, 173, 174
  gas flaring in, 169
  Home Information Packs and Energy Performance Certificates, 237
  mandatory market green certificates in, 212
  metro systems in, 370
  nuclear power reactors in, 311
  nuclear weapons developed in, 317, 430
  renewable energy certificates in, 168, 390
  REVA electric car used in, 462
  Royal Institute of Chartered Surveyors in, 237
  tram public transportation in, 369
  Wave Hub wave farm, Cornwall, 443
  Zero Carbon Britain UK energy strategy, 77
United Nations (UN)
  renewable energy issue and, 265
  See also International Renewable Energy Agency (IRENA); Kyoto Protocol; United Nations Framework Convention on Climate Change (UNFCCC); specific conference, convention, programme

United Nations Conference on Environment and Development, 216, 269, 413
United Nations Conference on the Human Environment, 413
United Nations Convention on the Law of the Sea (UNCLOS), 322–323
United Nations Conventions on the Laws of the Sea, 69
United Nations Environment Programme, 25, 74, 258
United Nations Framework Convention on Climate Change (UNFCCC), 25–26
  carbon trading issue and, 62
  Kyoto Protocol and, 74–75, 216, 269
  See also Berlin Mandate; Intergovernmental Panel on Climate Change (IPCC); Kyoto Protocol
United States (U.S.)
  acid rain in, 412
  Americans with Disabilities Act (ADA, 1990), 368
  appliance energy standards in, 9, 10
  Atomic Energy Commission, 98
  bicycle transportation in, 30
  Big Oil political influence in, 330–331
  biodiesel produced in, 33–34
  bioelectricity used in, 38
  biomass energy produced in, 38, 39, 132
  Canadian oil sands and, 332, 333
  cap-and-trade emissions trading system in, 58, 84, 121, 133, 144, 207, 216
  carbon equivalency measure in, 290–291
  coal-fired power generation in, 362
  coal reserves in, 80
  CO_2 emissions produced in, 50
  Cold War and, 98–99
  Columbia River, 85–87
  commuter rail transportation in, 370
  compressed natural gas buses in, 462
  concentrated solar power (CSP) systems in, 410
  dioxin emissions in, 103
  domestic crude oil produced in, 325
  Economic Stabilization Program, Nixon administration, 340
  electricity loss in transmission and, 337
  electricity use data in, 406
  electric trolleys and buses in, 115, 462
  electric vehicles in, 7
  Elwha Ecosystem Restoration Project, 87
  emissions reductions in, Berlin Mandate and, 26

Energy Independence and Security Act
    (2007), 34
energy mortgages in, 237
energy policies of, 132–133
environmental preservationism in, 146–147
ethanol from corn in, 156, 194, 223
farm windmills in, 444 (photo), 445
Federal Energy Regulatory Commission, 98
federal government purchasing power
    and, 141
Federal Power Commission, 98
feed-in tariffs in, 166
flex fuel vehicles in, 170–172
food miles in, 173
fuel efficiency standards in, 19
gas flaring in, 169
Geological Survey (USGS), 16, 17
geothermal energy produced in,
    199, 200 (table)
government utility ownership in, 373
green pricing programs in, 225–226
grid-connected system legislation in, 229
Home Energy Rating Systems (HERS),
    235–238
household energy audits in, 124
hybrid vehicles in, 462
hydroelectric power usage in,
    7, 223, 241, 383
inertial confinement fusion research in, 188
international energy distribution and,
    261–262
Kern River gas transmission pipeline, 353
Kyoto Protocol rejected by, 123, 269, 270
landfill methane in, 276
lead in gasoline elimination in, 122
lignite coal in, 79
Love Canal, Niagara Falls, New York, 104
maize agriculture in, 139
mandatory market green certificates in, 212
Mandatory Oil Import Quota (MOIP)
    program (1959), 339
maximum achievable control technology
    in, 440
metro systems in, 370
Midwestern Regional Greenhouse Gas
    Reduction Accord, 58
National Appliance Energy Conservation
    Act (NAECA, 1987), 10, 11 (table)
National Energy Policy plan (2001), 325
natural gas use in, 298
natural gas vehicles in, 301
nonpoint source water pollution in, 305

nuclear bomb developed by, 430
nuclear fission policies in, 131
nuclear power reactors in, 222, 311–312
Nuclear Regulatory Commission,
    98, 164, 423
nuclear weapons in, 317
oil consumption data in, 4, 17
oil production data and, 238
paratransit in, 368
photovoltaics (PV) green power in, 222
plutonium breeder reactor program
    of, 318
power grid and voltages in, 107
propane vehicle fuel source in, 21
public utilities data in, 372, 374
radioactive waste in, 315, 457
radium mining in, 430
recycling in, 379–381
Regional Greenhouse Gas Initiative, 58
renewable energy certificates in, 168
renewable energy-fired power plants
    in, 366
resource and environmental profile analysis
    in, 283
smart meters in, 288
Sonoran Desert, solar irradiation and, 404
sorghum (corn) ethanol fuel in, 20
soybeans as biofuel source in, 6
"splash and dash" biodiesel subsidies
    and, 34
Telsa electric sports car used in, 462
Times Beach, Missouri, 104
Title 40 Protection of Environment Part 41
    legislation, 118
tram public transportation in, 369–370
United Nations Convention on the Law of
    the Sea (UNCLOS) rejected by, 322
United States Enrichment Corporation, 318
uranium mining remediation in, 432
uranium resources in, 309–310, 431
weather contracts based on heating degree
    days in, 232
wind power in, 108, 219, 385, 448
wind power produced in, 222
Yucca Mountain, 315
See also **Arctic National Wildlife Refuge**
    (ANWR); Department of Energy, U.S.
    (DOE); Nuclear power; Renewable
    energy portfolio; *specific legislation,*
    organization, state
Universal Soil Loss Equation (USLE), 307
Uraninite, 431

Uranium, 429–433
decay products of, 429
enrichment technologies for, 318
"fertile" plutonium-239 process and, 429
fissioning chain reaction and, 429, 430, 458
geology of, 430–431
health effects from, 431–432
history of, 430
mining of, 432–433
neutrons absorbed by, 429
nuclear power and, 430, 432
nuclear weapons and, 318
phosphate mining and, 431
radioactive spent fuel rods issue and, 429, 430, 457
radioactive waste and, 458
sustainability aspects of, 432
uranium-235, 318, 429
uranium-238, 429
uranium deposit types and, 431
uranium enrichment process and, 429
versatility of, 430–431
See also **Chernobyl; Nuclear power; Three Mile Island**
URENCO group, 318
Urry, Lewis, 23
U.S. Department of Agriculture, 151, 156
U.S. Department of Energy. *See* **Department of Energy, U.S. (DOE)**
U.S. Environmental Protection Agency (EPA). *See* Environmental Protection Agency (EPA)
U.S. Fish & Wildlife Service. *See* Fish & Wildlife Service
U.S. Geological Survey (USGS), 16, 17
U.S. National Wildlife Refuge System, 14
U.S. Synthetic Fuels Corporation Act, 100
Utah
oil shale in, 334
renewable energy portfolio of, 390, 392 (table)
Skull Valley Goshute Indian Reservation, 459
Sonoran Desert, 404
Uzbekistan, 346

Vanadium redox reaction fuel cells, 184, 431
Vapor barriers, proper insulation, 255
Vegetable oil, as biofuel source, 20, 32–33, 39
Venezuela
gas flaring in, 169
oil reserves in, 346

OPEC membership of, 338, 339
Petroleos de Venezuela S. A. oil company in, 329
United Nations Convention on the Law of the Sea (UNCLOS) rejected by, 322
Vermont, 44–45, 392 (table)
Vertical axis wind turbines, 448
Vertical nuclear proliferation, 316
Very-large-scale photovoltaic systems (VLS-PV), 351–352
Vietnam
anaerobic digestion system used in, 219
microhydropower used in, 292, 293
Vinasse, 195
Virginia
coal mining in, 294
Dominion Virginia Power and, 110–111
renewable energy portfolio of, 390, 392 (table)
solar irradiation in, 404
VOC. *See* **Volatile organic compound (VOC)**
Volatile organic compound (VOC), 435–437
biofiltration removal of, 436
in building environments, 436
definition of, 435
electronic devices and, 436
ethanol production and, 436
gasoline source of, 435–436
green energy production and, 435, 436
methane as, 435
ozone affected by, 435
"sick building syndrome" and, 436
thermal *vs.* nonthermal generation of, 435
Volga River, 67
Volta, Allesandro, 23
Volt (V) electrical potential measure, 107
Voluntary market green certificates, 211–212, 226

Ward, Barbara, 413
Washington (state)
appliance efficiency standards in, 10
Columbia River and, 85
Hanford nuclear site, 86
irrigation from Grand Coulee Dam and, 86
radioactive waste disposal in, 424
renewable energy portfolio of, 392 (table)
ShoreBank Pacific eco-banking practices in, 210
Washington, D.C., 96, 390, 391 (table)

Waste incineration, 439–441
  biomass energy and, 39
  carbon dioxide emissions and, 440
  of commercial and industrial waste, 441
  cultural and religious factors in, 440
  dioxin production from, 102, 103
  emissions and residues from, 439
  farming soil enrichment and, 439
  feedstock sources evaluation and, 440–441
  "garbage in, garbage out" concept
    and, 439
  garbage *vs.* waste terms and, 439
  landfill gas collection systems and, 440
  maximum achievable control technology
    and, 440
  of medical waste, 440–441
  nuclear waste and, 441
  regulatory regimes for, 441
  syngas production and, 440
  technology advances in, 439, 440
  as waste management option, 439
  waste-to-energy technology and, 441
  See also **Life cycle analysis**
Water Quality Act (1987), 306
Water vapor
  from biomass energy, 40
  fossil fuel burning and, 180, 309
  as GHG, 378
  as hydrogen fuel component, 91, 114, 249
  as natural gas component,
    168, 169, 179, 298
  as natural GHG, 73, 213, 378
  parabolic dishes and, 401, 410
  short residence time of, 290
  wood burning and, 450
Watson, Robert K., 278
Watts, Michael J., 343–344
Wave power, 441–443
  definition of, 442
  Department of Energy work regarding,
    99–100
  disadvantages of, 442
  environmental measures regarding, 145
  as green power renewable energy source,
    83, 441–443
  IPCC studies of, 261
  Pelamis Wave Power machine,
    Aguçadora Wave Park, Portugal,
    442 (photo), 443
  power station electrical conversion
    and, 109
  reliability advantage of, 442

  research grants in, 131
  Salter's Duck model of, 442
  solar heating element of, 131
  wave farms and, 442 (photo), 443
WCED. *See* **World Commission on
  Environment and Development**
  (WCED)
Weather derivatives financial
  instrument, 232
"Weighted Average Source Distance" food
  miles calculation method, 173
Wells Fargo green rewards card, 210
West Virginia
  coal mining in, 294, 295
  Samples Mine (coal) in, 294 (photo)
Wet biomass biodiesel feedsource, 34
Wet natural gas, 298, 299
Wet-spray cellulose insulation, 256
White, Richard, 85
Wild and Scenic Rivers Act, 14
Wilderness Act, 14
Williams, James C., 46
Wind power, 443–447
  advantages of, 446
  costs associated with, 243
  electrical energy infrastructure for, 108
  energy payback calculation and, 127
  energy storage, smart grid and, 398
  environmental stewardship and,
    148, 386, 387
  European Wind Energy Association
    and, 108
  feed-in tariffs and, 166, 167
  as green power renewable energy source,
    83, 207, 217, 218, 219–220, 222, 223,
    252, 330, 366, 385–386, 443–447
  history of, 445
  intermittent nature of, 387–388
  limitations of, 220
  magnitude calculation and, 446
  power station electrical conversion
    and, 109
  process explained, 443–444
  solar heating element in, 131, 384,
    405, 443
  subsidies incentives and, 131
  Tararua Wind Farm projects,
    New Zealand, 273
  technology of, 445–446
  wind farms and, 445
  windmills, Halladay Standard Company,
    Batavia, Illinois, 444 (photo)

windmill technology and, 445
wind turbine energy process and, 385
Xcel Energy wind power facility,
    Minnesota, 107
See also **Wind turbines**
Wind turbines, 447–449
advantages *vs.* disadvantages of, 448–449
as alternative energy, 1 (photo)
components of, 447–448, 447 (photo)
energy production process of, 2, 8
as green power renewable energy source,
    217, 219–220, 330, 366
vertical *vs.* horizontal axis types of, 448
wind farms and, 8
See also **Wind power**
Wisconsin
Evergreen Renewable Energy Program of
    Dairyland Power Cooperative, 37, 277
renewable energy portfolio of, 392 (table)
smart meters used in, 288
Wood energy, 449–453
advantages and disadvantages of, 449,
    451–452
as biomass energy source, 449
Btus energy content measure and, 451
carbon dioxide (CO_2) and, 449–450
chunkwood (firewood), 449
combustion process of burning wood and,
    450–451
cord unit of firewood measure and, 450
firewood *vs.* wood pellets and, 452–453
forest products industry waste and, 450
as green power renewable energy source,
    449–453
heat value range of, 449, 451
installation and maintenance issues
    and, 451
local purchase importance and, 451
masonry heaters and, 451
outdoor wood boilers, 452
seasoned *vs.* green wood, 450
stumpage firewood, 450
sun and carbon roles in, 1
sustainable forest management practices
    element in, 449, 450
wood chips, 449
wood components and, 449
wood furnaces and boilers, 451–452
wood pellets, 449, 452
wood waste source of, 38 (photo)
Woodland management, 39
Wood pellets, 449, 452

Wood products, 56
Work breakdown structure, 28
World Bank
carbon credit trades and, 63
microhydro power data of, 292, 293
natural gas flaring data of, 170
renewable energy issue and, 265
World Commission on Dams, 244
**World Commission on Environment and
    Development** (WCED), **453–455**
Brundtland Commission name of, 453
Brundtland Report of, 413–414
common and integrative approach of, 455
global environmental degradation
    focus of, 453
international cooperation, multilateral
    action work of, 455
key issues for analysis by, 454
objectives of, 453–454
"Our Common Future" report of,
    454–455
public hearings trademark feature of, 454
sustainable development definition of,
    454–455
tragedy of the commons issue and, 455
World Conservation Strategy, 413
World Council for Renewable Energy, 266
World Energy Council (WEC), 142–145, 153
World Health Organization (WHO), 461
World Meteorological Organization,
    25, 74, 258
World Summit on Sustainable
    Development, 413
Worster, Donald, 46
Wyoming
bituminous coal in, 79
heating degree day measures in, 232
Kern River gas transmission pipline
    in, 353
oil shale in, 334

Xcel Energy wind power facility, Minnesota,
    107, 390
Xiluodo Dam, Jinsha River, China, 421
Xylene, 169, 178

Yucca Mountain, 457–459
alternatives to, 459
human health issues and, 458
mechanics of waste storage and, 458
millirem radiation measure and, 458
Barack Obama's policy regarding, 315

radioactive waste storage in, 315, 457
safety issues and, 459
tunnel excavation at, 457 (photo)
See also **Nuclear power**

Zaire, 334
Zero Carbon Britain UK energy strategy, 77
Zero-emission vehicle (ZEV), 461–464
    advanced-technology-partial-zero-emission
        vehicles (AT PZEVs) and, 461
    alternative-fueled vehicles and, 462
    BMW Hydrogen 7 prototype and,
        461 (photo)
    California automobile standards and, 463
    definition of, 461
    electric buses, 462
    electric cars, 113, 461, 462

emission-reducing technology
    and, 462
GHGs reduced by, 461–463
human powered vehicles, 463
hybrid electric vehicles, 462
hydrogen fuel cell vehicles, 462–463
partial-zero-emission vehicles (PZEVs),
    461–462
REVA electric car, 462
Telsa Roadster electric sports car, 462
Zero-net energy commercial buildings, 101
ZEV. *See* **Zero-emission vehicle** (ZEV)
Zinc-air fuel cells (ZAFC), 185
Zinc/bromine fuel cells, 184
Zoé atomic cell, 311
Zonaend, François, 314
Zone of Alienation (Chernobyl Zone), 70